PSYCHOLOGY LIBRARY EDITIONS:
COMPARATIVE PSYCHOLOGY

Volume 8

EVOLUTION OF BRAIN AND BEHAVIOR IN VERTEBRATES

EVOLUTION OF BRAIN AND BEHAVIOR IN VERTEBRATES

Edited by
R.B. MASTERTON, M.E. BITTERMAN,
C.B.G. CAMPBELL AND NICHOLAS HOTTON

Routledge
Taylor & Francis Group

LONDON AND NEW YORK

First published in 1976 by Lawrence Erlbaum Associates, Inc.

This edition first published in 2018
by Routledge
2 Park Square, Milton Park, Abingdon, Oxon OX14 4RN

and by Routledge
711 Third Avenue, New York, NY 10017

Routledge is an imprint of the Taylor & Francis Group, an informa business

© 1976 by Lawrence Erlbaum Associates, Inc.

British Library Cataloguing in Publication Data
A catalogue record for this book is available from the British Library

ISBN: 978-1-138-50329-8 (Set)
ISBN: 978-1-351-12878-0 (Set) (ebk)
ISBN: 978-0-8153-7146-5 (Volume 8) (hbk)
ISBN: 978-0-8153-7151-9 (Volume 8) (pbk)
ISBN: 978-1-351-24658-3 (Volume 8) (ebk)

Publisher's Note
The publisher has gone to great lengths to ensure the quality of this reprint but
points out that some imperfections in the original copies may be apparent.

Disclaimer
The publisher has made every effort to trace copyright holders and would welcome
correspondence from those they have been unable to trace.

EVOLUTION OF BRAIN AND BEHAVIOR IN VERTEBRATES

Edited by

R. B. MASTERTON   C.B.G. CAMPBELL
M. E. BITTERMAN   NICHOLAS HOTTON

 LAWRENCE ERLBAUM ASSOCIATES, PUBLISHERS
1976   Hillsdale, New Jersey

DISTRIBUTED BY THE HALSTED PRESS DIVISION OF

JOHN WILEY & SONS

New York   Toronto   London   Sydney

Lawrence Erlbaum Associates, Inc., Publishers
62 Maria Drive
Hillsdale, New Jersey 07642

Distributed solely by Halsted Press Division
John Wiley & Sons, Inc., New York

Library of Congress Catalogue Card Number: 76-6499

Printed in the United States of America

Contents

Preface

The idea behind this volume and its companion, *Evolution, Brain, and Behavior: Persistent Problems*, began with an interchange of letters with G. Gaylord Simpson in the spring of 1969. At that time we spoke only of updating a previous volume on the same subject, *Behavior and Evolution*, which had been edited by Roe and Simpson and published by Yale University Press in 1958. However, as the number of people who expressed interest increased, enthusiasm for the project also increased and its goals were gradually extended. Eventually, an entirely new effort seemed most appropriate—one directed toward presenting the facts and arguments of paleontology, neurology, and behavior and specifically written in a form usable to students in each other's field.

Accordingly, a meeting was eventually convened to plan the volume and a conference that was to produce it. The planning meeting was held at Tallahassee in February, 1972, under the sponsorship of the Psychobiology Program at Florida State University. It included representatives from the fields of paleontology, comparative neurology, and behavior: M. E. Bitterman, C. B. G. Campbell, William Hodos, Nicholas Hotton, Harry Jerison, J. M. Warren, James C. Smith, Stan Olsen, and Eliott Valenstein.

At the planning meeting the form of the conference was decided: it would consist of four days of meetings divided into morning and afternoon sessions. The morning sessions were to be devoted to the generation of a descriptive overview of what is now known about the evolution and variety in structure and behavior of vertebrates. This overview would be published as the first of two volumes. The afternoon sessions were to be devoted to discussions of the most persistent problems confronting students of the comparative neurosciences regardless of their particular discipline, and would be published as the second of the two volumes.

Subcommittees were then formed for each discipline in the morning sessions and for each discussion topic in the afternoon sessions. Later, these subcommittees selected and invited the participants and still later, edited the manuscripts. The chairmen appear as editors of the volumes themselves.

The conference itself was held in Tallahassee late in February, 1973. It was sponsored jointly by the Psychobiology Program of the Florida State University, the National Science Foundation, and the W. T. Grant Foundation.

Participants in the conference included:

Mark Berkeley	Giles MacIntyre
M. E. Bitterman	W. A. Mason
Kenneth Brookshire	Bruce Masterton
C. B. G. Campbell	Malcolm McKenna
William Clemens	Glenn Northcutt
William Corning	Stan Olsen
Judith Cruce	E. C. Olson
William Cruce	J. M. Petras
Donald Dewsbury	David Premack
Irving Diamond	Leonard Radinsky
Sven Ebbesson	W. A. Riss
Ford Ebner	M. J. RoBards
Carl Gans	Bobb Schaeffer
Michael Ghiselin	L. C. Skeen
Karen Glendenning	Joseph Sidowski
Mitchell Glickstein	Elwyn Simons
S. J. Gould	James C. Smith
Jack P. Hailman	William Stebbins
William C. Hall	N. S. Sutherland
William Hodos	Donald Tucker
Nicholas Hotton, III	J. M. Warren
John A. Jane	Douglas Webster
Harry Jerison	W. I. Welker
Jack Johnson	E. G. Wever
Harvey Karten	Martha Wilson
Richard Kay	William Wilson
Eric Lenneberg	William Woodard

During the conference, the presentations and discussions of each topic were tape-recorded and the tape given to the contributor for use in modifying his manuscript. After the participants returned home and rewrote their contributions the several editors collected, edited, and reedited the manuscripts until they took their present form.

Throughout the time from the original interchange of letters to the publication of the finished volumes, many people whose names do not appear in the volumes played important, even crucial, roles. Not the least of these were the Psycho-biology students at Florida State who inspired the conference in the first place, attended to the myriad of details and finally, played the role of host to the contributors and guests both at the planning meeting and the conference.

BRUCE MASTERTON

Introduction to Volume 1

By the turn of the twentieth century, some forty years after Darwin's *Origin of Species,* many neurologists and psychologists had grasped the importance of evolution for helping to understand brain–behavior relationships. Elliot Smith, Ariëns-Kappers, and the two Herricks in comparative neurology, and Thorndike, Yerkes, and Jennings, in comparative psychology, marked this period with an enthusiastic outburst of activity. Conscious effort was expended toward cooperation between the two fields. Correspondence between men like C. J. Herrick and Yerkes flourished and by 1908 Yerkes, Jennings, and Watson were members of the editorial board of the Journal of Comparative Neurology.

Over the next fifty years, however, the extensiveness of the inquiry and the energy behind it slowly disappeared. The scientific journals that had been begun in the early 1900s to foster comparative and evolutionary neurology and psychology had become, by 1930, journals of animal neurology or animal behavior, almost devoid of comparative argument and completely devoid of evolutionary conclusions.

It is now possible to see that this frustration of goal was probably the joint result of unexpected difficulties in neurological and behavioral comparison and the dissolution of some of the oldest and most fundamental concepts in each field (e.g., "complexity" in neurology, and "intelligence" in psychology). Except for Herrick and Yerkes, few would even venture to say exactly what it was about brains and minds that had evolved. Comparative neurology and psychology texts eventually became recitals of discontinuous and unexplained "variations" or "species differences" effectively chilling the enthusiasm of several generations of new students.

Now, more than a century since Darwin and almost a century since the first wave of activity, a serious renaissance of strictly evolutionary and comparative methodology in neurological and behavioral analysis seems to be underway.

Although it is impossible to pinpoint the events or persons to be credited for this revival, some landmarks lie in the volumes by G. G. Simpson (1949), C. J. Herrick (1956), W. E. LeGros Clark (1959), and J. Z. Young (1962) and certainly *Behavior and Evolution* (1958) edited by Roe and Simpson. Throughout the 1960s, these volumes proved to be important sources of information and inspiration for evolutionary thought in several neurological and psychological laboratories—particularly those of Irving T. Diamond at Duke University and Walle, J. H. Nauta at Walter Reed Hospital.

But through the joint interests of many, the evolution of the nervous system and the evolution of behavior are considered topics for serious inquiry once more. Certainly, the phylogeny of animals, the evolution of brains, and the evolution of brain–behavior relationships are becoming increasingly popular topics in journals and at scientific meetings.

Partly because of the activity generated over the last few years, a marked change has occurred in the subject matter of evolutionary neurology and behavior. New and promising points of view have been invented and old ones rediscovered. What once appeared to be serious gaps in the evolutionary history of nervous systems are now filled or at least bridgeable in the imagination. But along with this renewed activity, a number of real problems have appeared and a much larger number of old problems, pseudo-problems, and misunderstandings have reappeared.

The object of this volume is to present a relatively up-to-date overview of what is now known, what is suspected, and what remains to be discovered concerning the general question of the evolution of the vertebrate brain and behavior, and to present a list of references for those who want to delve deeper into one or another aspect of the problem. Accordingly, it contains chapters by paleontologists (1, 12, 13, 16), sensory morphologists and physiologists (2, 3, 4, 5, 6), comparative neurologists (7, 8, 14, 15, 17), and comparative psychologists (9, 10, 11, 18, 19).

The chapters are arranged in a sequence loosely approximating the order in which the various animals, brain structures, or behavior first appeared. Therefore, the chapters fall naturally into sections, each section directed to a group of vertebrates, beginning with those which have very remote common ancestry and progressing to those with more recent common ancestry with mankind. Each chapter within a section describes either the phylogenetic history, nervous system or behavior of the members of that group.

For ease in reference each chapter's bibliography is placed at its end, and the entire volume is supplied with a subject index which focuses almost entirely on particularly important animals, neurological structures, behavioral patterns, or evolutionary concepts.

EVOLUTION OF BRAIN AND BEHAVIOR IN VERTEBRATES

1

Origin and Radiation of the Classes of Poikilothermous Vertebrates

Nicholas Hotton III

Smithsonian Institution

INTRODUCTION

When arranged in order of their earliest appearance in the fossil record, vertebrate classes strikingly reflect a complex sequence of increasingly elaborate mechanisms, and with greater or lesser confidence later groups can be derived from earlier on morphological grounds. This is the primary basis of traditional phylogenetic treatments of vertebrate history (Figure 1). In addition, however, it has long been evident that the higher the taxonomic category, the clearer the adaptive significance of the characters which define it (cf. Simpson, 1953). Therefore, the historical sequence of the appearance of vertebrate classes can also be interpreted in terms of different adaptive themes that, when examined in relation to established phylogenetic patterns, provide a functional rationale for class origins.

The purpose of this chapter is to review the morphology and history of the lower vertebrates in terms of current opinion of their phylogenetic relationship to each other and to living forms, with special attention to the sequence of appearance in the fossil record. Adaptive aspects of this history are emphasized because of their value in determining possible modes of origin of classes. Because of the pervasiveness of adaptation in the evolution of nonnervous structure, it is suggested that evolution of the central nervous system may be more directly adaptive than is generally accepted at the present time.

Six classes of poikilothermous animals are currently recognized among the vertebrates that have lived since their earliest certain record in the Middle Ordovician: Agnatha, Placodermi, Osteichthyes, Chondrichthyes, Amphibia, and Reptilia. This arrangement, like all other biological classifications, is in some degree arbitrary. For example, each class includes morphological subdivisions

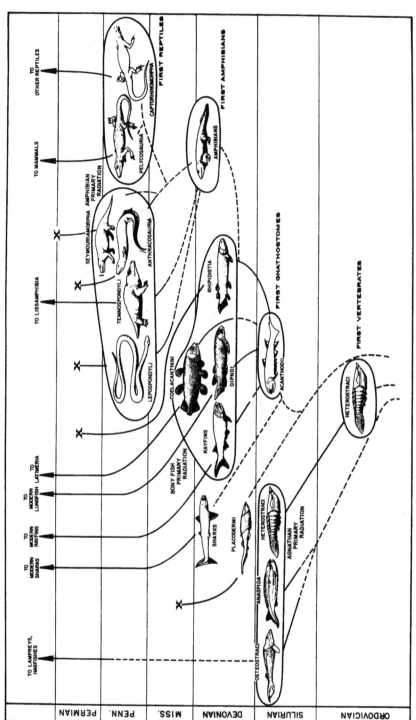

FIGURE 1 Phylogeny of vertebrates, showing first appearance and primary adaptive radiation of classes that arose during the Paleozoic. Diagrams not to scale. Sources as for subsequent figures except: rayfinned fishes (Gregory, 1951); lungfishes, coelacanth, nectridian and anthracosaurian amphibians, and *Seymouria* (Romer, 1966); Middle Pennsylvanian synapsid reptile (Reisz, 1972).

that are already distinct from each other in the oldest record of the class to which they belong, and because their distinctness implies independent origin, a case can be made for regarding most such subdivisions as classes unto themselves. However, these are taxonomic questions that modify neither the logic nor the conclusions of the present review but merely require consideration of systematics in greater detail. In the interest of conciseness it is more useful to the purpose at hand to follow the broader taxonomic arrangement.

All six classes have undergone primary adaptive radiation before the end of the Pennsylvanian, about 280 million years ago, but the Chondrichthyes (sharks) and Osteichthyes (higher bony fishes), currently or recently in the midst of intense radiations beyond the primary ones, are still well represented in the living fauna. The Agnatha are represented only by the living cyclostomes (lampreys and hagfishes), and the Placodermi have been extinct since the end of the Devonian, about 345 million years ago. The Reptilia are but a shadow of their former selves, having declined radically since their second phase of abundance and diversity during the Middle and Late Mesozoic. The Amphibia never have amounted to much except as a transitional group; their Late Paleozoic radiation has been overshadowed by contemporary bony fishes and reptiles, as is their present-day representation by such a relatively insignificant group as living reptiles.

MORPHOLOGY AND ADAPTATION IN POIKILOTHERMOUS VERTEBRATES

Agnatha

Members of the class Agnatha lack grasping jaws and must ingest their food by some sort of sucking mechanism. They also lack the controllable paired limbs characteristic of jawed vertebrates.

Fossil agnathans provide the earliest record of vertebrates. The earliest certain fragments of bone, identifiable as the remains of heterostracan agnathans, date from the Middle Ordovician, about 460 million years ago, but the class as a whole is best known from the record of its primary adaptive radiation during the Late Silurian and Early Devonian, about 400 million years ago. Most of the deposits in which these animals are found indicate an environment of fresh or brackish water.

Ancient agnatha were generally small, from 10 to 20 cm in length, although some forms attained a length of about 100 cm. The mouth was small and of limited mobility; details of the feeding mechanism would indicate that agnathans fed on finely particulate matter, sucking water in through the mouth and expelling it through the gill openings. In some fossil agnathans the head and trunk were expanded, flattened in varying degree, and covered by a heavy armor of dermal bone; a short tail covered with heavy scales projected from the rear of the trunk shield. In these animals the overall shape of the body resembled that of a frog tadpole, which together with the lack of paired fins would suggest that the mode of swimming was one of wriggling from point to point on the substrate, without much capacity for active maneuver.

In the Osteostraci (Figure 2C) the dermal armor covered the head and trunk as a solid shield and was underlain by perichondral bone composing the neurocranium and visceral skeleton (Stensiö, 1927); both kinds of bone were cellular. Branchial arches were unjointed and were continuous with the neurocranium. The head shield was pierced dorsally by orbits of modest size, pineal opening, and a single, median, conjoined nasohypophyseal opening like that of living lampreys and hagfishes. In the Anaspida (Figure 2B), the head and trunk were

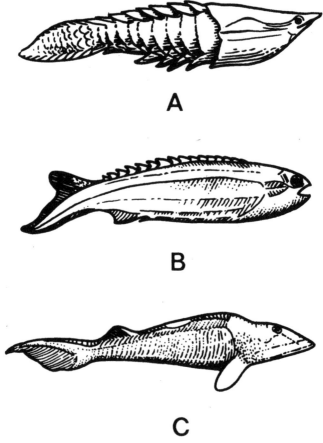

FIG. 2 Agnathan body form. (A) Heterostraci; (B) Anaspida; (C) Osteostraci. (After Romer, 1969.)

not expanded laterally and were covered with scalelike dermal plates that did not form a distinct shield. Apparently no perichondral bone was formed, for no internal structures were preserved. As in osteostracans, the dermal bone was cellular, and orbits, pineal opening, and nasohypophyseal opening were present. In the Heterostraci (Figure 2A), the head and trunk were expanded as in ostracoderms, but the shield consisted of a complex assemblage of larger dermal plates composed of a peculiar noncellular material that resembled dentine more than bone. Such internal structures as gill pouches, pineal body, and semicircular canals were partially preserved as impressions on the internal surfaces of dorsal plates (White, 1935). Orbits were small and in at least one form absent altogether, the pineal opening was usually absent, and there was no structure comparable to the nasohypophyseal opening of other agnathans. A nasal organ is believed by some to open near the mouth but its nature is obscure.

In osteostracans the head and trunk shield was very flat ventrally but was arched or crested dorsally; mouth and gills opened on the ventral surface (Figure 3), whereas the organs of special sense lay on top of the arch on or near the midline. These features would suggest that osteostracans were the most persistently bottom dwelling of the Agnatha. They probably fed simply by ingesting

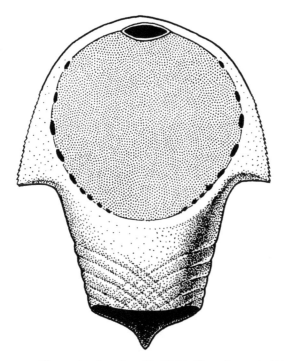

FIG. 3 Osteostracan (*Kiaeraspis;* after Stensiö, 1927). Ventral aspect of head and trunk shield to show mouth and gill openings.

bottom mud that was rich in organic material, living and dead; as it passed through the gut useful elements were digested automatically and inorganic material was expelled with the feces.

Anaspids differed from other Agnatha in the fusiform shape of the body; the mouth was terminal and the gill openings and orbits lateral, all of which would suggest that anaspids were somewhat more active than osteostracans. Parrington (1958) suggested that they fed in the manner of "tailing" trout, maintaining the body at a steep angle with the head plowing through the muck on the substrate, but there was no indication that they were any more selective in their choice of food than were the osteostracans.

In heterostracans the head and trunk shield was generally more rounded in cross-section than that of osteostracans and ranged in shape from nearly fusiform to very flat. Orbits were lateral and the gills had a common excurrent opening to the rear, like a jet exhaust, on either side. Some of the less flattened forms sported a prominent rostral beak above the subterminal mouth. The base of the beak was covered with fine pits that might indicate organs of chemical sense, implying that the beak might have been a means of locating small invertebrates on which the animal fed, rather in the manner of a living sturgeon or paddlefish. In the actual ingestion of food, however, heterostracans could not have been comparable to these fishes, for the mouth was tiny and its margins were nearly immobile. If the heterostracans did in fact have a more selective mode of feeding, the gills might have functioned as strainers, retaining food particles while permitting water to pass through. This mechanism is reminiscent of the true filter feeding of lower chordates, such as tunicates, and it may be significant that heterostracans predate other agnathans in the fossil record by 20–30 million years. In most flatter forms the mouth was ventral, suggesting a mode of life more nearly comparable to that of osteostracans.

Fossil Agnatha, therefore, were directly dependent on the substrate as the main source of food, were able to ingest only relatively small particles, and were of very limited swimming ability. The adaptive theme of the class could be characterized in terms of bottom dwelling, more or less sedentary habit, and detritus or filter feeding.

The lampreys and hagfishes are the only survivors of this strange group. There is some evidence (Stensiö, 1968) that they are independently derived, from Osteostraci and Heterostraci respectively, but they differ from all fossil agnathans in their eel-like body form and in the fact that their skeletons are completely cartilagenous and their skin scaleless. They also differ radically from fossil forms in the size of the organisms on which they feed, for lampreys feed by rasping and sucking at the bodies of living bony fish, and hagfishes by burrowing into the bodies of dead or dying bony fish. Appropriately to their feeding habits, they are much more maneuverable swimmers than any of the fossil forms can have been; the eel-like habitus appears to be a successful solution to the problem of attaining maneuverability without paired fins.

Placodermi

The Placodermi (Figure 4) are characterized by dermal bony armor that covers the head and trunk, by grasping jaws, and by paired fins. The earliest placoderms are in general the most heavily armored. The paired fins consist of bony spines, which in early members of the class are fixed to the trunk shield but which become jointed and movable in the course of evolution.

Definitive placoderms first appear in the Early Devonian and disappear, except for a few strays, at the end of the period. The earliest forms occur in sediments that indicate a freshwater environment, but most later ones are found in marine deposits.

Placoderms average much larger than agnathans, ranging from 30 cm to about 10 m in length. They resemble agnathans in having head and trunk shields and in being often broad and flat; in placoderms, however, the head and trunk shields are always distinct and are connected only by joints.

In a number of marine placoderms with short jaws and subterminal mouths the jaws are armed with flat-crowned teeth, presumably having a crushing function; these animals are thought to have been mollusc eaters. In some strongly flattened freshwater forms the small mouth is ventral and the short jaws are unarmed and weak; these features probably indicate redevelopment of a mud-ingestion mode of feeding similar to that postulated for many agnathans.

In one large group of placoderms in which the body was more or less fusiform, although very flat ventrally, the mouth was terminal and the jaws were long; although the jaw margins were toothless they were sharp and constituted a formidable shearing apparatus. These animals must have been predators of some sort, but the weight of their armor and the rigidity that it imposed on the trunk must have limited their activity. They probably fed by lying in wait for prey and capturing it by means of a short rush and a quick snap.

Jaws and paired fins express a predatory behavioral aspect of the adaptive theme of placoderms, in contrast to the detritus feeding of agnathans. The retention of armor and a strong tendency toward flattening of the body, however, indicate that ecologically the placoderms remained dependent in large measure on the substrate, and for want of a better term their adaptive theme can be characterized as one of "sedentary predation."

The substrate of marine environments is often very rich in shelled invertebrates, much more so than that of freshwater environments. The presence of jaws gave bottom-dwelling placoderms the potential of exploiting this source of food, and

FIG. 4 Placoderm body form (*Coccosteus;* after Romer, 1966).

when placoderms became able to cross the physiological threshold between fresh and salt water they evolved a variety of mollusc eaters. In contrast, the rich source of food in organic detritus and nanoorganisms in bottom muds might have predisposed bottom dwellers that remained in fresh water to reassume a detritus-feeding habit.

Subclass Acanthodii

As far as the fossil record shows, the placoderms are not the earliest jawed vertebrates. That honor goes to the Acanthodii, a long-lived group of generally small animals of beautifully generalized gnathostome structure. Their earliest members are preserved in deposits of Late Silurian age, about 400 million years old, and their latest in deposits of the Early Permian, a span of about 140 million years. Their relationships are at present the subject of controversy, in which some workers emphasize similarities to sharks in the visceral skeleton (e.g., Nelson, 1968) and other similarities to bony fishes in jaw suspension (e.g., Miles, 1968). Current assignment of acanthodians is based on the close resemblance of their scales to those of bony fish and on jaw suspension and other characters.

Acanthodians (Figure 5) have a fusiform "fishlike" body with a heterocercal tail; the head is short and blunt, with very large eyes, paired terminal nostrils, and a large terminal mouth supported by jaws that bear teeth in most forms. There are paired fins that consist of spines jointed to skeletal elements in the body wall; the spines act as cutwaters for webs of skin attached broadly to the body wall behind them. The number of pairs of fins had evidently not been stabilized at two, for some acanthodians had as many as seven pairs.

Most acanthodians lived in fresh water and ranged in length from 2 to 8 cm, although in marine rocks of Mississippian age, spines of acanthodian pattern were found that indicated an animal about 2 m long. During their history they were remarkably uniform in structure, differing primarily in size, number of paired spines, and amount of scaly covering.

Their fusiform shape and well-developed paired fins indicate that these animals have been active swimmers, the jaws show predatory habit, and the large eyes suggest a level of exteroception commensurate with active hunting. In short, the acanthodians show the earliest really effective step toward a habit of active predation that has later been exploited so successfully by both sharks and bony fish.

FIG. 5   Acanthodian body form (*Euthacanthus*; after Gregory, 1951).

Acanthodii and the Origin of Gnathostomes

Because of their generalized structure and time of earliest occurrence, acanthodians have been regarded as ancestral to all other jawed vertebrates. More detailed anatomical considerations confirm the idea that they are ancestral to later bony fish without barring them from shark ancestry but tend to confute any such relationship to placoderms. Such differences as the absence of head and trunk armor in acanthodians and its persistence in placoderms, for example, suggest that the two groups must have arisen independently from agnathan ancestors. Placoderms came into being as their detritus-feeding ancestors moved into a mode of sedentary predation, and acanthodians as their ancestors moved directly into a mode of active predation. The adaptive theme of sedentary predation, expressed by placoderms, therefore does not appear in the ancestry of bony fish.

Some of the less sedentary agnathans have been suggested as gnathostome ancestors, but all known members of the class Agnatha have morphological peculiarities that preclude them from such a role. The common ancestor of acanthodians and placoderms therefore must have been less specialized than any known agnathan and presumably very old. On theoretical grounds it is probable that such an animal may have had an unossified skeleton and no teeth or scales, which partially accounts for its being unknown and takes the problem out of the purview of paleontology.

Osteichthyes

In the Osteichthyes (bony fishes), the internal skeleton is well ossified, the neurocranium is covered by thin plates of dermal bone, and the body is covered by a coat of small scales. In generalized forms the large mouth is terminal and the jaws are long and armed with teeth. There are two pairs of fins that are very flexible because their point of attachment is narrow, allowing a high degree of maneuverability in swimming. The bony fishes have perfected the habit of active predation, initiated by their presumptive acanthodian ancestors, by narrowing the fin bases and perhaps by reducing the number of paired fins to two.

From their earliest record in the Early Devonian, the bony fishes have been represented by two subclasses, the Sarcopterygii, or lobefins, and the Actinopterygii, or rayfins. The lobefins are dominant in the Early Devonian but are soon surpassed in numbers and diversity by rayfins. The latter eventually radiate into the sea in dazzling variety and for the past 200 million years or so have dominated nearly all aqueous environments. However, the lobefins include the ancestry of tetrapods and are therefore the more appropriate to the present discussion.

Lobefins are so called because a part of the basic fin skeleton, clothed with the muscles that operate the fin, protrudes from the body wall as a narrow, scale-covered, fleshy lobe, forming an axis from which spring the short, slim bony rays that support the fin proper. Lobe-finned fishes range from about 25 to 150 cm in length; the body is cylindrical and the head is rather long and somewhat flattened. Two orders, the Crossopterygii (Figure 6) and the Dipnoi, or lungfishes, are distinguishable from their first appearance in the Early Devonian on the bases of dental armament and jaw suspension and of patterns of dermal skull bones. The jaws of crossopterygians are long and armed with large conical teeth; in lungfishes the jaws are shorter and the teeth consist of flattened, strongly ridged plates. The upper jaws of crossopterygians are movable on the braincase, as in most bony fishes, whereas those of lungfishes are firmly fused to the braincase. The crossopterygians are obviously highly predaceous, and in their jaws and teeth and stout, cylindrical bodies suggest the habits of living pike (*Esox*) or freshwater dogfishes (*Amia*), which stalk their prey from cover in quiet water and capture it by a sudden dash. Lungfishes are said to be adapted to a durophagous habit because of the firm support of the upper jaw and the platelike morphology of the teeth. However, the ridges give the teeth a shearing capacity in many forms, and the aggressive disposition and catholic taste of living lungfishes indicate that behaviorally they can be hardly less predaceous than any crossopterygian. Although body form is elongate in two of three living genera of lungfishes, most fossil lungfishes are little different from crossopterygians in this respect.

Lobefin structure indicates adaptation to life in quiet water, in which the lobate fins may be used in "walking" on the substrate in search of food, as illustrated by the behavior of living lungfishes. In crossopterygians the walking habit may have been analogous to the fin-controlled hovering of *Esox* and *Amia*, as used during the stalking phase of the hunt. The presence of lungs, by which lobefins can breathe air, enables them to survive when the water becomes foul, as it does in the highly wet–dry seasonal climate of Africa and South America where living lungfishes are found. Many of the deposits in which fossil lobefins of both orders occur together appear, on external evidence, to have been laid down in fresh water under cyclic climatic conditions in which the water became stagnant at intervals, presumably during dry seasons.

In structural detail, crossopterygians are far better candidates for tetrapod ancestry than are lungfishes. In functional terms, however, neither feeding habit nor physical environment accounts for the origin of tetrapods from one order rather than from the other, for both have been active predators in the same environment. The main functional determinant seems to have been a differential

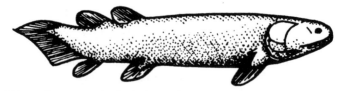

FIG. 6   Crossopterygian body form (*Sterropterygion;* after Thomson, 1972).

response to extreme fouling or drying of the water courses in which they lived. When conditions reach an extreme in the environment of the living African lungfish, it responds by burrowing into the mud and going into estivation. There is clear evidence that a variety of Late Paleozoic lungfishes also estivated but none that crossopterygians did so. However, some crossopterygians probably solved the problem of fouling and crowding by using their lobate fins and walking habit to struggle from one pool to another, presumably downstream, until they reached a pool big enough to allow survival until the rains came again. For a part of their lives they were therefore exposed to some of the selective effects of a land environment, which lungfishes had avoided altogether; tetrapods eventually appeared as descendants of the survivors of such conditions.

Chondrichthyes

The class Chondrichthyes, which includes sharks, skates, rays, and ratfishes, gets its name from the fact that the internal skeleton is completely cartilagenous; this is but one of a galaxy of features by which sharks can be distinguished from higher bony fishes. In habitus, however, sharks, like bony fishes, originate as active predators, as indicated by the usual combination of fusiform body, large terminal mouth with strong jaws and sharp teeth, and narrow-based, flexible paired fins.

The record suggests that sharks did not appear in this role until about 50 million years after the bony fishes had established themselves in it, and if this interpretation is valid, the success of sharks is somewhat surprising. A possible explanation is that sharks became active predators in marine waters, into which bony fishes had not yet penetrated in significant numbers. The first sharks, such as cladodonts (Figure 7), appear in marine deposits of Late Devonian age; they differ from later forms in that the fins are broad based, the head is short and blunt, the eyes are very large, and the mouth is terminal. In these respects cladoselachians resemble acanthodians and have probably borne a relationship to later sharks comparable to that which acanthodians bear to more advanced bony fishes.

The origin of primitive sharks is an open question. They might have been derived from among the placoderms (Stensiö, 1963), in which many marine

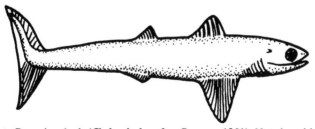

FIG. 7   Late Devonian shark (*Cladoselache;* after Gregory, 1951). Note broad-based fins.

forms were becoming increasingly sharklike by reduction and loss of armor during the late Middle Devonian. In contrast, they might be derived from an acanthodian source, perhaps from one of the groups that had already entered the sea. In the latter case they may have no closer relationship to the placoderms than do the Osteichthyes, leaving the placoderms isolated as a "sister group" to other gnathostomes (Schaeffer, personal communication, 1973). In any case, however, sharks must have been isolated from the line of other gnathostomes at least since the Devonian.

Amphibia

Present discussion of the class Amphibia can be restricted for the most part to the subclass Labyrinthodontia, and within the subclass to the relatively few forms that exhibit a terrestrial habitus as adults. Such animals are in general the earliest members of each of the three orders into which the labyrinthodonts are divided, and it is among them or their immediate derivatives that the best annectant forms are found. Some examples are *Ichthyostega* (Figure 8), of Late Devonian age, the earliest tetrapod (order Ichthyostegalia); the earliest members of the order Anthracosauria, which includes the ancestors of reptiles; and at least one line of the order Temnospondyli (Figure 9), also of very early appearance, which probably includes the ancestors of living amphibians.

It must be noted in passing that most labyrinthodonts retained an aquatic habit throughout the life cycle and that the most characteristic feature of labyrinthodont adaptive radiation has been the evolution of at least four distinct types of aquatic habitus. However, the priority of terrestrial adaptation indicates that the primary thrust of amphibian origin has been toward life on land and that adaptation to permanent life in water has been secondary. For this reason, and because later tetrapods appear to have sprung from forms of terrestrial habitus, aquatic labyrinthodonts need not be further considered.

Paleozoic labyrinthodonts as a whole range from the size of modern salamanders to more than 5 m in length, averaging much larger than living forms. They resemble a living salamander, such as *Ambystoma*, only insofar as they and *Ambystoma* differ from fish: the head is supported on a flexible neck because the bony gill covers of fish are lacking and gills (when present) are external, and the paired limbs are legs rather than fins. They differ from living amphibians in that the skin is covered with small bony scales, thin and overlapping in some and

FIG. 8 *Ichthyostega* body form (after Romer, 1966).

FIG. 9 Temnospondylous amphibian, Early Permian, terrestrial habitus. (*Cacops;* after Romer, 1966).

stout and articulated in others. Whether, in spite of the scales, the skin was glandular and slimy as in fish and living amphibians cannot be determined. However there is little question that labyrinthodonts are amphibians in the classic sense of the term, for larval forms complete with external gills, finned tail, and bodily proportions comparable to larval *Ambystoma* are known that are clearly referable to a variety of big, scaly adults.

The terrestrial habitus of labyrinthodonts is distinctive and is quite uniform throughout the subclass. In such animals the trunk is short, with from 24 to 27 presacral vertebrae; the limbs are stout, with massive processes for muscle attachment; the tail is of modest length; and the snout is not elongate, the orbits lying about halfway along the length of the skull. The structure of trunk and limbs clearly reflects the requirements of a land environment, in which the effect of gravity is not counteracted by the buoyancy of water. The small number of presacral vertebrae is part of an anatomical complex that is able to maintain vertical rigidity of the trunk while permitting a degree of lateral flexion, and the stout, well-muscled structure of the limbs permits a true walking locomotion. There are numerous trackways of amphibians of this general build which, although they may include tail drag marks, show no indication that the belly has dragged. Such amphibians therefore walked—they did not crawl.

Living forms that most closely approximate this sort of habitus are such turtles as the snapper, *Chelydra.* They have the same short "wheelbase," stout limbs with blunt feet, and moderately flattened profile. Like snapping turtles, Paleozoic amphibians of terrestrial habitus probably spent a lot of time in and near water; the point is that they were not adapted to living permanently in water as fishlike or eel-like swimmers, and they could move overland for long distances if necessary.

In *Ichthyostega*, the presence of many characters held over from fish ancestry merely emphasizes a primitiveness consistent with the early appearance of the animal. It in no way minimizes the level of terrestrial adaptation that is clearly evident in both general habitus and in finer anatomical detail of the axial skeleton. The first anthracosaurs, of Late Mississippian age, are also quite primitive but in some of them body form expresses a terrestrial habitus as clearly as it does in *Ichthyostega.* In these animals, however, detail of vertebral structure is essentially comparable to that of the earliest reptiles, although the skull is no different from that of any other primitive anthracosaur. This peculiar combination of "reptilian" postcranial structure and anthracosaur skull continues, in the suborder Seymouriamorpha, from the Mississippian to the end of the Permian, long after true reptiles have become well established and diversified. It evidently represents a line of anthracosaurs that have found a good thing in being terrestrially adapted as adults but are unable to shed the amphibian larval stage that must be passed in water.

Elimination of an aquatic larval stage is accomplished with the origin of the amniote egg, which must still be the definitive criterion of a reptilian level of organization. The fossil record can never show directly how this has come about, but some hints are available in certain tetrapod faunas of Middle Pennsylvanian age (Carroll, 1969).

The animals in this fauna are all quite small, not more than 20 cm or so in length. Their habitus is most closely comparable to that of a lizard; the body is slim and elongate, although there are no more than 27 presacral vertebrae, and the limbs are very long and slender, as are the digits, which are clawed. The skulls are also lizardlike, being short, narrow, and high, with large lateral orbits placed halfway along the length of the skull. This habitus is more completely terrestrial than that of terrestrial labyrinthodonts so far discussed. The fauna was from a coal forest, and Carroll suggests that some of its members were actually arboreal.

Most of these animals, by the criterion of skull structure, are true reptiles, the first of their kind (Figure 10). One, however, the genus *Gephyrostegus*, has a skull of anthracosaur structure and dermal bone pattern and is therefore technically an amphibian. In noting the close approximation of *Gephyrostegus* to a reptilian level of terrestrial adaptation, Carroll emphasizes the small size of the animals in question. The significance of size, he suggests, lies in the fact that problems of support, movement, and water conservation in a terrestrial environment are less difficult for small animals than for large. Finally, he notes that

FIG. 10 Middle Pennsylvanian anapsid reptile, lizardlike habitus. (*Hylonomus;* after Carroll & Baird, 1972).

small amphibians of living taxa, many of them arboreal, have repeatedly devised means of laying eggs on land, thus shortening or skipping the larval stage. He suggests that it was in such circumstances that the amniote egg evolved, sometime during the first half of the Pennsylvanian Period.

Origin of the subclass Lissamphibia (frogs, toads, and salamanders) is less well documented than that of reptiles. On the basis of skull structure, dermal bone pattern, and embryogeny of the vertebral column, modern amphibians are fairly readily derived from among the temnospondyls. Within that order, animals of terrestrial habitus (Bolt, 1969) are the best candidates for lissamphibian ancestry. More detailed consideration of this point is unwarranted; however, there is sufficient evidence to indicate that the Temnospondyli as ancestors of Lissamphibia, and the Anthracosauria as ancestors of Reptilia, are distinct from the time of their earliest appearance in the fossil record and that the living derivatives of these groups have arrived independently at the level of terrestrial adaptation that they show at the present time.

Reptilia

The adaptive theme of reptiles appears at the outset to have been an enhancement of the trend toward increasing terrestriality set by the earliest amphibians. Reptiles soon surpass most amphibians, as the latter retreat to various aquatic habitats, and ultimately come to dominate land faunas in a great variety of adaptive modes over a period of some 200 million years. During most of this time, however, the reptiles that are dominant are not those that give rise to mammals and need not be considered further.

The reptilian subclass Synapsida, from which mammals ultimately spring, is already recognizable, on the basis of skull anatomy, in one of the small forms of lizardlike habitus of the Middle Pennsylvania (Reisz, 1972). Others in this fauna are referable, on the same basis, to the subclass Anapsida (Carroll & Baird, 1972), which can be recognized with varying degrees of confidence as the source of most other reptilian lines. The Synapsida and Anapsida are perhaps the two best-documented reptilian subclasses and their Middle Pennsylvanian representatives are so similar in nearly all respects that they must lie close to their common ancestry.

Synapsids dominated the terrestrial environment from the Late Pennsylvanian to the end of the Early Triassic, an interval of about 85 million years. The first half of this time marks their primary adaptive radiation, which is manifest primarily as increase in size, with minor diversification of form, and is represented by the order Pelycosauria. Early Permian pelycosaurs ranged in length from about 30 cm to 4 m; predatory, fish eating, and other specializations appeared, but general habitus remained lizardlike.

Dimetrodon (Figure 11) is an Early Permian predaceous pelycosaur in which the dorsal neural spines are enormously elongate. The spines supported a skin-covered membrane, which would have increased the animal's surface area

FIG. 11   Early Permian pelycosaurian reptile (*Dimetrodon;* after Hopson, 1967).

greatly. It probably served as a heat exchanger, permitting *Dimetrodon* to warm up or cool off quickly. A similar elongation of dorsal neural spines supporting a "fin" appeared in the genus *Edaphosaurus,* which, whatever it was, was not a predator and was not closely related to *Dimetrodon.* Such fins were rare among pelycosaurs, however, and there were predators much like *Dimetrodon* and "whatnots" much like *Edaphosaurus* without them. They probably indicate some sort of modification of the temperature-control mechanism in the animals in which they occur but are too scarce to warrant broad generalizations about temperature control in pelycosaurs.

During the second half of the time in which they were dominant, essentially the Late Permian and Early Triassic, the synapsids underwent a more extensive secondary adaptive radiation, represented by the order Therapsida. Therapsids evolved a much more elaborate ecological interrelationship, producing a variety of large and small plant-eating forms, the first of their kind among tetrapods; large and small predators; insectivorous animals; and possibly a few fish eaters. In South Africa, the ratio of putative herbivores to all other kinds in numbers of individuals is about ten to one, which suggests that the South African deposits record the oldest ecologically balanced terrestrial fauna in which vertebrates are the primary consumers.

The habitus of therapsids was no longer lizardlike: legs were longer although still heavily muscled; feet were shorter, heads larger, and tails shorter. The new habitus probably reflected a trend toward greater activity, with which, especially among predators, went hints of a real start toward a constant-temperature metabolism, such as differentiation of teeth, proportionately larger heads accommodating longer dental batteries, and appearance of secondary palates and ultimately of nasal turbinals. These developments were not closely integrated, however, and did not proceed at a constant rate. For example, the skull of a gorgonopsian, such as *Lycaenops* (Figure 12), is but little advanced over the

FIG. 12 Late Permian therapsid reptile (*Lycaenops;* after Colbert, 1948).

condition of *Dimetrodon* but the limbs of *Lycaenops* are longer, slimmer, and more "mammal-like" in orientation than anything else that appears in the synapsid–mammal line before placentals approach a cursorial habit in the Paleocene, some 165 million years after the end of the Permian. In cynodonts (Figure 13), however, the skull is very mammal-like in dentition, secondary palate, details of the nasal passage, and proportion of the dentary bone, whereas the limb bones are reminiscent of those of *Dimetrodon.* Ultimately, however, it was some of the cynodonts in the Middle Triassic that gave rise to the mammals.

ADAPTIVE ORIGIN OF CLASS DIFFERENCES

The characterization of each class of poikilothermous vertebrates by a distinctive adaptive theme does not of itself demonstrate that class differences have originated by adaptation. There is, in fact, a contrary argument, based on such observations as the finely honed adaptations among insects and the flowers they pollinate, or the almost universal specificity of parasites, holding that the only effect of adaptation is an ever-narrowing specialization. In this view, adaptation is not innovative or creative and the origin of higher taxonomic categories must result from a process different in kind from adaptation.

FIG. 13 Early Triassic cynodont reptile (*Thrinaxodon;* restored by L. B. Isham and N. Hotton from Brink, 1956, and from a specimen in the Smithsonian Institution, catalogue number USNM 22812).

This conclusion is erroneous, in my opinion, for two reasons: one, it does not give sufficient weight to the evidence of the fossil record; and two, it expresses too rigid and narrow a concept of the function of specialized structures. When the fossil record and the role of function are properly taken into account, it becomes apparent that vertebrate classes do not merely express distinctive adaptive themes but that they actually come into being as a consequence of adaptive processes.

It is true that in detail the fossil record will always be full of gaps and biases. It is also true, however, and more importantly, that in sequences and relative abundance of forms the fossil record has yielded a remarkably consistent story over the past 100 years of study and that the increasingly numerous discoveries in this interval have gone far to document biological transitions in a manner consistent with evolutionary theory. Therefore, by the principle of Occam's razor, it must be assumed that the fossil record provides a pragmatically reliable measure of the actual history of the vertebrates—that the earliest record and the first record of abundance of a group indicate, respectively, the times of origin and of primary adaptive radiation of that group, within a practical margin of error. On this assumption, the fossil record also indicates that when radiating into a new environment or life style in the absence of significant competition, organisms tend to occupy the new environment to its very periphery in relatively short order.

These assumptions permit functional interpretation of the origin and primary adaptive radiation of the bony fish. Bony fishes originated as small active predators in fresh water. During primary radiation the ancestral habitat of well-aerated fresh waters remained well populated by rayfins and most acanthodians, even as such peripheral aqueous environments as brackish or salt waters, were being penetrated by some acanthodians and as fresh waters subject to periodic fouling or drying were by lobefins.

Crossopterygian lobefins are customarily conceived of as generalized fishes and their evolutionary potential is usually dealt with in terms of the subsequent history of tetrapods. In fact, however, they have been specialized for life in a rather specific environment and the process by which their evolutionary potential has been realized (as opposed to the descriptive history of this realization) must be clarified in functional terms; for no structure, however specialized to a specific function, serves that function alone nor is restricted to a single function in all circumstances.

In crossopterygians, lungs, shortness of the "presacral" vertebral column, and the lobate structure of the paired fins are all features that anticipate the adaptive theme of tetrapods. As a consequence their adaptive significance as fish structures is often underemphasized. Lungs are probably very basic gnathostome equipment, for they are known to have been present, with connection to the pharynx, in some placoderms (Denison, 1941). In most bony fish they are present in modified form as a swim bladder of hydrostatic function, which in one superorder of rayfins also serves as part of a sound detecting and transmitting mechanism, and in a variety of others serves very well for respiration. Even in lungfishes the function of the lungs is as much hydrostatic as respiratory. Shortness of the "presacral" vertebral column in crossopterygians may represent a primitive character of bony fish generally, but it also appears comparable to shortness of body form in a variety of rayfins inhabiting still waters at the present time. Similarly, the lobate fins function as an accessory locomotory mechanism analogous to one present in a variety of still-water fishes. Under ordinary circumstances all of these features can function in a totally fishlike way. For example, living crossopterygians, the Coelacanthini, inhabit deep waters of the sea. In bodies of water that became foul or dried up, however, these structures could assume functions other than those they ordinarily served; it was because they had such structures that some crossopterygians could survive in such places, but it must be emphasized that they survived as fishes and continued to do so long after the appearance of amphibians.

In the crossing of the adaptive threshold from fish to amphibian, two broad selective factors can be distinguished: stress of the peripheral aquatic environment, and opportunity combined with lack of competition in the terrestrial environment. Stress was the consequence of regular but partial and temporary exposure to terrestrial conditions but, as noted, the adaptations of the fish ancestors of amphibians accommodated them to it strictly as fish. The role of stress was merely one of setting the stage, therefore, and without the incentive of opportunity and lack of competition in the new environment, the crossopterygians would have continued as fishes until they became extinct or until the present, as lungfishes did.

The opportunity offered by the terrestrial environment was almost certainly a new source of food. By Late Devonian time, land surfaces had been supporting a diversified flora for at least 60 million years. Insects are recorded from the Mississippian, and it is extremely likely that some arthropods, which are admirably preadapted for life on land, have been living on land plants almost from the beginning. A biota of plants and arthropods lacking any significant competition for vertebrates therefore must have existed for some time before the origin of amphibians. Once certain crossopterygians began to feed on land, they became more completely subject to selective pressure toward adaptation to the terrestrial environment, and with no competition facing them it was a relatively short time before they shifted over to being amphibians.

The first appearance of amphibians coincides with the primary adaptive radiation of bony fishes. This coincidence, together with the similarity of the first amphibians to certain crossopterygians in structure and habitat, suggests that the appearance of the new class has been essentially a part of the primary radiation of the parent class. From the first appearance of acanthodians to the peak of primary radiation of the bony fishes is about 50 million years, a measure of the time required for the process to be completed. The origin of reptiles exhibits a similar pattern: the first appearance of reptiles coincides with the peak of primary radiation of amphibians, and from *Ichthyostega* to the earliest reptiles is also about 50 million years. In spite of a certain lack of faunal detail, the same process seems to be at work in the origin of acanthodians: the first appearance of acanthodians coincides with the peak of agnathan primary radiation, and from the oldest certain agnathan record to the first appearance of acanthodians is not more than 60 million years.

To recapitulate briefly, the origin of vertebrate classes during the Paleozoic seems to correspond to the following pattern. Once into a new environment or new mode of life, organisms radiated rapidly and soon impinged on the limits of that environment or life style; functional multipotentiality of structure permitted some forms to survive not merely near but right at the limits. The environment outside of these limits already had the potential of supporting vertebrates but contained no competitors for them, and when marginal forms became capable of exploiting this potential they crossed the limits and the process began again.

However, this pattern is not applicable to the origin of mammals, first because the origin of mammals is not as clearly definable as the origin of earlier classes; second, because the time from the first appearance of synapsid reptiles to the first appearance of mammals is much longer than the comparable interval in earlier transitions; and third, because the origin of mammals, however defined, does not correspond to any adaptive radiation of the parent class.

The Late Permian record shows synapsid reptiles to be evolving a variety of mammal-like features at different rates (Olson, 1944). Moreover, the record above a strictly synapsid level of organization indicates a morphological spectrum ranging from an advance over the synapsid level (by definition) only by virtue of a squamoso–dentary jaw joint in the Late Triassic, to marsupial and placental levels in the Late Cretaceous. The differences between this transition and earlier ones are not merely functions of the quality of the record, for whereas the Late Permian record is one of the best available, the Mesozoic record is no better than that of earlier transitions. Nevertheless, the pattern of Mesozoic evolution of the mammalian line is clearly a continuation of the pattern established during the Late Permian.

If the first mammals are defined at the lowest reasonable level of organization (Late Triassic), the interval from the origin of reptiles to the first mammals is about 100 million years; if the first mammals are defined at a marsupial or placental level (Late Cretaceous), the interval is about 165 million years. By the Late Triassic the secondary radiation of synapsid reptiles is past and the group is in its final decline, probably being outcompeted by the rising dinosaurs; by the Late Cretaceous the synapsids have long been extinct except for their derivatives, the mammals, and the dinosaurs themselves are probably declining.

The striking divergence of the pattern of mammalian origin was apparently created by increasingly extensive exploitation of available conditions of life. By the time therapsid ancestors of mammals had come on the scene, they were merely tetrapods among a variety of tetrapods (including other therapsids). The evolution of specialized mammalian features, primarily a matter of increasing mobility and refinement of physiological processes, did not offer the potential of freedom from competition, because other tetrapods were also experimenting successfully with improvement of mobility and physiology. In these circumstances, the only way in which new conditions would become accessible to the immediate ancestors of mammals was through changing ecologies involving the decline of competing or dominant forms, for example when the therapsids declined as a fauna before the end of the Triassic or when dinosaurs went into decline in the Late Cretaceous. Because such conditions did not exist at the time of the primary or secondary adaptive radiation of synapsids but appeared only long afterward, the synapsid ancestors of mammals had to wait until they came into existence. This probably accounts, at least in part, for the long drawn out nature of the transition from reptiles to mammals.

HERITAGE AND HABITUS IN THE CENTRAL NERVOUS SYSTEM

Ideally, the evolutionary changes of any organ system are traced through phylogenies, which are based for the most part on data and procedures of the complementary disciplines of comparative anatomy and vertebrate paleontology. Unfortunately, these disciplines do not complement each other fully, for both are biased against primitive and annectant forms, vertebrate paleontology because annectant forms have apparently been short lived as such, and comparative anatomy because its data base is restricted to the living fauna. The bias of comparative anatomy is by far the more pronounced, for the vertebrate fauna of the present day is absurdly lopsided when compared with the fossil record: one group of rayfins and a few chondrichthyans dominate the waters, both fresh and salt; the specialized descendants of only two of some 18 orders of reptiles dominate the land. From a strict phylogenetic viewpoint there are only a handful of animals that provide a window to the myriad extinct forms: for agnathans, the highly atypical lampreys and hagfishes; for placoderms, none (Schaeffer, personal communication, 1973), elasmobranchs (Stensiö, 1963), or ratfishes (Patterson, 1965); for primitive ray-finned fishes, the sturgeon, paddlefish, bowfin, gar, and *Polypterus,* each atypical of early forms in its own way (for additional detail see Schaeffer, this volume). For the ancestors of amphibians and reptiles what little exists is unreliable; in the former category are *Latimeria* and three kinds of lungfishes and in the latter are the living amphibians, none phylogenetically close to the stipulated forms. Finally, there are no living reptiles close to mammalian ancestry unless the egg-laying monotremes are considered synapsids.

Nervous structure suffers especially severely from the biases of the data base because, like most "soft parts," it is very poorly represented in the fossil record. Acceptable phylogenies are of limited value because functionally significant anatomical detail of the nervous system is unknown for most of each line and can be determined only in end forms. Because of broad similarities of nervous structure in end forms, it is often concluded that the central nervous system is evolutionarily conservative. To some, the limitations of established phylogenies means that the evolution of the central nervous system is inaccessible; to others, the presumed conservatism of the central nervous system means that details of its evolution are not significant.

Neither of these viewpoints is fully warranted, because evolutionary plasticity of the central nervous system has not been adequately explored. Adaptive plasticity of the skeletomuscular system and the organs of special sense certainly appears to be a dominant theme in vertebrate evolution, and it is inescapable that the central nervous system should in some way reflect this plasticity because of its very high functional correlation with these structures. Evidence of such reflection may eventually be identifiable in lower vertebrates, in which basic central nervous structure is not complicated by elaboration of higher centers as in birds and mammals.

The situation in living lobefins may be just such a case. At one time amphibians were thought to have been derived from lungfishes and the similarity of telencephalic structure in the two groups was part of the evidence for this conclusion. As the somatic structure of crossopterygians and fossil amphibians became better known, however, derivation of amphibians from lungfishes fell into disrepute, and the similarity of telencephalic structure came to be interpreted as a reflection of similarity of mode of life. This idea has been given added credence by the structure of the telencephalon of *Latimeria* (Millot & Anthony, 1965), the only living crossopterygian, which is more suggestive of rayfin than of lungfish structure. The mode of life of *Latimeria,* a deep-water marine fish, is much more reminiscent of the life style of rayfins than that of lungfishes or amphibians.

Sharks and bony fishes provide adequate subjects on which to test this notion more extensively, for they are not at all close to each other phylogenetically, and each group contains living forms that are comparable in habitat specialization and emphasis of exteroceptive and locomotor equipment. Comparison of the central nervous system in terms of habitus similarities in these classes should at the very least give a better perspective of the role of convergence in the evolution of the nervous system than is available at present. It may also actually make it possible to adduce central nervous system structure of fossil forms from living animals of similar habitus without as rigid an adherence to putative phylogenies as is currently required. Moreover this approach may also be useful outside the limits of sharks and bony fishes. For example, most ancient agnathans fed at the substrate and must have had a tadpolelike mode of swimming; in habitus they were radically different from modern agnathans. The living animals most like them in habitus and in position, if not mode, of feeding are frog or toad tadpoles, most of which, besides being larvae of tetrapods, are also free-living organisms functioning in an aqueous environment. It is not too farfetched to suggest that a study of the brains of tadpoles from this point of view, with due considerations of the brains of larval lampreys, may be at least as valuable to the understanding of the central nervous system of the earliest known vertebrates as is the study of the brains of adult lampreys.

REFERENCES

Bolt, J. R. Lissamphibian origins: Possible protolissamphibian from the Lower Permian of Oklahoma. *Science,* 1969, **166**, 888–891.

Brink, A. S. Speculations on some advanced mammalian characteristics in the higher mammal-like reptiles. Palaeontologica Africana, 1956, **4**, 77–96.

Carroll, R. L. Problems of the origin of reptiles. *Biological Reviews* (Cambridge), 1969, **44**, 393–432.

Carroll, R. L., & Baird, D. Carboniferous stem reptiles of the family Romeriidae. *Bulletin of the Museum Comparative Zoology* (Harvard), 1972, **143**(5), 321–363.

Colbert, E. H. The mammal-like reptile *Lycaenops. Bulletin of the American Museum of Natural History,* 1948, 89(6), 359–404.

Denison, R. H. The soft anatomy of *Bothriolepis. Journal of Paleontology,* 1941, **15**(5), 553–561.

Gregory, W. K. *Evolution emerging.* New York: Macmillan, 1951.

Hopson, J. A. Mammal-like reptiles and the origin of mammals. *Discovery,* 1967, **2**, 25–33.

Miles, R. S. Jaw articulation and suspension in *Acanthodes* and their significance. In T. Ørvig (Ed.), *Current problems of lower vertebrate phylogeny, Nobel Symposium.* Vol. 4. New York: Wiley, 1968. Pp. 109–127.

Millot, J., & Anthony, J. Anatomie de *Latimeria chalumnae.* 2. Systeme nerveux et organes de sens. Paris: Centre National Recherche Scientifique, 1965.

Nelson, G. J. Gill-arch structure in *Acanthodes.* In T. Ørvig, Ed., *Current Problems of Lower Vertebrate Phylogeny, Nobel Symposium.* Vol. 4. New York: Wiley, 1968. Pp. 129–143. 129–143.

Olson, E. C. Origin of mammals based upon the cranial morphology of the therapsid suborders. *Geology Society of America Special Papers,* 1944, **55**, 1–136.

Parrington, F. R. 1958. On the nature of the Anaspida. In T. S. Westoll, Ed., *Studies on fossil vertebrates.* London: University of London, Athlone Press, 1958. Pp. 108–128.

Patterson, C. The phylogeny of the chimaeroids. *Philosophical Transactions of the Royal Society* (London), 1965, **B249**, 101–219.

Reisz, R. Pelycosaurian reptiles from the Middle Pennsylvanian of North America. Bulletin of the Museum of Comparative Zoology, (Harvard), 1972, **144**(2), 27–61.

Romer, A. S. 1966. Vertebrate paleontology. (3rd ed.) Chicago: University of Chicago Press, 1966.

Romer, A. S. Vertebrate evolution with special reference to factors related to cerebellar evolution. *Neurobiology of cerebellar evolution and development: First international symposium.* Chicago: American Medical Association Education and Research Foundation, 1969. Pp. 1–18.

Simpson, G. G. *The major features of evolution.* New York: Columbia University Press, 1953.

Stensiö, E. A. The Downtonian and Devonian vertebrates of Spitsbergen. Part I. Family Cephalaspidae. *Skrifter om Svalbard og Nordishavet,* 1927, **12**, 1–391.

Stensiö, E. A. Anatomical studies on the arthrodiran head. 1. Preface, geological and geographical distribution, the organization of the arthrodires, the anatomy of the head in the Dolichothoraci, Coccosteomorphi, and Pachyostomorphi. Taxonomic appendix. *Kungliga Svenska Vetenskapsakademiens Handlingar Series* 4, 1963, 9, 1–419.

Stensiö, E. A. The cyclostomes with special reference to the Petromyzontida and Myxinoidea. In T. Ørvig, Ed., *Current problems of lower vertebrate phylogeny, Nobel Symposium*. Vol. 4. New York: Wiley, 1968, Pp. 13–71.

Thomson, K. S. New evidence on the evolution of the paired fins of *Rhipidistia* and the origin of the tetrapod limb, with description of a new genus of Osteolepidae. *Postilla, Peabody Museum of Natural History* (Yale University), 1972, 157, 1–7.

White, E. I. The ostracoderm *Pteraspis* Knerr and the relationships of the agnathous vertebrates. *Philosophical Transactions of the Royal Society (London)*, 1935, **B225**, 381–458.

2

Vertebrate Olfaction

Don Tucker
James C. Smith
Florida State University

INTRODUCTION

Olfaction does not appear to be very important to man as judged solely by the extent of anatomical development relative to that of the other senses, especially vision, audition, and somesthesis. Behaviorally, its greatest role seems to be in connection with eating, where its importance in the appreciation of flavor has been emphasized recently. There are also persistent suggestions that olfaction is important in sexual behavior, but the existing taboos render most such material rather anecdotal in character.

Familiarity with other animals, such as the bird or the dog, calls to mind different roles for olfaction. Since the days of Audubon, the bird has been considered as having poor, if any, sense of smell. The dog, in contrast, is an animal in which olfaction is thought to play a major role. In the literature regarding animal olfaction, the terms "microsmatic" and "macrosmatic" are often used. The classification of an animal into one of these two categories has traditionally been done on either anatomical or behavioral bases. The absolute or relative size of the olfactory bulbs, the number of olfactory ridges (conchae), area of olfactory mucosa, or color of the mucosa are some of the anatomical features used in such classifications. The behavior of an animal leading to a classification of either microsmatic or macrosmatic has been just as varied; i.e., whether tracking is observed, whether the nose is kept close to the ground, whether the animal digs for buried food sources, flies or swims toward an olfactory source, etc. In this review it will be seen that, on the basis of behavior, one vertebrate class (birds) has historically been classified as microsmatic. In the other classes, a wide variety of species has been described, ranging from micro- to macrosmatic.

The evolution of olfaction must be deduced mainly through study of living forms. Anatomy, physiology, and behavior are principal sources of information. Embryology yields valuable information about anatomy and to the extent that "ontogeny recapitulates phylogeny," embryology bears directly on the question. A taxonomic scheme is sometimes said to be analogous to a genealogical tree, indicating evolutionary relationships. Current methods permit examination of only a few terminal twigs on such a tree. However, we are convinced that the tree is exceedingly complicated and that the living forms have arrived at the present through parallel, convergent, and divergent lines from which great numbers have become extinct during the course of evolution. Therefore, we cannot tell an evolutionary story. Our type of information may be usefully integrated into the store of knowledge that specialists may possess, however. We shall present a resume of anatomy for purposes of orientation, examine a question about the origin of the olfactory organ of tetrapods, and then go into more detail by vertebrate class.

ANATOMY

The cerebral hemispheres are generally believed to have begun as a paired correlation center for the special sense of olfaction. The concept of the ancient smell brain, to which Herrick (1921) has certainly given impetus, has been enthusiastically demolished. The limbic system incorporates most of the neuro-anatomical structures that are in the former system. However, recent results suggest that olfaction does influence the functioning of some limbic structures. The suggestion has even been made that it may be profitable to regard the olfactory system as having been derived from the limbic system and that the nervus terminalis (see below) may be peculiar to the latter system (Riss, Halpern, & Scalia, 1969a).

Olfaction seems to have been a prominent feature of the sensory equipment of the earliest vertebrates. Because such animals have lived only in water, the anthropomorphic view of olfaction as being mediated by a stimulus conveyed through air must always be avoided. The structure of the olfactory receptors is remarkably constant in all the living forms in which they are to be found. A century ago it was just being determined that there was no sensory ganglion of the olfactory nerve analogous to the ganglia of the other sensory nerves. The nerve fibers originate from primitive neuroreceptor cells in the olfactory mucosa. They are primitive because this is the general scheme of invertebrate receptor cells in general. The origin of the vertebrate olfactory receptor may therefore have been from any of the invertebrate Metazoa. Many arthropods are used in studies of "olfaction" and "taste," although some workers designate the receptors as distance and contact chemoreceptors.

The olfactory nerve is transmitted from the nasal cavity to the olfactory bulb in the cranial cavity where the first synapse in the nervous pathway occurs. Commonly, there is a specialized subdivision of the olfactory system known as the "vomeronasal organ," or the "organ of Jacobson," that embryologically arises from the nasal placode, as does the olfactory organ proper. The vomeronasal nerve connects Jacobson's organ to the accessory olfactory bulb, which is typically located somewhere caudally relative to the principal olfactory bulb.

Jacobson's organ in man was discovered 270 years ago by Ruysch (Van Wijhe, 1919; Negus, 1958). Jacobson independently discovered the organ in various mammals. He also discovered the vomeronasal nerve and felt that the organ must have an olfactory function. The work became known because of a report made by Cuvier (1811), in whose laboratory some of the work was done, to the Institut de France of a note by M. Jacobson, a surgeon serving in the army of the king of Denmark.

The nervus terminalis (nerve of Pinkus) was discovered in the lungfish *Protopterus annectens* by Pinkus (1895) in 1894 and shortly thereafter was described by Locy (1905) in 20 genera of sharks and rays. At this time there was much confusion among the vomeronasal nerve, the terminal nerve, branches of the olfactory nerve, and even among branches of the trigeminal nerve, as discussed by Van Wijhe. The nerve has been found in many animals, especially in embryonic stages, where it appears only transiently in birds. The nerve has often been thought to have sensory and motor components but its distribution and function have never been adequately determined. The last major review is that by Larsell (1950).

The nasal cavity is also innervated by branches of the trigeminal nerve, which is generally believed to serve the modalities of tactile sense, temperature, pain, and proprioception. However, there are trigeminal fibers that respond well to odorants (Tucker, 1971). Parker (1922) popularized the concept of a common chemical sense mediated by "bare nerve endings" such as those of the trigeminal nerve fibers, being somewhat less sensitive in general than the gustatory sense, which of course is usually much less sensitive than the olfactory sense.

The olfactory (vomeronasal) sensory mucosa is recognized in almost all noses of tetrapods and the remainder of the nasal mucosa is commonly designated as "respiratory" mucosa. The respiratory mucosa typically has ciliated cells exhibiting a metachronal rhythm that propels a sheet of mucus. The trigeminal innervation seems to be denser in respiratory mucosa, although there are indications of trigeminal fiber endings in olfactory mucosa. The presence of nervus terminalis fibers has been claimed in both types of mucosa. The olfactory epithelium usually occupies a rather sheltered part of the nasal cavity and the respiratory epithelium may grade over into an indifferent nonmucosal type of epithelium. The geometry of the nasal cavity is extremely variable. The nasal cavity of tetrapods is connected to the oral cavity by the internal nares or choanae, which have figured prominently in evolutionary studies. In most fishes, of course, such an oronasal connection does not exist. The vomeronasal sensory mucosa is most often enclosed within a structure called the "organ of Jacobson," which may open within the nasal cavity (many rodents), into the nasopalatine duct transmitted by the incisive foramen just behind the incisor teeth (characteristic of mammals), or directly onto the palatine surface (snakes and lizards). Adult birds and some other forms do not have the organ. Fishes, in particular, appear not to have any vomeronasal sensory mucosa. An accessory olfactory bulb has been described in some forms of fishes, but apparently in no such instance has the peripheral anatomy been investigated.

Nasal anatomy has been invoked as an aid in the study of phylogenetic problems (Parsons, 1959a). However, Parsons (1967) has forthrightly decided that no definite conclusions can be reached about evolution of nasal structure in the lower tetrapods. His (Parsons, 1959b) large work on comparative embryology of the reptilian nose is more comprehensive than is suggested by the title and the bibliography represents an extensive literature review. Bertmar (1965) has studied the nasal embryology of lungfishes and notes the ganglion of the terminal nerve, first discovered in a lungfish. The relations between fish and tetrapod noses, choanal connections to the mouth, accessory nasal sacs, possible homologs of Jacobson's organ, etc., have been discussed further by Panchen (1967) and Bertmar (1969).

WAS THE TETRAPODAN VOMERONASAL ORGAN DERIVED FROM THE FISH OLFACTORY ORGAN?

As noted above, there is commonly a specialized subdivision of the olfactory system known as the vomeronasal organ, or Jacobson's organ, which embryologically arises from the nasal placode as does the olfactory organ proper. Histologists for a century referred to the olfactory cells of the sensory epithelium of Jacobson's organ. A diagnostic feature is that Bowman's glands, which are characteristic of most olfactory mucosae, are not present in any vomeronasal mucosae. The fact that there are no Bowman's glands associated with the olfactory organ of fishes has prompted Parsons (1970a) to revive the suggestion by Broman (1920) that the fish olfactory organ is the evolutionary precedent of the tetrapod vomeronasal organ. The medium for carrying stimulus molecules to the receptors would be glandular secretions, which would be transported by pumping movements in the mammalian organ and possibly by ciliary streaming in reptiles (Pratt, 1948; Bellairs & Boyd, 1950). Parsons' proposition is that the olfactory organ actually originated in connection with the transition to terrestrial conditions. This leads him to the conclusion that the olfactory bulb of fishes is missing and what is actually present is the homolog of the accessory olfactory bulb, to which the vomeronasal nerve connects the vomeronasal organ of tetrapods.

Broman (1920) tried to support his proposition that the Jacobson's organ was the old water smelling organ with the supposed association of the nervus terminalis with the vomeronasal organ of tetrapods and with the nasal sac of fishes. Although Kerkhoff (1924) supported Broman enthusiastically, Herrick (1921) rejected Broman's suggestion that Jacobson's organ was nothing other than the old water olfactory organ of vertebrates adapted for life on land; "an untenable position which he unfortunately attempted to support by reference to the nervus terminalis which reveals a total neglect of the recent contributions dealing with the innervation of this region." Herrick evidently accepted the proposition of Seydel (1895) that the organ of Jacobson developed in connection with the establishment of a choanal connection from the nose to the oral cavity, a proposition that was enthusiastically supported by Bruner (1914). Thus, a more intimate relation between gustation and olfaction during the act of feeding was established and the physiological integration of all the associated sensory information led to the appearance of a primitive amygdala (Herrick, 1921). The transition to terrestrial life was presumably an independent development.

Current knowledge about the nervus terminalis is still very scanty and is only somewhat better in respect to the nervus vomeronasalis (Tucker, 1971). The existence of a separate projection from the frog accessory bulb to the primordial amygdala has been confirmed (Scalia, 1972). However, there is every reason to believe that the gustatory system has persisted in parallel with the long existence of the olfactory system (Barnard, 1936). It seems likely that both are intimately involved in consummatory behavior of many representative types of living animals and that there must therefore be neural integration of the information arriving from the olfactory bulb at the front and from the bulb of the hindbrain, the medulla oblongata.

Parsons (1970a) indicated that the theories of Seydel and Broman were not necessarily in conflict, because Broman believed that the vomeronasal organ detected substances dissolved in glandular secretions, substances that likely came from the mouth. Parsons suggested that physiological evidence in the form of differential sensitivity to various kinds of chemical compounds might be decisive, because of the possibility that the vomeronasal organ of tetrapods represents the primitive nasal organ of fishes and still functions for chemoreception in an aqueous environment, whereas the olfactory organ of tetrapods is a new one adapted for chemoreception in air.

Another source of information that may bear on the question is ultrastructure. Until recently it has been thought that all olfactory receptors bear cilia, the olfactory hairs of the classical histologists, However, the vomeronasal receptors have been shown to be free of cilia (Tucker & Smith, 1969; Kleerekoper, 1969; Graziadei & Tucker, 1970; Moulton, 1971; Altner & Müller, 1968; Altner, Müller, & Brachner, 1970; Kolnberger, 1971), except for one report (Luckhaus, 1969) that seems questionable. The presence of cilia on lamprey and fish olfactory receptors is an embarrassment to the theory being considered if electron microscopy of more forms continues to show vomeronasal receptors without cilia. However, Reese and Brightman (1970) found that the olfactory receptors of the nurse shark *Ginglyomostoma cirratum* and the guitar fish *Rhinobatus lentiginosus* are nonciliated. The theory can be retained if it is posited that ancestral elasmobranchs have given rise to the tetrapods and that the other fishes have developed receptors like the tetrapodan olfactory receptors, proper. However, the report that the Australian lungfish does not have olfactory receptor cilia (Theisen, 1972) suggests the possibility that the early pattern of results may fall to pieces.

CYCLOSTOMES

The living Agnatha are presumably degenerate forms (many are parasitic) that diverged very long ago from ancestral stock common to the extinct ostracoderms. In adults there is a single nasal opening and a large olfactory organ, which exhibits bilateral symmetry in *Petromyzon marinus* (Kleerekoper, 1969). The nervus terminalis is presumed to be present (Riss *et al.*, 1969a). Taste buds and a lateral line system are present.

Shibuya (1960) has recorded electrically slow potential responses from the olfactory epithelium of *Lampetra japonica (Entosphenus japonicus)* in response to aqueous extract of dried silk worm pupae, responses that are fundamentally the same as those recorded from several species of fish. Kleerekoper (1969) and associates have experimented extensively with respect to the behavior of the North American Great Lakes form of the sea lamprey *P. marinus*. The lamprey can discriminate one of the components of a prey fish "body odor," which has been shown to be a complex mixture of free and bound amines, amino acids, and other compounds.

ELASMOBRANCHS

Sharks and rays are well known and their sense of smell is legendary. Their olfactory organs tend to be very well developed. Although the forebrain has been assumed to be dominated by olfaction, the olfactory bulb of the nurse shark has been shown to project to a restricted portion (Heimer, 1969). Electrical responses have been recorded from the forebrain of various sharks in response to nasal infusion of amino acids and tissue fluids of crabs, lobsters, fish, etc. (see Kleerekoper, 1969). Behaviorally, locomotor activity increases when such stimuli are presented in the tank.

PISCES

A tagged steelhead fingerling was released at the Alsea River fish hatchery on the coast of Oregon and 5 months later was caught 2000 miles away in Alaska. Seventeen months later, the fish returned to the Alsea hatchery (Wright, 1964). It is well known that several species of salmon spawn in fresh water, go to the sea for several years, and then home to the stream where they have hatched. Olfaction particularly has been implicated in homing behavior of salmon by observations of the effects of sectioning the olfactory nerves (Craigie, 1926). Plugging the apertures to the nasal sacs has yielded results supporting a home-stream odor theory (Wisby & Hasler, 1954). Salmon fingerlings are thought to be conditioned to a characteristic odor in home water. Jones (1968) has developed an elaborate sequential hypothesis for home-stream detection in salmon. Other sensory systems seem to be involved also in the remarkable performance of salmon, in which adults return to spawn at or near the sites where they themselves have been spawned (Hasler, 1966).

Another type of fish behavior mediated by olfaction has been discovered by von Frisch (1938). Injury of the skin of a European minnow *Phoxinus phoxinus* causes a fright reaction of the other members of the school. The fright reaction caused by release of alarm substance from specialized cells of the skin has been shown to be confined to the order Ostariophysi, a very large group composed of cyprinoid and siluroid (catfish) fishes, and the newly erected order Gono-rhynchiformes (Pfeiffer, 1962, 1967).

Schooling has also been shown to be partly mediated by olfaction in several fish species (Hemmings, 1966). The German workers have been very active in studying the chemical senses of fishes, especially with behavioral methods. Teichmann (1959) has found the olfactory sensitivity of the common European eel to be much superior to that of the European minnow. Enormous dilutions are required to reach behavioral threshold concentrations for chemically defined substances commonly used in the flavor and perfume industry. About 1800 molecules of β-phenylethyl alcohol, a constituent of rose oil, are required per cubic centimeter of water. The threshold for the minnow is about 10^{10} times greater. After determining the size of the nose and the rate at which cilia propel a water stream through it, Teichmann has calculated that at threshold as little as one molecule per second in the nose of the eel is sufficient. For a biologically meaningful odor, *Tubifex* worms have been homogenized and diluted more than 10^{17}-fold to reach threshold concentration.

Glaser (1966), Teichmann's student, determined the thresholds for *Phoxinus* of various sugars, saccharin, quinine hydrochloride, sodium chloride, and acetic acid with a behavioral technique. The taste thresholds were less than 1 μM for saccharin and 0.04 μM for quinine hydrochloride. The thresholds ranged from 24-fold to 2500-fold lower than those determined for man. The quinine taste threshold was lower than the β-phenylethyl alcohol olfactory threshold that had been determined for the minnow and elimination of the sense of smell had no influence on the performance in response to quinine and acetic acid stimulation. It was therefore concluded that the secondary chemoreceptive cells, i.e., the ancillary taste receptors, were as sensitive as are the primary chemoreceptors, the olfactory neuroepithelial cells.

This surprising result has been supported by a study in the goldfish *Carassius auratus* (Zippel, Von Baumgarten, & Westerman, 1970) in which the fish have been conditioned to associate *Tubifex* worms as food with coumarin or amyl acetate. Then the primary olfactory nerves are sectioned and a functionally specific regeneration occurs within 10–14 days. Test fish before regeneration and control fish in which the olfactory bulbs have been removed exhibit behavioral "taste" thresholds slightly more than one log unit higher than the 10^{-6} v/v amyl acetate threshold of normals. The human catagorization of substances as taste and odor stimulants does not seem to hold up well for fishes (Bardach & Todd, 1970).

The eel has been determined macrosmatic (Teichmann, 1959) by virtue of its great olfactory acuity. Anatomically, this condition is manifested by the relatively great development of the nose, which contains a bilateral arrangement of many folds, or lamellae, supporting the olfactory epithelium. The ratio of olfactory mucosal area to the retinal area is among the greatest for fishes. The flow of water through the nose is repeatedly directed between the lamellae by the so-called "repiratory" cilia. The microstirring by the propelling cilia and the intimate contact of the odorous medium with a large sensory area are theorized to enhance the filtering out of odorant molecules. However, that the development of the eel's nose is sufficiently greater than that of the minnow to account for a billionfold greater sensitivity is not intuitively evident, leaving the possibility of species differences at the receptor cell level.

Fishes from the family Polypteridae, found in Africa, have many olfactory lamellae in a radially symetric arrangement that Pfeiffer (1968, 1969a) has found resembles that of the living crossopterigian *Latimeria*. The noses of the lungfishes are much more like those of the Elasmobranchii and Actinoptergyii, however. The morphology of the soft nasal tissue of *Latimeria* and the Dipnoi therefore militates against their grouping in Choanichthyes and suggests an affinity between *Latimeria* and Polypteriformes. Whereas *Latimeria* has a large eye in addition to being a large fish, individuals of the Polypteridae have a ratio of olfactory to retinal areas equalling that of the eel. Accepting this as putative evidence for the macrosmatic condition, Pfeiffer (1969b) has tested behaviorally *Polypterus palmas*, *P. delhezi*, *Calamoichthys calabaricus*, and, for comparison, *Phoxinus laevis*. Responses to a standard beefheart extract have been obtained at dilutions of 10^{10}–10^{11} for *Polypterus*, 10^7–10^8 for *Calamoichthys*, and 10^3–10^4 for *Phoxinus*.

In long-term studies fishes exhibit seasonal variation of olfactory acuity. The behavioral threshold concentration of the eel increases by several orders of magnitude during late fall and early winter (Teichmann, 1959). The range cannot be so great, of course, for a relatively insensitive fish, such as *Phoxinus*. The seasonal variation of behavioral olfactory sensitivity is suspected to be under hormonal control. The olfactory system may feed information back to the neurohormone system of the hypothalamus. Kandel (1964) has shown that electrical stimulation of the goldfish olfactory tract synaptically activates neuro-endocrine cells that originate in the preoptic nucleus and terminate in the neurohypophysis. Perfusion of the gills, through the mouth, with tap water containing a fewfold increase of the 0.3 mequiv sodium ion per liter found causes suppression of neuroendrocrine cell baseline activity. These concentrations of sodium ions are near the beginning of the response range determined by electrical recording from taste and lateral line nerve preparations (e.g., Sutterlin & Sutterlin, 1970; Katsuki, Hashimoto, & Kendall, 1971). A quantitatively similar curve for sodium stimulation of the olfactory system has been obtained by EEG-type recordings from the olfactory bulb of the carp *Cyprinus carpio* (Satou, 1971). Olfactory bulb responses of the thyroidectomized lungfish *Protopterus annectens* are highly dependent on the level of injected thyroxine (Dupé & Godet, 1969). Unitary activity of cells in the olfactory bulb of goldfish has been recorded with stainless steel microelectrodes (Oshima & Gorbman, 1966). The nose is stimulated with 60 μM sodium chloride and pretreatment of the fish with estradiol, progesterone, testosterone, and thyroxine all cause changes in the types of responses recorded.

Unit activity from single-fiber recordings was obtained from the olfactory tract of goldfish (Nanba, Djahanparwar, & Von Baumgarten, 1966) and crucian carp *Carassius carassius* (Sato & Suzuki, 1969) in response to amyl acetate, coumarin, and morpholine in the former study and isoamyl acetate, butyl acetate, isoamyl alcohol, butyl alcohol, and sodium chloride in the latter. Many patterns of response were observed. In the goldfish mechanical sensitivity was seen in response to water flow or probing of the olfactory mucosa with a small bristle, and temperature sensitivity of some units was observed in the carp. Sodium chloride caused a tonic type of response and had no inhibitory effect from 6 to 500 μM. Similarly, single-fiber recordings were obtained from the olfactory tract of the burbot *Lota lota* (Döving, 1966). Twenty-three chemical compounds were used, all at the relatively high concentration of 1 mM. Of 281 stimulations, 32% of the responses were an increase in the rate of firing, 21% were inhibitory, and 47% were without effect.

EEG-type recordings from the olfactory bulbs of adult migrating salmon in response to stream water samples from various locations have been of much interest. Initially, it seemed that water from the locations nearer to the spawning site evoked larger responses with the culmination being that from the spawning site (Ueda, Hara, & Gorbman, 1967). A larger study of homing Pacific salmon from three spawning groups has not yielded such a simple pattern of results. The largest responses are evoked by the home waters, but a pattern of increasing response with approach to spawning location does not appear (Oshima, Hahn, & Gorbman, 1969). The nature of the stimulating compounds in river water is unknown, but there are some indications of nonvolatility. Olfactory receptor responses have been recorded directly from the nasal epithelium of young Atlantic salmon, *Salmo salar* (Sutterlin & Sutterlin, 1971). Amino acids as a class of compound are by far the most stimulatory of those tested. Suzuki (Suzuki & Tucker, 1971; Tucker & Suzuki, 1972) has remembered that catfish exhibit the fright reaction behavior and skin as well as other tissue extracts have been found to be highly stimulatory for the olfactory receptors. From this clue the common amino acids have been found to be extremely effective, with the lowest threshold concentration lying between 10^{-8} and $10^{-7}M$. The potency of amino acids for Pacific salmon *Oncorhynchus nerka* and *O. kisutch* and rainbow trout *Salmo gairdneri* has been confirmed by olfactory bulb EEG recording (Hara, 1972).

Recently, gustatory receptors of catfishes have been found to be at least as sensitive as olfactory receptors to amino acids (Caprio & Tucker, 1973; Tucker, 1973; Oakley, 1971). Perhaps catfishes are unique in this regard, in correlation with the great numbers of taste buds present on the external body surface, concentrated especially on the barbels. The distinction between taste and smell in fishes has been discussed (Bardach & Atema, 1971), with emphasis on taste. Similarly, Kleerekoper (1969) has reviewed the literature with emphasis on olfaction. A very broad range of roles of the chemical senses, olfaction, taste, and others, has been discussed for a variety of fishes (Bardach & Todd, 1970). Atema (1971) has concluded that, at least for the yellow bullhead *Ictalurus natalus,* food is found by taste only. The facial nerve portion of the taste system is thought to function as a true distance receptor, by virtue of its sensitivity, and also controls the picking up of food. The vagal and glossopharyngeal nerve portion of the taste system is essential for the swallowing of food. The sense of touch (trigeminal) combines with the sense of taste to control normal feeding behavior. The function of the olfactory system is essentially different and is involved in behavioral interaction with other yellow bullheads.

AMPHIBIA

Salamanders and newts, the Urodela, tend to have fairly simple noses. Frogs and toads, the Anura, have the most complicated noses, usually with three recognizable interconnecting cavities. Noses of the least familiar amphibians, Gymnophiona (= Apoda), are probably intermediate in complexity. A homolog of Jacobson's organ is usually recognized in all these three orders. Parsons (1959b) has discussed the literature on embryology and adult anatomy of the nose and Jacobson's organ of these forms, which cannot be arranged in a phylogenetic series. An accessory olfactory bulb with a more or less distinguishable vomeronasal nerve is commonly situated on the caudal, ventrolateral aspect of the olfactory bulb, which contrasts with the more familiar mediodorsal placement of mammals and reptiles. Bowman's glands are never found associated with the presumed vomeronasal epithelium and also are commonly missing from the olfactory epithelium of neotenous urodeles, i.e., those that do not undergo metamorphosis.

Homing behavior of the newt *Taricha rivularis* appears to be mediated largely through olfaction (Grant, Anderson, & Twitty, 1968). Blinded animals have performed successfully over mountainous terrain. Some salamanders make breeding migrations during rainy nights and therefore are not expected to orient visually, a feat of which some newts and frogs have been shown to be capable. Occasionally toads, and especially frogs of the genus *Rana*, have been very popular for electrical recording studies of olfaction and there has therefore been interest in the possibility of finding olfactorily mediated behavior, apparently without fruit. However, the well-known African clawed toad *Xenopus laevis* has been successfully conditioned to respond to odorants with food reward of *Tubifex* worms (Altner, 1962). The animals have been blinded by enucleation and the lateral line system of these highly aquatic anurans plays an important role in their behavior. The threshold concentration for β-phenylethyl alcohol is close to that found for the minnow by Teichmann (1959), but the threshold for the next higher homolog, γ-phenylpropyl alcohol, is about tenfold higher, being 3×10^{14} molecules per cubic centimeter (dilution ratio $1:1.5 \times 10^7$). The highest sensitivity has been found for β-ionone, 1.5×10^{13} molecules per cubic centimeter (dilution ratio $1:2 \times 10^8$). The value for citral is slightly higher and conditioning has not been successful with terpineol. Altner (1962) has examined the physiological anatomy of the nose of *Xenopus laevis* and has concluded that the olfactory mucosa, with Bowman's glands, is always covered by water and that the animal smells in an aqueous medium. The presentation of *Tubifex* extract on a brush near the nasal entrance during respiratory exchange of air elicits no search reaction in response to the airborne food odor. Typical frogs are thought to normally have air in their nasal and oral cavities. The mechanism of breathing has been analyzed (Gans, De Jongh, & Farber, 1969) and a comparison made with that of lungfishes. The buccal oscillation should be important in bringing stimuli to the olfactory organ, and even in the lungless salamanders buccal oscillations are commonly observed.

Since publication of Ottoson's (1956) monograph on the recording of slow potential changes recorded from the exposed olfactory mucosa of the frog in response to odorous stimulation, there has been much activity in recording the electroolfactogram, or EOG, as this type of response record is named. Müller (1971) has also obtained EOG's from the vomeronasal mucosa of frog in response to amyl acetate, propionic acid, propanal, and heptanal. The eminentia olfactoria in the floor of the horizontally expanded anuran nose is the usual recording site and is well illustrated in Figure 36b of Negus' (1958) book. Workers in laboratories scattered all over the world have published many papers on recording the EOG with macro— and microelectrodes from the frog olfactory mucosa, microelectrode recording of receptor unit activity from the olfactory mucosa, microelectrode recording of secondary or postsynaptic neural responses in the olfactory bulb, macroelectrode surface recordings from the olfactory bulb, and some other techniques that cannot be categorized this way. Much of this literature can be found in various symposium volumes, notable among these being the *Olfaction and Taste* series I–IV (edited by Zotterman, 1963; Hayashi, 1967; Pfaffmann, 1969; and Schneider, 1972). A review of the voluminous literature in this area is outside the purview of this chapter. Much of the work can be characterized as attempts to determine the nature of neural coding of odor qualities. The comparison of electrical recording results from frogs with human odor quality judgment must be based on the assumption of an essential similarity of the two odor "worlds." Actually, the results from both directions have been so variable that it appears that definitive statements cannot safely be made at this time. There seems to be growing agreement that electrically recorded data indicate a conspicuous lack of specificity at the receptor level and at secondary neural levels in the olfactory bulb (Tucker & Smith, 1969). Although in the gustatory field the idea of four primary qualities is strongly entrenched, many electrical recording data have been interpreted as supporting an across-fiber pattern theory rather than the older concept of specific taste qualities (see symposia cited above).

There have been questions about the degeneration and regeneration of olfactory mucosa for many years and retrograde degeneration of the receptors after sectioning of the primary olfactory nerve has been used as an experimental tool (Takagi, 1971). The use of tritiated thymidine, the pyrimidine precursor unique to DNA or deoxyribonucleic acid, has revealed that new olfactory receptors are continuously being formed in the frog's olfactory mucosa (Graziadei & Metcalf, 1971; Moulton & Fink, 1972). Autoradiography combined with electron microscopy leaves no doubt that in adult animals there is a continuous turnover of receptor cells and apparently a much slower turnover of supporting cells. Such findings are not confined to amphibia, of course; they have been reported in forms from lampreys (Thornhill, 1970) to mice (Moulton & Fink, 1972).

REPTILES

Turtles are the most primitive reptiles and their nasal structure differs considerably from that of the other reptiles. There is no nasolacrimal duct. The olfactory organ tends to be well developed and the vomeronasal mucosa is not housed in a separate cavity so as to form a distinct Jacobson's organ. Instead, the vomeronasal epithelium appears more or less ventrally within the principal nasal cavity. Tucker (1963a) has mistaken the medial nasal gland in the tortoise *Gopherus polyphemus* for the organ of Jacobson and has published a correction (Tucker, 1971). This problem has been discussed well by Parsons (1959b). A distinct accessory olfactory bulb is not clearly evident in many species of turtle. In histological sections of the olfactory bulb of three turtles, Riss, Halpern, & Scalia (1969b) have found that only in the painted turtle, *Chrysemys picta*, does the bulb appear clearly divided into a dorsal and ventral half. Noting the view of Crosby and Humphrey (cited by Riss *et al.*, 1969b) that the dorsal portion constitutes the accessory olfactory bulb, Riss *et al.* have stated that "if indeed true, the painted turtle offers an excellent opportunity to explore distinctive projections of the main bulb and the accessory bulb." The accessory bulbs are distinct and much smaller and are located on the dorsocaudal aspect of the main bulbs, and the vomeronasal nerves are quite distinct from the olfactory nerves in *Gopherus polyphemus* and *Terrapene carolina.* A paired Jacobson's organ opening onto the roof of the mouth is characteristic of the Squamata, the lizards and snakes, and it reaches its peak of development in many snakes with an accompanying elaboration of the accessory olfactory bulb. There tends to be a close association of the opening of the nasolacrimal duct and the opening of Jacobson's organ (Pratt, 1948; Bellairs & Boyd, 1950). Adult Crocodilia lack a Jacobson's organ and all the reptiles except the turtles have at least one true concha (Parsons, 1970b).

The red-eared terrapin, *Pseudemys scripta,* has been trained to associate the odor of amyl acetate with food. Animals that have had the olfactory nerve sectioned regain the discrimination in 5–6 weeks (Boycott & Guillery, 1962). Manton, Karr, & Ehrenfeld (1972) have developed an operant method for the study of chemoreception in the green turtle, *Chelonia mydas.* The young sea turtles have detected under water the odors of β-phenylethyl alcohol, isoamyl acetate, triethyl amine, and cinnamaldehyde at approximately 5×10^{-6} to $5 \times 10^{-5} M$. The amino acids serine and glycine are not effective at a concentration of about $10^{-4} M$. A reversible disruption of the behavioral performance has been obtained by intranasal perfusion with 0.35 M zinc sulfate solution, with recovery taking place in a variable period of 1–5 days. The discrimination may have been mediated by olfactory or vomeronasal receptors. Recordings from small twigs of the nerves have shown that olfactory, vomeronasal, and trigeminal receptors of the gopher tortoise and the box turtle respond to airborne and waterborne odorants and that there is much overlap in responsiveness of the different types of receptors (Tucker 1963a; Tucker & Shibuya, 1965; Tucker, 1971).

The box turtle and the gopher tortoise have been used for studying the relations between the electrically recorded neural response, unit receptor responses led off with microelectrodes, and the EOG or slow potential response of the olfactory mucosa. An "underwater" EOG has been routinely recorded in turtle preparations as has likewise been done in fish preparations (Tucker & Suzuki, 1972). The location of the turtle olfactory organ in a dorsal diverticulum suggests that an air bubble can be trapped in it when water is drawn into the naris, and because of its exposed, ventral location the vomeronasal mucosa should be in direct contact with the water. These differences have been suggested of possible importance in the behavioral utilization of these two kinds of olfactory receptors, which are freely exposed in the turtle's nasal cavity. Electron microscopy has revealed that the olfactory receptors proper are ciliated and that the vomeronasal receptors instead bear rather elaborate microvilli (Graziadei & Tucker, 1970). It may be of interest that when eating, these land turtles first touch their nose to the food, open their mouth, touch the food with their tongue, and then bite into it.

There is a large literature showing that lizards and particularly snakes can track prey and members of their own species. In some instances the olfactory organ or Jacobson's organ alone is sufficient, but in many both seem to be used. Wilde (1938) has shown that the garter snake *Thamnophis sirtalis* does not feed, that is, grasp and swallow, if the vomeronasal nerves are sectioned. The forked tongue tips seem to be relatively important in nerve-intact animals. Earlier experimenters had shown the importance of the tongue and suggested the probability that the tips were inserted into the openings to the paired Jacobson's organ (reviewed by Wilde, 1938; Burghardt, 1970; Tucker, 1971). Jacobson's organ has been implicated in recent studies of innate food preferences of newborn snakes (Burghardt, 1970).

Müller (1971) and co-workers have recorded EOG's from the exposed vomeronasal mucosa of lizards and snakes, using a wide variety of compounds at unspecified but probably rather high concentrations.

BIRDS

In 1826, John James Audubon wrote

as soon as, like me, you shall have seen the Turkey Buzzard follow, with arduous closeness of investigation, the skirts of the forests, the meanders of creeks and rivers, sweeping over the whole of extensive plains, glancing his quick eye in all directions, with as much intentness as ever did the noblest of Falcons, to discover where below him lies the suitable prey; when, like me, you have repeatedly seen that bird pass over objects calculated to glut his voracious appetite, unnoticed, because unseen; and when you have also observed the greedy Vulture, propelled by hunger, if not famine, moving like the wind suddenly round his course, as the carrion attracts his eye; then will you abandon the deeply rooted notion, that this bird possesses the faculty of discovering, by his sense of smell, his prey at an immense distance.

The controversy regarding avian olfaction existed before Audubon (Scarpa, 1789) and continues to the present (Wenzel, 1971). Owen (1837) has described the olfactory anatomy of the turkey vulture, although it has been claimed that Audubon's remarks reveal that he was often observing the black vulture *Coragyps* instead of *Cathartes* (Stager, 1967). Several other investigators have studied the olfactory anatomy of birds (e.g., Ariëns-Kappers, Huber, & Crosby, 1936), but it has not been until the work of Cobb (1960) and Bang (1960) that "the long overdue job of putting olfaction in context in evian ethology is now under way" (Bang & Cobb, 1968).

All birds appear to have the appropriate anatomy for odor perception, but the variation in that anatomy suggests the likelihood of significant differences among species. Bang and Cobb (1968) have measured the greatest diameter of the olfactory bulb and compared it to the greatest diameter of the cerebral hemisphere in 108 species of birds. They have found, for example, that the kiwi, *Apteryx australis*, has the largest bulb–hemisphere ratio at 38%; the turkey vulture *Cathartes aura* has a ratio of 28.7%; the domestic fowl *Gallus gallus*, 15%; and the house sparrow *Passer domesticus*, 4%. In all species studied, the anatomy of the avian olfactory bulb is similar to the mammalian forms, consisting of a glomerular layer, a mitral cell layer, and an internal granule cell layer (Wenzel, 1971). Not much is known about the neural projections from the olfactory bulb.

The nasal region in birds is sometimes said to be similar to that in reptiles (Portmann, 1961), except that no organ of Jacobson is present in adults. There are three conchae or turbinals in the nasal cavity, in which three chambers can be recognized. A transverse threshold separates the first from the second or main chamber, which extends to the choana. Above the main chamber is the olfactory chamber containing a prominent tubercle or a more highly developed turbinal system, such as is seen in *Cathartes*. True olfactory epithelium covers the olfactory tubercle or concha and may also lie along the roof and on the posterior and ventrolateral walls and the upper portion of the septum nasi of the olfactory chamber in some species (Bang, 1960). With the anatomy described, the question next arises of the physiology. Portmann (1961) has written that

the physiological evidence for the function of true olfactory structures is contradictory . . . the unsettled state of current knowledge invites a re-examination of the whole problem. The wide, open secondary choana provides a special pathway for olfactory stimuli; choanal smell is probably a fact in many birds and particularly in groups where the external nasal openings are small or even closed. The role, and interaction, of vision and smell in orientation must be more carefully studied. The morphological facts testify strongly against the simple conclusion that birds are completely anosmatic. The development of the olfactory part of the nasal cavity, and the well-marked variation in the proportion of the olfactory bulbs of the brain, are in favor of an opposite view.

A reexamination was accomplished with the publication of electrophysiological evidence for olfaction in 14 species of birds (Tucker, 1965) and the recording of receptor unit responses in vultures (Shibuya & Tucker, 1967). Among the birds used for olfactory nerve recording were the house sparrow, chicken, and turkey vulture with bulb–hemisphere ratios of 4, 15, and 28.7%. An interesting point was the lack of obvious species differences noted. Sieck and Wenzel (1969) recorded with chronically implanted macroelectrodes in the olfactory bulbs of pigeons. Thresholds were estimated for a variety of odorants, and bulbar responses were shown to cease if the olfactory nerve was sectioned, if the bird breathed through the mouth, or if the nostril was plugged. Bulbar responses were recorded from chicken, mallard duck *Anas platyrhynchos*, black-vented shearwater *Puffinus puffinus opisthomelas*, and black-footed albatross *Diomedea nigripes*, the latter two having a bulb–hemisphere ratio of 29% (Wenzel, 1971). Wenzel (1971) concluded that "regardless of the many gaps in our knowledge, however, it is perfectly clear that some birds, if not all, show the kind of activity in the olfactory system that is seen in other vertebrates whose olfactory ability has never been questioned."

The behavioral literature up to the middle of this century depicted a rather negative opinion about the sense of smell in birds; see Walter (1942) for a comprehensive review. Then a series of papers began to yield positive results more frequently (Wenzel, 1971). A response-contingent reinforcement procedure was used by Michelson (1959), working under the guidance of Stanley Cobb and B. F. Skinner, to train two pigeons to discriminate sec-butyl acetate or isooctane from air-only control trials. Calvin (1960) criticized this study on several grounds, the most important of which was the possibility that the discrimination was not truly olfactory but was instead mediated by the trigeminal system. Michelson (1960), in response to Calvin, reported that he performed bilateral sections of the olfactory nerves and that neither bird could discriminate isooctane postoperatively. sec-Butyl acetate was discriminated by one of the lesioned birds, indicating possible trigeminal activation. The postmortem examination revealed a total section of olfactory nerves. No histology was reported.

Pigeons were employed in a series of olfactory behavioral studies in which a modification of the conditioned suppression technique was used; see Smith (1970) for details. Henton, Smith, and Tucker (1966) demonstrated amyl acetate discrimination in three birds that lost the response when the olfactory nerves were bilaterally sectioned and regained the response when the concentration of amyl acetate was raised. Henton (1969) demonstrated thresholds to amyl acetate, butyl acetate, and butyric acid; demonstrated discrimination of amyl acetate from butyl acetate in normal pigeons; and demonstrated the odor quality discrimination in olfactory nerve sectioned birds at high concentrations, presumably through trigeminal sensitivity. Shumake, Smith, and Tucker (1969) showed that pigeons could make intensity discriminations with amyl acetate. In terms of vapor saturation, 7% was reliably discriminated from 2% but not from 3%. In summary, the only problem left appears to be whether birds employ olfaction in normal behavior and, if so, to what extent, which undoubtedly can be expected to vary greatly over species.

MAMMALS

Mammalian nasal anatomy is exceedingly variable but generally involves a maxilloturbinal invested with respiratory mucosa, a highly variable nasoturbinal of which part may bear olfactory mucosa, and an ethmoturbinal system of which most is covered with olfactory mucosa. The ethmoturbinals originate at the cribriform plate, through which foramina transmit the fila olfactoria (collectively, the olfactory nerve), and most often the anterior ends are connected together to form a continuous plate of bone covered with respiratory mucous membrane. These specialized projections of the ethmoid bone vary greatly in number among mammalian forms. For example, in man there is only one and in *Echidna,* the spiny anteater, there are 12. The nasal turbinates are conchal structures, of course, that have been variously homologized with the conchae of the other amniotes (Parsons, 1970b, 1971). Negus' (1958) book contains a wealth of illustrations of mammalian material and comparisons are made with other forms. Although this is an excellent source for gross and histological anatomy, many of the functional interpretations made are certainly questionable today. For example, Negus makes an argument to explain the greater acuity of macrosmatic mammals by describing the subethmoidal shelf or lamina transversa (= lamina terminalis?) that helps to recess the olfactory organ. This subethmoidal shelf is said to be well developed in typical carnivores, present in marsupials, and small in ungulates, rodents, and rabbits. His argument is difficult to follow, because recessing the olfactory organ excludes most of it from the air currents.

Jacobson's organ in mammals is typically a tubular structure with glands emptying into the caudal aspect. A small pore at the anterior end opens either within the nasal cavity near the infundibulum of the nasopalatine duct or directly into the duct somewhere over its course. The mystery has been how stimuli can be conveyed to the receptors within and whether stimulus compounds are necessarily volatile (Broman, 1920; Negus, 1958; Tucker, 1971). Ruysch, Jacobson, and Cuvier thought that the organ had a secretory function; yet the obvious innervation in nonprimate forms compelled consideration of an olfactory function. However, "quel agent extérieur pourroit [sic] aller se faire percevoir dans un réceptacle si caché, si profond, si peu accessible?"[1] (Van Wijhe, 1919, quoting Jacobson). Modern degeneration technique has revealed

---

[1] What outside agent could provoke perception from a receptacle so concealed, so deep, so inaccessible?

that the accessory olfactory bulb of rabbit projects to the mediocortical complex of the amygdala (Winans & Scalia, 1970). The old ideas of possible roles in feeding and sexual behavior are given new credence because of the projections of the cortical amygdaloid nucleus to the anterior medial hypothalamus and the ventromedial nucleus of the hypothalamus. The projections of the main bulb provide a separate, parallel route of influence to the hypothalamus. A large literature exists on the behavioral effects of olfactory bulb ablation and it is worth reemphasizing that in virtually all such experiments both of the olfactory bulbar systems as well as the nervus terminalis have been interrupted.

Comparison of the ratio of the largest cross-sectional area of Jacobson's organ to that of the olfactory organ suggests a great range for vertebrates, with the mammalian carnivore's ratio becoming very small, because of the large olfactory organ (Negus, 1958). Such a measure is reminiscent of the ratio of olfactory bulb to hemisphere diameters employed for birds (Bang & Cobb, 1968) or of the ratio of retinal to olfactory areas used for fish and mammals. Such relative measurements partly represent an effort to remove the confounding influence of the size of the animal. An elephant, for example, may not need an organ system any larger than that of the mouse in order to do the same job, although the elephant olfactory bulb is relatively large with multiple layering of the glomeruli. An interesting allometric approach relates brain structure volumes of simians, prosimians, and "progressive" insectivores to those of "basal" insectivores (Stephan & Andy, 1969). The only structure that clearly decreases on this basis with ascent of the primate scale is the olfactory bulb. There seems to be an enlargement of higher centers of the amygdaloid complex. The accessory olfactory bulb is well developed in prosimians, highly variable in new world monkeys, and very small or absent in old world monkeys. The human accessory olfactory bulb reaches its peak during fetal development at about the time movement begins and then regresses greatly (Humphrey, 1940). Still later, during fetal development, the main olfactory bulb seems to reach a peak and then to regress somewhat.

Electrophysiology of the mammalian olfactory system has been investigated extensively. Adrian (1956), a pioneer in this field, has written a highly readable account of methods of electrical recording from the rabbit olfactory bulb. Mitral cell (second-order neuron) responses are grouped according to the relative sensitivity displayed to the stimulating compounds, which continues to be the typical finding for both the receptor and secondary cell responses. A regional variation in the olfactory bulb of sensitivity to various kinds of compounds is also found. The theory that the bulbar waves (EEG-type recordings) originate in the peripheral organ has been retracted (Adrian, 1957).

Electrical recording from the receptor axons of responses to natural stimuli was first achieved in opossum and rabbit (Beidler & Tucker, 1955). Conventional methods of nerve recording were adapted to this smallest caliber of fibers and the response of a small population of receptors was shown to be determined primarily by the kind of odorant, odorant concentration, and inspiratory flow rate of the odorized air. This result was generalized to nonmammalian forms for both olfactory and trigeminal responses to odorants and to turtles for vomeronasal responses (Tucker, 1963b). Although responses were often recorded from rabbit vomeronasal receptors under unusual conditions, it was felt that anesthetization precluded the animal's exploitation of normal methods of stimulus transport to the interior of the long tubular organ (Tucker, 1963a).

Comparison across species of amphibian, reptile, bird, and mammal olfactory nerve twig responses indicates that if an odorant is efficacious for one it is for all the others. The lack of obvious species differences in olfactory responsiveness is puzzling when it is noted that in taste studies prominent differences are found within mammals and even within salts as stimuli. However, the olfactory recording results from fishes indicate a major difference between air- and water-breathing vertebrates. The odorants effective for air breathers can be applied in aqueous solution by nasal perfusion with similar effectiveness and the way is therefore open for testing air breathers with amino acids. However, the solutions must be carefully purified of "odorous" contaminants that come with the commercially available compounds, contaminants that appear to be quite ineffective in comparison with amino acids on fish olfactory receptors. Observations on newts that can live on land or in water can perhaps be taken up again with profit (Shibuya & Takagi, 1963). The point should be emphasized that comparisons of stimulatory effectiveness in electrical recording preparations are in general only relative, e.g., a ranking of compounds for a given preparation is obtained. The uncertainty derives from the unknown amount of stimulus attenuation between the input at the naris and the small population of receptors, localized somewhere in the olfactory organ, that is monitored. (Tucker, 1963b; Moulton & Tucker, 1964). Exceptionally, the effective stimulus concentration can be deduced at the receptor level. The curve fitted to such data for olfactory response of the gopher tortoise to amyl acetate extrapolates toward zero at about $10^{-5.5}$ of vapor saturation. Behavioral threshold determination with the conditioned suppression technique of the laboratory rat has yielded a value of $10^{-5.74}$ vapor saturation (Pierson, 1974), which is less than a factor of two smaller than the preceding value. However, the lowest amyl acetate threshold concentration found for the pigeon is $10^{-3.0}$ of vapor saturation.

Mammals are customarily credited with a good sense of smell. By far the most evidence regarding mammalian olfactory behavior is anecdotal and lacks objective demonstration. From the literature of the naturalists (e.g., Bedichek, 1960; Millen, 1960) information gathered from hunters, trappers, guides, pet owners, and kennel club devotees indicates that dogs, cats, foxes, sheep, rabbits, deer, hedgehogs, zebras, elephants, rhinoceroses, and bush babies are prominent among the mammals endowed with fine olfaction. Moncrieff (1967) has said "it is a matter of frequent observation that many of the mammals have a much keener sense of smell than man. . . . There is a correspondence between the amount and intensity of the nasal pigmentation and the acuteness of smell." Milne and Milne (1962) refer to the relative "odor blindness" of humans. Burton (1961), in explaining why "smell animals" get much more information from odors than man, has said "there are two reasons for this. The first is that their smelling-membrane is so much larger than ours. The second is that they have, as we say, a much finer discrimination for the various scents." He goes on to say "there is a third reason why we cannot use the sense of smell as efficiently as the smell-animals do. Our sense of smell readily tires." Matthews and Knight (1963) have said "hedgehogs are very discriminating in their sense of smell and will ignore ground beetles which have an unpleasant odour, while eating greedily those beetles and other insects which have not." Finally, Wilentz (1968) has said

the change that distinguishes man from these highly olfactory-oriented species occurred in the course of evolution. Man's ancestors moved from sea to land, and then to trees. It was here that vision and hearing took over as the prime distance senses. Monkeys are not good sniffers as rats or rabbits.... So man exists today, between the birds and four-footed beasts, far superior to the whales and porpoises that have no sense of smell at all, but still inferior to many others.

For man the sense is non-intellectual. We have great difficulty in describing smells— except as similes with other smells. But perhaps that very degree of weakness in the intellectual quality of smell explains why its emotional impact is so profound. Smell can conjure up the past or quicken the present; it can nauseate or excite, repel or entice us.

Although the above statements are based almost entirely on anecdotal evidence, there is good reason to believe that mammals do use olfaction to a great extent in their behaviors. The variety of mammalian scent glands, present in at least 15 of 18 orders (Mykytowycz, 1972), and the behavior known as "scent marking" (Ralls, 1971) quickly lead to such a conclusion. From hunters and dog trainers have come anecdotal records of fantastic feats of retrieval, tracking, and rescue work with a variety of canine species (see McCartney, 1968, p 15–77, for an excellent historical review). Perhaps the most quoted report is that of Romanes (1887), in which he refers to the "almost supernatural capabilities of smell in dogs" and concludes that his dog

distinguishes my trail from that of all others by the peculiar smell of my boots (1 to 6) and not by the peculiar smell of my feet (8 to 11). No doubt the smell which she recognizes as belonging distinctively to my trail is communicated to the boots by the exudations from my feet; but these exudations require to be combined with shoeleather before they are recognized by her. Probably, however, if I had always been accustomed to shoot without boots or stockings, she would have learnt to associate with me a trail made by my bare feet." (The parenthetical numbers refer to experiments.)

More objective studies were conducted on dog olfaction in the middle of this century by Neuhaus in Germany, Moulton and associates and Kalmus in England, and Becker and associates in the United States (see Wright, 1964, for a comprehensive review). Neuhaus (1953) found, on the one hand, that human subjects could identify the sheets of paper on which people had stepped and, on the other, that dogs were a millionfold more sensitive than humans to the fatty acids present in sweat from foot glands. Neuhaus reasoned that if only a thousandth part of the 16 cm^3 sweat released per day from a man's foot were to penetrate the sole of the shoe, there would be 2.5×10^{11} molecules left behind in each footprint. This quantity is so much in excess of the minimal amount required that there is nothing "supernatural" at all about the dog's ability to track humans, according to Neuhaus. Moulton, Ashton, & Eayrs (1960), however, have found the dog to be only about 100 times more sensitive than man to butyric acid. The differences in breeds tested, training methods, apparatus, and odor concentration estimation techniques in some way must account for the four log unit difference in the threshold results from Neuhaus' and Moulton's work. A series of laboratory experiments by Becker, King, and Markee (1962) has involved a method for testing olfactory thresholds in dogs in a "free-ranging laboratory environment." Dogs have been able to accurately select a "slightly fingerprinted" glass slide as long as 6 weeks after the print has been deposited (King, Becker, & Markee, (1964). They can even detect such a slide that has been weathered up to 1 week. These observations suggest the importance of care in handling odor stimulating equipment in such experiments with dogs. However, Lord Adrian (1956), who pioneered in electrical recording of olfactory responses from fishes, hedgehogs, and rabbits, said that "I do not think there is any direct evidence to show that the absolute sensitivity of the organ to various pure substances is greater in a dog than in a man" in his discussion of how a large olfactory organ may give better discrimination than a small one.

A large body of literature is available on many behavioral roles of olfaction in rodents, especially laboratory rats and mice. Pheromones of the inducer type have been implicated in the reproductive physiology of mice, in which three well-known effects are recognized (e.g., Dominic, 1969; Gleason & Reynierse, 1969). Pheromones are often thought to be involved in sexual attraction, alarm behavior, trail and territorial marking behavior, and individual recognition. Literature on the olfactory control of behavior in rodents has been reviewed by Schultz and Tapp (1973), with four main divisions being (1) movement in the living space, (2) feeding behavior, (3) rodent societies, and (4) fixed patterns and heredity. Reviewers are unanimous in emphasizing the importance of maintaining a heightened awareness of the importance of olfaction in behavioral and physiological studies.

The suggestion is frequently made that biologically significant or relevant odors should be used in the studies with electrical recording methods. After all, the insect sex pheromones have been shown to be highly specific for specialized receptor cells. Pfaff and Gregory (1971) have compared urine odors from rats of various endocrine status and nonurine odors, such as amyl acetate and phenylethyl alcohol. They have made single-unit recordings from neurons in the olfactory bulb and from the preoptic region of the hypothalamus, which has been shown to respond to olfactory input and which has been linked to hormonal control of reproductive function. They have found no evidence for a simple mechanism of coding in the sense of highly specific responding of any cells to a given type of odor. However, "differential response analyses, taking into account the direction and magnitude of response, showed that cells in the preoptic area gave more differential responses to female urine odors than cells in the olfactory bulb, even though the opposite was true for nonurine odors." The effects of testosterone injection have been nonspecific on preoptic units. As has been seen above, androgens and other hormones have been found to augment fish olfactory bulb responses.

ACKNOWLEDGMENTS

The authors' research was supported by the United States Public Health Service, grant NS-8814; by the United States Atomic Energy Commission, Division of Biology and Medicine, contract AT-(40-1)-2903; and by the Psychobiology Research Center, Florida State University, through the following grants: PHS NS-7468, PHS MH-11218, and NSF GU-2612.

REFERENCES

Adrian, E. D. The action of the mammalian olfactory organ. *Journal of Laryngology and Otology*, 1956, 70, 1–14.

Adrian, E. D. Electrical oscillations recorded from the olfactory organ. *Journal of Physiology*, 1957, **136**, 29P.

Altner, H. Untersuchungen über Leistungen und Bau der Nase des Südafrikanischen Krallenfrosches *Xenopus laevis* (Daudin, 1803). *Zeitschrift für Vergleichende Physiologie*, 1962, **45**, 272–306.

Altner, H., & Müller, W. Electrophysiologische und elektronenmikroskopische untersuchungen an der riechschleimhaut des Jacobsonschen Organs von eidechsen *(Lacerta)*. *Zeitschrift für Vergleichende Physiologie*, 1968, 60, 151–155.

Altner, H., Müller, W., & Brachner, I. The ultrastructure of the vomeronasal organ in reptilia. *Zeitschrift für Zellforschung*, 1970, **105**, 107–122.

Ariëns-Kappers, C. U., Huber, G. C. & Crosby, E. C. *The comparative anatomy of the nervous system of vertebrates including man*. New York: Macmillan Co., 1936, pp. 1358–1401.

Atema, J. Structures and functions of the sense of taste in the catfish (*Ictalurus natalis*). *Brain, Behavior and Evolution*, 1971, 4, 273–294.

Audubon, J. J. The black vulture or carrion crow, *Cathartes Jota*, Bonap., *In* J. J. Audubon, *Ornithological biography*, . . . , *or an account of the habits of the birds of the United States of America accompanied by descriptions of the objects represented in the work entitled the birds of America and interspersed with delineations of American scenery and manners*, Vol. II. Edinburgh: Adam & Charles Black, 1834, P. 33.

Bang, B. G. Anatomical evidence for olfactory function in some species of birds. *Nature* (London), 1960, 188, 547–549.

Bang, B. G., & Cobb, S. The size of the olfactory bulb in 108 species of birds. *Auk*, 1968, 85, 55–61.

Bardach, J. E., & Atema, J. The sense of taste in fishes. In L. M. Beidler, Ed., *Handbook of sensory physiology*, Vol. 4, Part 2, Springer-Verlag, New York: 1971, pp. 293–336.

Bardach, J. E., & Todd, J. H. Chemical communication in fish. In J. W. Johnston, Jr., D.G. Moulton, & A. Turk, Eds., *Advances in chemoreception.* New York: Appleton-Century-Crofts, 1970, pp. 205–240.

Barnard, J. W. A phylogenetic study of the visceral afferent areas associated with the facial, glossopharyngeal, and vagus nerves, and their fiber connections. The efferent facial nucleus. *Journal of Comparative Neurology,* 1936, **65**, 503–602.

Becker, R. F., King, J. E. & Markee, J. E. Studies on olfactory discrimination in dogs: II. Discriminatory behavior in a free environment. *Journal of Comparative and Physiological Psychology,* 1962, **55**, 773–780.

Bedichek, R. The sense of smell. Garden City, New York: Doubleday, 1960.

Beidler, L. M., & Tucker, D. Response of nasal epithelium to odor stimulation. *Science,* 1955, **122**, 76.

Bellairs, A. d'A., & Boyd, J. D. The lachrymal apparatus in lizards and snakes. II. The anterior part of the lachrymal duct and its relationship with the palate and with the nasal and vomeronasal organs. *Proceedings of the Zoological Society of London,* 1950, **120**, 269–310.

Bertmar, G. The olfactory organ and upper lips in Dipnoi, an embryological study. Acta Zoologica, 1965, **46**, 1–40.

Bertmar, G. The vertebrate nose, remarks on its structural and functional adaptation and evolution. *Evolution,* 1969, **23**, 131–152.

Boycott, B. B., & Guillery, R. W. Olfactory and visual learning in the red-eared terrapin, *Pseudemys scripta elegans* (Wied.). *Journal of Experimental Biology,* 1962, 39, 567–577.

Broman, I. Das Organon vomero-nasale Jacobsoni–ein Wassergeruchsorgan! *Anatomische Hefte,* 1920, 58, 141–191.

Bruner, H. L. Jacobson's organ and the respiratory mechanism of amphibians. *Gegenbaurs Morphologisches Jahrbuch,* 1914, 48, 157–165.

Burghardt, G. M. Chemical perception in reptiles, In J. W. Johnston, Jr., D. G. Moulton, & A. Turk, Eds., *Communication by chemical signals.* New York: Appleton-Century-Crofts, 1970, pp. 241–308.

Burton, M. *Animal senses.* London: Routledge and Kegan Paul, 1961, pp. 62–64.

Calvin, A. Olfactory discrimination. *Science,* 1960, **131**, 1263–1265.

Caprio, J., & Tucker, D. Amino acids as taste stimuli in the freshwater catfish, *Ictalurus punctatus. Federation Proceedings,* 1973, **32**, 328.

Cobb, S. Observations on the comparative anatomy of the avian brain. Perspectives in Biology and Medicine, 1960, **3**, 383–408.

Craigie, E. H. A preliminary experiment on the relation of the olfactory sense to the migration of sockeye salmon (*Oncorhynchus nerka,* Walbaum). *Transactions of the Royal Society of Canada,* 1926, **20**, 215–224.

Cuvier, G. Rapport fait à l'institut, sur un mémoire de M. Jacobson intitulé: Description anatomique d'un organe observé dans les mammifères. *Annals de le Muséum d'Historie Naturelle,* 1811, 18. (Cited by Van Wijhe in this reference list.)

Dominic, C. J. Pheromonal mechanisms regulating mammalian reproduction. *General and Comparative Endocrinology Supplement,* 1969, **2**, 260–267.

Döving, K. B. The influence of olfactory stimuli upon the activity of secondary neurones in the burbot (*Lota lota*L.) *Acta Physiologica Scandinavica,* 1966, **66**, 290–299.

Dupé, M., & Godet, R. Variations de la réponse électrique d'origine olfactive sous l'effet d'un traitement thyroxinen. *Comptas Rendus d'l Academie des Sciences (Paris),* 1969, **268**, 1314–1317.

Gans, C., De Jongh, H. J., & Farber, J. Bullfrog (*Rana catesbeiana*) ventilation: How does the frog breathe? *Science,* 1969, **163**, 1223–1225.

Glaser, D. Untersuchungen über die absoluten Geschmacksschwellen von Fischen. *Zeit-*

schrift für Vergleichende Physiologie, 1966, **52**, 1–25.

Gleason, K. K., & Reynierse, J. H. The behavioral significance of pheromones in vertebrates. *Psychological Bulletin*, 1969, **71**, 58–73.

Grant, D., Anderson, O. & Twitty, V. Homing orientation by olfaction in newts (*Taricha rivularis*). *Science*, 1968, **160**, 1354–1356.

Graziadei, P. P. C., & Metcalf, J. F. Autoradiographic and ultrastructural observations on the frog's olfactory mucosa. *Zeitschrift für Zellforschung*, 1971, **116**, 305–318.

Graziadei, P. P. C., & Tucker, D. Vomeronasal receptors in turtles. *Zeitschrift für Zellforschung*, 1970, **105**, 498–514.

Hara, T. J. Electrical responses of the olfactory bulb of Pacific salmon *Oncorhynchus nerka* and *Oncorhynchus kisutch*. Journal of the Fisheries Research Board of Canada, 1972, **29**, 1351–1355.

Hasler, A. D. Underwater guideposts. Madison, Wisconsin: University of Wisconsin Press, 1966.

Hayashi, T., Ed. *Olfaction and taste,* Vol II. New York: Pergamon Press, 1967.

Heimer, L. The secondary olfactory connections in mammals, reptiles and sharks. *Annals of the New York Academy of Sciences*, 1969, **167**, 129–146.

Hemmings, C. C. Olfaction and vision in fish schooling. *Journal of Experimental Biology*, 1966, **45**, 449–464.

Henton, W. W. Conditioned suppression to odorous stimuli in pigeons. Journal of the Experimental Analysis of Behavior, 1969, **12**, 178–185.

Henton, W. W., Smith, J. C. & Tucker, D. Odor discrimination in pigeons. *Science*, 1966, **153**, 1138–1139.

Herrick, C. J. The connections of the vomeronasal nerve, accessory olfactory bulb and amygdala in amphibia. *Journal of Comparative Neurology*, 1921, **33**, 213–280.

Humphrey, T. The development of the olfactory and the accessory olfactory formations in human embryos and fetuses. *Journal of Comparative Neurology*, 1940, **73**, 431–468.

Jones, F. R. H. *Fish migration.* New York: St. Martins, Press, 1968, pp. 259–271.

Kandel, E. R. Electrical properties of hypothalamic neuroendocrine cells. Journal of General Physiology, 1964, **47**, 691–717.

Katsuki, Y., Hashimoto, T. & Kendall, J. I. The chemoreception in the lateral-line organs of teleosts. The *Japanese Journal of Physiology,* 1971, **21**, 99–118.

Kerkhoff, H. Beitrag zur Kenntnis des Baues und der Funktion des Jacobsonschen Organs. *Zeitschrift für Mikroskopische Anatomische Forschung,* 1924, **1**, 621–638.

King, J. E., Becker, R. F. & Markee, J. E. Studies on olfactory discrimination in dogs: (3) Ability to detect human odour trace. *Animal Behaviour,* 1964, **12**, 311–315.

Kleerekoper, H. *Olfaction in fishes.* Bloomington, Indiana: Indiana University Press, 1969.

Kolnberger, I. Vergleichende Untersuchungen am Riechepithel, insbesondere des Jacobsonschen Organs von Amphibien, Reptilien und Säugetieren. *Zeitschrift für Zellforschung,* 1971, **122**, 53–67.

Larsell, O. The nervus terminalis. *The Annals of Otology, Rhinology, & Laryngology*, 1950, **59**, 414–438.

Locy, W. A. On a newly recognized nerve connected with the forebrain of selachians. *Anatomische Anzeiger,* 1905, **26**, 33–63, 111–123.

Luckhaus, G. Licht- und elektronenmikroskopische Befunde an der Lamina epithelialis des Vomeronasalorgans vom Kaninchen. *Anatomische Anzeiger*, 1969, **124**, 477–489.

Manton, M., Karr, A. & Ehrenfeld, D. W. Chemoreception in the migratory sea turtle, *Chelonia mydas. The Biological Bulletin*, 1972, **143**, 184–195.

Matthews, L. H., & Knight, M. *The senses of animals.* London: Museum Press Ltd., 1963, p. 107.

McCartney, W. *Olfaction and odours.* Berlin: Springer-Verlag, 1968, pp. 15–77.

Michelson, W. J. Procedure for studying olfactory discriminations in pigeons. *Science*, 1959, 130, 630–631.

Michelson, W. J. Olfactory discrimination. *Science*, 1960, 131, 1265–1267.

Millen, J. K. *Your nose knows*. Los Angeles, California: The Cunningham Press, 1960.

Milne, L. J., & Milne, M. The senses of animals and men. New York: Atheneum, 1962, pp. 119–134.

Moncrieff, R. W. *The chemical senses*. (3rd ed.) Cleveland, Ohio: Chemical Rubber Company Press, 1967, p. 361.

Moulton, D. G. Detection and recognition of odor molecules. In G. Ohloff & A.F. Thomas, eds., *Gustation and olfaction*. New York: Academic Press, 1971, pp. 1–27.

Moulton, D. G., Ashton, E. H. & Eayrs, J. T. Studies in olfactory acuity. 4. Relative detectability of *n*-aliphatic acids by the dog. *Animal Behaviour*, 1960, 8, 117–128.

Moulton, D. G., & Fink, R. P. Cell proliferation and migration in the olfactory epithelium. In D. Schneider Ed.,*Olfaction and taste*, Vol. IV. Stuttgart, Germany: Wissenschaftliche Verlagsgesellschaft MBH, 1972, pp. 20–26.

Moulton, D. G., & Tucker, D. Electrophysiology of the olfactory system. *Annals of the New York Academy of Sciences*, 1964, 116, 380–428.

Müller, W. Vergleichende elektrophysiologische Untersuchungen an den Sinnesepithelien des Jacobsonschen Organs und der Nase von Amphibien (*Rana*), Reptilien (*Lacerta*) und Saugetieren (*Mus*). *Zeitschrift für Vergleichende Physiologie*, 1971, 72, 370–385.

Mykytowycz, R. The behavioural role of the mammalian skin glands. *Die Naturwissenschaften*, 1972, 59, 133–139.

Nanba, R., Djahanparwar, B. & Von Baumgarten, R. Erregungsmuster einzelner Fasern des Tractus olfactorius lateralis des Fisches bei Reizung mit verschiedenen Geruchsstoffen. *Pflügers Archiv für Gesumte Physiologie*, 1966, 288, 134–150.

Negus, V. *The comparative anatomy and physiology of the nose and paranasal sinuses*. London: Livingstone, 1958.

Neuhaus, W. Über die Riechschärfe des Hundes für Fettsäuren. *Zeitschrift für vergleichende Physiologie*, 1953, 35, 527–552.

Neuhaus, W. On the olfactory sense of birds, In Y. Zotterman, Ed., *Olfaction and taste*, Vol. I. New York: Pergamon Press, 1963, pp. 111–123.

Oakley, B. Public communication at the Fourth International Symposium on Olfaction and Taste, Starnberg, Germany, August 2–4, 1971.

Oshima, K., & Gorbman, A. Influence of thyroxine and steroid hormones on spontaneous and evoked unitary activity in the olfactory bulb of goldfish. *General & Comparative Endocrinology*, 1966, 7, 482–491.

Oshima, K., Hahn, W. E. & Gorbman, A. Electroencephalographic olfactory responses in adult salmon to waters traversed in the homing migration. *Journal of the Fisheries Research Board of Canada*, 1969, 26, 2123–2133.

Ottoson, D. Analysis of the electrical activity of the olfactory epithelium. *Acta Physiologica Scandinavica*, 1956, 35(Supplement 122), 1–83.

Owen, R. Dissection of the head of the Turkey Buzzard and that of the common Turkey. *Proceedings of the Zoological Society of London*, 1837, 5, 33–35.

Panchen, A. L. The nostrils of choanate fishes and early tetrapods. *Biological Reviews*, 1967, 42, 374–420.

Parker, G. H. *Smell, taste, and allied senses in the vertebrates*. Philadelphia: Lippincott, 1922.

Parsons, T. S. Nasal anatomy and the phylogeny of reptiles. *Evolution*, 1959a, 13, 175–187.

Parsons, T. S. Studies on the comparative embryology of the reptilian nose. *Bulletin of the Museum of Comparative Zoology at Harvard College*, 1959b, 120, 101–277.

Parsons, T. S. Evolution of the nasal structure in the lower tetrapods. *American Zoologist*, 1967, 7, 397–413.

Parsons, T. S. The origin of Jacobson's organ. *Forma et Functio*, 1970a, 3, 105–111.

Parsons, T. S. The nose and Jacobson's organ, In C. Gans & T. S. Parsons (Eds.), *Biology of the reptilia*, Vol. 2. New York: Academic Press, 1970b, pp. 99–191.

Parsons, T. S. Anatomy of nasal structures from a comparative viewpoint. In L. M. Beidler, Ed., *Handbook of sensory physiology*, Vol. 4, Part 1. New York: Springer-Verlag, 1971, pp. 1–26.

Pfaff, D. W., & Gregory, E. Olfactory coding in olfactory bulb and medial forebrain bundle of normal and castrated male rats. *Journal of Neurophysiology*, 1971, 34, 208–216.

Pfaffmann, C., Ed. *Olfaction and taste*, Vol. III. New York: The Rockefeller University Press, 1969.

Pfeiffer, W. The fright reaction of fish. *Biological Reviews of the Cambridge Philosophical Society*, 1962, 37, 495–511.

Pfeiffer, W. Schreckreaktion und Schreckstoffzellen bei Ostariophysi und Gonorhynchiformes. *Zeitschrift für Vergleichende Physiologie*, 1967, 56, 380–396.

Pfeiffer, W. Das Geruchsorgan der Polypteridae (Pisces, Brachiopterygii). *Zeitschrift für Morphologie der Tiere*, 1968, 63, 75–110.

Pfeiffer, W. Das Geruchsorgan der rezenten Actinistia und Dipnoi (Pisces). *Zeitschrift für Morphologie der Tiere*, 1969a, 64, 309–337.

Pfeiffer, W. Der Geruchssin der Polypteridae (Pisces, Brachiopterygii). *Zeitschrift für Vergleichende Physiologie*, 1969b, 63, 151–164.

Pierson, S. C. Conditioned suppression to odorous stimuli in the rat. *Journal of Comparative and Physiological Psychology*, 1974, 86, 708–717.

Pinkus, F. Die Hirnnerven des *Protopterus annectens. Morphologische Arbeiten*, 1895, 4, 275–346.

Portmann, A. Olfaction. In A. J. Marshall, Ed., *Biology and comparative physiology of birds*, Vol. 2. New York: Academic Press, 1961, pp. 42–48.

Pratt, C. W. McE. The morphology of the ethmoidal region of *Sphenodon* and lizards. *Proceedings of the Zoological Society of London*, 1948, 118, 171–201.

Ralls, K. Mammalian scent marking. *Science*, 1971, 171, 443–449.

Reese, T. S., & Brightman, M. W. Olfactory surface and central olfactory connexions in some vertebrates. In G. E. W. Wolstenholme & J. Knight, Eds., *Taste and smell in vertebrates.* London: J. and A. Churchill, 1970, pp. 115–149.

Riss, W., Halpern, M. & Scalia, F. Anatomical aspects of the evolution of the limbic and olfactory systems and their potential significance for behavior. *Annals of the New York Academy of Sciences*, 1969a, 159, 1096–1111.

Riss, W., Halpern, M. & Scalia, F. The quest for clues to forebrain evolution—The study of reptiles. *Brain, Behavior & Evolution*, 1969b, 2, 1–50.

Romanes, G. J. Experiments on the sense of smell in dogs. *Nature*, 1887, 36, 273–274.

Sato, Y., & Suzuki, N. Single unit analysis of the olfactory tract of the crucian carp. *Journal of the Faculty of Science, Hokkaido University*, 1969, 17, (Series 6), 208–223.

Satou, M. Electrophysiological study of the olfactory system in fish I. Bulbar responses with special reference to adaptation in the carp. *Cyprinus caripo* L. *Journal of the Faculty of Science, University of Tokyo*, 1971, 12(Section IV), 183–218.

Scalia, F. The projection of the accessory olfactory bulb in the frog. *Brain Research*, 1972, 36, 409–411.

Scarpa, A. *Anatomicae disquisitiones de auditu et olfacto. Ticinum* (Italy). 1789; cited by: W. Neuhaus. On the olfactory sense of birds. In Y. Zotterman, Ed., *Olfaction and taste*, Vol. I. New York: Pergamon Press, 1963, p. 111.

Schneider, D., Ed. *Olfaction and taste*, Vol. IV. Stuttgart, Germany: Wissenschaftliche Verlagsgesellschaft MBH, 1972.

Schultz, E. F., & Tapp, J. T. Olfactory control of behavior in rodents. *Psychological Bulletin*, 1973, 79, 21–44.

Seydel, O. Über die Nasenhöhle und das Jacobson'sche Organ der Amphibien. *Gegenbaurs Morphologisches Jahrbuch*, 1895, 23, 453–543.

Shibuya, T. The electrical responses of the olfactory epithelium of some fishes. *Japanese Journal of Physiology*, 1960, 10, 317–326.

Shibuya, T., & Takagi, S. F. Electrical response and growth of olfactory cilia of the olfactory epithelium of the newt in water and on land. *Journal of General Physiology*, 1963, 47, 71–82.

Shibuya, T., & Tucker, D. Single unit responses of olfactory receptors in vultures. In T. Hayashi, Ed., *Olfaction and taste*, Vol. II. New York: Pergamon Press, 1967, pp. 219–234.

Shumake, S. A., Smith, J. C. & Tucker, D. Olfactory intensity-difference thresholds in the pigeon. *Journal of Comparative and Physiological Psychology*, 1969, 67, 64–69.

Sieck, M. H., & Wenzel, B. M. Electrical activity of the olfactory bulb of the pigeon. *Electroencephalography and Clinical Neurophysiology*, 1969, 26, 62–69.

Smith, J. C. Conditioned suppression as an animal psycho-physical technique. In W. C. Stebbins, Ed., *Animal psychophysics*. New York: Appleton-Century-Crofts, 1970, Pp. 125–159.

Stager, K. E. Avian olfaction. *American Zoologist*, 1967, 7, 415–419.

Stephan, H., & Andy, O. J. Quantitative comparative neuroanatomy of primates: An attempt at a phylogenetic interpretation. *Annals of the New York Academy of Sciences*, 1969, 167, 370–387.

Sutterlin, A. M., & Sutterlin, N. Taste responses in Atlantic salmon (*Salmo salar*) parr. *Journal of the Fisheries Research Board of Canada*, 1970, 27, 1927–1942.

Sutterlin, A. M., & Sutterlin, N. Electrical responses of the olfactory epithelium of Atlantic salmon (*Salmo salar*). *Journal of the Fisheries Research Board of Canada*, 1971, 28, 565–572.

Suzuki, N., & Tucker, D. Amino acids as olfactory stimuli in freshwater catfish, *Ictalurus catus* (Linn.). *Comparative Biochemistry and Physiology*, 1971, 40(A), 399–404.

Takagi, S. F. Degeneration and regeneration of the olfactory epithelium. In L. M. Beidler, Ed., *Handbook of sensory physiology*, Vol. 4, Part 1. New York: Springer-Verlag, 1971, pp. 75–94.

Teichmann, H. Über die Leistung des Geruchssinnes beim Aal. *Zeitschrift für Vergleichende Physiologie*, 1959, 42, 206–254.

Theisen, B. Ultrastructure of the olfactory epithelium in the Australian lungfish *Neoceratodus forsteri*. *Acta Zoologica*, 1972, 53, 205–218.

Thornhill, R. A. Cell division in the olfactory epithelium of the lamprey, *Lampetra fluviatilis*. *Zeitschrift für Zellforschung*, 1970, 109, 147–157.

Tucker, D. Olfactory, vomeronasal and trigeminal receptor responses to odorants. In Y. Zotterman, Ed., *Olfaction and taste*, Vol. I. New York: Pergamon Press, 1963a, pp. 45–69.

Tucker, D. Physical variables in the olfactory stimulation process. *Journal of General Physiology*, 1963b, 46, 453–489.

Tucker, D. Electrophysiological evidence for olfactory function in birds. *Nature* (London), 1965, 207, 34–36.

Tucker, D. Nonolfactory responses from the nasal cavity: Jacobson's organ and the trigeminal system. In L.M. Beidler, Ed., Handbook of sensory physiology. Vol. 4, Part 1. New York: Springer-Verlag, 1971, pp. 151–181.

Tucker, D. Rapid decline of olfactory and gustatory receptor sensitivities of wild catfish after capture. *Journal of the Fisheries Research Board of Canada*, 1973, 30, 1243–1245.

Tucker, D., & Shibuya, T. A physiologic and pharmacologic study of olfactory receptors. *Cold Spring Harbor Symposia on Quantitative Biology*, 1965, 30, 207–215.

Tucker, D., & Smith, J. C. The chemical senses. *Annual Reviews of Psychology*, 1969, 20, 129–158.

Tucker, D., & Suzuki, N. Olfactory responses to schreckstoff of catfish. In D. Schneider, Ed., *Olfaction and taste*, Vol. IV. Stuttgart, Germany: Wissenschaftliche Verlagsgesellschaft MBH, 1972, pp. 121–127.

Ueda, K., Hara, T. J. & Gorbman, A. Electroencephalographic studies on olfactory discrimination in adult spawning salmon. *Comparative Biochemistry and Physiology*, 1967, 21, 133–143.

Van Wijhe, J. W. On the nervus terminalis from man to amphioxus. *Koninklijke Akademie van Wetenschappen te Amsterdam (Proceedings of the Section of Sciences)*, 1919, 21, 172–183.

Von Frisch, K. Zur Psychologie des Fisch-Schwarmes. Die Naturwissenschaften, 1938, 26, 601–606.

Walter, W. G. *Some experiments on the sense of smell in birds*. 's-Gravenhage, Nederland: N.V. Boek-en Kunstdrukkerij Voorheen Mouten and Co., 1942.

Wenzel, B. M. Olfaction in birds. In L. M. Beidler, Ed., *Handbook of sensory physiology*, Vol. 4, Part 1. New York: Springer-Verlag, 1971, Pp. 432–448.

Wilde, W. S. The role of Jacobson's organ in the feeding reaction of the common garter snake, *Thamnophis sirtalis sirtalis*. (Linn.). *Journal of Experimental Zoology*, 1938, 77, 445–465.

Wilentz, J.S. *The senses of man*. New York: Thomas Y. Crowell Co., 1968, Pp. 114–115.

Winans, S. S. & Scalia, F. Amygdaloid nucleus: new afferent input from the vomeronasal organ. *Science*, 1970, 170, 330–332.

Wisby, W. J., & Hasler, A. D. Effect of olfactory occlusion on migrating silver salmon (*Oncorhynchus kisutch*). *Journal of the Fisheries Research Board of Canada*, 1954, 11, 472–478.

Wright, R. H. *The science of smell*. New York: Basic Books, 1964, Pp. 57–62, 71–79.

Zippel, H. P., Von Baumgarten, R. & Westerman, R. A. Histologische, funktionelle und spezifische Regeneration nach Durchtrennung der Fila olfactoria beim Goldfisch (*Carassius auratus*). *Zeitschrift für Vergleichende Physiologie*, 1970, 69, 79–98.

Zotterman, Y., Ed., *Olfaction and taste* I. New York: Pergamon Press, 1963.

3

The Vertebrate Eye

Mitchell Glickstein

Brown University

I should like to discuss the structure of the vertebrate eye in relation to two questions: first, are there suggestions from comparative study of the eye about the origin of vertebrates? Second, what factors in the environment are correlated with structural differences in the eyes of different animals? Although these questions can be answered only in part, comparative study is a solid foundation for understanding the structural basis of vision. Walls' monograph on the vertebrate eye (Walls, 1942), in my opinion, stands above all other work in this century in its scholarly breadth and insights. Many of the ideas in this chapter derive from Walls, and the reader desiring a more extended discussion of any of the topics covered is referred to this outstanding book.

The vertebrate eye is a device that focuses an image onto an array of receptors and ultimately transmits some derivative of this visual information to the brain. All vertebrate retinas, with the exception of those animals in which the eye is degenerated (for example, the hagfish *Eptatretus* or the blind cave fish *Amblyopsidae*), are composed of three cellular layers: an array of photoreceptors that transduces light energy into neural signals; a set of cells with relay and lateral connections, the cell bodies of which lie in the inner nuclear layer; and an array of ganglion cells that transmits a modified image to the brain (Ramón y Cajal, 1972).

Eyes of vertebrates and invertebrates are analogous in many respects but they are not homologous. Certainly the common ancestors of cephalopods and vertebrates have not had an image-forming eye (see discussion of homology p. xxx), yet some cephalopods have eyes that are optically similar to those of the vertebrates. For example, in the octopus, whose eye forms a good image, there is only one cell layer in the retina. Photoreceptors are oriented opposite in direction to those in the vertebrate eye; in octopus the tips of the photoreceptors point toward the pupil, whereas in all vertebrates photoreceptors point away from the pupil.

The uniformity of retinal structure among all vertebrates suggests that the answer to my first question is "No." Clues from retinal structure to understanding the origin of vision are hard to find. Froriep is quoted by Gordon Walls as saying that the sudden appearance of the vertebrate eye is similar to the birth of Athena, full grown and fully armed from Zeus' brow. Comparative study of vision can provide only modest evidence toward understanding vertebrate evolution. Although the basic connectionist plan is identical, however, eyes do differ among vertebrates. It is the main thesis of this chapter that such differences relate principally to the nature of the light with which the eye must deal in the normal life of the animal.

I shall now return to the second question, of how the makeup of the vertebrate eye depends on the visual environment of the animal. This question occupies the rest of this chapter. Figure 1 is a low-power view of the monkey (*Macaca mulatta*) retina through the fovea. The retina is made up of three obvious layers of cells with fibrous zones between them. The three-layer plan is pervasive, although there are regional exceptions to it. For example, in the fovea of man and the macaque the ganglion and inner nuclear layers give way to a single layer of receptors and their nuclei.

Despite great variation in diameter of the eye in vertebrates, there is comparatively little variation in thickness of retina. The largest mammalian eyes, for example, are about 50 times the diameter of the smallest, and yet the thickness of the retinas never vary by more than two to one. The relatively constant thickness of the retina can be related to an old puzzle in the study of vision. Comparative ophthalmologists have reported refractive errors in many vertebrates. Animals, especially small ones, are often reported to be hypermetropic. In such eyes, accommodation is necessary to focus distant objects. This is puzzling because it is hard to conceive of any selective advantage resulting from hypermetropia. When Michel Millodot and I studied refractive state in a number of animal species, we found that the apparent refractive error seemed to vary in an orderly way as a function of the size of the animal's eye (Glickstein & Millodot, 1970). Refractive error seemed to be greatest in animals with the smallest eyes. We wondered, however, whether the measurements themselves were at fault. What would be the error in retinoscopic determinations if the light used in retinoscopy were reflected from a plane other than that of the rods and cones? If the reflection were from a plane behind the receptors, the eye would be spuriously myopic. If the reflection were from a plane in front of the receptors, the eye would appear spuriously hypermytropic. Because the retina of the mammals is of a relatively constant thickness, we have calculated that the error in retinoscopy should be inversely proportional to the square of the focal length of the eye. We performed retinoscopy on representative subjects and plotted these data along with evidence in the literature. We concluded that the

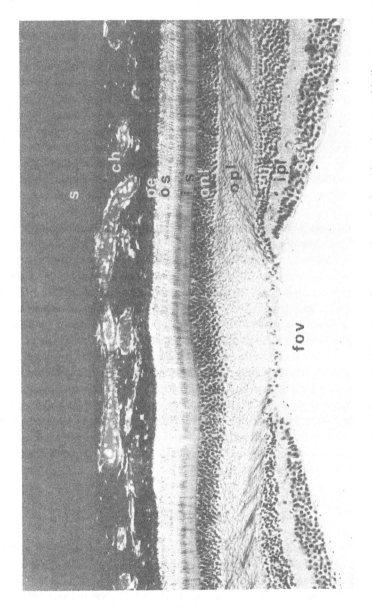

FIG. 1 Low-power view of monkey retina (*Macaca mulatta*) through the fovea. s, sclera; ch, choroid; p, pigment epithelium; os, receptor outer segments; is, receptor inner segments; onl, outer nuclear layer; opl, outer plexiform layer; inl, inner nuclear layer; ipl, inner plexiform layer; gcl, ganglion cell layer; fov, fovea. Note the basic three-cell layer plan of the retina and its change at the fovea.

reflection seen with the retinoscope is not from the plane of the receptors but is instead from the interface between the retina and the vitreous. This plane of reflection gives a spurious hypermetropia. The animals studied were in fact shown by other tests to be emmetropic; the eye when relaxed was in focus for distant objects. These results comforted us: hypermetropia could only reduce the usefulness of the eye by limiting the range of accommodation.

In most animals, then, a well-focused image is formed at the level of the receptors. What is the relationship between the density of receptors on which the image falls and the resolving capacity of the eye? Although the density of receptors may set limits to acuity, human foveal receptors are so densely packed that acuity is actually limited equally by the physical nature of light (Riggs, 1965). When the pupil is smaller than 4 mm, acuity is limited by diffraction; when the pupil is 4 mm or larger, spherical and chromatic aberration of the eye limit visual resolution. The ability to distinguish fine detail cannot be much better in the human eye even if the packing density of foveal cones is greatly increased.

Visual resolution of lines and gratings, or letters and symbols, is also limited by receptor density—humans can distinguish a pair of lines as a pair if the lines are separated by roughly one cone width. However, forms of visual acuity are much better than this and pose a major challenge for understanding vision. Under optimal conditions people can detect whether two lines are aligned or offset, even if the offset is as low as 2–4 sec of arc. Similarly, the displacement of an image on the two retinas of about 2–4 sec of arc may be used as a binocular disparity cue for depth. Foveal cones subtend an angle of about 30 sec of arc, however, and humans are therefore actually capable of resolving vernier offset and retinal disparities that subtend only one-tenth of the visual angle of a single cone. In the case of depth disparity this information must be carried through the lateral geniculate nucleus and decoded in the visual cortex, the first point in the brain at which there is binocular interaction. The mechanism for such a remarkable acuity is not well understood. Depth and vernier acuity are even more remarkable because involuntary eye movements are at least ten times as great as the magnitude of vernier or disparity thresholds.

RODS, CONES, AND ACUITY

The human eye is a splendidly designed machine that can operate over a very wide range of light intensities. It can work in the pale light of the moon or the midday sun. Not all vertebrates can use their eyes over such a wide range of light intensity. Indeed, anatomists have distinguished early between those eyes that seem to be adapted primarily for vision under dim illumination and those that are adapted for bright illumination. Schultze (1873) first distinguished two basic types of receptors about 100 years ago. He named these receptors "rods" and "cones" and pointed to their association with vision under conditions of dim or bright illumination. Schultze's classification of visual receptors has been extended and modified but it is still a most useful dichotomy. Rods and cones exist, are widely distributed among vertebrates, and are usually related to vision under different conditions of lighting.

The strong correlation between eye structure and light environment is best seen in those animals that are relatively pure in their diurnality or nocturnality. The squirrel family and some of the tree shrews (*Tupaiades*) are strongly diurnal. Figure 2 is a picture of the receptor layer of a tree shrew, *Tupaia glis*. All of the receptors appear to be cones and there is a great deal of melanin pigment in the pigment epithelium. The outer segment of each cone is shrouded in a dense layer of pigment. The strongly diurnal tree shrew retina can be contrasted with that of a nocturnal animal, such as the Virginia opposum (*Didelphis virginianis*) seen in Figure 3. Note the dense packing of rods. The rod nuclei do not lie in a single layer but are stacked in many layers below the receptors themselves, about ten nuclei deep to the outer limiting membrane. Nocturnal animal eyes are dominated by rods. All-cone eyes are typical of diurnal animals. In addition to their domination by rods or cones, there are other differences. In animals that are strictly diurnal there tends to be about the same number of receptors as ganglion

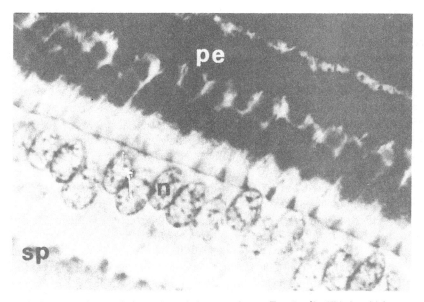

FIG. 2 Receptor layer of the retina of the tree shrew, *Tupaia glis*. This is a high-power photomicrograph that shows pigment epithelium (pe) at the top and synaptic pedicles (sp) of the receptors at the bottom. A single line of cone nuclei (n) of uniform size extending across the back of the eye can be seen. The inner segments (is) of the cones are seen clearly but the outer segments are largely hidden by the pigment epithelium.

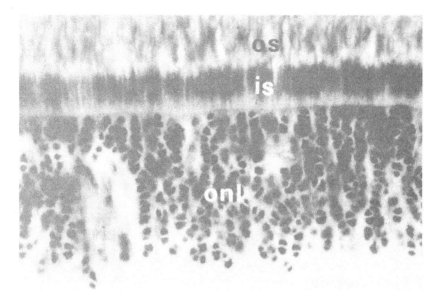

FIG. 3   Receptor layer of the Virginia opposum, *Didelphis virginianis*. Almost all of the receptors seen are rods. os, outer segments; is, inner segments; onl, outer nuclear layer.

cells. In contrast, nocturnal animals have eyes in which there is great convergence from many receptors down onto a single ganglion cell. This difference in convergence from receptors onto ganglion cells can be seen in Figure 4 and Figure 5, which show the retinas of the tree shrew (*Tupaia glis*) and the opposum (*Didelphis virginianis*). In the tree shrew there are roughly the same number of ganglion cells as there are receptors; in the opposum there are about 100 times as many receptors as there are ganglion cells. The eyes of nocturnal animals are so arranged that many receptors connect through cells in the inner nuclear layer onto a much smaller number of ganglion cells. Such an arrangement increases visual sensitivity at the expense of acuity. The ganglion cell are more readily activated but less likely to distinguish which receptor has been stimulated among the large set of receptors that contribute to its firing.

There are different shapes and types of rods and cones. The variance is great enough that some authors have even questioned whether the very concept of a rod and a cone is useful. I have retained the terms but have shown some examples of rods and cones in vertebrates to illustrate the variety of receptors that exist, and also to point out some of the functional problems that must be solved if a fuller understanding of vertebrate vision is to be gained. Figure 6 shows the retina of a frog (*Rana pipiens*), in which at least two types of receptors can be distinguished. The frog rod has a wide outer segment in contrast to a very narrow, tapering cone. There are actually two types of rod in the frog retina, but the ones that are easily seen in this figure are the so-called "red" rods.

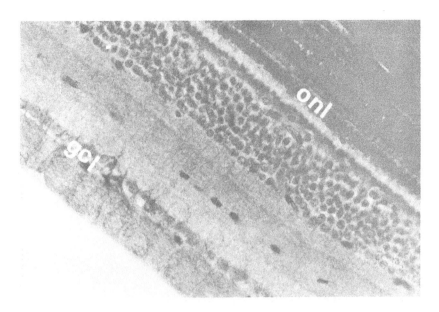

FIG. 4 Lower power view of the retina of the tree shrew, *Tupaia glis*. The number of receptor nuclei in the outer nuclear layer (onl) is roughly equivalent to the number of ganglion cells nuclei (gcl).

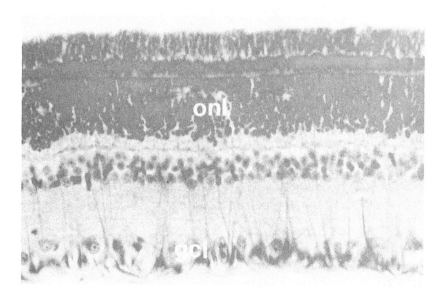

FIG. 5 Lower power view of the retina of the Virginia opposum, *Didelphis virginianis*. Note the relatively far greater number of receptor nuclei in the outer nuclear layer (onl) as compared to the number of ganglion cells in the ganglion cell layer (gcl).

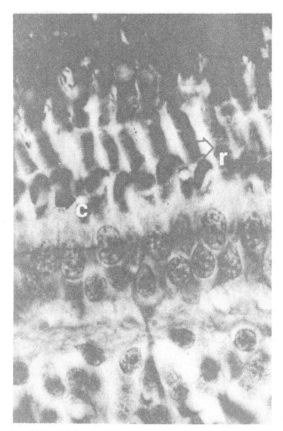

FIG. 6   Frog retina (*Rana pipiens*). Filled arrow points to a cone; unfilled arrow points to a (red) rod. Note the morphological difference between rod (r) and cone(c) and the presence of an oil droplet at the base of the outer segment of the cone.

If the cone is examined, it can be seen that at the beginning of the outer segment of the cone there is a small round structure, the oil droplet. In the frog eye these oil droplets are small and pale. In some animals oil droplets are larger and in many they are deeply pigmented. Because light that strikes the cone outer segment must go through the oil droplet, it must influence color vision in these animals. Figure 7 shows a whole mount of the retina of the turtle (*Pseudemys*) to show the array of oil droplets, which can be seen by simply looking at the unstained retina. Each oil droplet lies at the base of the cone outer segment and filters the light that reaches the receptor. Oil droplets are widely distributed in nature and are especially prominent in the retinas of turtles and birds. Something is known about the filtering properties of oil droplets, but their precise role in vision remains to be worked out.

Figure 8 is a high-power view of receptors in the frog (*Rana*) retina. The figure illustrates another basic puzzle of retinal function. A paired structure consisting

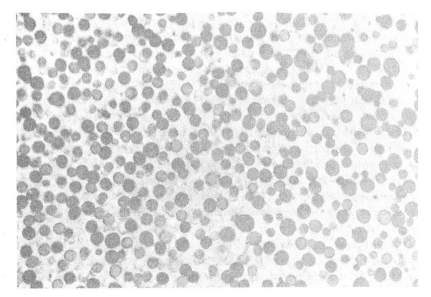

FIG. 7 Oil droplets of the turtle *Pseudemys* photographed from an unstained whole mounted retina.

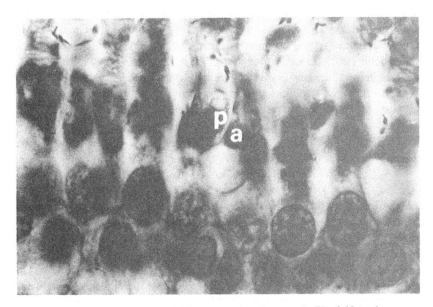

FIG. 8 Frog (*Rana pipiens*) retina, higher power than that seen in Fig. 6. Note the presence of a paired receptors. The principal cone (p) has an oil droplet and typical cone-shaped outer segment; the accessory cone (a) has an enormous parabaloid as part of its inner segment. The pair constitute a double cone.

61

of two cones can be seen; one rather small, the other quite large. This is the double cone. In the frog, one member of the pair has a prominent oil droplet and the other does not. The double cone is one example of a general type of paired receptor in vertebrate eyes. In the frog double cone there are two morphologically different elements. There are also paired but equal-sized cones—so-called "twin" cones—in the eyes of teleost fish. What is the function of this curious pairing? As a first step, it is necessary to know the spectral sensitivity of these paired elements. There are data on spectral sensitivity of the visual pigments in paired receptors of a few species. In the frog, for example, the spectral sensitivity for all receptor elements is known, thanks to the elegant work by Liebman and Entine (1968). These authors have determined by direct microspectrophotometry the peak sensitivity of the visual pigments of each receptor type in the frog eye. In the double cone, one element contains a cone pigment with maximum absorption at 575 nm. The accessory cone contains a 502 nm pigment that is indistinguishable in its absorption characteristics from rhodopsin, one of the major rod pigments. Each element of the paired receptors contains a different pigment and the pattern of pairing is entirely orderly. The data on cone sensitivity are a useful clue but the precise function of the double cone and paired receptors in general remains unknown.

The retina of many vertebrates contains both rods and cones. Often rods and cones can best be differentiated in histological structures by the staining properties of their nuclei. In most animals the cone nuclei lie nearest to the outer limiting membrane, with the rod nuclei below those of the cones. Figure 9 shows the outer nuclear layer of a leopard (*Pantherus pardus*) eye. There is a single layer of cone nuclei that lines the outer limiting membrane, with a deep layer of rod nuclei below. This arrangement of rod and cone nuclei is also seen in most of the human and monkey (*Mulatta*) retinas. Figure 10 shows the distribution of rod and cone nuclei in the retina of a rhesus monkey (*Macaca mulatta*) about $15°$ from the fovea.

REGULATION OF LIGHT INTENSITY

Light can be toxic—too much damages the eye. In the normal vertebrate eye ultraviolet light, the short-wavelength radiation, is well filtered by the lens. Infrared rays, however, can reach the retina. The image of the sun has enough energy to destroy a patch of retina. Human beings (as well as many other animals) have a built-in aversion to looking directly at very bright sources. In addition to built-in aversions to bright light there are several mechanisms for regulating the amount of light that reaches the receptors. The most familiar of these is the pupil. In bright lights the human pupil constricts; in dim light it expands. The human pupil at its largest can be 8 mm in diameter, and at its smallest 2 mm in diameter. This range of diameters gives a ratio of four, and the

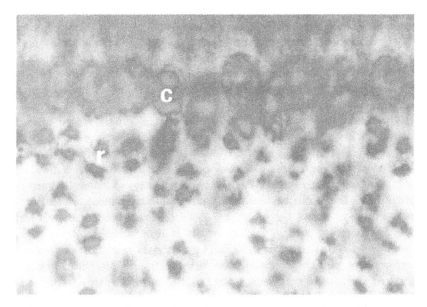

FIG. 9 Leopard eye, *Pantherus pardus*. Note the difference in size, shape, and distribution of rod (r) and cone (c) nuclei.

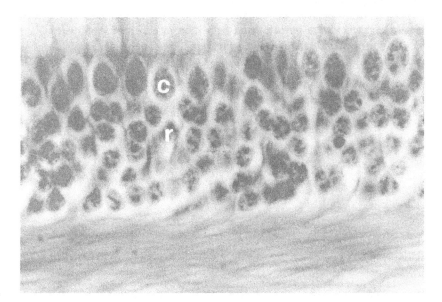

FIG. 10 Eye of monkey, *Macaca mulatta*. Note the similarity of the distribution of the rod (r) and cone (c) nuclei to that of the leopard seen in Fig. 9.

area of the largest pupil is therefore 16 times that of the smallest. However, a 16 : 1 ratio in area is hardly enough to regulate over the ten billionfold range of light intensities the eye may encounter. There are other, neural and photochemical, mechanisms in the human eye that serve to further regulate its sensitivity. Such mechanisms have been studied and a good deal is known, but they are far from being understood.

The potential toxicity of light can best be seen in certain special cases. In albinos even low levels of light may destroy receptors. Albinos have no light-absorbing pigment in the pigment epithelium, choroid, or iris. If an albino rat is subject to modest continuous illumination for about a month or two it will lose nearly all of its photoreceptors. Figure 11 shows the eye of a normal albino rat. Three cellular layers can be seen in the retina that are present in other vertebrates. Figure 12 shows the eye of an albino rat that has been exposed for 2 months to moderate but continuous intensity of light in the laboratory (Glickstein, Brown-Grant, & Raisman, 1972). There are only two layers in the retina. Receptors and pigment epithelium are both destroyed. Modest illumination carried on continuously for several weeks has been sufficient to destroy every receptor in the eye of this animal.

There is another way in which receptors can be screened from excessive illumination. In fish and birds, in some amphibians, and in a few reptiles, the pigment in the pigment epithelium can actively migrate. The relationship between receptors and pigment epithelium can be pictured as that of two hairbrushes which are pushed together. One set of the interlocking bristles represents the receptors, the other the processes of the pigment epithelium. In many animals the pigment can migrate from a position inside the cell body down and among the processes of the pigment epithelium. In so doing pigment can shroud

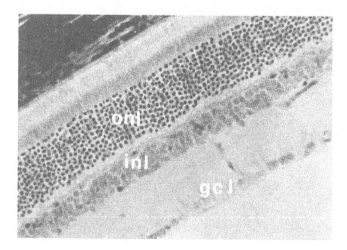

FIG. 11 Eye of normal albino laboratory rat. onl, outer nuclear layer; inl, inner nuclear layer; gcl, ganglion cell layer.

FIG. 12 Eye of albino rat exposed to 8 weeks of continuous light in the laboratory. Note absence of pigment epithelium and receptors. inl, inner nuclear layer; gcl, ganglion cell layer.

each individual cone. In the dark, the pigment migrates back into a position inside the cell body of the pigment epithelial cell.

SPECIAL STRUCTURES FOR INCREASING SENSITIVITY

Mechanisms for guarding against excessive light have been noted—there are parallel specializations for dealing with low levels of light. There are devices the main function of which seems to be to increase the probability that light is absorbed. Figure 13 shows one example of such a structure, the tapetum lucidum in the eye of a leopard, *Pantherus pardus*. This tapetum is situated between the pigment epithelium and the choroid, so that light traversing the retina without being absorbed by the receptors is reflected back to the receptors instead of being absorbed. Although the tapetum increases the visual sensitivity of the eye it does so at the expense of the acuity. Because the reflected light does not strike the receptors in exactly the same position as it does coming in, the position of the source of the light is less determined.

SPECIALIZED REGIONS IN THE EYE

Visual acuity is limited by the nature of light waves. Because of the diffraction properties of light waves, the smallest optical image that can be focused on the retina is a disk of light that has a considerably larger diameter than that of the

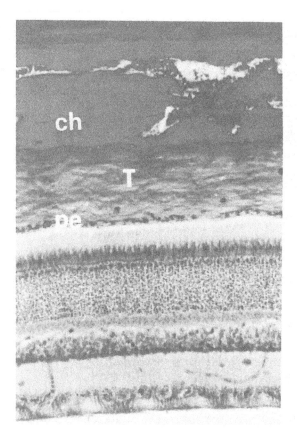

FIG. 13 Leopard, *Pantherus pardus.* Section of retina to illustrate position of the tapetum lucidum (t). ch, choroid; pe, pigment epithelium.

foveal cone receptor. In the human and monkey eye it appears as if receptors are packed in the center of the eye up to the point where the limits of acuity set by light are reached. Cones are long and slender in the center of the fovea being about 2 μm in diameter, which corresponds to about 30 sec of visual angle.

Why is the fovea shaped as it is? In the human and monkey eye the fovea has a parabolic shape with densely packed receptors (Figure 1). One suggestion (Weale, 1966) is that the fovea develops its shape secondarily to regional changes in the vasculature of the eye. In the human eye, the two inner layers of the retina are nourished by the retinal circulation. The receptor layers are nourished by the choroidal circulation. Some authors have argued that the primary reason for the special shape of the fovea is an advantage gained by displacing the retinal circulation away from the most acuitous region of retina. Because blood vessels are displaced, cells of the inner nuclear layer and ganglion cells must also be displaced such that no cell lies too far from its blood supply.

The argument that there is a selective advantage to displacing blood vessels from the most acuitous region of the eye is plausible and accounts for the shape and position of the human fovea. Another problem remains, however. Figure 14 shows the fovea in *Anolis*, the American chameleon. The walls of the fovea are far more steeply shaped than the gentle parabola seen in the human eye. An imaginative theory of the functions of this curious type of fovea is Pumphrey's (1948). Pumphrey suggested that these steep foveas are a device much like that of a sharpshooters lens, which helps to center the eye on the target of interest. A small target is distorted unless the center of the eye is precisely aligned with it.

Differences in their morphology and distribution of receptors and structure of the eye associated with increased or decreased sensitivity to light have been discussed. However, the eye is a three-layered structure. All of the action of the receptors must be transmitted to the brain via the axons of the ganglion cell, the fibers that make up the optic nerve. What sort of information is transmitted down the optic nerve? A crucial step in understanding the functions of ganglion cells has been the discovery by Kuffler (1953) of the spatial organization of the receptive field of a vertebrate eye. Kuffler has investigated the spatial distribution of points in which light can excite or inhibit the cell. The set of all such points is the receptive field. Although the receptive fields of ganglion cells in a cat are of two types, both types are small, roughly circular, and organized concentrically. In one type there is a small central area within which light excites the given retinal ganglion cell. Just outside that region is an annular area in which light has an inhibitory effect on the same cell. This is the so-called "on-center cell"; a light placed within a small circular region in the center of the receptive field causes an increase in the firing rate. When light is shone just outside the center of the ganglion cell's receptive field, the firing rate is slowed down. There is roughly an equal number of ganglion cells with the opposite

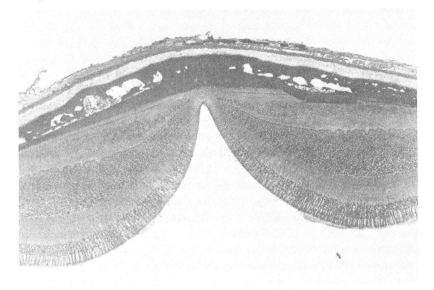

FIG. 14 Retina of the American chameleon, *Anolis,* showing fovea.

characteristics; light outside the center excites, whereas light inside it inhibits the response. The on-center and off-center types of receptive field are the only ones that Kuffler has found in the eye of the cat.

Knowledge of the spatial organization of the receptive field allows posing of an analytical question: how are the observed receptive fields formed? There is a spatially organized image of the visual world on the retina. Each receptor, then, is acted on by only a tiny pointlike region of the visual field. How is the retina organized so that this mosaic-like stimulation of rods and cones is put together to make up the observed receptive field of the ganglion cell? A major technical problem in the way of answering this question is posed by the fact that many retinal cells do not fire action potentials. Because such cells respond to stimulation only with a graded slow potential, the electrode must be placed inside the cell to understand its properties.

Several developments have helped make intracellular recording in the retina possible. One is the development of very small microelectrodes for use with a variety of vertebrate receptors; the other is the use of a preparation with relatively large cells in its retina—notably *Necturus,* the mudpuppy. Once penetrated and recorded from, retinal cells can be identified by intracellular dye marking.

Intracellular recording from fish cones has revealed a new and unsuspected property of these receptors (Tomita, Kaneko, Murakami, & Pautler, 1967). Unlike most sensory generator potentials, cones hyperpolarize when illuminated. Moreover, the response of cones to different wavelength of light shows that the peak sensitivity of these receptors is in good agreement with absorption measurements made on the cone pigments of the same species.

However, how do these receptors combine spatially? Rods and cones are connected both to horizontal and to bipolar cells. What are the response properties of these cell types? Bipolar cells are the first units to show a concentric receptive field organization (Werblin & Dowling, 1969). Illumination of the center and of the surround of receptive fields of a bipolar cell has opposite effects on the membrane potential.

Bipolar cells actuate both amacrine and ganglion cells. The role of amacrine cells in determining receptive field characteristics is not well understood, but they seem to respond principally with transient responses to changes of illumination. The receptive field of the ganglion cell is determined jointly by amacrine and bipolar cell inputs. Only the bare outline of ganglion cell receptive field construction is now known, but with the new intracellular recording techniques synaptic organization in the retina should be clarified within the next few years.

It is interesting that knowledge of retinal anatomy has been far ahead of its physiology. As early as 1892 Ramón y Cajal in his brilliant monograph summarized the structure of the retina from receptors to ganglion cells in all classes of vertebrates. Golgi staining methods permitted Ramón y Cajal to see the cell bodies and process of each cell type in the retina and describe their interconnections. Electron microscopic studies have added some details to Ramón y Cajal's picture, but the basic plan of synaptic organization is largely as Ramón y Cajal described it. Intracellular recording and dye marking have allowed physiologists to use fully Ramón y Cajal's anatomical description. It is satisfying to see these branches of science converge.

A most striking property of ganglion cells when stained with the Golgi method is the stratification of their dendrites. If the inner plexiform layer, the synaptic zone of the ganglion cells of the vertebrate retina, is examined with the appropriate staining methods, it is found that most ganglion cells do not spread their dendrite randomly in the inner plexiform layer. Instead, there are clear-cut sublaminae within the inner plexiform layer. Ganglion cells spread their dendrites preferentially in one, two, or more laminae. There are usually about five such sublayers in the inner plexiform layer of the eye. The ramifications of the dendrites of the ganglion cells are paralleled by a similar ramification in the cell layer that send processes into the inner plexiform layer, the amacrine and bipolar cells. Both cell types have preferential sublaminae in which their processes borize. What is the functional significance of this lamination? It seems likely that some specific properties of visual stimuli are segregated and organized by these dendritic ramifications. It seems that the inner plexiform layer constructs unique and specialized receptive fields. The electrophysiological studies by Lettvin, Maturana, McCulloch, and Pitts (1959) provide direct evidence of a specialized receptive field in the frog. Lettvin's experiments suggest that the cells in the frog are active filters of significant information about the visual world. The comparative physiology of the eye is a marvelous prospect for future study.

Study of the retina illustrates another simple and general principle of how the nervous system works. It is customary to think about the actions of neurons in terms of the spike potential. Spikes are prominent, relatively easily recorded, and therefore discovered early. However, it is now known that not all retinal cells fire action potentials. The ganglion cells certainly do and the amacrine cells may. Neither the receptors, nor horizontal cells, nor bipolar cells fire propogated spikes. This fact is easy to interpret if the function of action potentials is considered. The spike is an excellent device for coding the intensity of a stimulus as a frequency. However, there may be no need for such frequency coding when distances between cells are not great. The retina is never much more than 0.25 mm or so thick; the distance between any successive cells along the pathway within the retina is therefore small. The optic nerve, in contrast, can vary from a few millimeters to many centimeters in length in different vertebrates. Axons that are long tend to be thick; axons that are long or thick tend to have large cell bodies. This basic physiological principle is revealed beautifully in the structure of the eye. If animals with small eyes are examined, it is found that all of the cell elements, including the ganglion cells, are small. If animals with large eyes, such as the elephant (Figure 15) are considered, it is found that the receptors and inner nuclear cells are no larger than they are in the retina of the bat. Ganglion

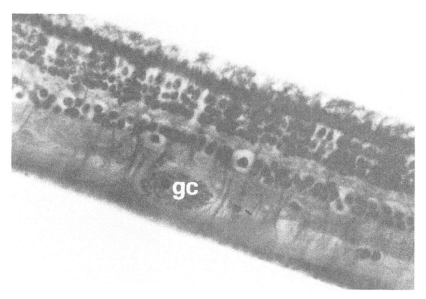

FIG. 15   Retina of the Asian elephant *Elephas*. Note large size of ganglion cell (gc).

cells of these large mammals are of an enormous size, however, suggesting that their size is associated with their function of carrying impulses for great distances. Ganglion cells must maintain a long and thick axon in order to communicate with the elephant brain.

This review has dealt with only a few salient points in the comparative anatomy of the eye. Much has been omitted, although some principles have emerged. The uniformity in cellular structure of vertebrate eyes has been seen. Variance, when it occurs, seems to relate more to control of light intensity and visual sensitivity than to taxonomic status of an animal. The structure of the eye tells more about the light environment of an animal than it does about the comparative history of the vertebrates.

ACKNOWLEDGMENTS

Most histological sections of the retinas illustrated have been prepared by Eilene LaBossiere. I also thank my colleagues who have read and criticized the manuscript, especially Dr. Lorrin Riggs.

REFERENCES

Glickstein, M., & Millodot, M. Retinoscopy and eye size. *Science,* 1970, **168,** 605–606.
Glickstein, M., Brown-Grant, K., & Raisman, G. Light-induced retinal degeneration in the rat and its implications for endocrinological investigations. Proceedings of the Anatomical Society of Great Britain, April 6–7, 1972; *Journal of Anatomy,* 1972, **111**(3), 515.

Kuffler, S. W. Discharge patterns and functional organization of mammalian retina. *Journal of Neurophysiology*, 1953, **16**, 37–68.

Lettvin, J. Y., Maturana, H. R., McCulloch, W. S., & Pitts, W. H. What the frog's eye tells the frog's brain. *Proceedings of the Institute & Radiological Engineering*, 1959, **47**, 1940–1951.

Liebman, P. A., & Entine, G. Visual pigments of frog and tadpole. *Vision Research,* 1968. 8, 761–775.

Pumphrey, R. J. The theory of the fovea. *Journal of Experimental Biology*, 1948, **25**, 299–312.

Ramón y Cajal, S. *The Structure of the Retina.* Translated by S. A. Thorpe and M. Glickstein. Springfield, Ill.: Charles C Thomas, 1972.

Riggs, L. A. Visual acuity. In *Vision and Visual Perception,* C. H. Graham, Ed. New York: John Wiley & Sons, 1965. pp. 321–349.

Schultze, M. The retina. In X. X. Stricker (Ed.), *Manual of Human and Comparative Histology.* Vol. 3. London: New Sydenham Society, 1873. pp. 218–292 (Chapt. 36).

Tomita, T., Kaneko, A., Murakami, M., & Pautler, E. L. Spectral response curves of single cones in the carp, *Vision Research*, 1967, 7, 519–531.

Walls, G. L. *The Vertebrate Eye and Its Adaptive Radiation.* Bloomfield Hills, Mich.: Cranbrook Institute of Science, Bulletin No. 19, 1942, pp. xiv and 785.

Weale, R. A. Why does the human retina possess a fovea? *Nature* (London), 1966, **212.**

Werblin, F. S., & Dowling, J. E. Organization of the retina of the mudpuppy, *Necturus maculosus.* II. Intracellular recording. *Journal of Neurophysiology*, 1969, **32**, 339–355.

4

Some Comments on Visual Acuity and Its Relation to Eye Structure

Mark A. Berkley

Florida State University

INTRODUCTION

It is possible to make some correlations between structure and function in the vertebrate visual system. In the previous chapter, Professor Glickstein elegantly demonstrated that certain aspects of an animal's lifestyle can be related to specific structural features of that animal's retina. For example, the presence of an exclusively cone retina is correlated with a diurnal lifestyle, whereas a retina dominated by rods is found in animals with a crepuscular or nocturnal lifestyle. It has been possible to make such correlations because: (1) there is a high degree of similarity between the retinas of various animals—a similarity convincingly shown in the previous chapter—and (2) the role of the receptors in vision is well understood.

This small measure of success achieved encourages attempts to make more specific statements about visual capacities and lifestyles in animals based on the structure of their eyes (Tansley, 1950). Although there have been numerous attempts, the results have been mixed. On the positive side, for example, the presence of cones indicates not only a diurnal lifestyle but a responsiveness to different wavelengths of light that give the animal the capacity for some degree of color vision. The complementary correlation, that eyes dominated by rod receptors can operate at low levels of illumination (they have a lower threshold for light), is also true. These correlations are more specific than those concerning diurnality or nocturnality and again are possible because absolute sensitivity to light and color vision has been physiologically linked with rod and cone receptors, respectively, in many animals.

Once these fundamental statements with regard to rod and cone function have been made, it becomes increasingly difficult to relate retinal or ocular structures with function. Part of this difficulty is because much of an animal's visual capacity is mediated by neural structures beyond the receptors, structures that are just now beginning to be understood (e.g., Dowling, 1970). Such early investigations, moreover, have revealed a much greater diversity of neural connections and properties than are found among the retinal receptors of different animals.

VISUAL ACUITY

One aspect of vision that has intrigued many investigators and that they have attempted to relate to structural properties of the eye is visual acuity, the capacity to resolve fine details (Wilcox & Purdy, 1933). The finest detail that an organism can detect under some standard set of conditions is usually taken as a measure of its visual acuity. Measures of visual acuity, however, are as numerous as the tests devised by experimenters to ascertain the limits of this visual capacity (Sloan, 1951). The size of the detail, however, is usually defined in minutes of visual angle subtended at the eye of the observer, the convention followed in this chapter. In the present discussion, I shall refer to visual acuity as the size of a single bar of a minimum resolvable grating. I have chosen this measure because of its wide use in animal acuity studies. Typically, at moderate levels of illumination, the finest grating that a human can distinguish from a nonstriped target of equal brightness has a bar width for each striation of about 1 min of visual angle.

Measures of visual acuity in various animals have shown that there is a wide range of acuity capacities among different animals. It is natural, therefore, to ask whether this range of acuities can be related to variations in ocular structures seen between species. Because visual acuity and spatial vision, in general, depend on the optical projection of an image of an object onto the retina, variations in some simple aspects of the geometry of the optics of the eye, the retinal image, and receptor mosaic are the most obvious factors to be considered in determining the limits of acuity. Other factors not optical in origin but that probably affect acuity include the neural connections within and beyond the retina. Some of these factors, which have been found to vary among animals, are described below.

Ocular Optics

An often neglected aspect of the eye is its optics. The eye is an optical instrument and obeys certain well-established physical laws. Such things as optical quality, aperture, and eye size can and do influence acuity. The quality of the optics of the eye sets certain limits to the image formation on the receptor mosaic; e.g., image formation depends on clear transmission media, minimal aberration, etc. A consideration of ocular optics also suggests that the length of the eyeball may be an important dimension. Although pupillary diameter is also an important optical variable, it usually is considered with optical quality measurements.

Receptor Mosaic

The image being cast by the optics of the eye onto the retina falls on a surface made up of small receptor units. If neural interconnections and eye size are disregarded for the moment, the finer this receptor mosaic (the smaller the diameter of the receptors), the greater the theoretical capacity for detecting fine targets.

Neural Connectivity

Finally, but equally important, are the limitations imposed by the neural connectivity within and beyond the retina. Probably the most confounding aspect of the relationship of the receptor mosaic to the image being cast on it is the extent of interconnection between the neural transmission elements and the receptor elements within the retina. For example, a receptor may or may not be the "functional unit" in the retina depending on the degree of convergence or interconnection that exists between it and the ganglion cells. A simplifying assumption made by many investigators is that a certain class of receptors, namely cones, have a simple one-to-one relationship with the neural output cells, ganglion cells, that are transmitting messages to the brain. In this case, the receptor unit is the functional unit. There are, however, many examples in which this is probably not the case, eyes in which the functional unit consists of a large number of receptors converging on the cell and sending its messages to the brain. In this instance, the spatial (anatomical) extent of the functional unit is usually not known. Simple light microscopic examination of the retina cannot and has not revealed the size and extent of these functional units.

MORPHOLOGY AND DETAIL VISION

Despite the warning of numerous early investigators, such as von Helmholtz (1867) and Hartridge (1922), that acuity estimates cannot be based on simple anatomical examination of the eye, such estimates continue to be made (Elliott-Smith, 1928; Duke-Elder, 1958). For example, Elliott-Smith (1928) suggests that the presence of a fovea (a depression within the receptor mosaic where vision is most acute) is a new evolutionary development and he suggests that animals possessing a fovea should have greater acuity than animals not possessing one. As will be seen below, examples can be found of animals with foveae that have as poor acuity as animals which do not have a fovea (see Table 3).

Another simple theory suggests that a cone-dominated eye has greater acuity than a rod-dominated eye. Here, as has been seen in Professor Glickstein's presentation, there is some evidence that the rod-dominated eye is one adapted for a nocturnal or crepuscular lifestyle, and the cone-dominated eye is adapted for a diurnal lifestyle. It is also true that cone-dominated eyes tend, on the average, to possess better acuity than rod-dominated eyes (see Walls, 1942, and Table 3).

One aspect of the structural organization of the retina that has not received its fair share of attention is the ganglion cell destiny of the retina. Ganglion cells, as will be recalled, are the innermost cells of the retina that receive information from the receptors and intervening cells and send their messages onward to the brain. They therefore receive all of the final, integrated input of the various elements within the retina and transmit that information to the brain. Ganglion cells and their interconnections must therefore represent what Lashley (1932) has called the "functional unit." That is, the ganglion cell and the extent of its interconnections represented spatially on the retina must ultimately set the limits of acuity. In the human and monkey eye, ganglion cell density and the cone density in the fovea are presumed to be one to one. In eyes that have mixed receptor mosaics in the region of acute vision, the relationship of rods and cones to the ganglion cells is not known. Therefore, acuity estimates based on receptor size tend to overestimate acuity in these animals and suggest receptor convergence (Lashley, 1932; Weymouth, 1958).

Finally, the dioptrics of the eye must set some sort of limits on its resolution. Of the simple dioptric features of the eye so far considered, eye size is the only one that has been suggested as an important factor (Grether, 1941; Walls, 1942). Such things as the quality of the lens and cornea, as well as the transmission media, must be considered because they are biological in origin and possess imperfections. These imperfections reduce image quality and must also set some limits on the resolution capacity of the eye. It may be thought that they set the final limits of acuity but, as will be seen, this is not the case.

I shall now turn to a somewhat more detailed discussion of these various factors and consider what evidence there is for and against each of them defining, or setting the limits for, acuity in the vertebrate eye.

Consider that because visual receptors among different animals are relatively similar in size, the limits on acuity may simply be determined by the size of the image cast on the receptor mosaic. For example, large diameter eyes should possess greater acuity than smaller diameter eyes. In an early paper by Grether (1941), comparisons are made between visual acuity (based on grating acuity) and the actual retinal image width of a stripe that a minimal resolvable grating cast on the retina, taking into account the different eyeball sizes. Grether (1941) concludes that the differences in acuity between the human and rhesus monkey, for example, can be attributed simply to the difference in eyeball size between these two animals, the monkey possessing a somewhat smaller diameter orb than man (see Table 1). However, other comparisons with such animals as pigeons and rats produce the conclusion that a simple explanation based on eye size is inadequate. This comparison is shown in Table 1. The behavioral acuity estimates are listed by investigators on the left. The retinal image widths as calculated by Grether (1941) are shown on the right. To make the comparisons more meaningful, the ratios between the visual acuity angles between animals and the ratio of the retinal image widths for each of these animals can be calculated. If, in fact, acuity is being limited by image size, these ratios should be the same. As can be seen from the table, they are not. However, some acuity reduction can be attributed to reduction in eyeball size among these various animals.

A second simple notion is that acuity may be limited by the size of the receptors in the central portion of the receptor mosaic. Support for this idea

TABLE 1

Retinal Image Widths at the Lowest Visual Acuity Thresholds Found for Several Species of Animals Tested by Means of Striae[a]

		Lowest threshold for each species		Ratio of visual angle to image width
Investigator	Animal	Visual angle (min)	Retinal image width (μ)	
Spence (1934)	Human (adult)	0.44	1.89	.23
Weinstein and Grether (1940)	Human (adult)	0.48	2.06	.23
Spence (1934)	Chimpanzee	0.47	1.86	.25
Weinstein and Grether (1940)	Rhesus monkey	0.67	2.33	.29
Johnson (1914)	Cebus monkey	0.95	3.31[b]	.29
Hamilton and Goldstein (1933)	Pigeon	2.70	4.89	.55
Chard (1939)	Pigeon	2.70	4.89	.55
Johnson (1914)	Gamecock	4.07	9.58[c]	.42
Lashley (1930)	Rat (albino)	86.0	79.2	1.08
	Rat (pigmented)	52.0	47.7	1.09

[a]From Grether (1941). Table lists the visual acuity and retinal image widths for various species and the source of the data. Visual acuity (third column) is expressed as the visual angle subtended by one bar of a threshold grating at the eye. The image width occupied on the retina by one bar of a threshold grating is listed in the last column.

[b]Calculation of this retinal image width based upon ocular dimensions for the rhesus monkey.

[c]Calculation of this retinal image width based upon ocular dimensions for the domesticated chicken.

may be found in correlations between human acuity and receptor size. For example, Green (1970), using a technique that bypasses the optics of the eye and is too complex to go into in this discussion, has found that the ocular optics actually reduce acuity very slightly and that the limits of acuity in humans are closely related to the cone size in the fovea. However, other examples may be found where this relationship may not hold. On the basis of anatomical examination of the eagle eye by Rochon-Duvingneaud, (Walls, 1942) the eagle possesses eight times as many cones in its fovea as does man and a somewhat shorter length eye. On the basis of this data, Walls (1942) suggests that the eagle may have eight times as great acuity as man, assuming also that he possesses a one-to-one relationship between the cones and the ganglion cells. Although this is a very attractive hypothesis, some recent examinations of the optics of the eagle eye suggest that its optics are incapable of forming an image that can produce an acuity eight times greater than man's (Schlaer, 1972). In duplex retinas, with both rods and cones present in the fovea, the receptor size argument also fails. In animals such as the cat and the owl, where the area centralis or the specialized central region or fovea are not completely populated by one type of receptor, acuity seems to be much poorer than either single-cone diameter or intercone distance suggests. For example, in the owl, Fite (1973) has found visual acuity to be poorer than the single-cone diameter in his fovea permits. This animal possesses both rods and cones in its fovea and its acuity is probably related not to cone diameter or intercone distance but to the inter-ganglion cell distance. Another example of this type may be found in the cat eye. Berkley and Watkins (1973), for example, have found the cat's visual acuity to be considerably poorer than its receptor mosaic (Steinberg, Reid, & Lacy, 1973) permits. Lashley (1932) comes to a similar conclusion in the rat based on his behavioral and ocular measurements. These studies all indicate the importance of knowing not the receptor size but the functional unit size. It is precisely here that current knowledge is so poor.

The urge to correlate ocular structures with the acuity has seized many investigators in vision. Walls (1942) and Elliott-Smith (1928) suggest albeit rather indirectly, that retinas which possess a fovea have better acuity than retinas which do not. Furthermore, rod-dominated eyes are poorer in acuity than cone-dominated eyes. Table 2 is a table taken from Walls (1942) in which are listed behavioral acuity estimates available to him at that time and the visual angle or the distance on the retina subtended by the minimum visible stripe in the target. As can be seen, there is a weak relationship between the diurnal eyes and acuity. However, there are numerous all-cone eyes that have yet to be examined for their acuity capacity, e.g., ground squirrels and eagles.

Another structural aspect that can be considered as setting a limit for acuity is the ganglion cell distribution in the retina. The ganglion cells, as will be remembered, are the final messengers from the retina sending their messages about the images cast onto the retina to the brain. Lashley (1932) suggests that the ganglion cell distribution is probably more closely related to the functional

TABLE 2

Visual Acuities for Parallel Lines (From Various Sources)[a]

	Visual angle (min)	Corresponding distance on retina (μ)	Visual angle corresponding to 1 mm distance along visual cortex
Diurnal animals			
Human adult	0.44	1.89	—
(different reports)	.048	2.06	—
	0.50	2.14	—
	0.80	3.43	—
	0.82	3.52	—
	0.83	3.56	—
Child	0.62	2.67	—
Chimpanzee	0.47	1.86	—
Rhesus monkey	0.67	2.33	—
Rhesus monkey, along visual axis	—	—	4'
Rhesus monkey, 7° from visual axis	—	—	20'
Cebus monkey	0.95	3.31	—
Pigeon	2.70	4.89	—
Pigeon, "homer"	0.38	.69	—
Gamecock (no fovea)	4.07	9.58	—
Nocturnal animals			
Cat, along visual axis	5.5	—	1°
Cat, 30° below axis	—	—	5°
Alligator	11.0	—	—
Opossum	11.0	—	—
Rat, pigmented	26.0	23.8	—
Rat, albino	52.0	47.7	—

[a]From Walls (1942, p. 207). Column labeled "visual angle" lists behavioral acuity estimates available to Walls for the species listed. The second column is a calculation of a distance of the retina that an image of a bar of the threshold grating would occupy. The upper portion of the table lists cone-dominated retinae and the lower portion lists rod-dominated retinae.

unit size distribution in the retina than is the receptor density and to a great extent this is a correct evaluation. For example, if acuity is measured at different positions along the horizontal meridian of the retina at different eccentricities away from central vision an excellent relationship is found between the decline of acuity with eccentricity and the decline in ganglion cell density, when moving away from central vision (Rolls & Cowey, 1970).

Finally I must turn to a consideration of the dioptric characteristics of the eyes in various animals and an evaluation of the dioptric limitations on acuity. In Table 3, I have presented, where available, the best behavioral estimate for grating acuity. (These numbers represent the interbar spacing in the minimum

TABLE 3

Table of Various Anatomical, Optical, and Behavioral Measurements
Related to Acuity for a Variety of Animals[a]

Animal	Eye size (PND) (mm)	Distance on retina (μ) per minute visual angle	Presence of fovea	Minimum interreceptor distance (min of angle)	Minimum interganglion cell distance (min of angle)	Optical resolution (min of angle)	Best behavioral estimate (min of angle)
Man	17.	4.9	Yes	0.4[1]	0.75[2]	0.5[3,4]	0.5[3,5]
Rhesus monkey	14.1[6]	4.1[6]	Yes[7]	1.08[6]	~.9[6]	—	0.65[6]
Squirrel monkey	10.2[6]	3.0[6]	Yes[6,7]	1.33[6]	~1.2[6]	—	0.74[6]
Stumptail monkey	—	—	Yes[8]	—	—	—	1.4[8]
Marmoset	—	—	Yes[7]	—	—	—	(.5–1.5)[9]
Lemur	—	—	No	—	—	—	(.5–1.5)[9]
Tree shrew	—	—	No	—	—	—	(.5–1.5)[9]
							(25)
African serpent eagle	15.5		Yes	~.16[10]	—	0.25[11]	—

	PND	Retinal distance per 1 min	Fovea	Single receptor	Interganglion cell	Optimal limits	Finest grating
Man (rod monochromat)	17.	4.9	(Yes)	—	—	0.5	6
Cat	12.5[12]	3.6[12]	No	1.7 (cone)[13]	*4.0[14]	3.5–5.5[15,16,17]	5[18]
Rat (pigmented)	2.93[19]	0.8[19]	No	~3.3[19]	~20–30[19]	30–61[19]	26–52[19]
Galago	—	—	No	—	—	—	3.5[20]
Aotes	—	—	No	—	—	—	8
Rabbit	9.8[21]	2.9[21]	No	—	—	—	10–20[22]
Great horned owl	—	—	Yes[23]	(.18, cone)[23]	(0.10)[23]	—	4–5[23]

[a]Eye size is represented by the posterior nodal distance (PND); second column lists the distance on the retina corresponding to 1 min of visual angle. The third column indicates the presence or absence of a specialized retinal depression (fovea). The fourth column gives visual angle subtended by a single receptor at the maximum receptor density region. The fifth column lists the visual angle subtended by the minimum interganglion cell distance. The sixth column lists an estimate of resolution at the optimal limits for that eye. The last column gives the finest grating (in interbar distance–visual angle) resolvable determined through behavioral tests that have been reported for that animal. (Superscript numbers refer to source of measurement listed in References.)

Numbered superscripts refer to the following references: [1] Østerberg, 1935; [2] Van Buren, 1963; [3] Campbell & Gubisch, 1966; [4] Gubisch, 1967; [5] Green, 1970; [6] Rolls & Cowey, 1970; [7] Woollard, 1927; [8] Yarczower, Wolbarsht, Galloway, Fligsten & Malcolm, 1966; [9] Ordy & Samorajsky, 1968; [10] Walls, 1942; [11] Schlaer, 1972; [12] Vakkur & Bishop, 1963; [13] Steinberg, Reid & Lacy, in press; [14] Stone, 1965; [15] Wassle, 1971; [16] Westheimer, 1962; [17] Morris & Marriott, 1961; [18] Smith, 1936; [19] Lashley, 1932; [20] Treff, 1967; [21] Hughes, 1972; [22] Van Hof, 1967; [23] Fite, 1973.

resolvable grating. Because of differences in testing conditions, target luminances, etc.. direct comparisons between animals are next to meaningless; thus these values must be considered as very approximate.) The eye-size estimate is in the first column (posterior nodal distance or PND). The second column lists the distance on the retina (in microns) subtended by 1 min of visual angle at the cornea. The third column indicates whether a foveal depression is present or not. The fourth column is the interreceptor distance in minutes of visual angle at the highest density of receptors in the eye. The fifth column gives the interganglion cell distance in minutes at the maximum ganglion cell density of receptors in the eye. The sixth column gives an estimate of the line resolvability based on the optical resolution measurements made by various investigators. In the last column are the behaviorally determined acuity measurements in these various animals. The lower portion of the table lists animals that are either crepuscular or nocturnal. As can be seen, cone-dominated eyes, in general, have better acuity than rod-dominated eyes. The absence or presence of a fovea appears to be a relatively poor predictor of acuity, as does the interreceptor distance for animals other than primates. The interganglion cell distance seems to have slightly better predictive value than the interreceptor distance, and the optical resolution measurements, although there are not many of them available, do appear to be reasonably well related to the acuity estimates. In summary then, it can be seen that there are numerous factors within the dioptric system of the eye and receptor mosaic that can set limits on the acuity of the organism. No single feature or aspect of the visual system seems to be simply related to visual acuity. If a choice must be made, however, optical resolution and interganglion cell distance are probably the best indicators.

EVOLUTION AND ACUITY

One puzzling aspect of visual acuity measurements in various animals is that many animals appear to possess acuity much greater than may be guessed necessary in their natural environment. For example, monkeys possess extremely good acuity, as do most birds. A suggestion made by a number of investigators for the high degree of acuity in many animals is for prey detection, for many predatory birds have excellent acuity. If it is assumed, for example, that as the eagle has eight times better acuity than man, this bird should be able to detect the presence of a rabbit at a distance of about 5 miles. I must point out that this distance is well beyond the striking range of the eagle. Although it is difficult to find any single aspect of the lifestyle of these diverse animals that is related to visual acuity, it is tempting to speculate on a number of possible factors. Walls (1942) points out, for example, that high acuity appears to be correlated with frontal placement of the eyes. I should like to suggest that this observation may represent the key for understanding the relatively high acuity of many animals.

If it is assumed necessary for many animals, both predatory and nonpredatory, to have acute depth perception, then these animals should have frontally placed eyes and accurate eye alignment. This would permit the use of parallax to detect small differences in depth (steropsis). To achieve the accurate alignment, the regions of acute vision (fovea) may act as a pair of crosshairs for the two eyes, permitting them to be aligned with great accuracy so that the small differences in retinal image position produced by targets either in front of or behind the plane of fixation can provide cues for depth. If this assumption is made, many animals with different lifestyles that require good depth perception also turn out to have quite outstanding visual acuity. Although it is not possible to obtain direct evidence in favor of this hypothesis, some data derived from the visual deprivation experiments may be relevant. In these experiments, if the developing animal, for example, a cat, is deprived of the simultaneous use of both eyes, it fails to develop a certain class of neural cells in its visual cortex that is required for stereopsis. In addition, its eyes are not aligned; that is, this animal has "cross-eyes." Therefore, both the ability (acuity) and the opportunity to use that ability (simultaneous use of both eyes) appear to be required for the development of a set of neurons capable of providing the depth information.

(For those interested in visual acuity in animals, a bibliography of experimental papers relevant to acuity is appended.)

ACKNOWLEDGMENTS

This chapter was prepared with support from NSF Grant GB 34166.

REFERENCES

Berkley, M., & Watkins, D. Grating resolution and refraction in the cat estimated from evoked cerebral evoked potentials. Vision Research, 1973, 13, 403–415.

Campbell, F., & Gubisch, R. Optical quality of the human eye. *Journal of Physiology (London)*, 1966, 186, 558–578.

Dowling, J. E. Organization of vertebrate retinas. *Investigative Ophthalmology*, 1970, 9, 655–680.

Duke-Elder, S. *System of Ophthalmology. Vol. 1. The eye in evolution.* Mosby Co., St. Louis, 1958, pp. 637–705.

Elliott-Smith, G. The new vision. Bowman Lecture. *Transactions of the Ophthalmology Society U.K.*, 1928, 28.

Fite, K. Anatomical and behavioral correlates of visual acuity in the great horned owl. *Vision Research*, 1973, 13, 219–230.

Green, D. G. Regional variations in the visual acuity for interference fringes on the retina. *Journal of Physiology (London)*, 1970, 207, 351–356.

Grether, W. F. Comparative visual acuity thresholds in terms of retinal image widths. *Journal of Comparative Psychology*, 1941, 31, 23–33.

Gubisch, R. Optical performance of the human eye. *Journal of the Optical Society of America*, 1967, 57, 407–415.

Hartridge, H. Visual acuity and the resolving power of the eye. *Journal of Physiology*, 1922, 57, 52–67.

von Helmholtz, H. *Physiological Optics.* 1867. (Translated by Southall.) New York: Dover Publications, 1962.

Hughes, A. A schematic eye for the rabbit. *Vision Research*, 1972, 12, 123–138.

Lashley, K. S. The mechanism of vision: V. The structure and image-forming power of the rat's eye. *Journal of Comparative Psychology*, 1932, 13, 173–200.

Morris, V., and Marriott, F. The distribution of light in an image formed in the cat's eye. *Nature* (London), 1961, 190, 176–177.

Østerberg, G. Topography of the layer of rods and cones in the human retina. *Acta Ophthalmologica* (Kbh.), 1935, suppl. 61, 1–102.

Rolls, E. T., & Cowey, A. Topography of the retina and striate cortex and its relationship to visual acuity in rhesus monkeys and squirrel monkeys. *Experimental Brain Research*, 1970, 10, 298–310.

Schlaer, R. An eagle's eye: Quality of the retinal image. *Science*, 1972, 176, 920–922.

Sloan, L. Measurement of visual acuity: A critical review. *Archiv fuer Ophthalmologie*, 1951, 45, 704–725.

Smith, K. U. Visual discrimination in the cat: IV. The visual acuity of the cat in relation to stimulus distance. *Journal of Genetic Psychology*, 1936, 49, 297–313.

Steinberg, R., Reid, M. & Lacy, P. The distribution of rods and cones in the retina of the cat. *Journal of Comparative Neurology*, 1973, 148, 229–248.

Stone, J. A quantitative analysis of the distribution of ganglion cells in the cat's retina. *Journal of Comparative Neurology*, 1965, 124, 337–352.

Tansley, K. Vision in Physiological Mechanism in Behavior. *SEB Symposium IV.* New York: Academic Press, 1950, pp. 19–33.

Van Buren, K. M. *The retinal ganglion cell layer.* Springfield, Ill.: Charles C Thomas, 1963.

Walls, G. *The vertebrate eye.* Bloomfield Hills, Illinois: Cranbrook Institute, 1942.

Wässle, H. Optical quality of the cat eye. *Vision Research*, 1971, 11, 955–1006.

Westheimer, G. Line-spread function of living cat eye. *Journal of the Optical Society of America*, 1962, 52, 1326.

Weymouth, F. W. Visual sensory units and the minimal angle of resolution. *American Journal of Ophthalmology*, 1958, 46, 102–113.

Wilcox, W. W., & Purdy, D. Visual acuity and its physiological basis. *British Journal of Psychology*, 1933, 23, 233–261.

Woollard, H. H. The differentiation of the retina in the primates. *Proceedings of the Zoological Society (London)*, 1927, 1, 1–17.

Yarczower, M., Wolbarsht, M., Galloway, W., Fligsten, K., & Malcolm, R. Visual acuity in a stumptail Macaque. *Science*, 1966, 152, 1392–1393.

BIBLIOGRAPHY

Baylor, E. R., & Shaw, E. Refractive error and vision in fishes. *Science*, 1962, 136, 157–158.

Beer, T. Die accommodation des Fischauges. *Pflügers Arch. ges. Physiol.*, 1894, 58, 523–650.

Berkley, M., & Watkins, D. Grating resolution and refraction in the cat estimated from evoked cerebral potentials. *Vision Research*, 1973, 13, 403–416.

Bingeli, R., & Paule, W. J. The pigeon retina: quantitative aspects of the optic nerve and ganglion cell layer. *Journal of Comparative Neurology*, 1969, **137**, 1–18.

Bishop, P. O., Kozak, W., & Vakkur, G. J. Some quantitative aspects of the cat's eye: axis and plane of reference, visual field coordinates and optics. *Journal of Physiology*, 1962, **163**, 466–502.

Block, M. T. A note of the refraction and image formation of the rat's eye. *Vision Research*, 1969, **9**, 705–712.

Blough, P. The visual acuity of the pigeon for distant targets. *J. Exp. Anal. Behav.*, 1971, **15**, 57–67.

Blough, P. Visual acuity in the pigeon II: Effects of target distance and retinal lesions. *Journal of the Experimental Analysis of Behavior*, 1973, **20**, 333–343.

Bonds, A., Enroth-Cugell, C., & Pinto, L. Image quality of the cat eye measured during retinal ganglion cell experiments. *Journal of Physiology (London)*, 1972, **220**, 383–401.

Catania, A. C. On the visual acuity of the pigeon. *Journal of the Experimental Analysis of Behavior*, 1964, **7**, 361–366.

Cavonius, C. R., and Robbins, D. O. Relationship between luminance and visual acuity in the rhesus monkey. *Journal of Physiology (London)*, 1973, **232**, 239–246.

Chard, R. D. Visual acuity in the pigeon. *Journal of Experimental Psychology*, 1939, **24**, 588–608.

Chard, R. D., & Gundlach, R. H. The structure of the eye of the homing pigeon. *Journal of Comparative Psychology*, 1938, **25**, 249–272.

Charman, W. N., & Tucker, J. The optical system of the goldfish eye. *Vision Research*, 1973, **13**, 1–8.

Coles, J. A. Some reflective properties of the tapetum lucidum of the cat's eye. *J. Physiology (London)*, 1971, **212**, 393–409.

Cowey, A., & Ellis, C. M. Visual acuity of rhesus and squirrel monkeys. *Journal of Comparative Physiology & Psychology*, 1967, **64**, 80–84.

Dawson, W., Birndorf, L. A., & Perez, J. M. Gross anatomy and optics of the dolphin eye (*Tursiops truncatus*). *Cetology*, 1972, **10**, 1–12.

Donner, K. O. The visual acuity of some passerine birds. *Acta Zoologica Fennica*, 1951, **66**, 1–40.

Farrer, D. N., & Graham, E. S. Visual acuity in monkeys: A monocular and binocular subjective technique. *Vision Research*, 1967, **7**, 743–747.

Fite, K. Anatomical and behavioral correlates of visual acuity in the great horned owl. *Vision Research*, 1973, **13**, 219–230.

Glickstein, M., & Millodot, M. Retinoscopy and eye size. *Science*, 1970, **168**, 605–606.

Graham, E. S. McVean, G. W., & Farrer, D. N. Near and far visual acuity in rhesus monkeys (*Macaca mulatta*). *Perceptual and Motor Skills*, 1968, **26**, 1067–1072.

Grether, W. F. Comparative visual acuity thresholds in terms of retinal image widths. *Journal of Comparative Psychology*, 1941, **31**, 23–33.

Gundlach, R. H. The visual acuity of homing pigeons. *Journal of Comparative Psychology*, 1933, **16**, 327–342.

Gundlach, R. H., Chard, R. D., & Skahen, J. R. The mechanism of accommodation in the pigeon. *Journal of Comparative Psychology*, 1945, **38**, 27–42.

Hamilton, W. F., & Goldstein, J. L. Visual acuity and accommodation in the pigeon. *Journal of Comparative Psychology*, 1933, **15**, 193–197.

Hands, A. R., Sutherland, N. S., & Bartley, W. Visual acuity of essential fatty acid-deficient rats. *Biochemical Journal*, 1965, **94**, 279–293.

Hester, F. J. Visual contrast thresholds of the goldfish *(Carassius auratus)*. *Vision Research*, 1968, **8**, 1315–1335.

Hughes, A. A schematic eye for the rabbit. *Vision Research*, 1972, **12**, 123–138.

Johnson, H. M. Visual pattern discrimination in the vertebrates. II. Comparative visual acuity in the dog, monkey, and chick. *Journal of Animal Behavior*, 1914, 4, 340–346.

Lashley, K. S. The mechanism of vision. III. The comparative visual acuity of pigmented and albino rats. *Journal of Genetic Psychology*, 1930, 37, 481–484.

Lashley, K. S. The mechanism of vision. V. The structure and image-forming power of the rat's eye. *Journal of Comparative Psychology*, 1932, 13, 173–200.

Leach, E. H., Marriott, F., & Morris, V. The distances between rods in the cat's retina. *Journal of Physiology (London)*, 1961, 157, 17P.

Lit, A. Visual acuity. *Annual Review of Psychology*, 1968, 19, 27–54.

Massof, R. W., & Chang, F. W. A revision of the rat schematic eye. *Vision Research*, 1972, 12, 793–796.

Meyer, D., Meyer-Hamme, S., & Schaeffer, K. Electrophysiological investigation of refractive state and accommodation in the rabbit's eye. *Pflügers Arch.*, 1972, 332, 80–86.

Meyer, D., & Schwassmann, H. Electrophysiological method for determination of refractive state in fish eyes. *Vision Research*, 1970, 10, 1301–1303.

Millodot, M., & Blough, P. The refractive state of the pigeon eye. *Vision Research*, 1971, 11, 1019–1022.

Morris, V., & Marriott, F. The distribution of light in an image formed in the cat's eye. *Nature* (London), 1961, 190, 176–177.

Muir, D., & Mitchell, D. Visual resolution and experience: Acuity deficits in cats following early selective visual deprivation. *Science*, 1973, 180, 420–422.

Nakamura, E. L. Visual acuity of two tunas, *Katserwonus pelamis* and *Enthynnus offinis*. *Copeia*, 1968, 1, 41–49.

Neuman, G. H. Visual learning ability of primitive mammals. *Zeitschrift für Tierpsychologie*, 1961, 18, 71–83.

Nye, P. N. The binocular acuity of the pigeon measured in terms of the modulation transfer function. *Vision Research*, 1968, 8, 1041–1053.

Nye, P. N. On the functional differences between frontal and lateral visual fields of the pigeon. *Vision Research*, 1973, 13, 559–574.

Ordy, J. M., Latanick, A., Samorajski, T., & Massopust, L. C., Jr. Visual acuity in newborn primate infants. *Proceedings of the Society of Experimental and Biological Medicine*, 1964, 115, 677–680.

Ordy, J. M., Massopust, L. C., Jr., & Wolin, L. R. Postnatal development of the retina, electroretinogram and acuity in the rhesus monkey. *Experimental Neurology*, 1962, 5, 364–382.

Ordy, J. M., & Samorajski, T. Visual acuity and ERG-CRF in relation to the morphological organization of the retina among diurnal and nocturnal primates. *Vision Research*, 1968, 8, 1205–1225.

Ordy, J. M., Samorajski, T., Collins, R. J., & Nagy, A. R. Postnatal development of vision in subhuman primate (*Macaca mulatta*). *Archives of Ophthalmology*, 1965, 73, 674–686.

Pepper, R. L., & Simmons, J. V., Jr. In-air visual acuity of the bottlenosed dolphin *Experimental Neurology*, 1973, 41, 271–276.

Perez, J. M., Dawson, W., and Landau, D. Retinal anatomy of the bottlenosed dolphin (*Tursiops truncatus*). *Cetology*, 1972, 11, 1–11.

Polyak, S. L. *The Retina*. Chicago: University of Chicago Press, 1941.

Protasov, V. *Vision and near orientation of fish* (Translated from russian 1970). I.P.S.T. cat. L5738. Jerusalem: Israel Program for Scientific Translations Ltd., 1900.

Rahmann, H., & Esser, M. Bestimmung der Sehschärfe sowie Dressurverhalten des skandinavischen Berglemmings (*Lemmus Lemmus L.*) *Zeitschrift für Sängetierkunde*, 1965, 30, 47–53.

Rahmann, H. Die Sehschärfe bei Wirbeltieren. *Naturwissenschaften Rundechan*, 1967, 20, 8–14.

Riesen, A. H., Ramsey, R. L., & Wilson, P. D. Development of visual acuity in rhesus monkeys deprived of patterned light during early infancy. *Psychonomic Science*, 1964, **1**, 33–34.

Rolls, E. T., & Cowey, A. Topography of the retina and striate cortex and its relationship to visual acuity in rhesus monkeys and squirrel monkeys. *Experimental Brain Research*, 1970, **10**, 298–310.

Rowley, J. B. Discrimination of pattern and size in the goldfish *Carassius auratus*. *Genetic Psychology Monographs*, 1934, **15**, 245–302.

Sadler, J. D. The focal length of the fish eye lens and visual acuity. *Vision Research*, 1973, **13**, 417–423.

Schlaer, R. An eagle's eye: Quality of the retinal image. *Science*, 1972, **176**, 920–922.

Schusterman, R. J. Visual acuity in pinniped. In H. E. Winn, & B. L. Olla, Eds., *Behavior of Marine Animals*. New York: Plenum Press, 1972, pp. 469–492.

Smith, K. U. The postoperative effects of removal of the occipital cortex upon visual acuity in the cat, as measured by oculocephalogyric responses. *Psychology Bulletin*, 1936a, **33**, 754.

Smith, K. U. Visual discrimination in the cat: IV. The visual acuity of the cat in relation to stimulus distance. *Journal of Genetic Psychology*, 1936b, **49**, 297–313.

Smith, K. U. Visual discrimination in the cat: VI. The relation between pattern vision and the visual acuity and the optic projection centers of the nervous system. *Journal of Genetic Psychology*, 1938, **53**, 231–272.

Spence, K. W. Visual acuity and its relation to brightness in chimpanzees and man. *Journal of Comparative Psychology*, 1934, **18**, 333–361.

Spong, P., & White, D. Visual acuity and discrimination learning in the dolphin (*Ragenorhynchus obliquideus*). *Experimental Neurology*, 1971, **31**, 431–436.

Steinberg, R., Reid, M., & Lacy, P. The distribution of rods and cones in the retina of the cat (*Felis domesticus*). *Journal of Comparative Neurology*, 1973, **148**, 229–248.

Stone, J. A. quantitative analysis of the distribution of ganglion cells in the cat's retina. *Journal of Comparative Neurology*, 1965, **124**, 337–352.

Suthers, R. A., & Wallis, N. E. Optics of the eyes of echolocating bats. *Vision Research*, 1970, **10**, 1165–1173.

Tamura, T. A study of visual perception in fish; especially on resolving power and accommodation. *Bulletin of the Japanese Society of Scientific Fisheries*, 1957, **22**, 536–557.

Treff, H. A. Tiefensehschärfe und sehscharfe beim galago (*Galago sinegalensis*). *Zeitschrift für Vergleichende Physiologie*, 1967, 26–57.

Vakkur, G. J., & Bishop, P. O. The schematic eye in the cat. *Vision Research*, 1963, **3**, 357–381.

Vakkur, G. J., Bishop, P. O., & Kozak, W. Visual optics in the cat, including posterior nodal distance and retinal landmarks. *Vision Research*, 1963, **3**, 289–314.

Van Hof, M. W. Visual acuity in the rabbit. *Vision Research*, 1967, **7**, 749–751.

Walk, R. D. A preference technique to investigate visual acuity in the rat. *Psychonomic Science*, 1965, **3**, 301–302.

Walls, G. *The vertebrate eye and its adaptive radiation*. Bloomfield Hills, Michigan: Cranbrook Institute, 1942. Pp. 207–209.

Ward, J., & Masterton, B. Encephalization and visual cortex in the tree shrew (*Tupaia glis*). *Brain, Behavior and Evolution*, 1970, **3**, 421–469.

Ware, C., Casagrande, V., & Diamond, I. T. Does the acuity of the tree shrew suffer from removal of striate cortex? *Brain, Behavior and Evolution*, 1972, **5**, 18–29.

Warkentin, J. The visual acuity of some vertebrates. *Psychology Bulletin*, 1937, **34**, 793.

Warkentin, J., & Smith, K. B. The development of visual acuity in kittens. *Psychology Bulletin*, 1936, **33**, 597.

Wässle, H. Optical quality of the cat eye. *Vision Research*, 1971, **11**, 995–1006.

Wässle, H., & Creutzfeldt, O. Spatial resolution in visual system: a theoretical and experimental study on single units in the cat's lateral geniculate body. *Journal of Neurophysiology*, 1973, **36**, 13–27.

Weinstein, B., & Grether, W. F. A comparison of visual acuity in the rhesus monkey and man. *Journal of Comparative Psychology*, 1940, **30**, 187–195.

Weiskrantz, L., & Cowey, A. Striate cortex lesions and visual acuity of the rhesus monkey. *Journal of Comparative Physiology & Psychology*, 1963, **56**, 225–231.

Weiskrantz, L., & Cowey, A. Comparison of the effects of striate cortex and retinal lesions on visual acuity in the monkey. *Science*, 1967, **155**, 104–106.

Westheimer, G. Line-spread function of living cat eye. *Journal of the Optical Society of America,* 1962, **52**, 1326.

Westheimer, G. Visual acuity. *Annual Review of Psychology*, 1965, **16**, 359–380.

White, D., Cameron, N., Spong, P., & Bradford, J. Visual acuity of the killer whale (*Orcinus orca*). *Experimental Neurology*, 1971, **32**, 230–236.

Wilcox, W. W., & Purdy, D. Visual acuity and its physiological basis. *British Journal of Psychology*, 1933, **23**, 233–261.

Wilson, P. D., & Riesen, A. H. Visual development in Rhesus monkey neonatally deprived of patterned light. *Journal of Comparative Physiology & Psychology*, 1966, **61**, 87–95.

Woodburne, L. S. Visual acuity of *Saimiri sciureus. Psychonomic Science*, 1965, **3**, 307–308.

Yamanouchi, T. The visual acuity of the coral fish, *Microconthus strigatus. Publications of Seto Marine Biological Laboratory*, 1956, **5**, 133–156.

Yarczower, M., Wolbarsht, M., Galloway, W., Fligsten, K., & Malcolm, R. Visual acuity in a stumptail macaque. *Science*, 1966, **152**, 1392–1393.

Young, F. A. Visual refractive errors of wild and laboratory monkeys. *Eye, Ear, Nose, Throat Monthly*, 1965, **8**, 55–70.

Young, F. A. Visual acuity and refractive errors in primates. Proceedings of the 77th Annual Convention of the American Psychology Association, Washington, D.C., 1969.

Young, F. A. Visual refractive characteristics and the subhuman primate. In *Non-Human Primates and Medical Research*. New York: Academic Press, 1973, pp. 353–379.

Young, F. A., & Farrer, D. N. Refractive characteristics of chimpanzees. *American Journal of Optometry*, 1964, **2**, 81–91.

Young, F. A., & Leary, G. A. Comparison of the optical characteristics of the human, ape, and monkey eye. Proceedings of the 75th Annual Convention of the American Psychology Association, Washington, D.C., 1967.

Zilbert, D. E., & Riesen, A. H. A comparison of the effects of infant and adult retinal lesions upon visual acuity in the rabbit. *Experimental Neurology*, 1971, **33**, 445–458.

5

Origin and Evolution
of the Ear of Vertebrates

Ernest Glen Wever

Auditory Research Laboratories
Princeton University

THE PREVAILING THEORIES

Toward the close of the nineteenth century, in a period of rapid development of evolutionary thinking, two hypotheses arose concerning the origin and evolution of the vertebrate ear. These two may be referred to as the "statocyst" hypothesis and the "acustico–lateralis" hypothesis, and both have persisted into modern times without serious conflict or critical evaluation.

The Statocyst Hypothesis

In a number of invertebrates, including coelenterates, ctenophores, echinoderms, and crustaceans, may be found small sacs containing sensory cells the cilia of which are in contact with a mass of calcareous or other dense materials. These sacs are now known as statocysts, and an equilibratory function is attributed to them largely because of their similarity to the macular organs of vertebrates. Earlier, however, before the true functions of the vertebrate labyrinth became known and the entire inner ear was considered to be auditory, these sacs were called "otocysts" and regarded as organs of hearing. The first theory of the origin of the vertebrate ear followed naturally from this erroneous conception: the ear was regarded as a development from an "otocyst" presumed to have been passed along the invertebrate series to the prevertebrates.

When later the invertebrate organ came to be recognized as a statocyst with an equilibrial function, the theory was modified correspondingly, and this organ was considered as the forerunner of the vertebrate labyrinth, and directly or indirectly of the ear as well.

The Acoustico–Lateralis Hypothesis

According to a second theory, the labyrinth and ear are considered to have arisen within the vertebrate line itself, being derived from the lateral line organ of fishes. This organ consists in general of a canal deep in the tissues along each side of the fish's body, with extensive branching on the head and numerous openings to the surface. Within the canals are many sensory endings, each containing ciliated cells and a cupula and therefore bearing some resemblance to the crista endings of the semicircular canals.

This theory had its first formal development at the hands of Ayers in 1892, although the basic idea had appeared earlier, and especially in the writings of Mayser (1882) and Beard (1884). Mayser saw a similarity between the endings in the lateral line canals and those of the labyrinth and indicated a functional relationship. The lateral line endings, he thought, might represent a crude sort of auditory receptor. Beard, after an extensive study of the "segmental sense organs" of fishes, concluded that the various sense organs of the head were derived from endings on the gill bars that originally gave warning of nearby objects that might damage the delicate gills beyond. Among these special sense organs is the ear, derived from a portion of this sensory system.

Ayers was more explicit in his formulation and conceived of the ear as elaborated from a portion of the canal system of the lateral line that had sunk below the surface.

Ayers' hypothesis was advanced at a time when the lateral line was attracting a large amount of attention among biologists, and a great variety of sensory functions were attributed to it. At different times this system was regarded as serving the senses of taste, smell, and touch, and as contributing to the development of the eye. Functions later ascribed to it, on more substantial grounds, include equilibrium, general chemical sensibility, and the perception of water motion. This theory of the ear's origin was widely accepted then, and since that time has been repeated and elaborated in numerous formulations. It was accepted by van Bergeijk and developed in masterly fashion in two essays on the evolution of the vertebrate ear appearing in 1966 and 1967.

As suggested above, the statocyst and acoustico–lateralis hypotheses are not in conflict and can be reconciled and combined, with the first accounting for the origin of the labyrinth in prevertebrates and early vertebrates and the second continuing the development within the vertebrate line. Several authors (e.g., Lurie (1933), Guggenheim (1948)) have adopted this combined theory, at least implicitly, but no one has gone into the matter in sufficient detail to trace the transition from statocyst to lateral line.

In the statocyst and acoustico–lateralis theories, and also in the combined form of the theory, it is usually stated or implied that a certain continuity has existed in the development of the statocyst through the invertebrate series, and that the development of the ear from the early stage observed in fishes has likewise proceeded in a continuous and progressive fashion through the vertebrates. This continuity is traced from fishes through amphibians to the reptiles, and then the development is considered to take separate paths to the birds and mammals, thus following the accepted course of the evolution of the vertebrates as a whole.

Often these authors have not gone into the details of their theories, but have taken it for granted that continuity exists in the development of the ear and have contented themselves with describing the several forms of organs to be found at successive stages along the animal series. At times, particular steps of the evolutionary processes have been emphasized, such as the development of the advanced type of ear out of the ear of amphibians.

One of the most frequent statements, in recent as well as in older literature, is that the basilar papilla of amphibians is the progenitor of the cochlea of higher forms. This choice of the basilar papilla seems curious at first glance, for the more elaborate papilla amphibiorum, which is always present in amphibians whereas the other ending may be lacking, would seem the more reasonable choice for this role. The matter becomes clear when the early history of the amphibian endings is considered. It turns out that the favored position given to the basilar papilla is simply the result of its earlier discovery and the name applied to it.

This naming was quite fortuitous. The auditory ending now known as the basilar papilla was the first to be discovered. When Deiters saw it in 1862, he called it the cochlea by analogy with this receptor in mammals and considered a membrane found near by to correspond to the basilar membrane of the mammalian cochlea. It is doubtful that Deiters obtained any clear view of the second ending, now known as the amphibian papilla, but Hasse (1868), who studied the structures further, saw it well enough to regard it as an accessory to the other structure and called it "pars initialis cochleae"—the initial part of the cochlea. What was then regarded as the principal auditory ending came to be called the "pars basilaris cochleae" (by Retzius, 1881, among others) and in more modern times simply the "papilla basilaris" (de Burlet, 1935).

This episode in the development of the evolutionary theory of the ear is presented in detail as an interesting example of the tyranny of terminology, and the extent to which an established prejudice can guide the course of thinking for many decades. The evidence bearing on the above theories is reviewed below.

A Consideration of The Evidence

The statocyst organs of invertebrates take a variety of forms and appear somewhat sporadically among the different groups. There is no good reason for thinking that continuity existed in the development of these organs among the various invertebrate phyla in which they occur, or that any of these organs gave rise to the lateral line system of fishes.

The appearance of the static receptors in numerous invertebrate groups is of interest, however, in indicating the capability of organisms to generate a special receptor of this kind.

The acoustico–lateralis theory was ably presented by van Bergeijk (1966, 1967), together with an assembly of arguments in its support. His outline is used as a basis for the following discussion. He indicated three lines of evidence for the theory, as follows:

1. In the process of embryonic development, the lateral line and ear arise from the same mass of ectodermal tissue, which is the acoustico-lateral placode.
2. Both lateral line and ear contain the same type of sensory cells, which are the hair cells.
3. Both systems are innervated in a similar manner, by nerve fibers arising from the same cochlear nuclei.

The Developmental Pattern

The course of embryonic differentiation becomes significant if the recapitulation hypothesis is accepted, which states that individual development follows a sequence that corresponds to phyletic evolution. The critical evidence comes from the study of embryological development in the fishes, in which a relation is observed in the early growth of lateral line and labyrinth, as indicated in the discussion of the work of Mayser (1882) and Beard (1884).

Beard was mainly concerned with the development of the sensory ganglia and nerves and only briefly mentioned the growth of sensory structures. Others a little later added details concerning the differentiation of these structures. Priority for this detailed observation is probably due to Mitrophanow, who published first in Russian in 1888 and then presented a brief summary of his work before the congress of natural sciences in Warsaw in 1889. He reported that the labyrinth and lateral line of selachian embryos are both developed from a lateral placode, a line of epithelial cells running lengthwise along each side of the developing embryo. He saw many similarities between the lateral line system and the labyrinth and suggested a functional relation between them. The labyrinth he considered to serve both for hearing and for the "sixth sense," by which he meant the organs for perception of bodily position and motion.

More accessible, and therefore better known, is a report by Wilson (1891) of work that was probably carried out independently of that of Mitrophanow. Wilson studied embryos of the sea bass, and then a short time later he collaborated with Mattocks (1897) in a similar study on salmon embryos. The results obtained on the different species of fishes are essentially the same with only minor variations.

In salmon embryos, during the development of the neural keel along the dorsal midline about the twelfth to thirteenth days of the egg's incubation, a thick layer of cells is left on either side, covering about two-thirds of the embryo's

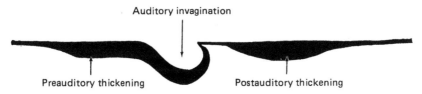

FIG. 1 Lateral placode of the salmon embryo at the end of the thirteenth day of development. (After Wilson & Mattocks, 1897, p. 659.)

length, and is identified as the lateral sensory placode. Near the end of the thirteenth day this placode begins to invaginate in its midportion and takes on an appearance as shown in Figure 1.

Three portions may now be distinguished: an anterior portion, designated as the preauditory thickening, which is destined to give rise to a number of superficial sense organs of the head and gill region; the invaginated middle region that becomes the auditory vesicle; and a posterior portion that develops into the lateral line system. On the fourteenth day the posterior portion separates off from the remainder. The middle or auditory portion continues to invaginate and sinks deeper in the head but retains its connection with the anterior portion. Finally, around the twenty-fifth day, the situation is as shown in Figure 2. The auditory vesicle has separated from the anterior portion and its open side has closed over to give a spherical form. The lateral line anlage meanwhile has moved posteriorly and has become fairly well defined, although it is still not completely isolated from the thin neural ectoderm of the surrounding area.

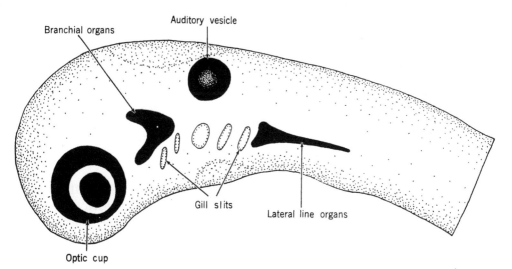

FIG. 2 Anterior portion of the salmon embryo at the twenty-fifth day of development. (After Wilson & Mattocks, 1897, p. 660.)

This evidence must be carefully regarded. It shows indeed that three groups of organs arise from the lateral placode: the branchial sense organs, the labyrinthine endings, and the lateral line system. The evidence does not indicate that these three sets of organs arise in common, from the same embryonic cells, or that any one set is prior to the others, or that one is transformed into another. The areas are adjacent and at first contiguous; then they separate and develop independently. The frequent assertion that the labyrinth arises from the lateral line anlage is obviously unjustified.

It may be noted in this connection that the lateral line anlage separates off early from the remaining masses of cells, and the association of the labyrinthine or "auditory" anlage is always closer to the anterior branchial portion. Moreover, this latter relation is maintained much longer in the course of development.

The statement made by van Bergeijk (1967) about this evidence, as given in the first argument outlined above, is more cautious than many others encountered in discussions of the ear's origin. He said merely that the lateral line and ear arise "from the same mass of tissue." Yet this formulation also is objectionable: it implies that a number of embryonic cells, at first indistinguishable from one another, become separated into two groups, whereas in fact the cells are strung out in a linear array and separable as soon as any structural distinctions in the region can be determined.

Cell Similarities

The second argument for a derivation of labyrinth and ear from the lateral line system rests on a similarity between the hair cells and supporting cells of these sensory structures. There are indeed many points of resemblance among these cells, both in their general forms and in their microstructure. However, many of these features are characteristic of all cells, and others are common to epithelial cells. The presence of cilia on the exposed ends of the hair cells is often emphasized, but this is not an exclusive feature of labyrinthine and lateral line sensory cells. Ciliated cells are of wide distribution, especially in epithelial tissue, and if the common characteristics of the labyrinthine and lateral line cells signify anything of importance it is that these cells are of ectodermal origin and come from the same general region of the embryo. No further relationships or dependencies are indicated.

The Innervation Argument

Van Bergeijk pointed to the extensive work of Herrick (1897, 1899, 1901) on the courses of the nerve fibers arising in the nuclei of the lateral region of the hindbrain. Herrick observed in teleost fishes that a number of sensory endings, including those from lateral line and labyrinth, are served by fibers that enter the tuberculum acusticum and cerebellum and appear to mingle there. Within the nuclei he was unable to separate these fibers or to identify the different kinds.

From this evidence van Bergeijk argued that the innervation of lateral line and labyrinth was in a sense common and interchangeable, and the labyrinth and ear could therefore be regarded as arising out of the lateral line.

It is often stated further that the nerve fibers passing through the seventh, ninth, and tenth nerves, and supplying the endings of the lateral line in its various regions, are actually "auditory" fibers derived from the cochlear nuclei. I do not find clear evidence for this assertion in Herrick's studies or those of others. Herrick said that he was unable to trace and identify these fibers.

The intermingling of nerve fibers does not necessarily imply an interchange or transfer of function. Here also the close physical relations between the innervation patterns of labyrinth and lateral line reflect the early contiguity of the anlagen of these systems and their connections to the lateral wing of the neural plate.

The Pore-Canal System

A further argument for the acoustico–lateralis hypothesis comes from a consideration of the pore-canal system found in certain fossil fishes.

The ostracoderms were primitive vertebrates of the Silurian period, now bracketed with the cyclostomes in the class Agnatha. Their bodies were covered with a bony armor, and in some, the Osteostraci, this armor contained a network of fine canals—the pore canals.

Although the pore canals have usually been considered as mucous secreting organs, it has been suggested that they may have had a primitive sensory function and may have been the forerunners of the lateral line and labyrinth. It is conjectured that the lateral line was first derived from the pore-canal system, and that later on the labyrinth arose from the lateral line.

Further study, as Denison (1966) has pointed out, fails to support this speculation. The ostracoderms already had semicircular canals along with their pore canals, and they lacked the lateral line system. Denison suggested that a likely progression would be from pore canals to semicircular canals to lateral line, but this order does not assist the argument for priority of the lateral line.

From this consideration of the evidence it is clear that the acoustico–lateralis hypothesis lacks any firm foundation. The labyrinth in its embryonic growth arises along with other senses—which include the lateral line and a number of sensory structures in the head region—but the evidence is that these various sensory endings appear independently, coming from different embryonic cells, and one part of the system is no more fundamental than another.

If this process is regarded as representing the course of phylogenetic development, the various endings must be placed on the same footing. To say that the labyrinth arises from the lateral placode is quite proper, but to say that it had its phyletic origin in the canal endings of the lateral line is unjustified. This latter idea of the development of the ear has arisen from loose statements about the actual embryological picture, and has gained weight and credibility largely through constant repetition.

THE LABYRINTH HYPOTHESIS

As indicated above, early in the development of the statocyst and acoustico—lateralis hypotheses the only function considered was auditory, and only later did it become known that the labyrinth has postural and motional sensing capabilities also. Rather late in the development of the acoustico—lateralis hypothesis—and most notably at the hands of van Bergeijk—two significant steps of development were conceived, one in which the lateral line was said to give rise to the labyrinth (strictly, the nonauditory labyrinth), and another in which the labyrinth gave rise to the inner ear.

If the first of these two steps is abandoned, as I think the evidence requires, the second may still be retained as a valid conception.

The evidence in favor of this labyrinthine theory, or at least consistent with it, has often been pointed out. The labyrinthine and auditory endings are contained in a common capsule; are bathed by a single fluid, the endolymph; and are more or less surrounded by another fluid, the perilymph. The endings themselves are made up of sensory cells, supporting cells, and other tissues with many points of resemblance. All are innervated by branches of the eighth nerve.

In addition to the anatomical similarities just mentioned there are functional correspondences. All these organs are types of mechanoreceptors, responding to physical energies in the form of motion, though the sorts of motion differ.

There is a further line of evidence, with both structural and functional aspects, that has only recently come to attention (Wever, 1973a, b). All the labyrinthine receptor organs, including saccular macula, utricular macula, lagenar macula, the cristae of the semicircular canals, and the papillae or cochlear organs, display the same general arrangement for the stimulation of their hair cells. In all these sense organs a tectorial structure lies over the outer ends of the hair cells and makes contact with their ciliary tufts. The tectorial structure varies in form, and includes perforated and solid plates, thick-walled and thin-walled canals, and fibrous networks of simple and complicated kinds, but always some form of tectorial tissue makes the connection to the ciliary tuft and is the immediate agency for the cell's stimulation. Sometimes the tectorial tissue transmits a motional stimulus to the ciliary tufts while the cell body remains at rest, and this produces the relative motion between cilia and cell body that constitutes a stimulation of the cell. More often, the tectorial connection with the ciliary tuft applies a restraint, and when the cell body is caused to move by a stimulus involving the foundation on which the cell rests there is likewise a relative motion that is stimulating.

The above evidence is not proof that the inner ear is a derivative of the labyrinth, but merely shows that the different sense organs are closely related and should be classed together as mechanoreceptors of a particular type. More pertinent indications of the ear's derivation must be sought, together with the specific ways in which this derivation has come about.

The Evidence from Primitive Vertebrates

The first critical evidence is the observation that the earliest true vertebrates had labyrinths but lacked any auditory organs. These were the agnathans, and included a group now living, the cyclostomes (such as the hagfishes and lampreys) and another group, the ostracoderms, known only as fossils. The labyrinths of hagfishes and lampreys contain both macular organs with an otolithic overlay and crista organs surmounted by a cupula, but these animals appear to lack a specific auditory organ. They respond to low-frequency sounds probably by means of their lateral line and skin organs.

Also to be mentioned in this connection are the Chondrichthyes (elasmobranchs, including the sharks and rays), which make up another and higher class of vertebrates. These fishes possess a well-developed labyrinth, containing most of the general features that continue without much change throughout the entire vertebrate group, but an ear seems to be absent. Many of these animals, and especially the sharks, have been found to respond readily to sounds, but there is still doubt about the sensory endings involved. It is likely that the sensitivity of these animals to sounds is mediated by their well-developed lateral line system, but the question must be left open whether some other organ is involved also.

Because animals exist, and have ancestral lines that go far back in history, in which labyrinths are present without inner ears, whereas the reverse relation is not known, it is reasonable to suppose that the labyrinth is the forerunner of the ear. Or, more strictly, it may be said that ears have arisen by a transformation of labyrinthine organs or through mutations occurring in certain early embryonic cells that originally were labyrinthine. Two means by which ears have arisen are implied by this last statement, as presently will be made clear.

The Appearance of the Teleostean Ear

The highest among the four classes of fishes belonging to the superclass Pisces are the Osteichthyes or bony fishes, of which the teleosts are the most abundant and most important in many ways. Among these, without any doubt, there are true ears and, in some members of the group, a fairly high level of auditory sensitivity over a moderate range of frequencies.

This teleostean ear arises as a transformation of a previously existing macular organ. The transformation takes place in two distinct ways. In most of these fishes the saccular macula, and probably the lagenar macula also, is employed for this new purpose, and in a few, the clupeids and their relatives, it is the utricular macula that is involved.

There is no indication so far that in either transformation any structural modification of the sensory ending is involved; the change appears to be in the neural connections and in the central pathways through which the actions at the sense organ are transmitted and interpreted. These maculae, as far as anyone has been able to see, have the standard form that macular endings exhibit everywhere, with a patch of sensory cells borne on a solid layer of connective tissue, and in contact with these cells a tectorial structure that leads to an otolithic mass beyond.

Auditory stimulation occurs when sound waves are produced in the water, enter the body of the fish, and set in vibratory motion the plate of tissue on which the hair cells lie. The bodies of these cells are set in motion, but their cilia, attached to strands or ribbons of tectorial tissue, are partially restrained. The restraint is communicated to the tectorial layer by the otolithic mass above it, the calcareous particles of which have a specific gravity of about 1.7 and therefore an inertia greater than that of the general tissue of the fish. The otolithic mass tends to lag in its response to the imposed vibrations, and it therefore moves over a shorter amplitude and probably in a different phase relation from the surrounding tissues (including the bodies of the hair cells). The differential motion between cell bodies and ciliary tufts causes a stimulation of the hair cells.

The stimulation of macular hair cells by sounds no doubt occurs in all animals that possess these organs, but the effects are interpreted as auditory only when the proper neural arrangement is provided. When the macular organ is employed as an ear, no doubt the arrangement also includes a means of suppressing the effects of stimulation by bodily position and motion—the normal manner of stimulation of a macular organ.

Many of the teleost fishes have developed additional mechanisms (Weberian ossicles, for example) that operate in conjunction with their transformed macular organs and considerably enhance the sensitivity to sounds.

The Amphibian Ear

In amphibians there are two separate types of auditory receptor, both of which appear as outpocketings of the walls of the saccule and are known as the amphibian and basilar papillae. In a few species only the first of these is present.

As shown in Figure 3, these two receptors lie in a path of vibratory fluid flow that extends from the footplate of the columella at the oval window and passes first through the saccular region and then through two portions of the perilymphatic duct system to the round window near the base of the labyrinth. The fluid circuit is continued from the round window through a region containing

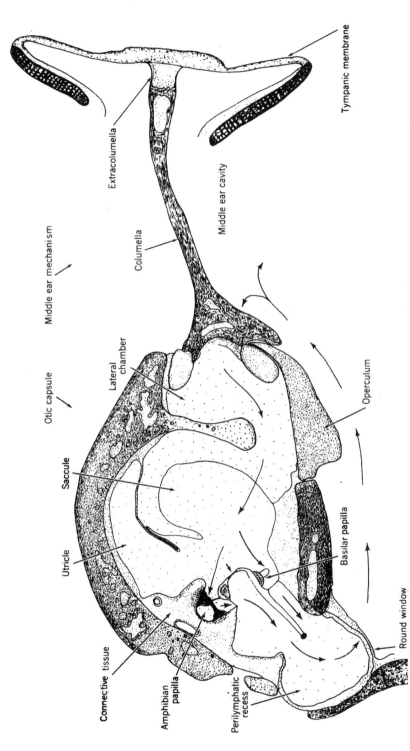

Tympanic membrane

Extracolumella

Columella

Middle ear cavity

Middle ear mechanism

Otic capsule

Lateral chamber

Saccule

Operculum

Utricle

Basilar papilla

Connective tissue

Amphibian papilla

Perilymphatic recess

Round window

FIG. 3 A schematic representation of the ear of a frog (*Rana pipiens*) in cross section to show the paths of vibratory fluid flow through the auditory papillae.

fluid and tissue to the middle ear cavity, which is filled with air. An inward thrust of the footplate, representing one phase of the alternating pressures exerted by a sound wave, sets up a displacement of fluids in the continuous paths through both papillae and the round window to the middle ear region, where the pressure is relieved either by a compression of the air of the middle ear cavity or by expansion into the space left by the ingoing surface of the footplate. A series of sound waves thus sets up a back and forth motion of the fluids and flexible membranes in this path, and also involves the sensory structures of the two papillae.

In the amphibian papilla the auditory hair cells lie on the solid roof of the capsule with their ciliary tufts in close contact with a complex tectorial structure below. This structure consists largely of long, thin-walled canals in a sort of honeycomb arrangement that hangs down and partly obstructs the passage through the organ. Also this structure connects to a thin membrane, also made up of tectorial tissue, that continues the obstruction of the passage and acts as a "sensing membrane." The fluid movements involve the tectorial mass to which the ciliary tufts of the hair cells are attached, and the movements therefore are communicated to these tufts, whereby the cells are stimulated.

The basilar papilla is similar in construction, except that the tectorial mass is greatly reduced and consists mainly of the sensing membrane. This membrane is stretched across one side of a short duct in which this ending is contained, partially occluding it. Again, vibratory fluid motion through the duct is communicated to the sensing membrane and thus to the ciliary tufts of the hair cells that lie in an arc along the inner surface of the duct.

This stimulating arrangement evidently provides a fairly sensitive mode of response to aerial (or aquatic) sounds, for some species of frogs hear very well as measured by the cochlear potential method.

The Reptilian Ear

Among the reptiles appears a new type of structure with a greatly improved method for the sensing of fluid vibrations. This structure is found throughout this class of vertebrates and in all the birds and mammals.

In this higher type of ear the hair cells are borne on a flexible membrane, the basilar membrane, which lies athwart the fluid path from stapedial footplate to round window. When vibratory movements occur in this fluid path the basilar membrane is carried along with the fluid and the motion involves the hair cells on its surface. The ciliary tufts of these hair cells are restrained, however, and do not follow the movements exactly. The manner of restraint varies with species and often varies in different regions of a given papilla, but the effect is to produce a differential of motion between ciliary tuft and cell body, which stimulates the cell. In most ears the ciliary tufts of some or all the hair cells are restrained by an attachment to a tectorial membrane that itself is anchored to a stationary wall of limbic tissue near by.

In most orders of reptiles, and also in the birds and mammals, a standard structural arrangement is present for this ciliary fixation. In snakes, for example, the ciliary tufts are connected to the underside of a tectorial plate, from which a tectorial membrane runs to an attachment on the limbus. In alligators and birds the tectorial membrane arises from its anchorage on the limbus and then splits up into numerous small fibers, with one of these fibers going to the tip of each hair tuft. In the lizards, however, there is an amazing variety of arrangements, varying from one family to another and varying also in the different regions of a single papilla. There are not only variations in the structures involved, but also in the physical principles brought to bear, from direct restraint to inertia, reactive force, and viscosity. And yet, whatever the method, the effect is the same: a relative motion is established between cilia and cell body through which the cell is stimulated.

The particular processes by which the auditory endings may have arisen among the vertebrates are now to be considered.

THE SPECIFIC PROCESSES OF AUDITORY DEVELOPMENT

The first two processes, occurring in the teleost fishes, have already been identified. They involved the transformation into sound receptors of previously existing macular organs, as has been seen. Because different macular organs were involved in different groups of fishes, there were two distinct events, though the processes were essentially the same.

In the amphibians the matter of specific origin presents more of a problem. There are some points of resemblance between the papilla amphibiorum and a crista organ. In both types of ending there is a special tectorial structure in contact with the layer of hair cells, though in the crista this structure is a simple network and in the papilla amphibiorum it takes a variety of forms from a perforated plate to an array of thick-walled canals. Beyond this structure in both endings is a large body of thin-walled, parallel canals, known as the cupula in the crista organs and forming the complex sound-receptive mass in the papilla amphibiorum.

Despite this structural correspondence, however, it seems unlikely that the amphibian organ is a transformed crista. The three cristae of the semicircular canals continue to be found in this labyrinth in their accustomed locations. The papilla might better be considered as an adaptation of a supernumerary crista. The basilar papilla then could be regarded as a simplified form of another such crista or, perhaps more likely, the result of a reduction of a second amphibian papilla.

My best suggestion is that the papillae of amphibia represent two new and distinct developments of auditory endings out of the basic labyrinthine epithelium, developments from elements with genetic characters closely similar to those that produced the crista organs, or mutant derivatives of the elements responsible for these organs.

The reptilian ear is clearly a new and separate development, distinct from both the macular ears of fishes and the papillar organs of amphibians. I think it represents a new mutation of the basic labyrinthine anlage.

The evidence for the acceptance of the reptilian ear as a distinct derivation from the labyrinthine epithelium I consider to be practically compelling. There are many differences of structure but the outstanding difference, and the one that I regard as determinative, is the difference in the stimulating mechanisms. These mechanisms vary profoundly in form and operate by different principles.

In the transformed macular organ of teleosts, as already described, the hair cells rest on a solid base and in the presence of sounds their cell bodies execute the same vibratory motions as the surrounding tissues. Their cilia, however, are connected to a layer of tectorial tissue that in turn is attached to an otolithic mass whose high inertia causes it to respond differently to the imposed vibrations. A differential action on cilia and cell bodies is thereby established.

This situation is represented schematically in Figure 4. Pictured here are the four tissue layers, with the connective tissue base and the epithelial layer taking part in the general vibratory motion of the body as a whole, and the otolithic mass above remaining relatively stationary, so that a transitional zone lies between these two regions. In this zone the tectorial tissue makes the connections to the ciliary tufts of the hair cells, exposing these cells to the discontinuity that is stimulatory.

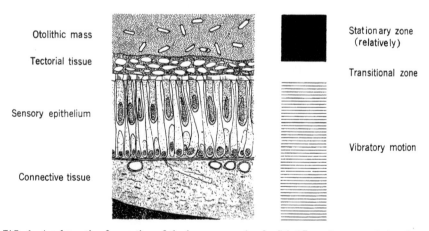

Otolithic mass

Tectorial tissue

Sensory epithelium

Connective tissue

Stationary zone (relatively)

Transitional zone

Vibratory motion

FIG. 4   A schematic of a portion of the lagenar macula of a fish (*Carassius auratus*) showing the four tissue layers and their relations.

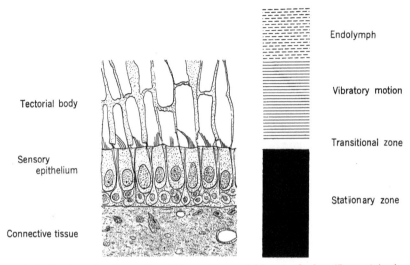

FIG. 5   A schematic of a portion of the papilla amphibiorum of a frog (*Rana pipiens*).

In amphibians the hair cells likewise rest on a solid base, but in this instance the cells remain stationary in the presence of sound, and their ciliary tufts are made to move through the action of the sound on the tectorial tissues in contact with them. Again a differential motion is established, but in a different manner.

This situation is shown in Figure 5, where the endolymph is to be thought of as in vibratory motion, so that this motion is communicated to the light and

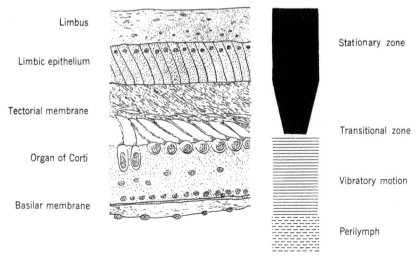

FIG. 6   A schematic of a portion of the cochlea of a bird (*Bubo virginianus*). Two types of hair cells are shown, inner (on the left) and outer, all with individual fiber connections with the body of the tectorial membrane.

tenuous strands of tectorial tissue. Connections between this tissue and the ciliary tufts transmit these movements and stimulate the cells.

In the reptiles, and also in all the higher animals, the hair cells are borne on a flexible membrane and are set in motion by the action of sound while their ciliary tufts are restrained. Here is still a third manner of stimulation of the hair cells, pictured in Figure 6. Vibratory motion transmitted through the cochlear fluids involves the basilar membrane and the bodies of the hair cells supported on it. The ciliary tufts of these cells, however, are connected to the tectorial membrane, which is more or less securely anchored to the epithelial and limbic layers as shown.

The bird cochlea is represented here, but reptilian and mammalian ears operate in essentially the same manner, though they vary in the ways in which the hair tufts are restrained.

It is almost inconceivable that one of the above stimulation systems, of teleosts, amphibians, and reptiles, could have been converted into another. From the reptiles on to both the birds and mammals the stimulating arrangement remains the same in basic form and principle. There is good reason to believe, therefore, that these higher forms of inner ear represent a single development.

In summary, five distinct developments of auditory receptor organs can be recognized: two in the teleostean fishes, two in amphibians, and one in reptiles.

REFERENCES

Ayers, H. Vertebrate cephalogensis, *Journal of Morphology*, 1892, 6, 1–360.

Beard, J. On the segmental sense organs of the lateral line, and on the morphology of the vertebrate auditory organ, *Zoologischer Anzeiger*, 1884, 7, 123–126, 140–143.

de Burlet, H. M. Vergleichend anatomisches ueber endolymphatische und perilymphatische Sinnesendstellen des Labyrinthes, *Acta Oto-laryngologica*, 1935, 22, 287–305.

Deiters, O. Über das innere Gehörorgan der Amphibien, *Archiv für Anatomie und Physiologie*, 1862, 277–310.

Denison, R. H. The origin of the lateral-line sensory system, *American Zoologist*, 1966, 6, 369–370.

Guggenheim, L. *Phylogensis of the ear*. Calver City, California: Murray and Gee, Inc., 1948.

Hasse, C. Das Gehörorgan der Frösche, *Zeitschrift für wissenschaftliche Zoologie*, 1868, 359–420.

Herrick, C. J. The cranial nerve components of teleosts, *Anatomischer Anzeiger*, 1897, 13, 425–431.

Herrick, C. J. The peripheral nervous system of the bony fishes, *Bulletin of the United States Fish Commission for 1898*, 1899, 18, 315–320.

Herrick, C. J. The cranial nerves and cutaneous sense organs of the North American siluroid fishes, *Journal of Comparative Neurology*, 1901, 11, 117–249.

Lowenstein, O. The concept of the acusticolateral system. In P. Cahn (Ed.), *Lateral line detectors*. Bloomington: Indiana University Press, 1967. Pp. 3–12.

Lurie, M. H. Phylogenetic development of the cochlea, *Annals of Otology, Rhinology and Laryngology*, 1933, 42, 1069–1080.

Mayser, P. Vergleichend anatomische Studien über das Gehirn der Knochenfische mit besonderer Berücksichtigung der Cyprinoiden, *Zeitschrift für wissenschaftliche Zoologie*, 1882, **36**, 259–364.

Mitrophanow, P. Sitzungsprotokolle der biologischen Sektion der Warschauer Naturforschergesellschaft, Sept. 27–Oct. 9, 1889, *Biologisches Centralblatt*, 1890–1891, **10**, 190–191.

Retzius, G. *Das Gehörorgan der Wirbelthiere*. Vol. 1. Stockholm: 1881. Pp. 151–213.

van Bergeijk, W. A. Evolution of the sense of hearing in vertebrates, *American Zoologist*, 1966, **6**, 371–377.

van Bergeijk, W. A. The evolution of vertebrate hearing. In W. D. Neff (Ed.), *Contributions to sensory physiology*. Vol. 2. Berlin and New York: Springer-Verlag, 1967. Pp. 1–49.

Wever, E. G. The mechanics of hair-cell stimulation, *Transactions* of the *American Otological Society*, 1971, **59**, 89–107; *Annals of Otology, Rhinology and Laryngology*, 1971, **80**, 786–804.

Wever, E. G. The labyrinthine sense organs of the frog, *Proceedings of the National Academy of Sciences USA*, 1973, **70**, 498–502. (a)

Wever, E. G. Tectorial reticulum of the labyrinthine endings of vertebrates, *Annals of Otology, Rhinology, and Laryngology*, 1973, **82**, 277–289. (b)

Wever, E. G. The ear and hearing in the frog, *Rana pipiens*, *Journal of Morphology*, 1973, **141**, 461–478. (c)

Wever, E. G. The evolution of vertebrate hearing. In W. D. Keidel & W. D. Neff (Eds.), *Handbook of sensory physiology*, Vol. 5, Auditory System. Berlin and New York: Springer-Verlag, 1974. Pp. 423–454.

Wilson, H. V. The embryology of the sea bass *Serranus atrarius*, *Bulletin of the United States Fish Commission for 1889*, 1891, **9**, 209–277.

Wilson, H. V., & Mattocks, J. E. The lateral sensory anlage in the salmon, *Anatomischer Anzeiger*, 1897, **13**, 658–660.

6

Comparative Hearing Function in the Vertebrates

W. C. Stebbins

University of Michigan

In his description of the evolving vertebrate ear in this book, Wever has presented a detailed, precise, and fascinating account of the varying structures and modes of operation of the ear in the vertebrate series. I hope the following may serve as a brief addendum to his presentation. In so far as possible I will describe what is now known about the nature of vertebrate hearing based on the behavioral evidence which, after all, is the "what for" of the structure and function so aptly discussed by Professor Wever. The greater part of this chapter is directed to the mammals (see also Masterton, Heffner, & Ravizza, 1969)—particularly the primates.

Reliable behavioral measures of hearing for the lower vertebrates are sparse. For two classes, amphibians and reptiles, there are almost no behavioral data, and we must depend on electrophysiological evidence together with what behavioral data exist for neighboring taxa. These animals have proved difficult to condition in the laboratory for purposes of behavioral audiometry. There are relatively uniform behavioral data for several avian species and, if the bird's ear is accepted as representing an advanced form of the reptilian ear, i.e., the crocodile's, it is perhaps justifiable to use the bird's hearing function as an approximation to the more advanced reptile's. The situation is more hopeful for the bony fishes (Osteichthyes) and the mammals—evidence from a representative sample of species is available.

Most of the behavioral evidence for hearing in animals is in the form of threshold functions for pure tones. Minimum detectable energy levels for acoustic stimulation are determined over a wide range of frequencies. The results are presented graphically as a plot of threshold sound pressure level as a function of frequency (see Figure 1). The threshold functions are generalized; they have

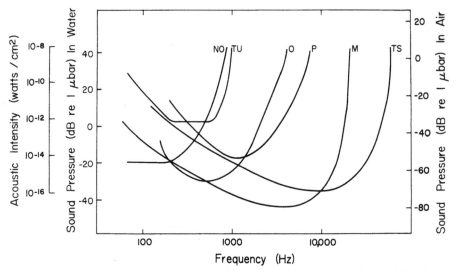

FIG. 1   Generalized auditory threshold functions for representative vertebrates: NO, the nonostariophysine fish, *Haemulon sciurus*, after Tavolga and Wodinsky (1965); O, the ostariophysine fish, *Ictalurus nebulosis*, after Weiss, Strother & Hartig (1969); TU, the turtle, *Pseudemys scripta*, after Patterson (1966); P, the pigeon, *Columba livia* after Stebbins (1970); M, man after Sivian and White (1933); TS, the tree shrew (Stebbins, 1969 unpublished data). The functions have been adjusted on the vertical axis to correct for the different acoustical properties of air and water.

been smoothed to indicate general form and placement on the frequency and sound pressure axes rather than the exact position of each experimentally obtained datum point. Two important characteristics of hearing emerge: the frequency range and the maximum sensitivity. These characteristics are used for comparisons in this chapter. Other measures of hearing are critical for an understanding of the evolution of hearing, but such measures are far more difficult to obtain from animal subjects. Questions about an animal's ability to localize sound or to distinguish minute differences in the frequency or intensity of acoustic stimulation have been directed only at the higher mammals. Answers to these questions must be sought if the behavior called "hearing" is to be meaningfully related to the evolution of structure.

FISHES

Of the four classes of fishes, two have been shown unequivocally to respond to sound. Of these, further discussion of the cartilaginous fishes (Chondrichthyes) must be deferred pending a more precise and detailed account of their hearing

(Kritzler & Wood, 1961). For the bony fishes (Osteichthyes) there is an abundance of data for both the ostariophysans (with Weberian ossicles) (Weiss, 1966;
Weiss, Strother, & Hartig, 1969) and nonostariophysans, including many marine
teleosts (Tavolga & Wodinsky, 1963). The auditory threshold function for the
blue-striped grunt (*Haemulon sciurus*) (after Tavolga & Wodinsky, 1965) is
presented in Figure 1. Maximum sensitivity is in the 100–300 Hz region and the
high-frequency limit is probably below 1000 Hz. There are nonostariophysans
sensitive to higher frequencies (see Tavolga & Wodinsky, 1963), but the grunt is
fairly typical of the taxon in having what may be called primitive hearing.

It may be that the Weberian ossicles, almost a prototype of the mammalian
ossicular chain, act as an amplifier to increase the sensitivity and perhaps also to
extend the frequency range of the ostariophysine ear. Tavolga and Wodinsky
(1963) have shown that there are nonostariophysans in which the coupling
between the sonic transducer, the swim bladder, and the inner ear may rival the
Weberian chain in conductive efficiency with resultant frequency range and
sensitivity to match the ostariophysans.

The auditory threshold function for the catfish (*Ictalurus nebulosis*) (Figure 1,
after Weiss *et al.*, 1969) provides an example of ostariophysan hearing. Peak
sensitivity is between 600 and 700 Hz, close to the high-frequency limit of the
blue-striped grunt. The upper frequency cutoff is near 4000 Hz at 40 dB μbar. It
is possible that a better conductive mechanism in the ostariophysans has produced an extension of the frequency range of hearing together with a shift in the
region of maximum sensitivity toward higher frequencies—a trend that continues
in the reptiles and mammals. Sensitivity at very low frequencies (below 1000
Hz) is comparable to that of all the other vertebrates. A review of the literature
by Tavolga (1971) has underscored some of the difficulties in both underwater
acoustic measurements and conditioning techniques in fishes that may have led
to some of the inconsistencies in recent findings.

AMPHIBIANS AND REPTILES

There is sufficient evidence based on cochlear microphonic recordings to ensure
that both these classes have functional hearing. Like the fishes, most amphibians
and reptiles are restricted in their hearing to a relatively narrow band of
frequencies and they are most sensitive in the 200–1000 Hz region. If the turtle
(*Pseudemys scripta*) (after Patterson, 1966) is accepted as an example of a
relatively primitive reptilian ear and the pigeon (*Columba livia*) (after Stebbins,
1970) as a representative of the birds and comparable to the most advanced
reptilian ear, we can imagine what reptilian hearing may encompass with respect
to frequency range and sensitivity (see Figure 1). It is not unreasonable to
speculate that within the evolving reptiles there has been an extension of the

frequency range, a tenfold increase in absolute sensitivity, and a shift upward in the frequencies to which the reptiles are most sensitive.

BIRDS

The pigeon has been treated above as an advanced reptile (see Figure 1). Some birds have an extended frequency range (to 15,000 Hz and possibly higher) (Schwartzkopff, 1955). The data that are available indicate that there is not great variance with respect to sensitivity or frequency range within the class. Low-frequency sensitivity in the ostariophysans, in the birds, and again by extrapolation in the more advanced reptiles is comparable to that of many of the mammals (Masterton et al., 1969). The behavioral evidence, then, indicates that both the Weberian ossicles in the ostariophysine fishes and the two-element columellar attachment between tympanic membrane and cochlea in birds and some reptiles rivals the mammalian ossicular chain at least in the transduction of low-frequency stimulation.

MAMMALS

In the eutherian mammals, a three-element bony ossicular chain couples the tympanic membrane, or ear drum, to the inner ear via the oval window of the cochlea. In the cochlea itself there are a comparatively large number of two types of receptor cells with an extensive neural network. The consequences of this plan for hearing have been dramatic. Profound increases in absolute sensitivity to higher frequencies and in the frequency range of hearing and considerable inter- and intraorder variance in hearing function characterize the mammals. Whereas in the lower vertebrates the major role of the ear is perhaps as a simple low-frequency sound detector, in the mammals, the ear, in addition to increasing in sensitivity as an energy detector, has become a much more powerful discriminator, able to respond differentially to small changes in the energy, spectral composition, and temporal properties of the signal. In addition I suggest that there is substantial improvement over the lower vertebrates in both sound localization and distance judgment in many of the mammals. The experimental evidence, however, is incomplete. In some species certain characteristics have been enhanced to a high degree of specialization yet unknown in the hearing of earlier forms—extremely high-frequency sensitivity together with the bat's outstanding ability to localize sound is one striking example.

Masterton et al. (1969) stressed the selective advantages for sound localization that accrued to the early small mammals as a consequence of the extension of their frequency range. The comparatively short wavelengths produced a significant intensity difference at the two ears, even in very small heads with a short interaural distance, thus affording an excellent cue for location of the sound source. Such remote sensing, particularly in the nocturnal mammals, the acuity and far vision of which were clearly inferior to that of the birds and the primates, helped to ensure an early warning system for prey and predator alike.

Because of the variation in hearing function within the mammals it is difficult to point to one taxon as representative of the class. The tree shrew (*Tupaia glis*) shows some fairly typical mammalian characteristics (see Figure 1): high-frequency hearing with an upper limit at about 60,000 Hz, absolute sensitivity at all frequencies except the lowest markedly superior to the other vertebrates, and an upward shift in the frequency region of maximum sensitivity.

There is not at the present time a clear indication of change in low-frequency sensitivity in the evolving vertebrates. The functions show good agreement below 1000 Hz between fish, turtle, pigeon, and tree shrew. Some mammals, however, such as man (see Figure 1, after Sivian & White, 1933) cat, and desert rat, possess exceptionally good low-frequency sensitivity not seen in other vertebrate taxa.

The basic design and structure of the middle and inner ear in the many living eutherian mammals are very similar (Pye & Hinchcliffe, 1968). There are some differences in the size of the ossicles and in the middle ear structure (Masali & Chiarelli, 1967; Pye & Hinchcliffe, 1968) and these may partially account for some of the variation in the frequency range of hearing among mammals (contrast the upper limit of man at 20 kHz and of the bat at 120 kHz) and in low-frequency sensitivity (compare thresholds at 1000 Hz: for the racoon, −89 dB μbar; and for the rat, −24 dB μbar). Variation in the mammals may also be referred to subtle differences in cochlear microstructure—in the number of hair cells and particularly in the complex neural network by which the hair cells transmit information to higher centers.

If the primates in general, and man in particular, have any acoustic superiority among the mammals it may be in their ability to resolve small differences in the frequency and sound pressure of auditory stimulation. A measure of this discriminative capacity is the ratio of the frequency difference threshold (i.e., the least discriminable difference between two frequencies) to the base frequency against which the comparison is made. The discrimination ratio (or Weber fraction) is plotted across frequency. (See Figure 2, after Heffner, Ravizza, & Masterton, 1969a, b, for *Tupaia* and *Galago;* after Stebbins, 1971, for *Macaca;* and after Filling, 1958, for *Homo.*) Recent findings for four genera are shown. Forced to speculate on the basis of these data and little else I must conclude that there is a hierarchy with respect to frequency resolving power at least in the primates and that the loss in high-frequency hearing in man and Old World monkey is perhaps compensated by the marked ability to discriminate between frequencies within that limited range. Limited findings for fishes (Tavolga, 1971) indicate difference thresholds at 500 and 1000 Hz greater than those obtained for the tree shrew (*Tupaia*).

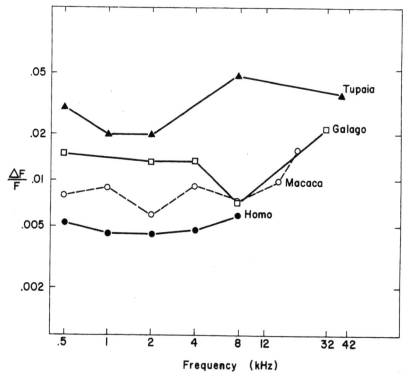

FIG. 2 Frequency discrimination functions at 40 dB above threshold for the tree shrew, *Tupaia*, after Heffner, Ravizza & Masterton (1969a); galago, *Galago*, after Heffner, Ravizza & Masterton (1969b); macaque, *Macaca*, after Stebbins (1970); and man, *Homo*, Filling (1958); from Stebbins (1971).

I believe that it may be possible to make a similar statement to the above for intensity (sound pressure) difference thresholds, but insufficient data preclude such a statement at the present time.

Finally, acoustic sensitivity has provided obvious additional ways by which prey may be obtained and predator eluded. Acoustic communication between conspecifics may be important even in the lowest vertebrates. In the mammals increases in sensitivity and in frequency range have added considerable versatility to the hearing function. Prey and predator are not only detected but located in space and at a distance. Better resolution of small differences in the physical dimensions of acoustic signals has opened up the possibility of extremely complex and varied intraspecies communication.

It is unfortunate that current knowledge of hearing, particularly in the amphibians and reptiles, is unable to match Wever's interesting and detailed account of the important evolutionary changes in morphology that have occurred in these taxa. A high degree of success has not yet been achieved with behavioral conditioning techniques in the lower vertebrates except in the fishes. Here I have tried to present an outline of some of the more obvious changes in hearing function that may have occurred in the evolving vertebrates. The picture is one of increasing complexity in acoustic function. An extension of the frequency range of hearing and a shift in the most sensitive frequency have been accompanied by advances in the capacity to localize sound even at a considerable distance. In the higher primates, although some diminution in the frequency range of hearing has occurred, this perhaps has been offset by a marked improvement in the ability to discriminate small changes in the frequency, intensity, and complexity of the acoustic signal.

REFERENCES

Filling, S. Studies on a series of normal subjects and on a series of patients from a hearing rehabilitation centre. *Difference limen for frequency.* 1 Odense, Denmark: Andelsbog- trykkeriet, 1958.

Heffner, H. E., Ravizza, R. J. & Masterton, B. Hearing in primitive mammals, III: Tree shrew *(Tupaia glis). Journal of Auditory Research,* 1969, 9, 12–18. (a)

Heffner, H. E., Ravizza, R. J. & Masterton, B. Hearing in primitive mammals, IV: Bushbaby *(Galago senegalensis). Journal of Auditory Research,* 1969, 9, 19–23. (b)

Kritzler, H., & Wood, L. Provisional audiogram for the shark, *Carcharhinus leucas. Science,* 1961, **133**, 1480–1482.

Masterton, B., Heffner, H. & Ravizza, R. The evolution of human hearing. *Journal of the Acoustical Society of America,* 1969, **45**, 966–985.

Masali, M., & Chiarelli, B. The ear bones of the old world primates. In D. Starck, R. Schneider, & H.-J. Kuhn (Eds.), *Progress in Primatology.* Stuttgart: Gustav, Fischer, Verlag, 1967. Pp. 145–149.

Patterson, W. C. Hearing in the turtle. *Journal of Auditory Research,* 1966, **6**, 453–464.

Pye, A., Hinchcliffe, R. Structural variations in the mammalian middle ear. *Medical & Biological Illustration,* 1968, **18**, 122–127.

Schwartzkopff, J. On the hearing of birds. *Auk,* 1955, **72**, 340–347.

Sivian, L. G., & White, S. D. On minimum audible sound fields. *Journal of the Accoustical Society of America,* 1933, **4**, 288–321.

Stebbins, W. C. Studies of hearing and hearing loss in the monkey. In W. C. Stebbins, (Ed.), *Animal Psychophysics: The Design and Conduct of Sensory Experiments.* New York: Appleton-Century-Crofts, 1970. Pp. 41–66.

Stebbins, W. C. Hearing. In A. M. Schrier & F. Stollnitz, Eds., *Behavior of Nonhuman Primates,* Vol. 3. New York: Academic Press, 1971, Pp. 159–192.

Tavolga, W. N. Sound production and detection. In W. S. Hoar & D. J. Randall (Eds.), *Fish Physiology: Sensory Systems and Electric Organs,* Vol. 5. New York: Academic Press, 1971. Pp. 135–205.

Tavolga, W. N., & Wodinsky, J. Auditory capacities in fishes. Pure tone thresholds in nine species of marine teleosts. *Bulletin of the American Museum of Natural History,* 1963, **126**, 179–239.

Tavolga, W. N., & Wodinsky, J. Auditory capacities in fishes: Threshold variability in the blue-striped grunt, *Haemulon sciurus. Animal Behavior,* 1965, **13**, 301–311.

Weiss, B. A. Auditory sensitivity in the goldfish. *(Carassius auratus) Journal of Auditory Research,* 1966, **6**, 321–335.

Weiss, B. A., Strother, W. F. & Hartig, G. M. Auditory sensitivity in the bullhead catfish. *(Ictalurus nebulosis). Proceedings of the National Academy of Sciences, U.S.,* 1969, **64**, 552–556.

7
Neurology of Anamniotic Vertebrates

Sven O. E. Ebbesson
University of Virginia

R. Glenn Northcutt
University of Michigan

INTRODUCTION

In introducing the topic of comparative neurology as it exists today it may be of interest to reflect on its history, which is closely related to the limitations of its methods. The study of nonmammalian brain structure was pioneered at the turn of the century. The founders included such dynamic personalities as Edinger, Ariëns-Kappers, Retzius, and Ramón y Cajal, in Europe; and Herrick, Crosby, and Johnston in the United States. Their contributions consist essentially of careful characterization of neurons and their distribution in a large number of vertebrate species (Ariëns-Kappers, Huber, & Crosby, 1936; Ramón y Cajal, 1894, 1918, 1922). Useful as these data are, they seldom provide accurate information about interneuronal connections, especially with respect to polarity. The importance of determining the connections of given neurons therefore remained.

Although these workers have attempted to determine interconnections of neurons, their techniques often failed them, especially in attempts to trace long connections. The Golgi (Nauta & Ebbesson, 1970) technique can at times reveal the extent of axonal trajectories over short distances, as Herrick (1948) often demonstrates in his classical work on the tiger salamander. The beauty of the Golgi method is that only a small percentage of neurons stain, and sometimes axons can be traced to their termination. Other silver methods have been developed by Ramón y Cajal and Bielshowsky in 1904 (Nauta & Ebbesson,

115

1970) and have been used extensively for many years by comparative anatomists. The problem with these methods however, is that all neurons and most neuronal processes are stained, resulting in an incredibly complicated microscopic picture in which it is virtually impossible to trace a given fiber for any distance. Therefore when these methods are used as hodological tools they often provide erroneous results. In fact, many concepts of nonmammalian brain organization based on such material have been proved wrong in the last 10 years. For example, the telencephalon of anamniotes has been considered primarily, if not solely, an "olfactory lobe" (Nieuwenhuys, 1967). Bäckström (1924), for example, thought he could trace olfactory tract fibers in silver preparations to all parts of the shark telencephalon. Better techniques have now revealed that this projection is restricted to a very small portion of the telencephalon (Ebbesson & Heimer, 1970).

The introduction of techniques for silver impregnation of experimentally selected pathways by Nauta and co-workers (Ebbesson, 1970a; Ebbesson & Rubinson, 1969; Nauta & Ebbesson, 1970) has resulted in a renaissance of comparative neurology in the last 10 years. Although the renaissance is a reality, at the present time relatively few laboratories use Nauta's method or the new histochemical, electrophysiological, and ultrastructural techniques in the study of anamniote brains. Therefore the data appear sketchy and the present contribution must be considered a preliminary attempt to reassess the organization of the central nervous system of anamniotes and its relationship to adaptive zones.

This discussion is restricted to (1) a brief consideration of living anamniote radiations and their adaptive zones, (2) documentation of the variation in anamniote nervous systems, and (3) correlation of this variation with changes in adaptive zones. This broad-ranging discussion purposely avoids anatomical details published elsewhere.

LIVING ANAMNIOTE RADIATIONS AND THEIR ADAPTIVE ZONES

About 55% of living vertebrate species are collectively grouped as anamniotes on the basis of their reproductive strategy. These vertebrates have evolutionary lineages that have possessed separate phylogenetic histories for the last 400 million years. At least four more or less distinct radiations can be recognized. Although the exact systematics and assignment of taxonomic categories is not complete, enough is known to suggest evolutionary relationships and to orient comparative neurological studies.

Living species referred to in this work, and their classifications, are listed in Table 1. We have primarily followed the recent classification of Miles (Moy-Thomas & Miles, 1971), but alternate opinions (Bertmar, 1968; Jarvik, 1968a, b; Romer, 1966; Schaeffer, 1968, 1969; Stensiö, 1968) must also be considered.

TABLE 1

The Major Living Groups of Anamniote Vertebrates and
Some of Their Species

Superclass Agnatha	Superclass Gnathostomata
Class Cephalospidomorphi	Class Teleostomi (*continued*)
Infraclass Petromyzonida	Infraclass Crossopterygii
Petromyzon marinus (sea lamprey)	*Latimeria chalumnae* (coelacanth)
Infraclass Myxinoidea	Infraclass Dipnoi
Myxine glutanosa (hagfish)	*Lepidosiren paradoxa* (South American lungfish)
Superclass Gnathostomata	*Protopterus annectens* (African lungfish)
Class Teleostomi	*Neoceratodus forsteri* (Australian lungfish)
Infraclass Actinopterygii	Class Amphibia
Superorder Chondrostei	Subclass Lissamphibia
Scaphirhynchus platorynchus (shovelnose sturgeon)	*Rana pipiens* (leopard frog)
Polypterus palmas (bichir)	*Rana catesbeiana* (bullfrog)
Calamoichthys calabaricus (reedfish)	*Bufo marinus* (giant toad)
Polydon spathula (paddlefish)	*Ambystoma tigrinum* (tiger salamander)
Superorder Holostei	Class Elasmobranchiomorphi
Lepisosteus osseus (longnose gar)	Subclass Chondrichthyes
Amia calva (bowfin)	Infraclass Elasmobranchii
Superorder Teleostei	*Ginglymostoma cirratum* (nurse shark)
Carassius auratus (goldfish)	*Galeocerdo cuvieri* (tiger shark)
Opsanus tau (toadfish)	*Negaprion brevirostris* (lemon shark)
Gymnothorax funebris (moray eel)	*Platyrhinoides trisserriata* (thornback ray)
Holocentrus sp. (squirrel fish)	Infraclass Holocephali
Eugerres plumieri (mojarra)	*Hydrolagus* (ratfish)

The four anamniote radiations are represented by the five classes listed in Table 1. The agnathans are the oldest of these radiations and are represented today by lampreys and hagfishes. These forms possess neither jaws nor paired fins and the size and kind of prey object and the efficiency of their locomotion are therefore far more restricted than those of the other anamniote radiations. Agnathan larval structure and habitat are reminiscent of protochordates. At metamorphosis agnathans cease life as filter feeders and transform to either parasitic predators or nonfeeders. The living agnathans have lost the dermal armor of their ancestors, have elongated the trunk, and have radically reorganized many of the head structures for this special sort of predatory existence. All of these changes indicate that these organisms are not "living fossils," but a very specialized part of the once flourishing agnathan radiation.

The teleostomes and elasmobranchiomorphs are represented today by bony fishes, and by sharks, skates, rays, and ratfishes, respectively. These two radiations comprise the vast majority of living anamniote vertebrates. Both radiations appear equally old and can be characterized by the development of jaws and paired fins. These acquisitions have allowed entry into new adaptive zones. The origin of jaws has increased the efficiency of feeding and allowed the utilization of new and larger prey. Concomitantly, the origin of paired fins has increased locomotive stability and maneuverability, resulting in new and more active predatory forms. Whereas teleostomes have evolved in both freshwater and marine environments, the elasmobranchiomorphs have evolved primarily in a marine environment.

Teleostomes comprise some 87% of all living anamniotes and are frequently divided into two major divisions, the actinopterygians and the sarcopterygians. Within the actinopterygian fishes, three grades of organization have traditionally been recognized: Chondrostei, Holostei, and Teleostei. These grades are certainly not monophyletic but represent repeated evolution of separate lineages that have reached the same grade of organization a number of times. These organizational grades represent phyletic trends toward more effective feeding and locomotive mechanisms (Schaeffer, 1968). Adaptations for more effective feeding include the forward displacement of the jaw articulation with the reduction and loss of several bony elements and the freeing of many of the remaining elements for more mobile and protrusive jaw mechanisms. Reduction in thickness of scales and development of homocercal tails and fusiform bodies have contributed to increased locomotive efficiency. These organizational grades may also represent a differential complexity of overall behavior. The living Chondrostei and Holostei represent relict populations of once flourishing radiations and are considered more generalized than the teleosts.

There are few surviving chondrosteans and holosteans. *Polypterus* and *Calamoichthys* inhabit freshwater rivers and lakes of tropical Africa and appear to be the most generalized living chondrosteans. The sturgeons and paddlefishes are the only other living chondrosteans, and their anatomy is more specialized than that of the polypterids. The gars and the bowfins, both found in American freshwater systems, are the only surviving holostean genera. The surviving teleosts form a group that surpasses all other living vertebrate groups in number of species and morphological diversity.

The second major teleostome division (sarcopterygians) is composed of the crossopterygians (lobe-finned fishes) and the dipnoans (lungfishes). Recently, many systematists have expressed the opinion that the lobe-finned fishes and the lungfishes are separate radiations equally distinct and no more closely related to each other than to the actinopterygian fishes (Moy-Thomas & Miles, 1971; Schaeffer, 1968; Schaeffer & Rosen, 1961). Whereas the sarcopterygians may not be a natural category, the term has been retained because of certain similarities in the nervous systems of lobe-finned fishes and lungfishes. The crossopterygians are divided into two major groups: the coelacanths and the rhipidistians. The coelacanths are thought to have evolved from Silurian rhipidistians and are represented today by a single species, *Latimeria chalumnae*. The rhipidistians became extinct during the Permian but prior to that time gave rise to the fourth major anamniote radiation, the amphibians.

The lungfishes (Dipnoi) are represented by three living genera and appear little changed from their Devonian ancestors, at least in regard to their skeletal specializations and endocast structure (Schaeffer, 1969). This does not mean that these forms are more generalized than other teleostomes. The dipnoans diverged from other teleostomes very early, and the living forms have occupied a very stable niche (tropical freshwater lakes and streams subject to periodic drought conditions) that has probably changed little since the Permian.

VARIATION IN ANAMNIOTE NERVOUS SYSTEMS

Even a casual gross examination of anamniote brains reveals marked differences in their total size and differential development (Figure 1). However, the real task of comparative neuroanatomy is to document the spectrum of brain variability, to establish the patterns and distribution of that variability, and to discover the biological principles of development, function, and evolution of the nervous system.

An interesting and not altogether fanciful analogy between brains and computers can be made. For instance, it may be asked how change in either can be accomplished. At least three ways are apparent: (1) the number of units can increase (or decrease) without a change in the capacity of existing units, (2) the information-carrying capacity or "sophistication" of the units can change, and (3) the wiring or circuitry can change. At present, the data are insufficient to evaluate the relative contributions of these mechanisms and to formulate an

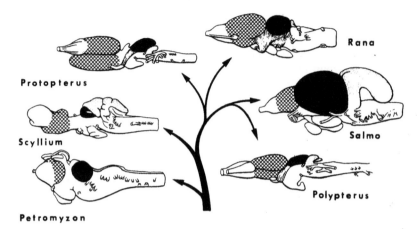

FIG. 1 Phenetogram illustrating some of the gross variation that exists in anamniotic brains. Represented are an agnathan, *Petromyzon* (sea lamprey); a chondrostean, *Polypterus* (bichir); a dipncan, *Protopterus* (African lungfish); an amphibian, *Rana* (frog); a teleost, *Salmo* (trout); and an elasmobranch, *Scyllium* (shark). The extent of the telencephalon is indicated by the hatching, the optic tectum by solid black, and the cerebellum by stippling. These brains have not been drawn to scale.

overall picture of vertebrate brain evolution. However, enough is presently known to establish that all three of the above mechanisms have operated and are operating to mold brain evolution.

Brain—Body Weight Ratios

Studies of changes in brain—body ratios probably reflect changes in the number of units. Recently Jerison (1970) has critically reviewed the problem of gross brain—body indices and their relationship to encephalization. He suggests that the philosophy of curve fitting, based on the assumption that the samples represent random deviations of a true mean caused by measurement error, should be replaced by a curve-fitting procedure that assumes the samples represent a region within which a set of brain—body data exist for a living taxon. Such a region can be represented by a principal axis defined for a set of points distributed rectangularly in the area within which they lie. This area and its principal axis can be enclosed by a minimum convex polygon, which then maps the area of the sample set.

In Figure 2 is redrawn Jerison's (1970) evaluation of the data collected by Crile and Quiring (1940). New data collected by our laboratories have been added. It is clear that the lamprey, *Petromyzon*, possesses a brain—body ratio that is lower than that of any other living vertebrate. Further examination of these data reveals that, with the exception of elasmobranchs, all anamniotes (bony fishes and amphibians) and reptiles possess comparable brain—body ratios.

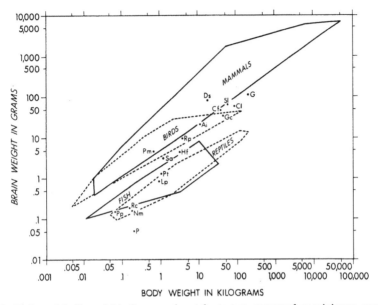

FIG. 2 Brain and body weights for several vertebrate taxa expressed as minimum convex polygons after Jerison (1970). The individual points, unless scored by an asterix, represent newly collected data. The asterixed points represent data collected by Crile and Quiring (1940). Ai, *Apriodon isodon* (finetooth shark); Cf, *Carcharhinus floridanus* (silky shark); Cl, *Carcharhinus leucas* (bull shark); Ds*, *Dasyatis sabina* (Atlantic sting ray); G*, *Galeocerdo cuvieri* (tiger shark); Gc, *Ginglymostoma cirratum* (nurse shark); Hf, *Heterodontus francisci* (horn shark); Lp, *Lepidosiren paradoxa* (South American lungfish); Nm, *Necturus maculosus* (mudpuppy); P, *Petromyzon marinus* (sea lamprey); Pm, *Potamotrygon motoro* (South American freshwater sting ray); Pp, *Polypterus palmas* (bichir); Pt, *Platyrhinoides triseriata* (thornback ray); Rc, *Rana catesbeiana* (bullfrog); Rp, *Rhinobatos productus* (guitarfish); Sa, *Squalus acanthias* (spiny dogfish); Sl, *Sphyrna lewini* (scalloped hammerhead shark).

Birds and mammals form a second group, with much higher brain–body ratios than the majority of anamniotes. Although the present data were insufficient to treat elasmobranchs by a nonnumerical analysis of convex polygons, it was clear that such a treatment would result in an overlay with some anamniotes as well as with amniotes. The extremely high brain–body ratios of elasmobranchs may be biased by the loss of the dermal skeleton in these forms. However, our measurements indicate that bone is only 21% more dense than cartilage and that a body correction factor of some 900% becomes necessary if elasmobranchs are to be placed on the same principal axis as other anamniotes. Our preliminary data therefore suggest that at least three levels of gross brain–body ratios exist in anamniotes. The first level may characterize agnathans, with a ratio of at least 500% less than other anamniotes; a second level may characterize the bony fishes and amphibians; and a third level may characterize elasmobranchs, with brain–body ratios as much as 400% higher than all other anamniotes.

Taking this analysis a step further, ratios analyzing the relative sizes of parts of the brain can be examined. In Figure 3 are plotted three divisions of the brains of 26 species representing most of the major vertebrate groups. These data are from our laboratories and are based on live adult specimens. An examination of the forebrain–body ratios reveals a striking similarity to the gross brain–body ratios seen in Figure 2. Lampreys possess the smallest forebrain–body ratio of all the measured species. A second group may be represented by bony fishes, amphibians, and reptiles; and a third group may be represented by birds, mammals, and elasmobranchs. The hindbrain–body ratios also possess a similar distribution, but the midbrain–body ratios are less similar. Notable exceptions are the urodele amphibians, which have reduced visual systems.

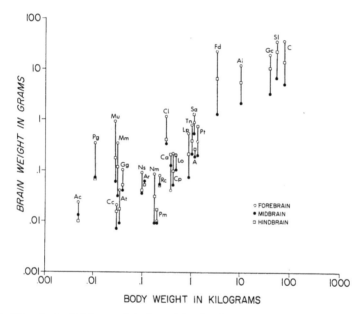

FIG. 3 Major brain divisions and body weights for 26 vertebrate species based on newly collected data. Note the similarity between the gross brain weight plots and the brain division plots, which in most instances demonstrates an allometric function. A, *Amia calva* (bowfin); Ac, *Anolis carolinensis* (green anole); Ai, *Apriodon isodon* (finetooth shark); Ar, *Ambloplites rupestris* (rock bass); At, *Ambystoma tigrinum* (tiger salamander); C, *Carcharhinus leucas* (bull shark); Ca, *Carassius auratus* (goldfish); Cc, *Calamoichthys calabaricus* (reedfish); Cl, *Columba livia* (domestic pigeon); Cp, *Chrysemys picta* (painted turtle); Fd, Felis domestica (domestic cat); Gc, *Ginglymostoma cirratum* (nurse shark); Gg, *Gekko gecko* (Tokay gecko); Lo, *Lepisosteus osseus* (longnose gar); Lp, *Lepidosiren paradoxa* (South American lungfish); Mm, *Mus musculus* (house mouse); Mu, *Melopsittacus undulatus* (budgerigar); Nm, *Necturus maculosus* (mudpuppy); Ns, *Natrix sipedon* (northern water snake); Pg, *Poephila guttata* (zebra finch); Pm, *Petromyzon marinus* (sea lamprey); Pt, *Platyrhinoides triseriata* (thornback ray); Rc, *Rana catesbeiana* (bullfrog); Sa, *Squalus acanthias* (spiny dogfish); Sl, *Sphyrna lewini* (scalloped hammerhead shark); Tn, *Tupinambis nigropunctatus* (tegu).

Our data do not support the traditional concept of encephalization. Instead, they suggest that several levels of brain–body ratios exist and that these levels have been reached independently by a number of vertebrate taxa at different times. These levels do represent changes in the numbers of units and probably also reflect levels of complexity of information processing and motor control. Although there has clearly been a general tendency toward increased neural complexity in phylogeny, there are also marked cases of atrophy of specific neural components. For example, the periventricular position of neurons in lungfish and urodele amphibians is almost certainly an example of secondary reduction rather than the retention of an ancestral vertebrate character. The loss of Mauthner cells, Müller cells, and lateral line systems in tetrapods is clearly another case of atrophy. Equally important are specific neural adaptations to particular niches. Such adaptations are sometimes correlated with hypertrophy of specific neural components, the complexity of which frequently reveals and often surpasses the neural complexity of taxa that have traditionally been regarded as "higher forms." Actinopterygian fishes represent an evolutionary epitome in utilization of diverse niches and the complexity of their medullary and tectal organization appears to be as complex as any other group of vertebrates, if not more so.

It is also clear that several independent variables determine size and numbers of neuronal elements. The size of neurons appears to vary as a function of neurite volume. The longer axons in a larger animal are reflected in larger perikarya and volume of glial elements. The positive correlation between volume of effector tissue and the number of neurons innervating such tissue is also well documented in certain systems; in other structures, e.g., the retinal ganglionic neurons, however, there is virtually no correlation with body size.

However, far more data are needed to evaluate these variables. Estimations of neuron population number and relative concentrations of neural enzymes and transmitters can aid substantially in the creation of a truly quantitative comparative neuroanatomy.

Evolution at the Neuronal Level

Changes in information-carrying capacity of units are best studied by such techniques as the Golgi method and electron microscopy studies of the distribution and types of synapses and by neurophysiological experimentation. Few studies have explored any part of this potentially powerful evolutionary mechanism for change in the central nervous system, (Ramón y Cajal 1922; Leghissa, 1962; Ramón-Moliner & Nauta, 1966; Heimer, 1969; Ramón-Moliner, 1969). Ramón-Moliner (Ramón-Moliner & Nauta, 1966: Ramón-Moliner, 1969) has examined the morphological complexity of dendrites in the medulla and thalamus and has suggested that dendritic organization becomes more complex as the type of information received by neurons becomes more restricted. In other words, the more restricted the type of information a neuron receives, the more structurally complex are that neuron's dendrites. Further examination of a number of vertebrate taxa has led him to speculate that this may be a major evolutionary mechanism. Lamination of neuronal populations is frequently associated with an increase in the type and degree of neuronal morphological complexity. These two features may really be part of a single mechanism the function of which is to increase informational specificity. An understanding of the organization and evolution of the vertebrate optic tectum should allow the functioning of this mechanism to be understood in greater detail. The degree of lamination in the optic tecta of different vertebrate taxa is considerable (see

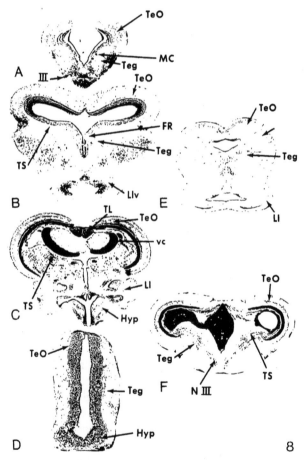

FIG. 4 Mesencephalon. A, lamprey; B, bichir; C, goldfish; D, lungfish; E, shark; F, frog. FR, fasciculus retroflexus; Hyp, hypothalamus; LI, lobus inferior; MC, Müller neurons; NIII, oculomotor nucleus; Teg, tegmentum; TeO, optic tectum; TL, torus longitudinalis, TS, torus semicircularis; VC, valvula cerebelli; III, oculomotor nerve.

Figures 4 and 5). There is a strong correlation between the degree of lamination and the number of distinct morphological types of neurons (Leghissa, 1962). The degree of segregation of afferent fiber systems also correlates well with the degree of lamination (see Figure 5). It appears likely that these architectural patterns are correlated with behavioral complexity. Similar mechanisms are probably operating in all suprasegmental structures, i.e., the cerebellum, the thalamus, and the pallium of the telencephalon.

The Range of Variation of Anamniote Neuronal Systems

One of the striking characteristics of the central nervous system of anamniotes is the hypertrophy of particular primary sensory areas in certain species. These are frequently related to greatly expanded peripheral sensory receptors of specific modality. The presence of these enlarged sensory areas in one species and their

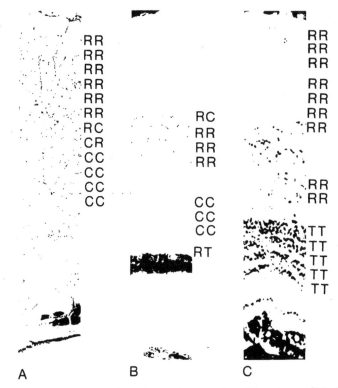

FIG. 5 Variation in appearance of Nissl-stained preparations (Fernstrom, 1958; Ebbesson, 1972a; Rubinson, 1968) and lamination of afferent fiber systems in the tectum of (a) shark, (b) squirrel fish, and (c) frog. R, retinal input; C, telencephalic input; T, tectal input. The tectal input in the frog was described by Rubinson (1968).

absence in others are correlated with characteristic niches. These variations clearly suggest that differences in brain function exist among different species. This variation is probably most striking in the organization of the teleost medulla. For example, many genera have developed an elaborate system of taste buds in the mouth, palatal structures, and pharynx that aid in the separation of food from mud taken into the mouth while feeding. These taste buds are innervated by nerves IX and X, which terminate in the much enlarged vagal lobes (Aronson, 1963). The lobes are enormous in the teleost families Cyprinidae and Siluridae yet barely exist in other fishes.

In some cyprinids (goldfish, carp, and suckers) the recurrent branch of the facial nerve is greatly enlarged and innervates large areas of the head, body, and fins. The sensory division of the facial nerve in these forms terminates in a huge facial lobe, located just rostral to the vagal lobes. Recent electrophysiological experiments have revealed a striking somatotopic organization within this lobe (Peterson, 1972).

Although forebrain organization is probably most interesting to comparative neurologists, psychologists, etc., the range of structural variation of the forebrain is so great among anamniotes that it is virtually impossible to evaluate the significance of this variation without an understanding of the organization of lower centers. Therefore, we have begun our comparative anatomical studies on the spinal cord, which appears to vary the least among vertebrates. Below is a summary of the most recently obtained data.

The spinal cord. Figure 6 illustrates some of the variation in the general morphology of the spinal cord in a sample of anamniotes. All vertebrate spinal cords possess the same basic configuration of white and gray matter. Some species have several large axons, called Müller or Mauthner fibers (Eccles, 1964; Rovainen, 1967a, b), that run the length of the spinal cord (Figure 6). These axons originate from cell bodies in the brain stem, but their precise terminations are not understood. Another interesting feature is the presence of intramedullary primary sensory neurons in *Petromyzon*, a feature not understood at this time.

Dorsal root projections have been studied only in *Bufo* and are similar to those described in mammals. However in *Bufo* some primary sensory fibers reach the cerebellum (Joseph & Whitlock, 1968). Most anamniote forms so far examined possess telencephalospinal fibers that extend as far caudally as the first or second spinal segment (Ebbesson, 1972a; Ebbesson & Schroeder, 1971; & Northcutt, unpublished observation). In all cases this projection is very sparse. The long corticospinal tracts appear to correlate with the appearance of mammalian laminated neocortex.

Ascending spinal cord projections are essentially the same in lampreys (Northcutt & Ebbesson, in preparation), sharks (Ebbesson, 1972a; Hayle, 1973a, b), and amphibians (Ebbesson, 1969; Nieuwenhuys & Cornelisz, 1971). However, only in sharks has a distinct spinothalamic connection been described (Figure 7).

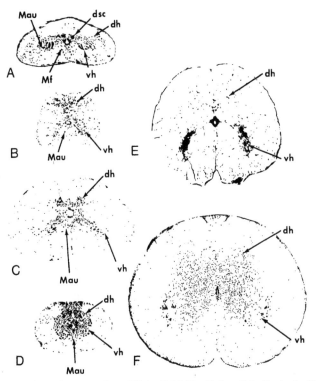

FIG. 6 Spinal Cord. A, lamprey; B, goldfish; C, bichir; D, lungfish; E, shark; F, frog. dh, dorsal horn; dsc, dorsal sensory cell; Mau, Mauthner fiber; Mf, Müller fiber; vh, ventral horn.

The evolution of substantial spinothalamic projections as well as the appearance of laminar somatosensory cell aggregates in the telencephalon may be correlated with the increasing importance of limbs.

The rhombencephalon. The general morphology of many genera displays considerable variations of the medulla oblongata and cerebellum. Some have enormous vagal and facial lobes; others have highly differentated cerebella (e.g., mormyrids) (Nieuwenhuys & Nicholson, 1969a, b). Others present a more general configuration (Figure 8) (Nicholson, Llinás, & Precht, 1969).

The input of the medulla oblongata and cerebellum is remarkably stable, and little or no variation is discernible at the light microcopic level of analysis. The inputs to the rhombencephalic reticular cell groups from the telencephalon are known in the nurse shark (Ebbesson, 1972a; Ebbesson & Schroeder, 1971) and in several amphibians. The tectal contribution to these cell groups is known in lampreys, sharks (Ebbesson, 1972a), teleosts (Ebbesson, unpublished), and amphibians (Ebbesson, Jane, & Schroeder, 1972; Rovainen, 1967b). The cerebellar projections are also known in sharks (Campbell & Ebbesson, 1971;

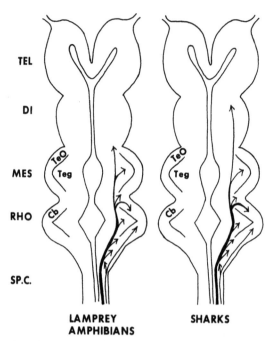

FIG. 7 Highly schematic representation of ascending spinal projections as revealed by recent Nauta studies. Cb, cerebellum; DI, diencephalon; MES, mesencephalon; RHO, rhombencephalon; SP.C., spinal cord; Teg, tegmentum; TEL, telencephalon; TeO, optic tectum.

Ebbesson & Campbell, 1973), holosteans (Northcutt, unpublished observation), teleosts (Braford, 1970), and amphibians. The spinobulbar projections are known in lampreys (Northcutt & Ebbesson, in preparation), sharks (Ebbesson & Schroeder, 1971), and frogs (Ebbesson, 1969). These inputs are very similar but require additional analysis.

The sensory root distribution of the trigeminal (Fuller & Ebbesson, 1972), vestibular (Fuller & Ebbesson, 1973), glossopharyngeal, and vagal (Friedman, Rubinson, & Colman, 1973) nerves has been described only in Anura. These are similar to those described in mammals except for the lack of differentiation of some of the recipient cell groups. Structural differences related to the fish–amphibian transition are found in the lateral line–auditory system (Maler, 1971).

Only the spinal cerebellar input has been examined in lampreys (Northcutt & Ebbesson, in preparation), sharks (Ebbesson & Schroeder, 1971), and amphibians (Ebbesson, 1969). As in other vertebrates both dorsal and ventral spinocerebellar tracts are recognized. A dorsal tract is located in the dorsal funiculus of the spinal cord, and in *Bufo*, at least, this fasciculus is partially composed of primary sensory fibers (Joseph & Whitlock, 1968). Both tracts reach the cerebellar nuclei and the granule cell layer of the corpus cerebelli as

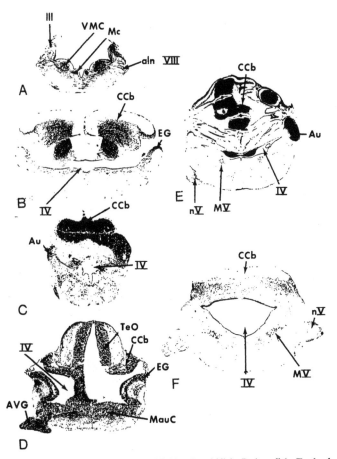

FIG. 8   Rhombencephalon. A, lamprey; B, bichir; C. goldfish; D, lungfish; E, shark; F, frog. aln, acustico-lateralis nerve; Au, auricle; AVG, acusticovestibular ganglion; CCb, corpus cerebelli; EG, eminentia granularis; lll, lateral line lobe; MauC, Mauthner neuron; Mc, Müller neuron; MV, trigeminal motor nucleus, nV, fifth cranial nerve; TeO, optic tectum; VMC, visceral motor column; IV, fourth ventricle.

they do in all other vertebrates (Ebbesson & Schroeder, 1971; Ebbesson, 1969; Larsell, 1967).

The efferents of the medulla oblongata have not been studied except for the projections of the vestibular nuclear complex in *Rana* (Fuller & Ebbesson, 1973). They appear similar to those reported in mammals.

The cerebellar cortical efferents may or may not extend beyond the cerebellar nuclei. In the nurse shark (Campbell & Ebbesson, 1971; Ebbesson & Campbell, 1973), we have found no evidence of Purkinje cells of the corpus projecting beyond these nuclei. In *Rana* some project to the vestibular nerve (Llinás, Precht, & Kitai, 1967). In teleosts it is not possible to selectively destroy the

Purkinje cells because the cells of the cerebellar nuclei and the Purkinje cells are located in the same layer (Braford, 1970). Therefore, the experiments give the impression that the cerebellar cortex of these forms projects to the reticular formation, thalamus, and nuclei of the oculomotor complex.

The deep cerebellar nuclei of the nurse shark project, via the brachium conjunctivum, to the thalamus, nucleus ruber, nuclei of the extraocular muscles, and reticular formation (Campbell & Ebbesson, 1971; Ebbesson & Campbell, 1973). A small contralateral descending component of the brachium conjunctivum distributes to the medial reticular formation, and an ipsilateral lateral descending pathway extends as far caudally as the first spinal segment. This arrangement of cerebellar efferents is strikingly similar to that reported in mammals (Figure 9).

The mesencephalon.   Little specific variation in general morphology is noted in the mesencephalic tegmentum. However, marked variation exists in the optic tectum (Figure 4).

The visual inputs to the mesencephalon are documented in lampreys (Northcutt & Przybylski, 1973), sharks (Ebbesson, 1967a; Ebbesson & Ramsey, 1968;

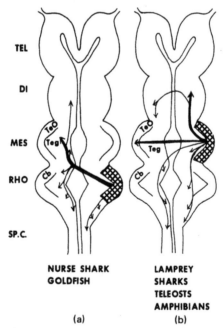

FIG. 9 Highly schematic representation of cerebellar efferent projections (a) and tectal efferent projections (b) as revealed by recent Nauta studies. Cb, cerebellum; DI, diencephalon; MES, mesencephalon; RHO, rhombencephalon; SP.C., spinal cord; Teg, tegmentum; Tel, telencephalon; TeO, optic tectum.

Graeber & Ebbesson, 1972a), chondrosteans, holosteans (Northcutt, unpublished observations), teleosts (Campbell & Ebbesson, 1969; Ebbesson, 1968, 1970b; Roth, 1969; Sharma, 1972; Vanegas & Ebbesson, 1973), lungfishes and amphibians (Ebbesson, 1970b; Riss, Knapp, & Scalia, 1963; Scalia, Knapp, Halpern, & Riss, 1968). The optic tectum receives a visual input in all these groups, and the retinal input to a basal optic nucleus exists in some. The spinomesencephalic projection is known and is similar in lampreys (Northcutt & Ebbesson, in preparation), sharks (Ebbesson, 1972a), and amphibians (Ebbesson, 1969). The projection, unlike that in mammals, is limited to the tegmentum. The cerebellomesencephalic projections are now known in sharks (Ebbesson & Campbell, 1973), holosteans (Northcutt, unpublished observations), and teleosts (Braford, 1970), and appear similar at the present level of analysis. Telencephalic and tectal projections to the mesencephalic tegmentum are also very similar in the few species studied (Figures 9b and 17).

The efferent pathways of the tegmental nuclei have not been studied in anamniotes, but the outflow from the optic tectum is well documented in lampreys (Northcutt, unpublished observation), sharks (Ebbesson, 1971, 1972a, b), teleosts (Ebbesson, 1970b), and amphibians (Llinás et al., 1967). The ascending tectal pathways characteristically impinge on several pretectal nuclei and one or more thamalic nuclei; they also consist of a crossed pathway that reaches the contralateral diencephalon via the supraoptic decussation. The descending tectal pathways in the four groups studied consist of an ipsilateral lateral tectoreticular pathway and a contralateral medial tectoreticular pathway. The latter pathway usually reaches the cervical spinal segments (Figure 9b).

The diencephalon. The range of morphological variation in anamniote diencephalons is considerable (Figure 10). Most anamniotes possess poorly differentiated diencephalons, and nuclear boundaries are therefore difficult to ascertain. Much of the variation is centered in the dorsal thalamus and pretectum. Vision appears to be the most dominant modality in the anamnoite thalamus.

The organization of the anamnoite hypothalamus has not been extensively studied with modern methods. Retinal inputs are known to this region in lampreys (Northcutt & Przybylski, 1973), sharks (Ebbesson, 1967a; Ebbesson & Ramsey, 1968; Graeber & Ebbesson, 1972a), teleosts (Campbell & Ebbesson, 1969; Ebbesson, 1968; Vanegas & Ebbesson, 1973) and amphibians (Jakway & Riss, 1969; Riss et al., 1963; Scalia et al., 1968). The number and configuration of the recipient cell groups vary as a function of the differentiation of the visual system (Ebbesson, 1972b). Most teleosts probably possess the most differentiated thalami and optic tecta, whereas lungfish and urodele amphibians probably possess some of the most poorly differentiated visual structures (Ebbesson et al., 1972) (Figure 11).

Afferents from the spinal cord to the thalamus have only been demonstrated in the nurse shark (Ebbesson, 1972a), although ascending spinal projections have

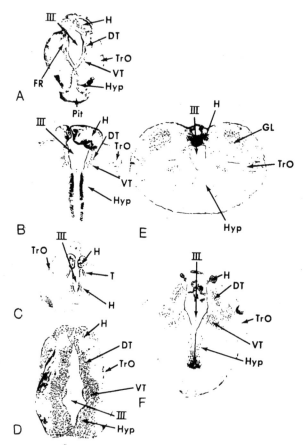

FIG. 10 Diencephalon. A, lamprey; B, bichir; C, goldfish; D, lungfish; E, shark; F, frog.
FR, fasciculus retroflexus; DT, dorsal thalamus; GL, lateral geniculate nucleus; H, habenula;
Hyp, hypothalamus; Pit, pituitary; T, thalamus; TrO, optic tract; VT, ventral thalamus; III,
third ventricle.

been traced to the mesdiencephalic border in lampreys (Northcutt & Ebbesson,
in preparation) and amphibians (Ebbesson, 1969; Ebbesson *et al.,* 1972). A
cerebellar input to thalamic cell groups has been documented in sharks (Ebbesson & Campbell, 1973) and teleosts (Braford, 1970) (Figure 12). Other
inputs have not been studied (Figure 12).

Hypothalamic projections are, as of this writing, not known in any anamniote
species. Thalamic projections have been studied in sharks (Ebbesson &
Schroeder, 1971; Schroeder & Ebbesson, 1971), chondrosteans (Northcutt,
unpublished observation), and amphibians (Northcutt, 1972; Vesselkin, Agayan,
& Nomokonova, 1971). The thalamotelencephalic pathways are primarily
crossed in sharks but predominantly ipsilateral in chondrosteans and amphibians

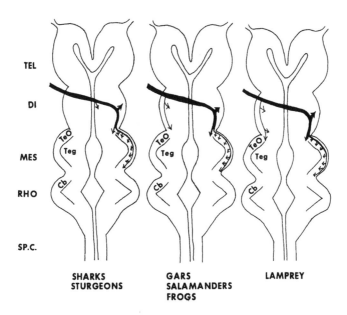

FIG. 11   Highly schematic representation of retinal projections as revealed by recent Nauta studies. Cb, cerebellum; DI, diencephalon; MES, mesencephalon; RHO, rhombencephalon; SP.C., spinal cord; Teg, tegmentum; Tel, telencephalon; TeO, optic tectum.

(Figure 13). Both behavioral (Graeber & Ebbesson, 1972b; Graeber, Ebbesson, & Jane, 1973) and physiological (Cohen, Duff, & Ebbesson, 1973) studies have confirmed that these pathways convey visual information in sharks and amphibians. In the latter group a somatosensory component has been demonstrated.

The telencephalon   The greatest morphological variation exists in the telencephalon of anamniotes and important misconceptions about its organization have been revealed in the last five years. The range of variation of the cell aggregates is enormous (Figure 14). Many of these misconceptions arise from differences in the embryonic development of the telencephalon in actinopterygian fishes and other anamniotes (Nieuwenhuys, 1962a, b, 1966). In actinopterygian fishes the telencephalic hemispheres are formed by a process of eversion, whereas in all other anamniotes the telencephalic hemispheres are formed by a process of evagination (Figure 15).

The last few years have produced striking new evidence that the olfactory apparatus does not completely dominate the telencephalon in anamniotes as previously thought (Ebbesson, 1972a; Ebbesson & Heimer, 1968, 1970; Ebbesson & Schroeder, 1971; Vesselkin *et al.*, 1971). Instead, it is now clear that relatively large regions are concerned with vision and probably other modalities

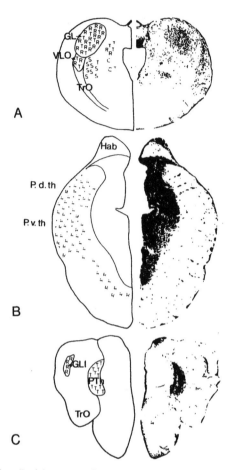

FIG. 12   Recently described inputs to the thalamus of the nurse shark (a), tiger salamander (b), and squirrel fish (c) (Ebbesson *et al.*, 1972). C, cerebellar input; GL, lateral geniculate nucleus; GLI, nucleus geniculatum laterale ipsum; Hab, habenula; PTh, prethalamic nucleus; P.d.th., pars dorsalis thalami; P.v.th., pars ventralis thalami; R, retinal input, S, spinal input; T, tectal input; TrO, optic tract; VLO, ventrolateral optic nucleus.

as well (Graeber, Ebbesson, Jane, & Best, 1971; Graeber, Schroeder, Jane, & Ebbesson, 1972; Kicliter, 1973). The data come from essentially three kinds of experiments, one dealing with the determination of the olfactory tract projections, the second with the delineation of the thalamotelencephalic projections, and the third with defining telencephalic efferents.

The olfactory tract connections have been studied in sharks (Ebbesson & Heimer, 1968, 1970), chondrosteans (Braford & Northcutt, 1974), teleosts (Scalia & Ebbesson, 1971), and amphibians (Royce & Northcutt, 1969). The pathway in all these species issues fascicles to a relatively small lateral zone

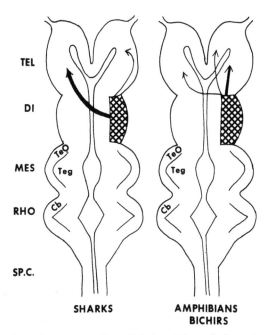

FIG. 13 Highly schematic representation of thalamotelencephalic systems as revealed by recent Nauta studies. Cb, cerebellum; DI, diencephalon; MES, mesencephalon; RHO, rhombencephalon; SP.C., spinal cord; Teg, tegmentum; Tel, telencephalon; TeO, optic tectum.

ipsilaterally and, in teleosts and amphibians, to a restricted portion of the contralateral hemisphere as well (Figure 16). Tertiary olfactory projections have also been examined in the nurse shark (Ebbesson, 1972a) and several amphibians (Northcutt, unpublished observation). These are also clearly limited, leaving a large volume of telencephalic tissue apparently void of olfactory input.

Thalamotelencephalic pathways are now known in sharks (Ebbesson & Schroeder, 1971), chondrosteans (Gruberg & Ambros, 1974; Kicliter & Northcutt, 1975; Northcutt, 1972; Vesselkin *et al.,* 1971), and amphibians (Northcutt, 1972; Vesselkin *et al.,* 1971) and are found to reach portions of the telencephalon not concerned with olfaction. In addition to the anatomical studies, electrophysiological studies in sharks (Cohen, Duff, & Ebbesson, 1973) and amphibians (Northcutt, 1970; Vesselkin *et al.,* 1971) have confirmed the anatomical data. Surprisingly, this pathway is almost completely crossed in sharks (Figure 13).

The total output patterns of telencephalic projections are known only in a few taxa of sharks (Ebbesson, 1972a, b; Ebbesson & Schroeder, 1971), teleosts (Ebbesson, unpublished observation), and amphibians (Halpern, 1973; Jane, unpublished observation; Kokoros, 1972; Northcutt, unpublished observation). The descending fibers extend as far caudally as the first spinal segment, im-

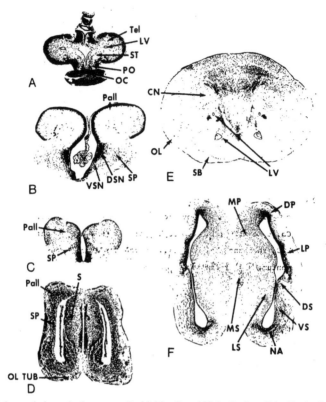

FIG. 14 Telencephalon. A, lamprey; B, bichir; C, goldfish; D, lungfish; E, shark; F, frog. CN, central telencephalic nucleus; DP, dorsal pallium; DS, dorsal striatum; DSN, dorsal subpallial nucleus; LP, lateral pallium; LS, lateral septal nucleus; LV, lateral ventricle; MP, medial pallium; MS, medial septal nucleus; NA, nucleus accumbens; OC, optic chiasma; OL, lateral olfactory area; Ol tub, olfactory tubercle; Pall, pallium; PO, preoptic area; S, septum; SB, area superficialis basalis; SP, subpallium; ST, striatum; Tel, telencephalon; VS, ventral striatum; VSN, ventral subpallial nucleus.

pingeing on thalamic and reticular nuclei in their course. In sharks, in contrast to teleosts and amphibians, the majority of the telencephalic efferents cross at diencephalic levels for distribution to contralateral thalamic, tectal, and reticular cell groups. Telencephalotectal fibers in the classical sense appear not to exist in *Rana* (Figures 5 and 17).

NERVOUS SYSTEM VARIATION AND ADAPTIVE ZONES

Although present understanding of nervous system variability is scanty, it should prove useful to examine that variability in relation to major adaptive zones. Can the development or loss of certain neural structures be correlated with the

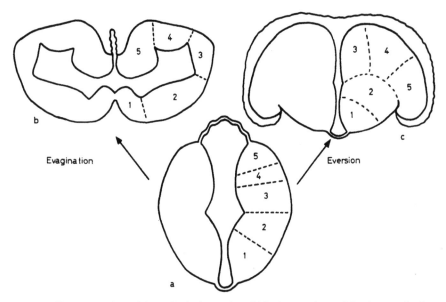

FIG. 15 Representation of hypothetical way in which the topology of fundamental sub-divisions of embryonic telencephalon (a) might be preserved after evagination (b), as in the amphibian, or eversion (c), as in the teleost. 1, septal region; 2, striatum; 3, pyriform lobe; 4, general cortex; 5, hippocampal region (Scalia & Ebbesson, 1971).

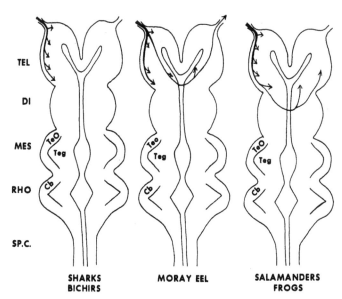

FIG. 16 Highly schematic representation of olfactory tract projections as revealed by recent Nauta studies. Cb, cerebellum; DI, diencephalon; MES, mesencephalon; RHO, rhom-bencephalon; SP.C., spinal cord; Teg, tegmentum; Tel, telencephalon; TeO, optic tectum.

137

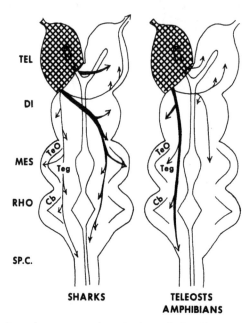

FIG. 17 Highly schematic representation of telencephalic efferents as revealed by recent Nauta studies. Cb, cerebellum; DI, diencephalon; MES, mesencephalon; RHO, rhombencephalon; SP.C., spinal cord; Teg, tegmentum; Tel, telencephalon; TeO, optic tectum.

occupation of a particular adaptive zone? In using the term "adaptive zone" we are following Simpson's usage to mean a way of life and not just a place where life is led (Simpson, 1953).

Unfortunately, little can be said about the modifications of the CNS as a result of changes in adaptive zones because the work hitherto has dealt mostly with attempts at delineating the range of neuronal systems that can be related to mammals. Therefore, most of the structural configurations specifically related to shifts in anamniote adaptive zones are poorly understood. Some of these aspects are readily identifiable and can be correlated with specific adaptations. The agnathan–gnathostome transition is characterized by the acquisition of true jaws and paired fins. An increase in size of all parts of the brain is correlated with this transition to a more active predatory life. The appearance of the mesencephalic nucleus of the trigeminal nerve correlates with the appearance of true jaws.

Present information on the organization and variation in the nervous system of teleostomes and elasmobranchs precludes the recognition of specific modifications of their nervous systems to their adaptive zones. In fact, so little is known about the biology of these organisms in general that it is difficult to even characterize the differences in their adaptive zones in a meaningful way.

The Water—Land Transition

Perhaps the most striking change in anamniote adaptive zones has been the invasion of the land. The nervous systems of terrestrial forms possess unique elaborate receptor organs, and CNS structures related to them, that give the animal power to react to many aspects of environmental change that cannot be discriminated by aquatic organisms. In contrast, many aquatic vertebrates have specializations of their own, such as electroreceptors, with concomitant specializations in the CNS. These structures have been lost with the invasion of land. Although some recent physiological studies have described various aspects of these specializations, the CNS anatomical details are still unknown.

There were probably few if any organ systems that remained unmodified in the transition of crossopterygian fishes to primitive amphibians. For example, the internal gills were lost; the opercular dermal bone series was reduced, with a caudal shift and reorganization of the pectoral girdle and limbs; there was a hypertrophy and change in articulation of the pelvic girdle and limbs; and the hyomandibular element shifted from a connecting skull component to a sound-conducting component. At present few of these transitions can be correlated with structural modifications in the nervous system. Even two of the most obvious morphological transitions, reorganization of limb and of auditory structures, cannot be analyzed in detail.

The crossopterygian—amphibian trend to develop more complicated limbs is probably reflected in the nervous system by (1) an increase in the number of neuronal elements of all levels that relate to the functioning of those limbs and (2) an increased complexity of the involved neuronal pools. An outstanding difference between the spinal cords of amphibians and most fishes is the presence in amphibians of an intumescentia cervicalis and an intumescentia lumbalis. These enlargements are correlated with the increasing development of limbs. Both the quantity and the complexity of neural elements increase with limb development. In anamniotes there is no fine motor control of distal limb elements, whereas in some mammals even individual digits have identifiable cell groups at spinal, medullary, thalamic, and cortical levels. Another trend seems to be that the more complex the limb, the greater is its representation rostrally.

The skeletal and muscular modifications that have evolved with increased utilization of the limbs have become even more elaborate since the limbs have taken an increased supportive postural role in tetrapods. It is one thing to use an appendage in an aquatic environment and quite another to support and loco-mote the body on land. It is evident that increased neural circuits must have evolved to insure required "postural tone" and locomotor mechanisms.

Except for the obvious enlargements of the spinal cord there is little information about changes in the CNS that relates to the evolution of limbs. The size and complexity of the dorsal column system reflects the limb development rather well. We have noted above that snakes have by far the smallest dorsal funiculus of the reptilian orders and the dorsal funiculi of fishes also appear poorly developed. However, it is also clear that many differences must exist in the spinocerebellar and spinobulbar connections in aquatic versus terrestrial forms. Only neurophysiological studies can elucidate these differences. Studies of muscle receptors in anamniotes can greatly increase current understanding of this transition.

The reduction of the lateral line system and the evolution of a terrestrial ear and auditory system capable of detecting and analyzing airborne sound is a hallmark of the crossopterygian–amphibian transition. Although urodeles and caecilians possess no tympanic membrane or amphibian and basilar papillae, most of the adult anurans have a well-developed auditory system. All amphibians possess a lateral line system of sense organs during larval stages but some genera (*Xenopus* and *Pipa*) in part retain, the lateral line system in the adult in addition to a well-developed acoustic system.

The remarkable transformation in the CNS of the acousticolateralis area is probably best studied in amphibians during metamorphosis, because only the presumed before and after conditions are seen in larval and adult anurans (Campbell & Boord, 1974). In larval stages, the dorsolateral cell group of the acousticolateralis area has an input from both lateral line and dorsal eighth root. During metamorphosis, lateral line connections are lost, whereas eighth-nerve connections are retained. The cells in the dorsolateral and medial parts of the acousticolateralis area do not degenerate but transform into dorsal and ventral eighth-nerve nuclei, respectively. Boord has recently shown that all auditory neurons, that is, those from the amphibian and basilar papillae, terminate within the dorsal acoustic nucleus (Boord & Eiswerth, 1972; Boord, Grochow, & Frishkopf, 1970), supposedly the homolog of the mammalian cochlear nuclei (Ariëns-Kappers, Huber, & Crosby, 1936). The dorsal acoustic nucleus in *Rana catesbeiana* projects to the central nucleus of the torus semicircularis in a manner strikingly reminiscent of the projections of the mammalian cochlear nuclei (Ebbesson, unpublished observation). It must be remembered that the latter observations are only preliminary in nature, and only detailed analysis can reveal the interspecific differences that no doubt exist.

SUMMARY

It is readily apparent from the above discussion that, at the present level of analysis with Nauta (Nauta & Ebbesson, 1970) and Fink and Heimer (1967) methods, many fiber systems show marked stability, particularly the ascending spinal projections, the tectal projections, and the cerebellar projections. In contrast, those neuronal connections that relate to specializations of either a peripheral sensory system or the forebrain show a high degree of variability.

These variations reflect the ability of neuronal systems to respond to changes in selective pressures. These modifications provide essential clues to form–function relationships. Much of the variation can be accounted for by differences in lamination or decussating versus nondecussating fiber systems. Correlation of these structural differences with physiological and behavioral differences should provide clues to their biological significance.

The range of variation of crossing versus noncrossing fibers in some fiber systems is considerable. The small sample of species studied makes speculation about its significance somewhat dangerous at this time. Nevertheless, certain patterns are emerging that suggest a working hypothesis: many systems may have been bilateral originally but selective pressures have resulted in the disappearance of either the ipsi- or contralateral component (or both), and once a component is lost, succeeding genera have been unable to reestablish the pathway. The presence or absence of the ipsilateral retinothalamic pathway apparently is not necessarily related to binocular vision as previously thought; e.g., the owl has considerable overlap of visual fields, yet all birds examined so far have a completely crossed optic tract. It is interesting that their closest reptilian relatives, the crocodilians, also lack this pathway (Braford & Northcutt, unpublished observations). The possibility then exists that knowledge of variations in location of tracts may provide important clues to evolutionary relationships.

ACKNOWLEDGMENTS

The authors' research reported herein was supported by Grant No. 5 RO1 EY00154, National Eye Institute, National Institutes of Health; Grant No. NGR47-005-186, National Aeronautics and Space Administration; Grant No. NS46292, Research Career Development Award, National Institute of Neurological Diseases and Stroke to Sven O. E. Ebbesson. Grant No. 7 RO1-NS11006, National Institutes of Health; Grant No. GB-40134, National Science Foundation to R. Glenn Northcutt.

REFERENCES

Ariëns-Kappers, C. U., Huber, G. C., & Crosby, E. C. *The comparative anatomy of the nervous system of vertebrates, including man.* New York: Macmillan, 1936.

Aronson, L. R. The central nervous system of sharks and bony fishes with special reference to sensory and integrative mechanisms. In P. W. Gilbert, Ed., *Sharks and Survival.* Boston: Heath, 1963.

Bäckström, K. Contributions to the forebrain morphology in selachians. *Acta Zoologica* 1924, *5*, 123–240.

Bertmar, G. Lungfish phylogeny. In T. Ørvig, Ed., *Current problems of lower vertebrate phylogeny.* Stockholm: Almqvist and Wiksells, 1968.

Boord, R. L., & Eiswerth, L. M. The central terminal fields of posterior lateral line and eighth nerves of *Xenopus. American Zoologist* 1972, *12*, 727.

Boord, R. L., Grochow, L. B., & Frishkopf, L. S. Organization of the posterior ramus and ganglion of the eighth cranial nerve of the bullfrog, *Rana catesbeiana*. *American Zoologist*, 1970, **10**, 555.

Braford, M. R., Jr. Projections of the corpus cerebelli in the goldfish. *Anatomical Record*, 1970, **166**, 282.

Braford, M. R., Jr. Ipsilateral retinal projections in non-mammalian vertebrates. *American Zoologist*, 1972, **12**, 728.

Braford, M. R., Jr., & Northcutt, R. G. Olfactory bulb projections in the bichir, *Polypterus*. *Journal of Comparative Neurology*, 1974, **156**, 165–178.

Campbell, C. B. G., & Boord, R. L. Central auditory pathways of nonmammalian vertebrates. In W. D. Keidel & W. D. Neff (Eds.), *Handbook of Sensory Physiology*. Berlin: Springer-Verlag, 1974. Pp. 337–362.

Campbell, C. B. G., & Ebbesson, S. O. E. The optic system of a teleost: *Holocentrus* re-examined. *Brain, Behavior & Evolution*, 1969, **2**, 415–430.

Campbell, C. B. G., & Ebbesson, S. O. E. On the organization of cerebellar efferent pathways in the nurse shark. *Ginglymostoma cirratum* (Bonnaterre). Society for Neuroscience, First Annual Meeting, 1971.

Cohen, D. H., Duff, T., & Ebbesson, S. O. E. Electrophysiological identification of a visual area in shark telencephalon. Science, 1973, **182**, 492–494.

Crile, G., & Quiring, D. P. A record of the body weight and certain organ and gland weights of 3690 animals. *Ohio Journal of Science*, 1940, **40**, 219–259.

Ebbesson, S. O. E. Retinal projections in two species of sharks (*Galeocerdo cuvieri* and *Ginglymostoma cirratum*). *Anatomical Record*, 1967, **157**, 238.

Ebbesson, S. O. E. Retinal projections in two teleost fishes (*Opsanus tau* and *Gymnothorax funebris*). An experimental study with silver impregnation methods. *Brain, Behavior & Evolution*, 1968, **1**, 134–153.

Ebbesson, S. O. E. Brain stem afferents from the spinal cord in a sample of reptilian and amphibian species. *Annals of the New York Academy of Science*, 1969, **167**, 80–101.

Ebbesson, S. O. E. Selective silver impregnation of degenerating axoplasm in poikilothermic vertebrates; In W. J. H. Nauta & S. O. E. Ebbesson, (Eds.), *Contemporary Research Methods in Neuroanatomy*. Berlin and New York: Springer-Verlag, 1970, Pp. 132–161. (a)

Ebbesson, S. O. E. On the organization of central visual pathways in vertebrates. *Brain Behavior & Evolution*, 1970, **3**, 178–194. (b)

Ebbesson, S. O. E. Projections of the optic tectum in the nurse shark (*Ginglymostoma cirratum*) (Bonnaterre). *Proceedings of the First Annual Meeting of the Society of Neuroscience*, 1971, p. 109.

Ebbesson, S. O. E. New insights into the organization of the shark brain. (Paper Presented at the Elasmobranch Biology Symposium, June 1971.) *Comparative Biochemistry & Physiology*, 1972, **42**, 121–129. (a)

Ebbesson, S. O. E. A proposal for a common nomenclature for some optic nuclei in vertebrates and the evidence for a common origin of two such cell groups. *Brain, Behavior & Evolution* 1972, **6**, 75–91. (b)

Ebbesson, S. O. E., & Campbell, C. B. G. On the organization of cerebellar efferent pathways in the nurse shark (*Ginglymostoma cirratum*). *Journal of Comparative Neurology*, 1973, **152**, 233–254.

Ebbesson, S. O. E., & Heimer, L. Olfactory bulb projections in two species of sharks (*Galeocerdo cuvieri* and *Ginglymostoma cirratum*). *Anatomical Record*, 1968, **160**, 469.

Ebbesson, S. O. E., & Heimer, L. Projections of the olfactory tract fibers in the nurse shark (*Ginglymostoma cirratum*). *Brain Research*, 1970, 17, 47–55.

Ebbesson, S. O. E., Jane, J. A. & Schroeder, D. M. A general overview of major interspecific variations in thalamic organization. *Brain, Behavior & Evolution*, 1972, 6, 92–130.

Ebbesson, S. O. E., & Ramsey, J. S. The optic tracts in two species of sharks (*Galeocerdo cuvieri* and *Ginglymostoma cirratum*). *Brain Research*, 1968, 8, 36–53.

Ebbesson, S. O. E., & Rubinson, K. A simplified Nauta procedure. *Physiology & Behavior*, 1969, 4, 281–282.

Ebbesson, S. O. E., & Schroeder, D. M. Connections of the nurse shark's telencephalon. *Science*, 1971, 173, 254–256.

Eccles, J. C. *The physiology of synapses.* New York: Springer-Verlag, 1964. Chapter XIV.

Fernstrom, R. C. A durable Nissl stain for frozen and paraffin sections. *Stain Technology*, 1958, 33, 175–176.

Fink, R. P., & Heimer, L. Two methods for selective silver impregnation of degenerating axons and their synaptic endings in the central nervous system. *Brain Research*, 1967, 4, 369–374.

Friedman, B. E., Rubinson, K., & Colman, D. R. Vagus nerve projections in Anurans. *Anatomical Record*, 1972, 172, 312.

Fuller, P. M., & Ebbesson, S. O. E. Central pathways of the trigeminal nerve in the bullfrog (*Rana catesbeiana*). *Anatomical Record*, 1972, 172, 312.

Fuller, P. M., & Ebbesson, S. O. E. Central connections of the vestibular nuclear complex in the bullfrog (*Rana catesbeiana*). *Anatomical Record*, 1973, 175, 325.

Graeber, R. C., & Ebbesson, S. O. E. Retinal projections in the lemon shark (*Negaprion brevarostris*). *Brain, Behavior & Evolution,* 1972, 5, 461–477. (a)

Graeber, R. C., & Ebbesson, S. O. E. Visual discrimination learning in normal and tectal-ablated nurse sharks (*Ginglymostoma cirratum*). *Comparative Biochemistry & Physiology*, 1972, 42A, 131–139. (b)

Graeber, R. C., Ebbesson, S. O. E., Jane, J. A., & Best, P. J. Visual discrimination and the effects of ablations of the central visual system in lemon and nurse sharks. Society for Neuroscience, First Annual Meeting, 1971.

Graeber, R. C., Ebbesson, S. O. E., & Jane, J. A. Visual discrimination in sharks without optic tectum. *Science,* 1973, 180, 413–415.

Graeber, R. C., Schroeder, D. M., Jane, J. A., & Ebbesson, S. O. E. The importance of telencephalic structures in visual discrimination learning in nurse sharks. Society for Neuroscience, Second Annual Meeting, 1972.

Gruberg, E. R., & Ambros, V. R. A forebrain visual projection in the frog (*Rana pipiens*). *Experimental Neurology,* 1974, 44, 187–197.

Halpern, M. Some connections of the telencephalon of the frog, *Rana pipiens. Brain, Behavior and Evolution*, 1973, 6, 42–68.

Hayle, T. H. A comparative study of spinal projections to the brain (except cerebellum) in three classes of poikilothermic vertebrates. *Journal of Comparative Neurology*, 1973, 149, 463–476. (a)

Hayle, T. H. A comparative study of spinocerebellar system in three classes of poikilothermic vertebrates. *Journal of Comparative Neurology,* 1973, 149, 477–496. (b)

Heimer, L. The secondary olfactory connections in mammals, reptiles, and sharks. *Annals of the New York Academy of Science,* 1969, 167, 129–146.

Herrick, C. J. *The brain of the tiger salamander.* Chicago: University of Chicago Press, 1948.

Jakway, J. S., & Ris,, W. Retinal projections in the tiger salamander. *Anatomical Record,* 1969, 163, 203.

Jarvik, E. Aspects of vertebrate phylogeny. In T. Ørvig, (Ed.), *Current problems of lower vertebrate phylogeny.* Almqvist and Wiksells, 1968. (a)

Jarvik, E. The systematic position of the Dipnoi. In T. Ørvig (Ed.), *Current problems of lower vertebrate phylogeny.* Almqvist and Wiksells, 1968. (b)

Jerison, H. J. Gross brain indices and the analysis of fossil endocasts. In C. R. Noback & W. Montagna (Eds.), *Advances in Primatology,* Vol. 1, *The Primate Brain.* New York: Appleton-Century-Crofts, 1970.

Joseph, B. S., & Whitlock, D. G. The morphology of spinal afferent-efferent relationships in vertebrates. *Brain, Behavior & Evolution,* 1968, **1,** 2–18.

Kicliter, E. Two visual systems in the frog. *Anatomical Record,* 1973, **175,** 357–358.

Kicliter, E., & Northcutt, R. G. Ascending afferents to the telencephalon of ranid frogs: an anterograde degeneration study. *Journal of Comparative Neurology,* 1975, **161,** 239–254.

Kokoros, J. J. Efferent projections of the telencephalon in the toad (*Bufo marinus*) and the salamander (*Ambystoma maculatum*). *Anatomical Record,* 1972, **172,** 349.

Larsell, O. In J. Jansen, Ed., *The comparative anatomy and histology of the cerebellum from Myxinoids through birds.* Minneapolis: University of Minnesota Press, 1967.

Leghissa, S. L'evoluzione del tetto ottico nei bassi vertebrati. *Archivio Italiano di Anatomia e di Embriologia,* 1962, **67,** 344–413.

Llinás, R., Precht, W., & Kitai, S. T. Cerebellar purkinje cell projection to the peripheral vestibular organ in the frog. *Science,* 1967, **158,** 1328–1330.

Maler, L. Projections of the lateral line nerves of *Gnathonemus petersii. Anatomical Record,* 1971, **169,** 374.

Moy-Thomas, J. A. & Miles, R. S. *Palaeozoic fishes.* Philadelphia: W. B. Saunders Co., 1971.

Nauta, W. J. H., & Ebbesson, S. O. E. (Eds.) Contemporary Research Methods in Neuroanatomy. Berlin and New York: Springer-Verlag, 1970.

Nicholson, C., Llinás, R., & Precht, W. Neural elements of the cerebellum in elasmobranch fishes: structural and functional characteristics. In R. Llinás (Ed.), *Neurobiology of cerebellar evolution and development.* Chicago: American Medical Association, 1969.

Nieuwenhuys, R. The morphogenesis and the general structure of the Actinopterygian forebrain. *Acta Morphologica Néerlando-Scandinavica,* 1962a, **5,** 65–78.

Nieuwenhuys, R. Trends in the evolution of the Actinopterygian forebrain. *Journal of Morphology,* 1962b, **111,** 69–88.

Nieuwenhuys, R. The interpretation of the cell masses in the teleostean forebrain. In R. Hassler & H. Stephan, Ed., *Evolution of the forebrain.* Stuttgart: Georg Thieme Verlag, 1966. Pp. 32–39.

Nieuwenhuys, R. Sensory Mechanisms. In Y. Zotterman (Ed.), *Progress in brain research,* Vol. 23. Amsterdam: Elsevier, 1967, p. 1.

Nieuwenhuys, R., & Cornelisz, M. Ascending projections from the spinal cord in the axolotl (*Ambystoma mexicanum*). *Anatomical Record,* 1971, **169,** 388.

Nieuwenhuys, R., & Nicholson, C. Aspects of the histology of the cerebellum of mormyrid fishes. In R. Llinás (Ed.), *Neurobiology of cerebellar evolution and development.* Chicago: American Medical Association, 1969, pp. 135–169. (a)

Nieuwenhuys, R., & Nicholson, C. A survey of the general morphology, the fiber connections, and the possible functional significance of the giganto-cerebellum of mormyrid fishes. In R. Llinás (Ed.), *Neurobiology of cerebellar evolution and development.* Chicago: American Medical Association, 1969, pp. 107–134. (b)

Northcutt, R. G. Pallial projections of sciatic, ulnar and trigeminal afferents in a frog (*Rana catesbeiana*). *Anatomical Record,* 1970, **166,** 356.

Northcutt, R. G. Afferent projections of the telencephalon of the bullfrog (*Rana catesbeiana*). *Anatomical Record,* 1972, **172,** 374.

Northcutt, R. G., & Ebbesson, S. O. E. Ascending spinal projections in the lamprey. Manuscript in preparation.

Northcutt, R. G., & Przybylski, R. J. Retinal projections in the lamprey, *Petromyzon marinus. Anatomical Record*, 1973, **175**, 400.

Peterson, R. H. Tactile responses of the goldfish (*Carassius auratus L.*) facial lobe. *Copeia*, 1972, **4**, 816–819.

Ramón-Moliner, E. The leptodendritic neuron: its distribution and significance. *Annals of the New York Academy of Science*, 1969, **167**, 65–70.

Ramón-Moliner, E., & Nauta, W. J. H. The isodendritic core of the brain stem. *Journal of Comparative Neurology*, 1966, **126**, 311–335.

Ramón y Cajal, P. *Investigaciones micrograficas en el encefalo de los batracios y reptiles. Cuerpos geniculodos y tuberculos cuadrigenios de los mamiferos.* La Dereche, Zaragoza, 1894. 88 pp.

Ramón y Cajal, P. Neuvo estudio del encefalo de los reptiles. *Trabajos Lab. Investigaciones Biologicas*, 1918, **16**, 308–333.

Ramón y Cajal, P. El cerebro de los batracios. *Libro en Honor de S. Ramon y Cajal*, 1922, **1**, 13–150.

Riss, W., Knapp, H. D., & Scalia, F. Optic pathways in *Cryptobranchus alleghiniensis* as revealed by the Nauta technique. *Journal of Comparative Neurology*, 1963, **121**, 31–43.

Romer, A. *Vertebrate Paleontology.* Chicago: The University of Chicago Press, 1966.

Roth, R. L. Optic tract projections in representatives of two fresh-water teleost families. *Anatomical Record*, 1969, **163**, 253–254.

Rovainen, C. M. Physiological and anatomical studies on large neurons of central nervous system of the sea lamprey *(Petromyzon marinus).* I. Müller and Mauthner cells. *Journal of Neurophysiology*, 1967, **30**, 1000–1023. (a)

Rovainen, C. M. Physiological and anatomical studies on large neurons of central nervous system of the sea lamprey (*Petromyzon marinus*). II. Dorsal cells and giant interneurons. *Journal of Neurophysiology*, 1967, **30**, 1024–1042. (b)

Royce, G. J., & Northcutt, R. G. Olfactory bulb projections in the tiger salamander (*Ambystoma tigrinum*) and the bullfrog (*Rana catesbeiana*). *Anatomical Record*, 1969, **163**, 254.

Rubinson, K. Projections of the tectum opticum of the frog. *Brain, Behavior & Evolution*, 1968, **1**, 529–561.

Scalia, F., & Ebbesson, S. O. E. The central projections of the olfactory bulb in a teleost (*Gymnothorax funebris*). *Brain, Behavior & Evolution*, 1971, **4**, 376–399.

Scalia, F., Knapp, H., Halpern, M., & Riss, W. New observations on the retinal projection in the frog. *Brain, Behavior & Evolution*, 1968, **1**, 324–353.

Schaeffer, B. The origin and basic radiation of the Osteichthyes. In T. Ørvig (Ed.), *Current problems of lower vertebrate phylogeny.* Stockholm: Almqvist and Wiksells, 1968.

Schaeffer, B. Adaptive radiation of the fishes and the fish-amphibian transition. *Annals of the New York Academy of Science*, 1969, **167**, 5–17.

Schaeffer, B., & Rosen, D. E. Major adaptive levels in the evolution of the actinopterygian feeding mechanism. *American Zoologist*, 1961, **1**, 187.

Schroeder, D. M., & Ebbesson, S. O. E. The organization of thalamic nuclei and their efferent pathways in the nurse shark, *Ginglymostoma cirratum* (Bonnaterre). Society for Neuroscience, First Annual Meeting, 1971.

Sharma, S. C. The retinal projections in the goldfish. An experimental study. *Brain Research*, 1972, **39**, 213–223.

Simpson, G. G. *The Major Features of Evolution.* New York: Columbia University Press, 1953.

Stensiö, E. A. The cyclostomes with special reference to the diphletic origin of the Petromyzontida and Myxinoidea. In. T. Ørvig (Ed.), *Current problems of lower vertebrate phylogeny*. Stockholm: Almqvist and Wiksells, 1968.

Vanegas, H. & Ebbesson, S. O. E. Retinal projections in the perch-like teleost, *Eugerres plumieri. Journal of Comparative Neurology*, 1973, **151**, 331–358.

Vesselkin, N. P., Agayan, A. L., & Nomokonova, L. M. A study of thalamo-telencephalic afferent systems in frogs. *Brain, Behavior & Evolution,* 1971, **4**, 295–306.

8

The Forebrain of
Reptiles and Mammals

Ford F. Ebner

Brown University

The most striking differences between anamniote and amniote nervous systems are related to their forebrain organization. Although these differences have been described many times in general terms, exactly what is different is still a matter of debate. Early attempts to formulate the general principles underlying forebrain evolution have lacked the detailed results that have been produced during the last 10 years. As a result a number of tenets of comparative neurology that had been generally accepted only 20 years ago have had to be revised because of new information acquired during the last few years. One such tenet is that most of the forebrain is exclusively olfactory in nonmammalian vertebrates. In recent years the discovery that limbic system structures in mammals are polysensory and not exclusively olfactory has diminished the apparent extent of the "rhinencephalon" in this vertebrate class. At the same time the unexpectedly restricted distribution of olfactory bulb projections in the forebrain of nonmammalian vertebrates studied to date suggests that the olfactory portion of the forebrain includes only a fraction of the areas traditionally related to olfaction. The expected corollary of this is that nonolfactory sensory modalities preempt major portions of the forebrain in all vertebrates, and indeed, recent studies have demonstrated, for example, that both the retina and the tectum project via the thalamus to the forebrain in representatives from all classes of vertebrates. This finding, together with the demonstration of direct thalamocortical projections in reptiles, has lead to renewed interest in the possibility that homologies can be established between some parts of the reptilian cortex and some parts of mammalian neocortex. Such homologies, when firmly established, aid in reconstructing the characteristics of common ancestors and lead to more accurate general models of forebrain organization.

The purpose of this chapter is to discuss some of the issues related to the organization of thalamus and cortex that bear on the topic of homologies between structures in reptilian and mammalian forebrains.

The patterns of afferent and efferent connections of the thalamus, especially thalamotelencephalic projections, have been pivotal in the quest to establish homologies between reptilian and mammalian forebrains. LeGros Clark's early paper on the structure and connections of the thalamus (LeGros Clark, 1932) identified many cytoarchitectural subdivisions of the sensory thalamus and raised many questions about homologous thalamic cell groups, that is, in comparisons among mammals showing various grades of thalamic differentiation, and between reptiles and mammals. Rose and Woolsey (1949) have also compared the brains of a number of mammalian species and have emphasized several important organizational features that vary in different species with the degree of differentiation of the forebrain. One of their general conclusions is that only nuclei in the dorsal thalamus (with the exception of the reticular nucleus of the ventral thalamus) show degenerative cell changes following removal of the entire endbrain. Not all nuclei in the dorsal thalamus are altered, however, by removal of the cortex alone. In particular, the results derived from retrograde cell degeneration studies in many mammalian species support the general conclusion that midline and intralaminar nuclei project to the basal regions of the telencephalon but not directly to the cortex. Among those dorsal thalamic cell groups that have been associated with neocortex, the retrograde cell changes suggest an orderly, point-to-point spatial relationship between thalamus and cortex which is still accepted as a general characteristic of mammalian thalamocortical projections.

Several major modifications of these conclusions were subsequently introduced by Rose and Woolsey (1958) because in some cases retrograde cell responses were too subtle and complex to be explained by a simple one-to-one correspondence of adjacent thalamic nuclei onto cortical areas. These results led Rose and Woolsey to propose the concept of two different types of thalamocortical projections: essential projections and sustaining projections.

Essential thalamocortical projections, as defined operationally by Rose and Woolsey (1958), can be identified by the great severity and extreme localization of the thalamic cell responses following even very small cortical lesions. These responses are consistent with the type of response normally described in studies of retrograde cell changes.

The concept of sustaining thalamocortical projections derived from the finding that a lesion in the primary auditory cortex (AI) in the cat would produce a severe chromatolytic change in a restricted group of cells in the principle medial geniculate nucleus, plus less severe changes in a second thalamic nucleus, the magnocellular medial geniculate nucleus. Only very large lesions of the auditory cortex that involved several different cytoarchitectural areas would be adequate

to produce extensive retrograde cell changes in the magnocellular medial geniculate nucleus. This was an important observation because it implied that more
than one thalamic nucleus could project in an overlapping manner to a single
cortical area, and that one thalamic nucleus may project to more than one
cortical area.

This concept led to the question of whether the cortex in various species varies,
in the relative amount of sustaining projections it receives from the thalamus.
For example, from Herrick's generalization that specific systems evolve out of
the nonspecific neuropil (Herrick, 1934, 1948), it can be predicted that some
species of primitive mammals may show a type of cortical organization in which
all nuclei in the dorsal thalamus send widespread sustaining projections to the
cortex. In an effort to test this prediction, Diamond and Utley (1963) and Hall
and Diamond (1968) have studied the oppossum and hedgehog, respectively, to
see the extent to which essential and sustaining thalamocortical projections exist
in these species. When it is remembered that these types of projection are
operational definitions for the extent and severity of cell changes in the thalamus following cortical removal, the retrograde cell degeneration results provide
good evidence for the interpretation that thalamic nuclei, including cells in the
lateral geniculate nucleus of these primitive mammalian forebrains, respond as if
they give rise only to the sustaining type of thalamocortical projection. It is
important to this interpretation, however, to note that in these studies the
definition of a sustaining projection has been widened to include the possibility
that a single thalamic cell group can project to different parts of a single
cytoarchitectural area and still be considered a sustaining projection. This is an
important modification of the original definition because when the presumed
collaterals can distribute in the same area (e.g., striate cortex) the criteria for
identifying sustaining projections overlap with the definition for rather widespread essential projections. The severity of retrograde cell changes resulting
from a cortical lesion of a given size certainly varies widely among mammalian
species, and it is important to note that Diamond and his colleagues have
compared the severity of degenerative changes following cortical lesions in the
hedgehog and oppossum with similar results in species with a highly differentiated visual cortex, such as the tree shrew and gray squirrel (Diamond, Snyder,
Killackey, Jane, & Hall, 1970). In these latter species, the border between
normal and degenerated zones of retrograde cell changes is extremely sharp; it
can therefore be concluded, at the very least, that the hedgehog and oppossum
react in a qualitatively different manner in these comparisons.

The results just described, which are all based on the method of retrograde cell
degeneration, indicate that thalamic nuclei differ in their mode of projection
onto the cortex. However, on the basis of this technique alone it is not possible
to convert the useful, but necessarily operational, definition of essential and
sustaining thalamocortical fiber systems into a more direct concept based on the

morphology of the axons and the actual location of the terminations of each type of thalamocortical fiber system.

Anterograde axonal degeneration has been traced to Layer IV (plus adjacent Layer III) of at least the first sensory area in all sensory modalities of all mammals studied to date, where the thalamic fibers repeatedly branch and terminate in a restricted area with very little lateral dispersion (Benevento & Ebner, 1970c; Colonnier & Rossignol, 1969; Cragg, 1969; Ebner, 1967; Garey, 1970; Garey & Powell, 1967, 1971; Glickstein, King, Miller, & Berkley, 1967; Hall, 1972; Hand & Morrison, 1970, 1972; Harting, *et al.* 1973a; Hubel & Wiesel, 1969, 1972; Jones, 1967; Jones & Powell, 1969, 1970a, b; Killackey, 1973; Killackey & Ebner, 1972, 1973; Lund, 1973; Nauta, 1954; Petrovcky & Druga, 1972; Polley & Dirkes, 1963; Rossignol & Colonnier, 1971; Tobias & Ebner, 1973; Wilson & Cragg, 1967, 1969). This terminal arborization mimics the appearance of specific thalamic afferent fibers that have been drawn by Lorente de No (1938) from Golgi preparations of mouse parietal cortex. Almost without exception, these studies show a few fibers that continue through Layer IV to Layer I in the region of the main projection field. This type of thalamocortical projection shows a relatively high degree of topologic localization in all mammals. The "Layer IV" projection system is therefore the most probable system of thalamocortical fibers to correspond to the "essential" thalamocortical projections.

Whether there is a second Layer IV system that provides the basis for sustaining thalamocortical projections, or whether, on the other hand, sustaining systems project to different cortical layers has been more difficult to establish. There is, however, evidence that structures in the thalamus do project to additional cortical layers.

The strongest single line of evidence for a system to layers other than Layer IV arises from looking at the total pattern of thalamic fiber terminals in cortex following the removal of all thalamic nuclei. Such experiments lead to the strong conclusion that there is a thalamocortical fiber system that terminates densely in Layer I throughout all areas of the neocortex (Ebner, 1967a, b, 1969). Such experiments to date, however, have two major limitations. One is that they have been done mainly in primitive, nonprogressive mammalian brains with an extraordinarily well-developed Layer I and hence the conclusion that a well-developed Layer I thalamocortical fiber system is present in all mammals may be incorrect. The other is that large lesions of the thalamus do not specify which nuclei give rise to any of the thalamocortical projections, and the Layer I projections may arise from specific thalamic nuclei.

The analysis of the influence of intralaminar nuclei on cortex by Jasper and his colleagues over 20 years ago (see, e.g., Hanberry *et al.*, 1954) led them to postulate that a system of cells project to Layer I of cortex in a parallel and independent fashion from the projections of the specific thalamocortical projections. Considerable recent evidence supports the idea that some intralaminar nuclei do project directly to the cortex in addition to the basal telencephalic nuclei. Murray (1966) has described changes in the cell bodies of neurons constituting the cat's intralaminar nuclei following large cortical removals and long periods of survival. These results suggest that the axons of intralaminar cells project to wide areas of the cortex, although some rostral to caudal topography has been detectable in the organization of the fiber projections (i.e., in the location of the cells showing degeneration). Similar retrograde cell changes have also been reported in other mammals (for example, in *cat*, Waller & Barris, 1937, Powell & Cowan, 1967; in *monkey*, Walker, 1938; Peacock & Combs, 1965; and others). Further support for the concept of direct projections from intralaminar nuclei to the cortex has come from studies of the distribution of intralaminar cell axons. Fibers from intralaminar cells have been traced to the orbitofrontal cortex, for example, in rapid Golgi preparations in the mouse (Scheibel & Scheibel, 1967). Similar projections from the intralaminar nuclei to orbitofrontal cortex have been demonstrated in the cat by experimental techniques (Nauta & Whitlock, 1954), and more recent studies of thalamocortical connections using similar anterograde degeneration techniques in the opossum and hedgehog have provided results that clearly corroborate the concept of direct projections from intralaminar nuclei to all areas in the neocortex (for example, see Killackey & Ebner, 1973). These anterograde degeneration studies also indicate that the axons arising from cells in intralaminar nuclei terminate in a different combination of cortical layers than the Layer IV system, namely in all Layers from VI to I. Intralaminar thalamocortical projections can therefore be identified in axonal degeneration studies by the fact that they show a widespread distribution in the cortex and by the layers of termination of intralaminar cell axons in the cortex. The intralaminar nucleus projections are always associated with a localized terminal field in the basal telencephalon but this may not be unique to projections from intralaminar nuclei. Unfortunately, intralaminar nucleus projections to the cortex have not been demonstrated directly in mammals with more differentiated sensory cortex, such as the cat and monkey, although physiological results have presented strong arguments for the existence of such projections (for example, Hanbery & Jasper, 1953; Hanbery *et al.*, 1954; Jasper, Naquet, & King, 1955).

To summarize then, although two types of thalamocortical projections can be recognized on the basis of differences in the severity of degenerative changes in thalamic neurons which follow restricted cortical lesions, it is not clearly established how these types relate to the morphology and distribution of axonal arborizations revealed by Golgi and anterograde degeneration techniques. In some cases, essential and sustaining thalamocortical projections may correspond to the specific and nonspecific afferent fiber types described originally by Lorente de No. The intralaminar nuclei may be one example of a sustaining system that is constituted of nonspecific afferent fibers which project to widespread cortical areas. On the other hand, the correlation is not perfect, since several instances are known of sustaining projections which project as "specific afferents" to Layer IV and the adjacent part of Layer III. The connectional organization of the latter fiber system is unknown.

THALAMOTELENCEPHALIC PROJECTIONS IN REPTILES

There has been a considerable increase in knowledge of the connections of the reptilian forebrain since Huber and Crosby (1926), LeGross Clark (1932), and Ingvar (1923) postulated that the reptilian nucleus rotundus was homologous to the ventral group of nuclei of the mammalian thalamus based on its topography and cytoarchitecture. New information has shifted the focus from somatic sensory to visual interconnections for the nucleus rotundus, but many of the nagging questions of homologies between structures in mammalian and reptilian telencephali still remain.

The gross subdivisions of the reptilian diencephalon and telenceophalon are similar to those of mammals, and although different names have been given to equivalent thalamic nuclei in turtles, lizards, snakes, and crocodilians, similar nuclei can be identified in Nissl-stained sections of almost all reptilian thalami.

The study of reptilian thalamocortical projection systems is complicated by the fact that no existing reptile brain has a neocortex, *in sensu strictu*; that is, a well-differentiated "six"-layered telencephalic cell array that receives an input from the dorsal thalamus. The main question about reptiles has been where do the thalamotelencephalic projection fibers terminate, and how do these sites of termination correspond to the layers of termination in mammalian neocortex? Whatever the basis for the difficulty in interpreting subtle retrograde cell changes in primitive mammals, the problem is even more difficult in reptiles, in which no detectable retrograde cell change occurs following removal of only dorsal (general) cortex in several different reptilian species (Kruger & Berkowitz, 1960). Thalamic cell changes occur only in the nucleus rotundus and in the dorsal medial anterior nucleus following very extensive telencephalic lesions that involve the dorsal cortex, the dorsal ventricular ridge, and the basal telencephalic nuclei. However, anterograde axonal degeneration can be traced to all three regions following large thalamic lesions in the turtle, *Pseudemys* (Hall & Ebner, 1970b), and the lizard, *Iguana* (Butler, 1972). At the same time, there are several striking differences between the total pattern of thalamic projections in the two reptilian species that have been studied.

These differences suggest that reptiles, like mammals, may show many grades of organization and that differentiation in reptiles may have evolved in a different direction from mammals. For example, in the general cortex alone there are marked differences in thalamocortical fiber patterns; the thalamic fibers enter from the rostral and lateral edge of the cortex in *Pseudemys*, whereas they enter the cortex from a posteromedial direction in *Iguana*. The thalamocortical projection fibers in Pseudemys are fine caliber and terminate exclusively in the outer 100 μm of the cortex among the apical dendrites of the underlying cell layer. In *Iguana* the majority of the thalamic fibers in Layer I are also small, but there is an additional large fiber component. In addition, *Iguana* shows a definite terminal field among the cortical cell bodies.

In order to establish probable homologies between thalamic and telencephalic cell groups in reptiles, it is necessary to examine the characteristics of individual thalamic nuclei. It is well known that two thalamic nuclei in reptiles receive an input from components of the visual system; the dorsal lateral geniculate nucleus (labeled with various names in different groups) which receives a major input from the retina (Armstrong, 1950, 1951; Burns & Goodman, 1967; Butler & Northcutt, 1971a; Hall & Ebner, 1970a; Knapp & Kang, 1968a, b; Northcutt & Butler, 1973), and the nucleus rotundus, which receives a major input from the optic tectum (Braford, 1972; Butler & Northcutt, 1971b; Hall & Ebner, 1970a; Pritz, 1973). Of these two nuclei only the lateral geniculate nucleus projects to general cortex, and these fibers have few, if any, collaterals in the thalamus, basal nuclei, or dorsal ventricular ridge (Hall & Ebner, 1970b). In contrast, the nucleus rotundus projects to a portion of the dorsal ventricular ridge, with some terminal collaterals in the basal nuclei (Hall & Ebner, 1970b; Pritz, 1973). Unfortunately, the visual projections have not been studied in detail in reptiles, such as *Iguana*, that show several components in the total thalamocortical projection to general cortex. It is therefore not possible to predict the range of variation that can be expected in different reptiles in the manner of termination of, for example, the lateral geniculate nucleus fibers.

The auditory pathways show considerable specialization in some reptiles. In the crocodilian *caiman*, for example, the thalamic nucleus that receives the ascending auditory projections, the nucleus reuniens (Pritz, 1972), is an expanded midline cell group that in turn projects to the dorsal ventricular ridge but not to the general cortex. This is consistent with reports of physiological studies in which visual and somatic sensory but not auditory stimuli can evoke responses in general cortex (Belekhova & Kosareva, 1971; Kurger & Berkowitz, 1960; Orrego, 1961). In contrast, cells in the dorsal ventricular ridge do respond to click stimuli, and it is this area that receives the auditory projections (Pritz, 1972). The evoked responses to somatic sensory stimulation suggest that somatic sensory information has a cortical representation in reptiles, but neither the thalamic relay nucleus nor its manner of projection onto the cortex have been described anatomically (see Belekhova & Kosareva, 1971).

COMPARISON OF TRENDS IN DIFFERENTIATION OF THALAMOCORTICAL
CONNECTIONS IN REPTILES AND MAMMALS

The fine caliber fibers and superficial position of the thalamic axon terminals in Layer I of the turtle cortex might indicate that this fiber system is equivalent to the mammalian projection system from the intralaminar nuclei to the outer half of Layer I. However, turtle thalamocortical fibers arise in large part from the thalamic visual relay nucleus, the dorsal lateral geniculate nucleus. As previously mentioned, the lizard *Iguana* appears to have several components to the thalamocortical fiber system, with large- and small-caliber fibers in Layer I and what appears to be a third system of thalamic inputs that terminates among the cortical cell bodies. Cortical differentiation in reptiles therefore appears to take the form of increasing the number and complexity of inputs to a system of cortical cells that appears rather stable in number and intracortical complexity. Unfortunately, not enough is known of the connections of the thalamus and cortex to make a detailed comparison among groups of reptiles. It is not possible, therefore, to extract the features of these systems that are common to all reptiles.

Cortical differentiation in mammals takes quite a different form; namely, the number of thalamic inputs to particular cortical areas appears to decrease and the number of cortical cell layers and areas increases. For example, the striate cortex is relatively undifferentiated in both opossum and hedgehog and both visual systems have three inputs from three separate thalamic nuclei (Benevento & Ebner, 1970c; Hall, 1971). One input to the cortex is from the lateral geniculate nucleus and these fibers terminate in Layer IV and adjacent Layer III, with very few fibers that continue out to Layer I. A second input is from the lateral nuclear group of the thalamus and although these fibers project outside of striate cortex, they also terminate in Layer VI and some in Layer I, of the striate cortex proper. A third input is from the intralaminar nuclei, and these fibers terminate sparsely over a wide area of the visual cortex, with terminals in all layers but characteristically ramify as very fine fibers in the outer part of Layer I. In contrast to this rather complicated multiple input from the thalamus to undifferentiated striate cortex, more highly specialized visual cortex receives fibers from one nucleus, the lateral geniculate, but apparently none from a second source such as the lateral nuclear group. No direct projections from intralaminar nuclei to the striate cortex in highly specialized systems have ever been conclusively demonstrated, although some indirect evidence predicts that they exist.

This difference in mode of differentiation of cortex with more thalamic inputs to a stable cellular arrangement in reptiles, on the one hand, and toward fewer thalamic inputs to an increasingly complex cortical analyzer on the other, has not been emphasized in the past. More data are obviously needed to support this supposition as a general case for either reptiles or mammals. However, in the present context it is interesting to speculate on the physiological limitations and advantages that each type of differentiation may impose on each group. One supposition is that what a reptile's eye tells its forebrain is more like what a frog's eye does than what a cat's eye does. That is, if visual information is already complexly integrated at the retinal level in reptiles it may not be possible to recombine elementary units of visual information at the cortical level. This produces a system in which increasing the number of different inputs becomes the way to increase the integrative capacity of that cortex. In mammals, the trend is to project an increasingly resolved, but minimally integrated, representation of the visual world onto the striate cortex and to increase the integrative capacity of cortical mechanisms. What the selective advantage of these different types of specialization is for individual species in each class remains to be studied.

HOMOLOGIES IN THALAMOCORTICAL PROJECTIONS SYSTEMS IN REPTILES AND MAMMALS

Recognizing that both reptiles and mammals show thalamic inputs to a visual area in the cortex raises the question of which systems are homologous in each group. The mammalian dorsal lateral geniculate nucleus is the thalamic nucleus that receives direct input from the retina and in turn projects to visual Area I without evidence of collateral branching or subcortical terminations. These characteristics, plus the absence of *direct* inputs from the mesencephalic reticular formation, suggest that the mammalian geniculocortical projections share more features with the reptilian geniculocortical pathway than does any other mammalian projection system. One obvious difference between these two systems is that the lateral geniculate projection in adult mammals is mainly a Layer IV system (see Laemle, Benhamida, & Purpura, 1972 for changes in layers with age), whereas in reptiles the same projection terminates mainly in Layer I. In order to discuss this apparent discrepency, it is necessary to look more closely at the concept of cortical layers.

In Nissl-stained sections of normal brains, one of the most striking differences between reptilian general cortex and mammalian neocortex is the difference in the number of cells constituting each structure. Reptilian cortex is relatively thin, only about .5 mm thick, and in most areas contains only a single prominant row of neuronal cell bodies located at varying distances from the underlying ependymal cells that line the lateral ventricle. There is a considerable number of neuronal cell bodies above and below the main tier, but these scattered neurons are rarely dense enough to look like or be counted as a cellular lamina in Nissl preparations.

Mammalian neocortex, in contrast, is 1–3 mm thick, and even in the most undifferentiated mammalian brains a minimum of four to five cellular laminae can be distinguished. In highly differentiated cortex, eight or nine laminae are commonly present even though they are conventionally numbered so that every mammal ends up with six layers in most areas. One reason that this degree of difference has not been given equal weight in comparative neurology is that the difference between reptiles and primitive mammals has seemed a qualitative difference, whereas that between primitive and differentiated mammals has seemed a quantitative difference. This is one historical reason that information about the cortex of any mammal has been accepted as contributing to the understanding of all "mammalian neocortex," whereas studies on reptiles, especially of forebrain organization, have been thought to contribute little, if anything, to an understanding of the more layered and complex mammalian neocortex.

The five or more perikarial layers of neocortex can be conveniently thought of in terms of three organizational subzones; the granule cell Layer IV and the laminae above and below, which are the supragranular and infragranular layers, respectively. In mammals, the geniculocortical fibers terminate on populations of "stellate" or "granular" neurons and the portions of infragranular cell apical dendrites that are located in Layer IV.

Thalamocortical fibers in reptiles, such as turtles, terminate just under the cortical surface where only the apical dendritic spines and shafts of the single lamina of cortical neurons form the only obvious source of postsynaptic membranes.

There is a striking difference between Layer IV of the opossum or hedgehog striate cortex and that of, for example, a *Macaca* monkey. All primates have large numbers of densely packed, very small neurons in the fourth layer where the thalamic fibers terminate. Most of these cells are short-axon or Golgi type II cells, the axon and dendrites of which are restricted to a very local area. If these small "granule" neurons are thought of as "local modulator units" (LMUs) that constitute one of two main targets of the thalamic input and that modulate the discharge rate of other neurons in the vicinity, then it can be seen that the trend is from apparently no LMUs in the reptile, to relatively few in primitive mammals, to very many LMUs in differentiated neocortex wherever it is found, not just in primates. This condition was noted by Ramón y Cajal, who said that "the increase in small cells is the anatomical reflection of delicacy of function of the human brain," and, it may be added, of all highly differentiated cortex.

However, the conventional interpretation requires an important assumption be accepted; that the local modulator unit is a seperate cell, with inputs to dendrites, outputs via an axon, and metabolic machinery in the immediate area. Yet considerable evidence has accumulated in the last few years that dendrites of some neurons possess the membrane specializations characteristic of both inputs and outputs which are far from the cell body and which potentially operate by local potentials rather than through spikes down an axon. Systems that may operate in this fashion are the anaxonal amacrine cells of the retina (Kidd, 1962); Mitral cell dentrites in the olfactory bulb (Hinds, 1970; Rall, Shepherd, Reese, & Brightman, 1966); and Golgi II cells in the lateral geniculate (Famiglietti & Peters, 1972), medial geniculate (Morest, 1971), and ventrobasal nuclei (Ralston & Herman, 1969), to mention just a few of the possible examples.

In reptilian cortex, there are no apparent alternative candidates for the thalamic fiber terminations except the dendrites of the underlying cell bodies, the axons of which constitute the output fibers from the cortex. However, there are two cell types in the molecular layer of reptilian cortex that can receive a portion of the thalamic fibers; the scattered neurons with horizontally coursing dendrites and the processes of ependymal cells that run from the ventricle perpendicularly to the pial surface even in adult reptiles (and amphibians and some fishes, but not adult mammals). The ependymal cells may possibly be involved in the synaptology of the reptile cortex, because neuroependymal contacts exist in the turtle cortex that are indistinguishable from synaptic contacts between neurons (Ebner & Colonnier, 1973, 1975). It is not yet known whether thalamic fibers are among those axons terminating on the ependymal cells, but if this can be demonstrated, then the elaborate ependymal cells of nonmammalian vertebrates may function in part as an extensive population of local modulator units. Herrick (1948) has suggested that the elaborate ependymal cells in adult amphibian brains may act like neurons in some locations in the amphibian CNS (based on the appearance of the cells in Golgi preparations).

In recent Golgi studies of the embyological development of the opossum cortex, developing cells have been described that are very similar to ependymal cells in adult reptiles; that is, they have a cell body near the ventricle and a radiating process that extends to the outer limiting layer (Morest, 1970). However, in the opossum and all other mammals these cells migrate out to the pial surface during ontogenetic development, the deep process retracts from the inner limiting layer (at the ventricle), and the cell subsequently differentiates into a separate neuron (Berry & Rogers, 1965; Morest, 1970; Rakic, 1972). Autoradiographic studies have shown that the first wave of nuclei forms the deepest layer of neocortex (Layer VI). Successive waves of migrating nuclei form Layers V, IV, III, and II in that order (Angevine & Sidman, 1961; Berry & Rogers, 1965; Hicks & D'Amato, 1968; Shimada & Langman, 1970). Unfortunately, the number and types of genetic regulators of this process are not well understood, but there remains the long-standing morphological observation that all reptiles have only one predominant cell layer and all mammals have at least five cell layers. There are few intermediate stages, no good transition forms, and little sign of gradual gradients. Likewise, there are no neuropathological conditions or congenital anomalies in mammals in which two, or three, or four well-developed laminae accur throughout an otherwise normal neocortical mantle. However, there are single-gene mutations in mice where the disturbance in the neocortex consists of nearly normal numbers of cortical neurons, but they differentiate without any appearance of lamination throughout the neocortex (Sidman, 1968). It is possible that a very small number of mutations, for example, a change in a regulatory or timing gene that stops migration after a certain number of layers are formed, can produce a germinal matrix in which the cells do not stop until five waves of migrating cells have moved out from the germinal layer and have differentiated into neurons. If the pressure in cortical differentiation is always to select for systems with more modulator units, then mutations with this effect have very likely already been present in the animals that have survived in the niches occupied by reptilelike mammals.

These results from developmental biology speak to another question at the cellular level that has occupied comparative neurologists for decades; namely, which layer, if any, of the mammalian neocortex is homologous to the single-cell layer in the reptile cortex? The above sequence of cortical development might support the idea that if only one layer had developed in the mammalian cortex, it would have been the deeper Layers VI and V, because Layers IV, III, and II developed in sequence and at a later time. Recent cytoarchitectural studies, however, support the idea that Layer II shows the greatest similarity in cell type to reptilian neurons and is therefore also a possible homolog (Sanides & Sanides, 1972).

The connections of the mammalian cortex strongly support Layer V–VI (infragranular layers) as the cell laminae homologous to that of reptilian cortex. These cells appear to receive thalamic input, for example, from the lateral geniculate nucleus, in part directly on their dendritic spines, and in part indirectly through local modulator units or stellate cells in Layer IV. The lamina of reptilian cortical neurons and mammalian Lamina V–VI cells both project back to the brainstem. From the point of view of circuitry rather than cell migration into laminae or cytoarchitecture, therefore, both reptiles and mammals may show the organization of mammalian Layers VI–V and IV, whereas it appears that only mammals display the connections of Layer III–II. Recently it has become clear that the neurons in Layers II and III of the neocortex mainly interconnect various areas of the cortex within layers of the same area they occupy, and between areas of the same sensory modality. That is to say, Layers II and III of a sensory area, such as the striate cortex, project to the underlying Layer V cells in the striate cortex and to at least Layers II and III of at least one other visual area (Area 18), and in some mammals also to further visual areas of the cortex (Benevento & Ebner, 1970b; Butler & Jane, 1973; Garey, Jones, & Powell, 1968; Spatz, Tigges, & Tigges, 1970). Reptiles, in contrast, have very limited corticocortical connections, and these appear to be mostly collateral processes of the same cells that project back to the brainstem. However, studies in several different reptilian species indicate that there are few interhemisperic connections between the dorsally located general cortex in reptiles, even though such connections exist between regions of the archicortex and more basal telencephalic areas (Hall, 1971; Lohman & Mentink, 1972; Voneida & Ebbesson, 1969). Current ideas about the details of reptilian telencephalic connections are sure to change considerably in the next few years, but no reptile is likely to be described that shows the complexity of corticocortical connections that is the hallmark of the supragranular layers of the mammalian neocortex. Therefore, the most parsimonious working hypothesis at the present time is that the neurons in the reptile cortex are homologous to the neurons located in the infragranular layers (V and VI) in the mammalian neocortex, and that the geniculocortical projection systems in reptiles and mammals are homologous, despite the differences in the complexity of the cortical system they project onto in each group.

The second visual pathway to the telencephalon is less well established in reptiles, but the parallel connections suggest that the thalamic cells receiving a major input from the optic tectum and projecting to the telencephalon are homologous in the two classes of vertebrates. These thalamic cells constitute some or all of the lateral posterior nucleus (pulvinar) in mammals and some or all of the nucleus rotundus in reptiles. Both systems show terminations in the basal telencephalic nuclei, as mentioned above, and in a further visual area in the telencephalon. In studies of reptiles to date, the main rotundal projection has been to the core nucleus of the dorsal ventricular ridge, in a region spatially quite separate from the geniculocortical terminations. In the hedgehog, the cortical target of the lateral posterior nucleus is strongly to Layer VI of striate cortex. Strict interpretation by this connectional criterion alone leads to the position that the rotundal-recipient dorsal ventricular ridge is homologous to Layer VI of the striate cortex in the hedgehog. However, in other mammals the tectal recipient lateral posterior (pulvinar) nucleus projects to VA II and the more distal cortex outside of the striate (Cluner & Campos-Ortego, 1969; Glendenning, Hall, & Hall, 1972; Graybiel, 1972; Harting et al., 1973a, b). This has led to the more commonly held opinion that the visual part of the dorsal ventricular ridge is homologous to extra striate visual areas. Much more evidence is needed before either of these interpretations, or other alternative conclusions, can be seriously entertained.

The remaining thalamocortical projection system found in mammals is even more of a puzzle. What nuclei and projections in reptiles are homologous to the intralaminar nuclei of mammals? This is really two questions; first, are there reptilian thalamic nuclei that show the cellular organization and connections of mammalian intralaminar nuclei? Second, is there a reptilian Layer I system equivalent to that of mammals?

In line with Herrick's theory of brain specialization, it is logical a priori that the thalamus of the common ancestors of reptiles and mammals contain only a small number of ill-defined and diffusely organized nuclei that represent the rostral part of what is now called the "reticular formation" or "isodendritic core" of the brainstem. These dorsal thalamic cells would receive inputs from several sensory modalities and would project with extensive overlapping to wide areas of the telencephalon. Mammalian intralaminar nuclei appear as a specialized version of such an ancestral dorsal thalamus, in the sense that there is convergence of several modality-related structures onto one thalamic subarea and that this subarea projects to all of the cortex that is functionally related to that sensory modality, plus some part of the basal nuclei.

It is not apparent, however, that the reptilian thalamus contains this type of organization. Most dorsal thalamic nuclei in existing reptiles are quite well differentiated and where afferent fiber connections have been studied with experimental techniques, they have been found to terminate in quite localized and separate thalamic zones. There are many channels through which the ascending reticular formation can influence dorsal thalamic cells, and the most obvious route is via the midbrain tegmentum, similar to the condition found in mammals. However, no such afferents have been described to date. More detailed study in a variety of reptiles may lead to the identification of cell groups that satisfy the criteria that have been useful in identifying the intralaminar nuclei in mammals. At present, however, there is little evidence to support the idea that nuclei homologous to the mammalian intralaminar nuclei are well developed in the reptilian thalamus, if they exist at all. If new information does not provide strong evidence for the presence of intralaminar nuclei in reptiles, then these nuclei might be best considered as a specialization of the mammalian thalamus that was very likely not present in the common ancestors of reptiles and mammals.

THE PROBLEM OF THE COMMON ANCESTOR

It is evident from the comparisons just made between reptilian and mammalian brains that there are a number of outstanding differences between them. These differences presumably reflect, in part, the evolutionary trends followed by ancestral reptiles in their divergent specialization into the existing species of these two classes. Clues to the significance of these specializations may become apparent if an attempt is made to correlate paleontological evidence with differences in the central nervous systems of existing reptiles and mammals.

The organization of the auditory system constitutes one dramatic divergence between reptiles and mammals. In mammals, three articular jaw bones have been modified to perform complicated amplifying and damping functions in the link between the tympanic membrane and the oval window of the inner ear. In reptiles, only one ossicle has separated from the mandibular articulation and this columnella forms an unhinged link between the two membranes. Although both reptiles and mammals show some similarities in the basilar membrane organization of the cochlea, it is more extensive, coiled, and tuned to a broader frequency spectrum in mammals (see Webster, this volume). These differences in the peripheral receptor predict differences in the degree of auditory system differentiation in reptiles. In the above discussion of the central connections of the auditory and visual systems in reptiles, it has been pointed out that there is one striking difference between the two modalities; there are two ascending visual pathways from the thalamus, one to the dorsal ventricular ridge and one to the cortex, whereas there is only one known ascending auditory pathway to the dorsal ventricular ridge (Pritz, 1972). There is no known auditory input to the dorsal cortex equivalent to the reptilian lateral geniculocortical pathway or the mammalian medial geniculocortical pathway. It becomes interesting to know whether there are limitations in the processing of auditory information caused by the absence of auditory representation in the general cortex, but no physiological comparisons have been reported that have been designed to answer this question.

Another attribute suggested by paleontological evidence is that mammal-like reptiles have had small-sized bodies and that the majority have been nocturnal. Mammals with primitive brains frequently possess nearly pure-rod retinas and are behaviorally most active at night. In contrast, the majority of existing reptiles are diurnal and have nearly pure-cone retinas. This suggests that the ancestors of present-day reptiles have been dependent on visual acuity and perhaps color in their adaptation, whereas mammal-like reptiles have selected for movement detection at low light intensities. These behavioral differences suggest that if reptiles and mammals evolved from a common gene pool, then one of the earliest divergences has been development of specializations for daytime and nightime activities in subgroups of the population, with mammal-like reptiles selecting biological equipment for nightime activity. This difference in activity cycles may have some bearing on the puzzling fact, mentioned above, that mammals appear to have an internal cortical activating system through the brainstem reticular formation–intralaminar nuclei-nonspecific thalamocortical projection system. According to the argument presented in the previous sections, reptiles appear to have no comparable reticular activating system which projects to the cortex; instead, the system that appears at first glance to be wired similarly to the mammalian reticular formation at the cortical level turns out to share many characteristics with the light-activated geniculostriate system of mammals. Therefore, daylight may have been the primary stimulus to cortical activation in the reptilian common ancestors, whereas nightfall may have reset the internal activity programmer in crepuscular or nocturnal mammalian common ancestors. To fill many diverse ecological niches, mammals are then required only to reset such an internal circadian pacemaker for maximum activity at any part of the day or night. Although most reptiles (and birds) have remained diurnal there are some nocturnal forms, and it will be interesting to look for adaptations that parallel the mammalian intralaminar nuclei in these particular species.

Other marked differences in peripheral sense organs are related to the unusual development of the olfactory system in mammals, with the parallel differentiation of a soft and movable nose. Tactile sensation of the nose and vibrissae is highly developed in many so-called "macrosmatic" mammals, with a corresponding increase in the number of fibers in the trigeminal nerve. The trigeminal nerve nucleus, in turn, is usually quite differentiated and gives rise to ascending projections mainly to the thalamus and from there to the somatic sensory cortex. In reptiles (and birds) this ascending trigeminal system appears to project densely to the basal telencephalon, in a location near the olfactory cortex (Hall & Ebner, 1970; Karten, this volume). However, much more information is needed to specify what these differences reflect in the use and integration of olfactory with snout-tactile information.

A final divergence between reptiles and mammals is suggested by the paleontological interpretation that early mammals have shown considerable refinement in patterns of locomotion and presumably in control of the limbs. Quadrapedal perambulation with the belly up off the ground and with intricate control of the distal limb musculature is particularly well developed in mammals, and these motor skills are correlated with the development of a direct neural pathway from the neocortex to the spinal cord. In birds (Karten, this volume) and even sharks (Ebbesson & Northcutt, Chapter 7, this volume), however, certain telencephalic cell groups have been shown to project directly to the spinal cord, suggesting that present concepts of descending telencephalic projections in nonmammalian vertebrates may be far from complete.

These few comparisons between reptiles and mammals raise more questions about their common ancestors than they answer. The questions have perhaps become more specific and the criteria for establishing homologies slightly more rigorous, but the lack of comparable information about various groups of reptiles remains a major impediment to reconstructing this phylogenetic history. More subtle mechanisms, such as those involved in controlling body responses to changes in environmental temperatures, have led to differences in thermoregulatory mechanisms, but these more subtle differences are only beginning to be studied with newer techniques. Many questions require comparisons of the ultrastructure of neurons and single-unit responses, and very little is known about different areas of the reptilian central nervous system at these levels of analysis. The main change in perspective during the last decade has not been caused by the volume of new information available, although a steady progress is apparent, but has been because sufficient similarities have been demonstrated between the forebrains of existing reptiles and mammals to encourage the idea that homologies and therefore history may be reconstructed with considerable certainty.

ACKNOWLEDGMENTS

I wish to thank Dr. W. C. Hall for his careful reading of the manuscript and many detailed suggestions for improvements.

REFERENCES

Angevine, J. B., Jr., & Sidman, R. L. Autoradiographic study of cell histogenesis of cerebral cortex in the mouse. *Nature (London)*, 1961, **192**, 766–768.

Armstrong, J. A. An experimental study of the visual pathways in a reptile *(Lacerta vivipara)*. *Journal of Anatomy (London)*, 1950, **84**, 146–167.

Armstrong, J. A. An experimental study of the visual pathways in a snake *(Natrix natrix)*. *Journal of Anatomy (London)*, 1951, **85**, 275–288.

Belekhova, M. G. Posttetantic potentiation and recruitment in cerebral cortex of the turtle. *Neuroscience Transactions*, 1967, **2**, 204–212.

Belekhova, M. G., & Kosareva, A. A. Organization of the turtle thalamus: Visual, somatic, and tectal zones. *Brain, Behavior, & Evolution*, 1971, **4**, 337–375.

Benevento, L. A., & Ebner, F. F. The contribution of the dorsal lateral geniculate nucleus to the total pattern of thalamic terminations in striate cortex of the Virginia opossum. *Journal of Comparative Neurology*, 1970, **143**, 243–260. (a)

Benevento, L. A., & Ebner, F. F. The areas and layers of corticocortical terminations in the visual cortex of the Virginia opossum. *Journal of Comparative Neurology*, 1970, **141**, 157–190. (b)

Benevento, L. A., & Ebner, F. F. Pretectal, tectal, retinal and cortical projections to thalamic nuclei of the Virginia opossum in stereotaxic coordinates. *Brain Research*, 1970, **18**, 171–175. (c)

Berry, M., & Rogers, A. W. The migration of neuroblasts in developing cerebral cortex. *Journal of Anatomy*, 1965, **99**, 691–709.

Braford, M. R., Jr. Ascending efferent tectal projections in the South American spectacled *Caiman*. *Anatomical Record*, 1972, **172**, 275–276.

Burns, A. H., & Goodman, D. C. Retinofugal projections of *Caiman sklerops*. *Experimental Neurology*, 1967, **18**, 105–115.

Butler, A. B. Thalamotelencephalic projections in the lizard *Iguana*. *Anatomical Record*, 1972, **172**, 282.

Butler, A. B., & Jane, J. A. An ultrastructural study of interlaminar connections of rat supragranular visual cortex. *Anatomical Record*, 1973, **175**, 282.

Butler, A. B., & Northcutt, R. G. Retinal projections in *Iguana iguana* and *Anolis carolinensis*. *Brain Research*, 1971 **26**, 1–13. (a)

Butler, A. B., & Northcutt, R. G. Ascending tectal efferent projections in the lizard *Iguana iguana*. *Brain Research*, 1971, **35**, 597–600. (b)

Cluner, P. F. de V., & Campos-Ortego, J. A. The cortical projection of the pulvinar in the cat. *Journal of Comparative Neurology*, 1969, **137**, 295–308.

Colonnier, M., & Rossignol, S. On the heterogeneity of the cerebral cortex. In: H. H. Jasper, A. A. Ward, & A. Pope (Eds.), *Basic mechanisms of the epilepsies*, Boston: Little, Brown, 1969. Pp. 29–30.

Cragg, B. G. The topography of the afferent projections in the circumstriate visual cortex (CVC) of the monkey studied by the Nauta method. *Vision Research*, 1969, 9, 733–747.

Diamond, I. T., Snyder, M., Killackey, H., Jane, J., & Hall, W. C. Thalamocortical projections in the tree shrew *(Tupaia glis)*. *Journal of Comparative Neurology*, 1970, **139**, 273–306.

Diamond, I. T., & Utley, J. D. Thalamic retrograde degeneration study of sensory cortex in opossum. *Journal of Comparative Neurology*, 1963, **120**, 129–160.

Ebner, F. F. Afferent connections to neocortex in the opossum *(Didelphis virginiana)*. *Journal of Comparative Neurology*, 1967, **129**, 241–268. (a)

Ebner, F. F. Medial geniculate nucleus projections to telencephalon in opossum. *Anatomical Record*, 1967, **157**, 238–239. (b)

Ebner, F. F. A comparison of primitive forebrain organization in metatherian and eutherian mammals. *Annals of the New York Academy of Science*, 1969, **167**, 241–257.

Ebner, F. F., & Colonnier, M. Interneuronal and neuroependymal synaptic relationships in "visual" cortex of adult turtle *(Pseudemys scripta)*. *Anatomical Record*, 1973, **175**, 312.

Ebner, F. F., & Colonnier, M. Synaptic patterns in the visual cortex of the turtle: An electron microscopic study. *Journal of Comparative Neurology*, 1975, **160**, 51–80.

Famiglietti, E. V., Jr., & Peters, A. The synaptic glomerulus and the intrinsic neuron in the dorsal lateral geniculate nucleus of the cat. *Journal of Comparative Neurology*, 1972, **144**, 285–334.

Garey, L. S. The termination of thalamocortical fibers in visual cortex of the cat and monkey. *Journal of Physiology (London)*, 1970, **210**, 15–17.

Garey, L. S., Jones, E. G., & Powell, T. P. S. Interrelationship of striate and extrastriate cortex with the primary relay sites of the visual pathway. *Journal of Neurology, Neurosurgery, & Psychiatry*, 1968, **31**, 135–157.

Garey, L. S., & Powell, T. P. S. The projection of the lateral geniculate nucleus upon the cortex in the cat. *Proceedings of the Royal Society*, 1967, **B169**, 107–126.

Garey, L. S., & Powell, T. P. S. An experimental study of the termination of the lateral geniculocortical pathway in the cat and monkey. *Proceedings of the Royal Society,* 1971, **B179**, 41–63.

Glendenning, K. K., Hall, J. A., & Hall, W. C. The connections of the pulvinar in a primate *(Galago senegalensis)*. *Anatomical Record*, 1972, **172**, 316.

Glickstein, J., King, R. A., Miller, J., & Berkley, M. Cortical projections from the dorsal lateral geniculate nucleus of cats. *Journal of Comparative Neurology*, 1967, **130**, 55–76.

Graybiel, A. M. Some ascending connections of the pulvinar and nucleus lateralis posterior of the thalamus in the cat. *Brain Research*, 1972, **44**, 99–125.

Hall, J. A. Efferent projections from general cortex in the turtle *(Pseudemys scripta)*. Masters thesis, Brown University, 1971.

Hall, W. C. Visual pathways to the telencephalon in reptiles and mammals. *Brain, Behavior, & Evolution*, 1972, **5**, 95–143.

Hall, W. C., & Diamond, I. T. Organization and function of the visual cortex in hedgehog: I. Cortical cytoarchitecture and thalamic retrograde degeneration. *Brain, Behavior, & Evolution*, 1968, **1**, 181–214.

Hall, W. C., & Ebner, F. F. Parallels in the visual afferent projections of the thalamus in the hedgehog *(Paraechinus hypomelas)* and the turtle *(Pseudemys scripta)*. *Brain, Behavior, & Evolution*, 1970, **3**, 135–154. (a)

Hall, W. C., & Ebner, F. F. Thalamotelencephalic projections in the turtle *(Pseudemys scripta)*. *Journal of Comparative Neurology*, 1970, **140**, 101–122. (b)

Hanbery, J., Ajamone Marsan, C., & Dilworth, M. Pathways of non-specific thalamocortical projection system. *Electroencephalography & Clinical Neurophysiology*, 1954, **6**, 103–118.

Hanbery, J., & Jasper, H. H. Independence of diffuse thalamocortical projection system shown by specific nuclear destructions. *Journal of Neurophysiology*, 1953, **16**, 103–118.

Hand, P. J., & Morrison, A. R. Thalamocortical projections from the ventrobasal complex to somatic sensory areas I and II of the cat. *Experimental Neurology*, 1970, **26**, 291–308.

Hand, P. J., & Morrison, A. R. Thalamocortical relationships in the somatic sensory system as revealed by silver impregnation techniques. *Brain, Behavior, & Evolution*, 1972, **5**, 273–302.

Harting, J. K., Diamond, I. T., & Hall, W. C. Anterograde degeneration study of the cortical projections of the lateral geniculate and pulvinar nuclei in the tree shrew *(Tupaia glis)*. *Journal of Comparative Neurology*, 1973, **150**, 393–440. (a)

Harting, J. K., Glendenning, K. K., Diamond, I. T., & Hall, W. C. Evolution of the primate visual system: Anterograde degeneration studies of the tecto-pulvinar system. *American Journal of Physical Anthropology*, 1973, **38**, 383–392. (b)

Harting, J. K., Hall, W. C., & Diamond, I. T. Evolution of the pulvinar. *Brain, Behavior, & Evolution*, 1972, **6**, 424–452.

Herrick, C. J. The amphibian forebrain X. Localized functions and integrating functions. *Journal of Comparative Neurology*, 1934, **59**, 239–266.

Herrick, C. J. *The brain of the tiger salamander*. Chicago: University of Chicago Press, 1948.

Hicks, S. P., & d'Amato, G. J. Cell migrations to the isocortex in the rat. *Anatomical Record*, 1968, **160**, 619–634.

Hinds, J. Reciprocal and serial dendrodendritic synapses in the glomerular layer of the rat olfactory bulb. *Brain Research*, 1970, **17**, 530–534.

Hubel, D., & Wiesel, T. N. Anatomical demonstration of columns in the monkey striate cortex. *Nature (London)*, 1969, **221**, 747–750.

Hubel, D., & Wiesel, T. N. Laminar and columnar distribution of geniculocortical fibers in the Macaque monkey. *Journal of Comparative Neurology*, 1972, **146**, 421–450.

Huber, G. C., & Crosby, E. C. On the thalamic and tectal nuclei and fiber paths in the brain of the American alligator. *Journal of Comparative Neurology*, 1926, **40**, 97–154.

Ingvar, S. On thalamic evolution. *Acta Medica Scandinavica*, 1923, **59**, 696–745.

Jasper, H. H., Naquet, R., & King, E. Thalamocortical recruiting responses in sensory receiving areas in the cat. *Electroencephalography & Clinical Neurophysiology*, 1955, **7**, 99–114.

Jones, E. G. Pattern of cortical and thalamic connections of the somatic sensory cortex. *Nature (London)*, 1967, **216**, 704–705.

Jones, E. G., & Powell, T. P. S. The cortical projection of the ventroposterior nucleus of the thalamus in the cat. *Brain Research*, 1969, **13**, 298–318.

Jones, E. G., & Powell, T. P. S. An electron microscopic study of the laminar pattern and mode of termination of afferent fiber pathways in the somatic sensory cortex of the cat. *Philosophical Transactions of the Royal Society of London*, 1970, **B257**, 45–62. (a)

Jones, E. G., & Powell, T. P. S. Connections of the somatic sensory cortex of the rhesus monkey. III. Thalamic connections. *Brain*, 1970, **93**, 37–56. (b)

Kidd, M. Electron microscopy of the inner plexiform layer of the retina in the cat and the pigeon. *Journal of Anatomy (London)*, 1962, **96**, 179–187.

Killackey, H. P. Anatomical evidence for cortical subdivisions based on vertically discrete thalamic projections from the ventral posterior nucleus to cortical barrels in the rat *Brain Research*, 1973, **51**, 326–331.

Killackey, H. P., & Ebner, F. F. Two different types of thalamocortical projections to a single cortical area in mammals. *Brain, Behavior, & Evolution*, 1972, **6**, 141–169.

Killackey, H. P., & Ebner, F. F. Convergent projection of three separate thalamic nuclei onto a single cortical area. *Science*, 1973, **179**, 283–285.

Knapp, H., & Kang, D. S. The visual pathways of the Snapping turtle *(Chelydra serpentina)*. *Brain, Behavior, & Evolution,* 1968, **1,** 19–42. (a)

Knapp, H., & Kang, D. S. The retinal projections of the side-necked turtle *(Podocnemis unifilis)*. *Brain, Behavior, & Evolution,* 1968, **1,** 369–404. (b)

Kosareva, A. A. Projection of optic tract fibers to visual centers in a turtle *(Emys obicularis)*. *Journal of Comparative Neurology,* 1967, **130,** 263–276.

Kruger, L., & Berkowitz, E. C. The main afferent connections of the reptilian telencephalon as determined by degeneration and electrophysiological methods. *Journal of Comparative Neurology,* 1960, **115,** 125–141.

Laemle, L., Benhamida, C., & Purpura, D. P. Laminar distribution of geniculo-cortical afferents in visual cortex of the postnatal kitten. *Brain Research,* 1972, **41,** 25–37.

LeGros Clark, W. E. The structure and connections of the thalamus. *Brain,* 1932, **55,** 406–470.

Leonard, C. M. The prefrontal cortex of the rat. I. Cortical projection of the mediodorsal nucleus. II. Efferent connections. *Brain Research,* 1969, **12,** 321–343.

Lohman, A. H. M., & Mentink, G. M. Some cortical connections of the tegu lizard *(Tupinambis teguixin)*. *Brain Research,* 1972, **45,** 325–344.

Lorente de No, R. Cerebral cortex: Architecture, Intracortical connections, motor projections. In J. F. Fulton (Ed.), *Physiology of the nervous system.* London: Oxford University Press, 1938.

Lund, J. S. Organization of neurons in the visual cortex, area 17, of the monkey *(Macaca mulatta)*. *Journal of Comparative Neurology,* 1973, **144,** 455–496.

Morest, D. K. A study of neurogenesis in the forebrain of opossum pouch young. *Zeitschrift für Anatomie Entwicklungsgeschichte,* 1970, **130,** 265–305.

Morest, D. K. Dendrodendritic synapses of cells that have axons: The fine structure of the Golgi type II cell in the medial geniculate body of the cat. *Zeitschrift für Anatomie Entwicklungsgeschichte,* 1971, **133,** 216–246.

Morrison, A. R., Hand, P. J., & O'Donoghue, J. Contrasting projections from the posterior and ventrobasal thalamic nuclear complexes to the anterior ectosylvian gyrus of the cat. *Brain Research,* 1970, **21,** 115–121.

Murray, M. Degeneration of some intralaminar nuclei after cortical removals in the cat. *Journal of Comparative Neurology,* 1966, **127,** 341–368.

Nauta, W. J. H. Terminal distribution of some afferent fiber systems in the cerebral cortex. *Anatomical Record,* 1954, **118,** 333.

Nauta, W. J. H., & Whitlock, D. G. An anatomical analysis of the non-specific thalamic projection system. In J. F. Delafresnaye (Ed.), *Brain mechanisms and consciousness.* Oxford: Blackwell, 1954. Pp. 81–116.

Northcutt, R. G., & Butler, A. B. Retinal projections in the northern water snake *Natrix sipedon sipedon* (L.). *Journal of Morphology,* 1973, **142,** 117–135.

Orrego, F. The reptilian forebrain, I. The olfactory pathways and cortical areas in the turtle. *Archives Italiennes de Biologie,* 1961, **99,** 425–445.

Orrego, F., & Lisenby, D. The reptilian forebrain, IV. Electrical activity in the turtle cortex. *Archives Italiennes de Biologie,* 1962, **100,** 17–30.

Peacock, J. H., & Combs, C. M. Retrograde cell degeneration in diencephalic and other structures after hemidecortication in rhesus monkeys. *Experimental Neurology,* 1965, **11,** 367–399.

Petrovcky, P., & Druga, R. Peculiarities of the cytoarchitectonics and some afferent systems of the parietal cortex. *Folio Morphologica,* 1972, **20,** 161–163.

Polley, E. H., & Dirkes, J. M. The visual cortical (geniculocortical) area of the cat brain and its projections. *Anatomical Record,* 1963, **145,** 345.

Powell, T. P. S., & Cowan, W. M. The interpretation of the degenerative changes in the intralaminar nuclei of the thalamus. *Journal of Neurology, Neurosurgery, & Psychiatry,* 1967, **30,** 140–153.

Pritz, M. B. Ascending projections of some auditory nuclei in *Caiman crocodilia. Anatomical Record*, 1972, **172**, 386–387.

Pritz, M. B. Connections of the alligator visual system; telencephalic projections of nucleus rotundus. *Anatomical Record*, 1973, **175**, 416.

Rakie, P. Mode of cell migration of the superficial layers of fetal monkey neocortex. *Journal of Comparative Neurology*, 1972, **145**, 61–84.

Rall, W., Shepherd, G. M., Reese, T. S., & Brightman, M. W. Dendrodendritic synaptic pathway for inhibition of the olfactory bulb. *Experimental Neurology*, 1966, **14**, 44–56.

Ralston, H., & Herman, J. The fine structure of neurons and synapses in the ventrobasal thalamus of the cat. *Brain Research*, 1969, **14**, 77–97.

Ravizza, R. J., & Diamond, I. T. Projections from the auditory thalamus to neocortex in the hedgehog *(Paraechinus hypomelas). Anatomical Record*, 1972, **172**, 390–391.

Rose, J. E., & Woolsey, C. N. Organization of the mammalian thalamus and its relationship to the cerebral cortex. *Electroencephalography & Clinical Neurophysiology*, 1949, **1**, 391–404.

Rose, J. E., & Woolsey, C. N. Cortical connections and functional organization of the thalamic auditory system of the cat. In J. P. Harlow & C. N. Woolsey (Eds.), *Biological and biochemical basis of behavior*. Madison, Wisconsin: University of Wisconsin Press, 1958. Pp. 127–150.

Rossignol, S., & Colonnier, M. A light microscopic study of degeneration patterns in cat cortex after lesions of the lateral geniculate nucleus. *Vision Research Supplement*, 1971, **3**, 329–338.

Sanides, F., & Sanides, D. The "Extraverted Neurons" of the mammalian cerebral cortex. *Zeitschrift für Anatomie Entwicklunsgeschichte*, 1972, **136**, 272–293.

Sheibel, M. E., & Scheibel, A. B. Structural organization of non-specific thalamic nuclei and their projection toward cortex. *Brain Research*, 1967, **6**, 60–94.

Shimada, M., & Langman, J. Cell proliferation, migration, and differentiation in the cerebral cortex of the golden hamster. *Journal of Comparative Neurology*, 1970, **139**, 227–244.

Sidman, R. L. Abnormal cell migrations in developing brains of mutant mice. In: G. Z. Jervis (Ed.), *Expanding concepts in mental retardation* (A symposium from the Joseph P. Kennedy, Jr., Foundation). Springfield, Illinois: Charles C Thomas, 1968, Pp. 40–49.

Spatz, W. B., Tigges, J., & Tigges, M. Subcortical projections, cortical associations, and some intrinsic interlaminar connections of the striate cortex in the squirrel monkey (*Saimiri*). *Journal of Comparative Neurology*, 1970, **140**, 155–174.

Tobias, T. J., & Ebner, F. F. Thalamocortical projections from the mediodorsal nucleus in the Virginia opossum. *Brain Research*, 1973, **52**, 79–96.

Valverde, F. Structural changes in the area striata of the mouse after enucleation. *Experimental Brain Research*, 1968, **5**, 274–292.

Voneida, T. J., & Ebbesson, S. O. E. On the origin and distribution of axons in the pallial commissures in the tegu lizard *(Tupinambis nigropunctatus). Brain, Behavior, & Evolution*, 1969, **2**, 467–481.

Walker, A. E. *The Primate Thalamus*. Chicago: University of Chicago Press, 1938.

Waller, W. H., & Barris, R. W. Relationship of thalamic nuclei to the cerebral cortex of the cat. *Journal of Comparative Neurology*, 1937, **67**, 317–339.

Wilson, M. E., & Cragg, B. G. Projections from the lateral geniculate nucleus in the cat and monkey. *Journal of Anatomy (London)*, 1967, **107**, 677–692.

Wilson, M. E., & Cragg, B. G. Projections from the medial geniculate body to the cerebral cortex in the cat. *Brain Research*, 1969, **13**, 462–475.

9
Vertebrate Learning: Common Processes

M. E. Bitterman

William T. Woodard

University of Hawaii

By "common processes" of vertebrate learning, is meant processes to be found at least in a wide range of living vertebrates, if not in all. It should be said at once that there is no very good evidence for the existence of such processes. The argument from common ancestry is not compelling because marked transformations in process may have occurred in the course of evolution. Communality has been inferred also from certain gross similarities in the results of presumably analogous experiments with animals drawn from diverse groups—goldfish, pigeons, rats, dogs, rhesus monkeys—but the similarities may be only superficial.

Communality of process is more difficult to establish than perhaps may be evident at first. Learning processes are not given directly by learning experiments but are inferred from behavioral outcomes, and the relation of process to outcome may be quite complex. Recent experience shows that different processes may be responsible for what may seem to be identical outcomes of analogous experiments with animals of different species, or even with animals of the same species studied under different conditions. Nor do unlike results obtained in such experiments necessarily imply the operation of different learning processes, because behavior in learning situations depends to a considerable extent on processes other than learning. Whether there are common processes and, if so, what they are can only be known from the systematic study of a wide range of vertebrates, but there has been no such study. Although exploratory work has been done with many different species at one time or another, there has been systematic work with few.

Despite the lack of compelling evidence, the existence of common processes has long been taken for granted. In fact, it has been the assumption of leading theorists that there are only common processes—that intellectual differences among animals in general are (to use Darwin's phrase) differences of degree and not of kind (Bitterman, 1960). Consider, for example, the view of Thorndike (1911), who perhaps more than any other man can be said to have brought the study of animal intelligence into the laboratory. On the basis of his early experiments with chickens, cats, dogs, and monkeys, Thorndike concludes that "all" animals are "systems" of sensory–motor "connections" (some innate, others formed in the course of experience), and that (however much animals differ in what connections they form, or in the number of connections they are capable of forming) the connecting principles are the same for all. A like view was expressed by Pavlov (1927), another pioneer in the study of animal intelligence, who confidently asserted the phyletic generality of the principles emerging from his work with dogs. Although there was some substantial disagreement at first, the communality assumption prevailed, and interest centered on the discovery of the principles that were assumed to hold for all animals. The phyletic scope of research on learning soon became rather narrow in consequence (Bitterman, 1965). With the laws of learning thought to be the same in all animals, it seemed perfectly reasonable to try to work them out in rats, or in dogs, or in whatever species came most conveniently to hand, just as geneticists worked out the laws of inheritance in peas or in fruit flies.

Although recent years have brought a resurgence of interest in comparative analysis, there is not yet much to say about common processes. All that can be done here is, first, to describe under four broad headings—(1) classical conditioning, (2) reward and punishment, (3) extinction, and (4) avoidance and omission—certain phenomena of learning the generality of which can for the moment be assumed on the basis of the diversity of species in which they are found, and, second, to speculate about underlying processes in the very small number of species on which there has been some systematic work. It will be evident that much remains to be done before general statements about vertebrate learning can be made with confidence.

CLASSICAL CONDITIONING

In classical conditioning, stimuli are paired. From time to time, for example, a dog may be exposed to a 10-sec tone the termination of which coincides with the presentation of food or of brief shock. What is sought in such an experiment is some alteration in the properties of the first stimulus (S_1) attributable to its pairing with the second (S_2). A tone paired with food comes to elicit salivation, whereas a tone paired with shock to a forelimb comes to elicit flexion of the forelimb. The change in behavior with successive pairings (trials) is gradual. The probability of response to S_1 and the magnitude of the response (for example, the quantity of saliva secreted) increase progressively to asymptotic levels, whereas the latency of response (time between onset of S_1 and initiation of the response) declines. The effect of the pairing may be evidenced not only by new responses that S_1 comes to elicit, but by other changes in behavior as well. For example, response-suppressing properties may be acquired by conditioning: shock disrupts the performance of a hungry rat that is pressing a lever for food, and when a tone is paired with the shock the tone itself soon comes to do the same (conditioned suppression). Motivational properties also may be altered by conditioning: a rat works to produce a stimulus that has been paired with food but leaves the vicinity of a stimulus that has been paired with shock or works to terminate it. In general, the first stimulus of the pair is found to acquire some of the properties of the second, but not the reverse. A tone that follows food does not come in consequence to elicit salivation.

Changes in the properties of S_1 that occur in the course of pairing with S_2 cannot safely, without proper controls, be attributed to the pairing, because the mere presentation of the stimuli may produce like effects. For example, feeding a dog repeatedly in some situation increases the probability of its salivating in response to stimuli never before encountered in that situation (pseudoconditioning). One way to distinguish the effects of pairing from the effects of the stimuli independently of pairing is to compare the performance of conditioned animals and of animals that have experienced the same stimuli equally often, but unpaired and in random sequence. A more economical procedure, because it requires only a single group of animals, is to use two readily discriminable stimuli, S_1 and S_1', pairing one consistently with S_2 and the other never; differential response to the two stimuli indicates the effectiveness of the pairing.

According to the "S–R contiguity theory" of classical conditioning, responses are "connected" (in some functional sense of the term) to contiguous stimuli, with each contiguous occurrence of stimulus and response increasing (to some asymptotic value) the strength of the connection between them. Traditionally, neural connections between brain "centers" have been contemplated (cf. Pavlov, 1927), in this case between a sensory center (the center for S_1) and any motor centers activated by the center for S_2. A limitation of S–R theory is that all of the properties acquired by S_1 must be conceived as motor. To account for conditioned suppression, for example, it has been assumed that S_1 becomes connected to emotional responses incompatible with feeding. To account for the acquisition of reward properties, it has been assumed that S_1 becomes connected to some fraction of the consummatory response elicited by the reward. Less awkward in this respect is the "S–S contiguity theory," according to which contiguous stimuli (or their brain centers) are "connected." When activated by S_1, the center for S_1 now activates the center for S_2, and thus activates indirectly any motor center or other center with which the center for S_2 is connected.

Evidence for the S–S theory is provided by experiments on "sensory preconditioning." For example, a light (S_1) and a tone (S_2), neither of which may have any apparent effect, are paired repeatedly while a rat is working for food, after which the tone is paired with brief shock until the tone itself disrupts the behavior (Prewitt, 1967). Now the light is also found to be disruptive, despite the fact that it never has been contiguous with the responses evoked by shock. The behavior of a control animal, to which light and tone are presented but not paired in the first stage of the experiment, is not so disrupted by the light in the third stage. These results indicate that the pairing of light and tone in the first stage of the experiment has produced some connection between them.

Evidence for the S–R theory has been sought in experiments of several designs. Attempts have been made to bypass the center for S_2. For example, pairing S_1 with electrical stimulation of a point in the motor cortex that evokes leg flexion may produce some tendency for the tone to elicit flexion (Doty & Giurgea, 1961). Unfortunately, however, this procedure does not rule out sensory contributions from directly activated sensory elements in the motor cortex, from adjacent sensory areas activated by spread of stimulation, or from proprioceptors activated by the evoked response. Better support for the S–R theory was provided by early experiments in which properties of S_2 were changed after they had been conditioned to S_1. In work with monkeys, for example, a bell was paired with display of a party favor (a "snake blowout"), which the animal feared, and soon the bell alone began to evoke an overt fear response (Harlow, 1937). When fear of the snake then was eliminated by adaptation and association with food, the bell itself continued to evoke fear; indeed, fear of the snake could be redeveloped by pairing it with the bell. These results seemed to indicate that the pairing of bell and snake in the first stage of the experiment produced some direct association between the bell and the emotional response evoked by the snake. Recent work with rats has failed, however, to produce analogous results (Rescorla, 1973).

After S_1 has come to elicit some response characteristic of S_2 in consequence of the pairing of S_1 and S_2, the response may be transferred to a third stimulus, S_s, by pairing it with S_1 alone. This phenomenon, which is known as "second order conditioning," may seem to provide another instance of the operation of the S–S contiguity principle, but recent experiments indicate that it may have an S–R component. Response to S_s has been found to persist after response to S_2 has been eliminated (Rizley & Rescorla, 1972). At best, however, evidence for the formation of S–R connections in classical conditioning is rather insubstantial. If it is to be taken seriously, connections of both kinds, S–S and S–R, must be assumed.

Conditioning theory should take account of the fact that the responses that come to be elicited by S_1 are not identical with those elicited by S_2. On the one hand, certain responses to S_2 do not seem to be conditioned to S_1: although a dog may salivate in response to a tone that has been paired with food, it usually does not make ingestive or chewing movements. On the other hand, S_1 may come to evoke responses that are not made to S_2: in the presence of the tone, the dog may show a certain restlessness—pawing, yawning—that it does not show in the presence of food. One explanation of the difference between "conditioned" responses (to S_1) and "unconditioned" responses (to S_2) is that the motor centers that produce the observed behavior are not activated in the same way by S_1 (either directly, via new S–R connections, or indirectly, via new S–S connections) as by S_2. Furthermore, certain aspects of the response to S_2 may depend on physical properties of S_2 that are not shared by S_1, whereas S_1 may have some properties (apart from those acquired by conditioning) that affect the form of the conditioned response.

The outcome of a conditioning experiment depends on the nature and intensity of the paired stimuli. Pavlov found, for example, that strong cutaneous shock paired with food does not come in consequence to elicit salivation, although mild shock does. In general, performance is better with more intense stimuli unless (as in the previous instance) innate response to S_1 is incompatible with response to S_2. Interstimulus interval is another important variable (Bitterman & Schoel, 1970). There is not much evidence of conditioning with strictly simultaneous presentation of S_1 and S_2—that is, when the interstimulus interval is zero—but conditioning usually is good even at very short intervals (.5–1.0 sec). The shape of the function varies markedly from species to species, and also within species depending on the stimuli employed and the responses measured. With long intervals, the latency of response tends to be long, although the magnitude of response may be substantial. At all but the shortest intervals, conditioning is much better when S_1 stays on until the appearance of S_2 (delay procedure) than when it terminates soon after onset (trace procedure). Experiments with complex stimuli point also to a selective process in conditioning (Kamin, 1969). After a tone and a light have together been paired with shock, for example, the tone alone may elicit an emotional response in rats and the light alone fail to do so, although the pairing of light alone with shock produces satisfactory evidence of conditioning in control animals. At the very least, the conditioned response to light may be substantially greater when it alone has been paired with shock than when tone has been present at the same time (overshadowing), and the difference is especially marked if the tone previously has been paired with shock (blocking). An indication of the generality of these phenomena, of which detailed study so far has been made primarily in mammals, is provided by experiments with goldfish in our own laboratory (Tennant & Bitterman, 1975).

REWARD AND PUNISHMENT

Experiments on what is called "instrumental conditioning" show what seems to be another quite general phenomenon of vertebrate learning—the control of behavior by its consequences. The distinctive feature of instrumental conditioning is response contingency. S_2 does not follow S_1 independently of the animal's behavior (as in classical conditioning) but is contingent on the occurrence of some defined response (R_i) to S_1. For example, if a rat presses a lever (R_i) in the presence of a tone (S_1), a pellet of food (S_2) is presented, but if the lever is not pressed no food is presented. In a hungry rat, the effect of the contingency is to strengthen the tendency of S_1 to elicit R_i, as indicated by a progressive decrease over trials in the latency of response (the time between the onset of S_1 and the occurrence of R_i) and a progressive increase in its probability (the likelihood that it occurs during the presentation of S_1 for some fixed period). Quite the opposite results are obtained, however, if shock is substituted for food, in which case the effect of the contingency is to weaken the tendency of S_1 to elicit R_i. In general, any S_2 that increases the likelihood of a response required to produce it is called a "reward," whereas one that decreases the likelihood of the response is called a "punishment" and characterized as "aversive." Note that what is rewarding or punishing for an animal of one species may not be rewarding or punishing for an animal of another species, or even for the same animal under other conditions. A drop of water is rewarding for a rat that has been deprived of water but not for one that has drunk its fill a moment before.

Just as it is necessary in classical conditioning experiments to control for all effects of experience apart from the temporal relation between S_1 and S_2 in order to isolate the effects of the pairing, so in instrumental conditioning it is necessary to control for the effects of all experience apart from the contingent relation between R_i and S_2 in order to isolate the effects of the contingency. A hungry rat that is rewarded with food or punished with shock for pressing a lever may show an increased or a decreased tendency to press the lever simply because its general level of activity has been increased by the food or decreased by the shock. One way to get at the role of the contingency is to compare the performance of instrumental animals with that of yoked control animals confined in identical environments and subjected to the same stimuli, except that the responses of the instrumental animals produce food or shock both for themselves and for the controls. In such an experiment, the effectiveness of the contingency is evidenced by a difference in the degree to which the probability of lever pressing changes in the two groups. If animals are trained in a choice situation that provides opportunity for two responses of comparable difficulty occurring with equal probability at the outset (say, a chamber with two levers, or a simple T maze at the choice point of which the animal may turn right or left), and if only one of the two responses is rewarded, then an increase in its probability relative to that of the alternative is evidence of the effectiveness of the contingency.

From the fact that the classical and instrumental conditioning procedures are different, it does not follow that the underlying processes are different. The response contingency in instrumental conditioning may serve simply to produce the S–S or S–R contiguities that seem to be responsible for the changes in behavior observed in experiments on classical conditioning. The effects of punishment can be explained rather easily in these terms. Local stimuli, such as the sight of the lever when R_i is lever pressing, acquire aversive properties through their pairing with the aversive S_2 produced by R_i, and the animal therefore may be expected to turn away from them, or to refrain from approaching them. Because proprioceptive feedback (S_p) from R_i—contributed by receptors in the participating muscles, tendons, and joints—also is paired with shock, competing responses (perhaps largely emotional) evoked by the shock and conditioned to S_p may be assumed to disrupt R_i as soon as it is initiated.

The effects of reward also may be explained in terms of S–S or S–R contiguity. A situation widely used for the study of instrumental learning in rats is the runway, a straight alley leading from a starting box to a goal box (the contents of which are not visible until after it has been entered). The increasing readiness with which a hungry animal runs from the starting box to a goal box that contains food may be explained in terms of the conditioning of an approach response, evoked originally by the food, to visual stimuli contiguous with the food (such as sight of the entrance to the goal box). The development even of such manipulative responses as lever pressing in rats and key pecking in pigeons may be understood in terms of classically conditioned approach: pigeons begin to peck at, and rats to lick at, objects that are introduced into the experimental situation (or, if already present, simply illuminated) several seconds before the noncontingent delivery of food (Brown & Jenkins, 1968; Peterson, Ackil, Frommer, & Hearst, 1972), and analogous results have been obtained in our own experiments with goldfish (Woodard & Bitterman, 1974). When a manipulandum must be operated in some specific way (such as a lever that must be moved to the right rather than to the left if reward is to follow), the sensory context in which the required form of the response is most likely to occur (such as a view of the left side of a lever that must be moved to the right) may be paired with the reward and elicit approach in consequence (cf. Bindra, 1972). There is some doubt, however, that certain other instrumentally conditionable activities, such as ear scratching or leg lifting in dogs, can be treated as approach responses to S_2 that are classically conditioned to S_1.

For a solution of this problem, proprioceptive feedback from R_i can be invoked, which, owing to the contingency, is closely and consistently followed by reward. If it is assumed that one of the properties acquired by stimuli paired with reward is to facilitate on-going behavior (as stimuli paired with punishment may be assumed to disrupt on-going behavior), then it follows that a previously rewarded response, once initiated to the point of producing distinctive feedback, is more likely to be continued than a response not previously rewarded. From this point of view, the effect of reward is to make more likely not the initiation of R_i but its continuation, and nothing more may be required to account for instrumental conditioning on the assumption that the aroused animal scans its repertoire of responses and that feedback sampling can take place very rapidly

(cf. Mowrer, 1960). The great speed with which even rather complex instrumental tasks may be performed after long training has suggested, however, that
the sampling process eventually is short circuited by direct S_1-R_i connections
(cf. Morgan, 1894; Sheffield, 1965). Feedback-induced facilitation increases the
frequency with which R_i occurs and therefore (by the S–R contiguity principle)
the strength of the S_1-R_i connection, which in turn increases the frequency
with which R_i is initiated, and so forth, until the connection becomes very
strong. In any case, the feedback-facilitation concept makes possible a perfectly
general interpretation of instrumental learning in terms of contiguity.

The formal advantage of such an interpretation is, of course, that no learning
process other than those which operate in classical conditioning need be assumed, but further assumptions as to the way in which the products of
conditioning are translated into instrumental performance seem to be required—
for example, the assumption that stimuli paired with reward facilitate on-going
behavior. An alternative interpretation is that a connection between a stimulus
and a contiguous response can be established and strengthened directly by
reward—the "S–R reward theory" (Thorndike, 1911). It also has been proposed
that the same process may operate in classical conditioning (Hull, 1943). The
tendency to salivate in response to an S_1 paired with food may be strengthened,
for example, only because the salivation heightens the taste of food, or with
weak acid as S_2 only because the salivation dilutes the acid. Flexion of a limb
may be conditioned to a stimulus paired with shock only because flexion
minimizes the pain of shock. There is, in fact, some evidence that the classical
conditioning of flexion evoked by electrical stimulation of the motor cortex
may represent nothing more than a postural adjustment to the disequillibrating
effect of the forced movement (Wagner, Thomas, & Norton, 1967). According
to S–R reward theory, it is not the contingent relation between response and
reward that is important for learning but the contiguity between them—the
contingency serves only to insure the contiguity, which is effective whether or
not it is intended by the experimenter (Hull, 1943).

It seems clear, however, that the S–R reward principle is insufficient to
account for all the facts either of classical or of instrumental conditioning. Not
only does the sensory preconditioning experiment point to the operation of a
pure contiguity process, but so also does the so-called "placement" experiment,
an instrumental analog of sensory preconditioning. Just as in sensory preconditioning a response is conditioned to S_2 only after S_2 has been paired with S_1, so
in a placement experiment reward properties are acquired by S_2 only after the
R_i-S_2 contingency has been experienced. For example, a rat traverses (R_i) a
runway (S_1) that leads to an empty but distinctive goal box (S_2). After
subsequent experience with the same goal box into which it is placed directly
and which then contains food, the animal is tested again in the runway. Its speed
now is found to be much greater than before, despite the fact that running never
has been rewarded. A control animal that is placed and fed in a goal box

different from the one previously found at the end of the runway does not show the same increase in speed on a subsequent test (Gonzalez & Diamond, 1960). These results are readily interpreted in terms of S–S contiguity theory. For the experimental group, stimuli associated with the entrance to the goal box are paired with S_2 in the first stage of the experiment, and S_2 is paired with food in the second stage.

The fact that performance is better for large reward than for small reward has suggested that larger rewards produce stronger S–R connections. Convincing evidence that rewards do not simply strengthen S–R connections but come to be anticipated by the animal and in that way facilitate instrumental responding is found in the so-called "depression effect" (Crespi, 1942). Rats trained in a runway with large reward run more rapidly than rats trained with small reward, but rats shifted to small reward after training with large reward become emotionally disturbed, and their running speed declines to a level substantially below that of control rats trained with small reward from the outset. The only reasonable explanation of the disruption yet proposed is that it is caused by discrepancy between "anticipated" and actually encountered magnitudes of reward; the greater the discrepancy, in fact, the greater is the disruption (Gonzalez, Gleitman, & Bitterman, 1962). The reward principle, according to which the large reward simply produces a stronger $S_1 - R_i$ connection, makes no reference to anticipation. According to contiguity theory, of course, anticipation of reward, based either on S–S or S–R linkages, is a central determinant of preshift performance—the S–R interpretation is that runway stimuli are connected to some "fraction" of the consummatory response—the magnitude of which increases with the size of the reward (Hull, 1931). It should be noted that the depression effect is not found in goldfish (Lowes & Bitterman, 1967) or in painted turtles (Pert & Bitterman, 1970), which may mean that contrast of present with previous reward is not a simple conditioning phenomenon but a reflection of some higher order process. Recently, however, we have discovered in goldfish another contrast effect, which appears also in rats (Bower, 1961) and in pigeons (Brownlee & Bitterman, 1968). Animals trained concurrently with small reward in one situation and large reward in a different situation perform more poorly for small reward than do control animals that are trained with small reward in both (Burns, Woodard, Henderson, & Bitterman, 1974). Here, then—whatever the relation between the two contrast effects may prove to be—is a more general indication that rewards do not simply strengthen S–R connections.

Evidence of this sort, of course, applies only to the sufficiency of the S–R reward principle, and not to its validity. There is no reason why contiguity and reward should not operate jointly, as various "dual process" theorists have proposed. It has been·suggested, for example, that S–S connections are established by contiguity alone and S–R connections by reward (Maier & Schneirla, 1942), or that autonomic responses are connected to stimuli by contiguity alone and skeletal responses by reward (Mowrer, 1950). Another possibility is that the same connections are strengthened both by contiguity and by reward (Thorndike, 1911). Given the connection of responses to contiguous stimuli, it is not difficult to imagine that the process may be facilitated by contiguous reward. The S–R reward principle surely does not go very far beyond the assumption (designed to bolster the contiguity interpretation) that stimuli associated with reward promote on-going behavior, and in some respects it is a good deal simpler. The validity of the reward principle, once widely accepted but now generally discounted, may profitably be examined further. For example, instrumental experiments in which feedback (proprioceptive and exteroceptive) is eliminated or minimized should provide some critical information (Taub, Bacon, & Berman, 1965).

If "reward" is defined as an event that strengthens the tendency of an animal to make a response that produces it, then "escape" experiments show the termination of an aversive stimulus to be rewarding. The S_1 in these experiments is the onset of an aversive stimulus, such as shock, and S_2 is the offset of S_1. The consequent strengthening of R_i is readily explained in terms of the S–R reward principle. A contiguity interpretation may begin with the assumption that exteroceptive stimuli other than those at the locus of R_i (if R_i is a localized response, such as jumping over a hurdle or pressing a lever) are more likely to be paired with shock, and the animal is more likely, therefore, to turn away from them. Proprioceptive feedback from responses other than R_i may come to disrupt those responses, once initiated, because of pairing with shock. As has been assumed in the case of conventional rewards, it is possible that conditioned approach to local stimuli paired with shock termination plays some role, and that the occurrence of R_i, once initiated, may be facilitated by its proprioceptive feedback, which also is paired with shock termination. Any of these factors can be expected to increase the relative frequency with which R_i occurs and to differentially strengthen its connection with S_1 in consequence. An important consideration, of course, is the compatibility of R_i with the powerful responses that may be evoked at the outset by S_1—some escape responses are therefore more easily acquired than others.

EXTINCTION

The effects of conditioning show considerable temporal stability—for example, goldfish trained to press a lever for food begin almost immediately to press the lever on being reintroduced into the training situation 1 year later—but a few minutes in the situation with lever pressing no longer rewarded produces a marked decrement in responding. This is "extinction," which must be distinguished from forgetting. In forgetting experiments, the effects on conditioned behavior of time between training and testing are examined. In extinction experiments, the effects of eliminating those features of the training procedure that produced the conditioned behavior in the first place are examined. If the presentation of S_2 contiguously with S_1 in classical conditioning, or contingently on R_i in instrumental conditioning, is called "reinforcement," then extinction experiments are concerned with the effects of nonreinforcement. The term "extinction" describes both the procedure and its behavioral consequences. In the classical case, S_1 is no longer followed by S_2 and loses the properties gained from the pairing. In the instrumental case, R_i to S_1 no longer controls the presentation of S_2, and the tendency of S_1 to elicit R_i is weakened if previously increased by reward or strengthened if previously decreased by punishment.

The biological significance of extinction is no less than that of conditioning in that it represents an adjustment to changing circumstances. Extinction is important also in relation to a phenomenon of conditioning known as "stimulus generalization," which has not so far been considered here: the properties, not only of S_1 but of stimuli similar to it are altered by reinforcement. A pigeon trained to peck at an orange key pecks also at keys of other colors (never previously presented) in proportion to the degree of their similarity to orange; that is, yellow produces less response than orange, but more response than green or blue. The animal is therefore prepared for variability in its environment and is not required to deal *de novo* with each stimulus discriminably different from S_1. If, however, response to yellow is not reinforced, the generalized response to it soon extinguishes. Because the effects of nonreinforcement also generalize to similar stimuli in proportion to the degree of similarity, response to orange diminishes somewhat in consequence, but differential reinforcement (reinforcement for response to certain colors and nonreinforcement for response to other colors) soon narrows the range of stimuli to which the response is made (within limits, of course, which are imposed by the differential sensitivity of the animal). The effects of reinforcement and of nonreinforcement interact in this way to produce behavior appropriate to the demands of the environment.

Extinction experiments call for control procedures analogous to those employed in conditioning experiments. If, in the classical case, S_2 simply is withheld, the resulting change in behavior cannot safely be attributed to the fact that S_1 no longer is paired with S_2. A better method is to continue to present both stimuli, but separately and in random sequence. If a discrimination procedure has been used to begin with (S_1 followed by S_2, but S_1' never), a useful procedure is now to pair S_1', but not S_1, with S_2, (discrimination reversal). In the instrumental case, again, it is inappropriate simply to withhold S_2. Where a yoked control group has been employed in training, the roles of experimental and control groups may be interchanged. Alternatively, S_2 may be introduced with the same frequency as before for the conditioned animals but randomly with respect to the occurrence of R_i. Reversal is appropriate after choice training in which the animal has been reinforced for making one of two alternative responses to the same stimulus.

The simplest explanation of extinction is that the connections produced by reinforcement are destroyed, but it does not seem to be an adequate one. That the effects of conditioning do not disappear in the course of extinction is suggested by the phenomenon of "spontaneous recovery": a lapse of time after extinction may restore response to S_1. Furthermore, conditioning even after a great many extinction trials usually is much more rapid than it was originally. There is the fact, too, of "disinhibition": a conditioned response may reappear after extinction if S_1 is accompanied by some unusual or distracting stimulus. Findings such as these led Pavlov (1927) to see in extinction not gradual destruction of the excitatory relation between brain centers assumed to have been produced by their contiguous activation, but the development of an overriding inhibitory process that is less stable than the excitatory process. From this point of view, excitatory and inhibitory processes sum algebraically in the control of behavior. Reinforcement produces excitation, nonreinforcement produces inhibition, and the passage of time or the occurrence of any distorting event that differentially interferes with the more fragile inhibitory process yields a net increase in the level of responding. Further evidence of an independent, nonreinforcement-produced suppressive process may be found in the phenomenon of "conditioned inhibition," also discovered by Pavlov. If S_2 is paired with S_1 but never with $S_1 + S_x$, not only does the animal stop responding to $S_1 + S_x$, but S_x suppresses responding to other conditioned stimuli with which it never before has been associated. Although the phenomena that have suggested the inhibitory theory of extinction have been discovered in experiments on classical conditioning, they are found also, it should be noted, in instrumental conditioning (Brimer, 1972).

Still a third explanation of extinction is that it is caused, not by the development of inhibition, but by the conditioning of new, incompatible responses (counterconditioning). A common explanation of the Crespi effect is that negative discrepancy between anticipated and actual reward generates an emotional response (Amsel, 1958), incompatible with R_i, that is connected by contiguity to S_1. The same explanation may be offered for the extinction of rewarded instrumental responses, as well as for the surprising fact that rats trained with large reward extinguish more rapidly than do rats trained with small reward (Wagner, 1961). The greater is the discrepancy between anticipated and actual magnitudes of reward (zero reward taken as a point on the magnitude continuum), the more intense the emotional response that is elicited and conditioned and the greater the competition with R_i. A counterconditioning explanation of extinction after conditioning with an aversive S_2 seems less feasible. It is difficult, furthermore, to find a counterconditioning explanation of such phenomena as disinhibition. If the mechanisms of conditioning and extinction are identical, why should the effects of nonreinforcement be so much less stable than those of reinforcement?

Both the inhibition and the counterconditioning theories of extinction are confronted with what may be referred to as the "asymptotic problem." Suppose that an animal is conditioned, then extinguished, then reconditioned, then extinguished again, and so forth (a conditioning–extinction series); or suppose that a discrimination between two stimuli is established, then reversed (the previously unreinforced stimulus now reinforced and the previously reinforced stimulus now unreinforced), reversed again, and so forth (a "reversal series"— which may be thought of as two, concurrent, oppositely phased conditioning– extinction series). If, as the counterconditioning interpretation suggests, reinforcement connects S_1 to R_a and nonreinforcement connects it to R_b, and if neither of these connections is weakened by events that strengthen the other, and if the strength of each connection (instead of increasing indefinitely) approaches some limiting value, then continued training should bring the animal to a point at which R_a cannot be replaced by R_b, or the reverse, depending on which response is dominant as asymptote. In fact, however, a great deal of flexibility is displayed after such training—much more, usually, than at the outset. The typical asymptotic pattern of performance shown by rats in conditioning–extinction series is to respond readily at the start of each session and then rapidly to stop responding if no reward is forthcoming. The inhibitory interpretation again is at an advantage here, for the instability of inhibition already has been assumed on other grounds. Reversal experiments with pigeons, as well as with rats, provide rather convincing evidence, in fact, that asymptotic performance is under the control of a separate inhibitory process that develops rapidly in each experimental session (suppressing response to the unreinforced stimulus of that session) and fades rapidly between sessions (Woodard, Shoel, & Bitterman, 1971). At the start of each session, the animals respond readily to both stimuli, suggesting that the effects of reinforcement are quite stable. Somewhat different results are obtained in analogous experiments with goldfish, the reason for which is not clear, but there is nothing in those results to support a counterconditioning interpretation.

It should be noted that extinction sometimes brings with it certain changes in the sensory antecedents of the response being extinguished. When conditioning trials follow each other in close succession (that is, when trials are "massed"), some trace of the S_2 with which Trial N ends may be present along with S_1 at the start of Trial $N + 1$. When S_2 is food, for example, particles of the food may remain for a time in the animal's mouth. Furthermore, any emotional response to the absence of food on Trial N may generate some feedback that is present along with S_1 on Trial $N + 1$. If S_R is the sensory carryover from reinforcement, and S_N is the sensory carryover from nonreinforcement, then response is conditioned to the stimulus complex $S_1 + S_R$, but the stimulus complex present in extinction is $S_1 + S_N$. Sensory carryover is particularly important in so-called "free-operant" training, which may be characterized as instrumental conditioning with an intertrial interval of zero—S_1 is presented for a fixed period during which, after each response, the animal is free immediately to respond again.

The carryover concept provides an explanation of the well-known "partial reinforcement effect," at least as it occurs in massed trials (Hull, 1952). The greater resistance to extinction found in animals that are only "partially" reinforced during instrumental training (that is, reinforced on some trials—selected at random—and not on others) as compared with animals that are reinforced consistently may be attributed to the fact that the partial animals are reinforced for response to $S_1 + S_N$ whenever a reinforced training trial follows one or more unreinforced training trials; the consistent animals never are reinforced for response to $S_1 + S_N$, which is, of course, the stimulus complex present in extinction. It should be noted, however, that the partial reinforcement effect is relatively rare in classical conditioning (Bitterman & Shoel, 1970), a fact that has called into question the assumption that the same associative processes are brought into play by the classical and the instrumental procedures. Furthermore, performance during conditioning does not seem to be sustained as well by partial reinforcement in the classical as compared with the instrumental case.

The concept of sensory carryover may help to circumvent the difficulty for counterconditioning theory that is presented by conditioning–extinction and reversal experiments. It has been proposed that adjustment to conditioning–extinction series is based on discrimination between the stimulus complexes $S_1 + S_R$ and $S_1 + S_N$, the former reinforced consistently and the latter never (Wickens & Miles, 1954). A similar interpretation has been offered for the results of reversal training on the assumption that traces of the stimuli to be discriminated (S_A and S_B) also may be carried over from trial to trial (Hull, 1952). Adjustment to reversal training then may be thought of as based on the differentiation of stimulus complexes consisting of present discriminanda, traces of preceding discriminanda, and after effects of preceding reinforcements or nonreinforcements; for example, the animal is consistently reinforced for response to S_A + Trace $S_A + S_R$, S_A + Trace $S_B + S_N$, S_B + Trace $S_B + S_R$, or S_B + Trace $S_A + S_N$, but never reinforced for response to S_A + Trace $S_A + S_N$, S_A + Trace $S_B + S_R$, S_B + Trace $S_B + S_N$, or S_B + Trace $S_A + S_R$. Because stimulus traces are known to be extremely short lived, this interpretation is feasible only when trials are highly massed. Furthermore, the carryover interpretation does not help to account in terms of counterconditioning for the differential stability of the effects of reinforcement and nonreinforcement that (as already noted) are found in such experiments.

If extinction is to be explained in terms of inhibition, the source of the inhibition must be specified. One suggestion is that its source is stimulation—that the action of S_1 generates inhibition at its center (Pavlov, 1927). Another suggestion, offered in conjunction with a reward principle of excitation, is that the source of inhibition is action (reactive inhibition)—the occurrence of R_i generates an inhibitory tendency specific to the further occurrence of R_i that becomes connected to S_1 even while reward is producing an opposed excitatory tendency (Hull, 1943). In both cases, inhibition is assumed to develop on reinforced as well as on unreinforced trials, and the probability of response is assumed to increase during conditioning only because inhibition develops more slowly than excitation. Pavlov's idea that inhibition is a general property of the S_1 center (rather than specific to a given response) is consistent with his finding that a conditioned inhibitor developed for one response also comes to inhibit other responses, and that continuing extinction beyond the point at which the animal ceases to respond may reduce spontaneous recovery. It is consistent also with the finding that conditioning is retarded by prior unreinforced presentations of S_1 (Lubow & Moore, 1959), which, explained in Hullian terms, requires the assumption that some receptor-orienting response to S_1, important for conditioning, is extinguished during the prior unreinforced presentations.

Extinction presents a special problem for S–R contiguity theory that it does not for S–S contiguity theory or for S–R reward theory. The contiguous occurrences of stimulus and response at the outset of extinction are not sufficient, apparently, to strengthen further or even to maintain the tendency for the stimulus to evoke the response. A like problem is presented by "habituation"—the progressive diminution in the magnitude of an unlearned response to a stimulus (such as the rat's startle response to a loud noise) in the course of repeated evocation by the stimulus—a phenomenon that may seem intuitively to have much in common with extinction (Groves & Thompson, 1970). The S–R contiguity principle can be salvaged, of course, with the aid of a variety of supplementary principles, but for the sake of simplicity it may better be dispensed with entirely. The evidence for the formation of S–R connections in classical conditioning certainly is rather scanty. Insofar as the contiguity interpretation of instrumental conditioning is concerned, the central assumption—that the selection of R_i may be traced to new properties acquired by stimuli contiguous with S_2—can rest as easily on the S–S as on the S–R principle. The S–R contiguity principle is invoked explicitly mainly to account for the speed and precision of asymptotic performance, on the assumption that the sampling of proprioceptive feedback derived from scanning by the animal of its response repertoire is inherently too slow a process and must in some way be short circuited. If the S–R contiguity principle is to be abandoned, some concept of central scanning can be turned to as a substitute for peripheral scanning, but a simpler alternative, and one more susceptible to experimental test, is provided by the S–R reward principle.

AVOIDANCE AND OMISSION

The consideration of instrumental conditioning here has been limited so far to cases in which the response contingency is positive—that is, in which R_i produces S_2 (whether reward or punishment). Now cases in which the contingency is negative—R_i prevents or avoids S_2—are considered. Most of the information about negative contingencies comes from so-called "avoidance" experiments, in which S_2 is aversive.

Suppose that whenever a dog flexes its right forelimb (R_i) in response to a tone (S_1), the tone immediately is terminated and a brief shock to the forelimb (S_2), which otherwise would have been presented 10 sec after the onset of the tone, is omitted. Because the tone does not evoke flexion at first, the tone is followed by shock, just as in classical conditioning, and it soon comes (as does the shock) to elicit flexion. When that happens, however, the shock is withheld and the new response tendency begins to extinguish, whereupon the tone is paired with shock once more. At the outset, then, there may be some pronounced oscillation in the strength of the tendency, but the oscillation becomes less evident as training continues, and the probability of flexion may be very high at asymptote despite the fact that frequency of shock must at the same time be very low. Although the origin of the tendency for the tone to elicit flexion can be attributed entirely to classical conditioning—before the first response to tone, the classical and avoidance procedures are, in fact, identical—the asymptotic performance cannot be, because low probabilities of reinforcement are not adequate to maintain high levels of response in classical conditioning. That the avoidance contingency plays an important role may be demonstrated by comparing the performance of avoidance animals with the performance of yoked controls. Although the two groups of animals have exactly the same experience with the two stimuli—the same number of presentations of S_1 alone and the same number of S_1-S_2 pairings—the level of response to S_1 is much greater in the avoidance group than in the control group.

How does the contingency work? It may be assumed that a stimulus complex in the presence of which the animal frequently has been shocked (S_A) is replaced, on the occurrence of R_i, by a stimulus complex in the presence of which the animal never has been shocked (S_0). In the simplest case, S_1 is a component of S_A but not of S_0 (because S_1 is terminated by R_i), whereas S_p (the proprioceptive feedback from R_i) is a component of S_0 but not of S_A. It should be noted that performance is better under such circumstances than when S_1 remains on for a fixed period independently of response—that is, when $S_A = S_1$ and $S_0 = S_1 + S_p$ (Mowrer, 1950). In the widely used "Sidman" procedure, there is no warning stimulus (S_1), but shock is scheduled by a fixed-interval clock that is reset by R_i (again, S_0 and S_A are distinguished only by the presence or absence of S_p). Performance in Sidman avoidance is improved when R_i is made to produce some distinctive stimulus, such as a click, which then becomes a component of S_0 but not of S_A (Bolles & Popp, 1964). In general, the efficiency of avoidance conditioning seems to be closely related to the discriminability of S_A and S_0. It follows from this analysis that avoidance conditioning involves classical conditioning, not with partial reinforcement but with differential reinforcement—shock is paired with S_A but not with S_0. The response conditioned to S_A extinguishes only very slowly despite the low probability of shock when the animal performs at a high level, perhaps because the unreinforced presentation of S_A is curtailed, and its activating properties therefore "conserved," by response (cf. Solomon & Wynne, 1954). The better performance of avoidance animals relative to that of yoked controls may be attributable to the fact that control animals do experience partial reinforcement. For a control animal that responds when the avoidance animal with which it is yoked does not, S_p is paired with shock; for a control animal that fails to respond when the avoidance animal does respond, S_A is not paired with shock.

In some avoidance experiments, R_i is a dominant feature of response to S_2 that is readily conditioned to S_1 by a purely classical procedure. Experiments with dogs in which S_2 is shock to a forelimb and R_i is flexion of the forelimb fall into this first category, and so also do experiments with goldfish in the "shuttlebox" (a two-compartment apparatus in which the animal avoids shock in one compartment by going to the other). Shuttling in the goldfish may be considered an expression of general activity evoked by shock and classically conditioned to S_1 (Woodard & Bitterman, 1971). In such experiments, the only problem for a contiguity interpretation is presented by the stability of performance at asymptote despite the fact that R_i prevents the shock that produces it in the first place. One solution to this problem, in terms of the principle of conservation, already has been noted. It is necessary, however, to consider also a second category of avoidance experiments in which R_i to S_1 cannot be either developed or maintained by noncontingent pairing of S_1 and S_2—for example, lever-press avoidance in rats (Davenport & Olson, 1968). Although a stimulus paired with shock from the floor of a compartment in which a rat is contained may come to elicit a variety of responses, lever pressing has a relatively low probability of occurrence. The problem here is to account for the fact that lever pressing becomes a dominant response in avoidance conditioning (that is, when lever pressing prevents shock) but not in classical conditioning.

A contiguity interpretation of these results may begin with the assumption that, on nonavoidance trials, local stimuli experienced at the time of shock acquire aversive properties and come therefore to elicit withdrawal. The validity of this assumption is suggested by experiments on "secondary escape," which show that (given the opportunity) rats go quickly from a place in which they have been shocked in the past to a place in which they have not (Bitterman & Shoel, 1970). The proprioceptive feedback from any response in progress at the time of shock also may be assumed to acquire aversive properties and therefore to suppress that response whenever it subsequently is initiated. The contiguity interpretation of second-category avoidance is directly analogous, then, to the contiguity interpretation of escape training. It may also be that stimuli contiguous with the termination of S_A (transition to S_0) play the role assigned in escape training to stimuli contiguous with shock termination, exteroceptive components eliciting approach and proprioceptive components facilitating R_i once it has been initiated. Here again, however, a somewhat more straightforward interpretation is provided by S–R reward theory, which applies as well to first-category avoidance. If S_A acquires aversive properties in consequence of its pairing with shock, then R_i to S_A is rewarded by the termination of S_A (transition to S_0) just as, in escape conditioning, R_i is rewarded by the termination of shock. Whether the contiguity or the reward interpretation is correct, of course, the outcome of avoidance training must depend on the compatibility of R_i with the responses that S_1 tends to evoke in consequence of its pairing with S_2. For example, if an S_1 paired with shock tends to evoke crouching, a manipulative response such as lever pressing becomes difficult to establish. The common finding that avoidance training is more effective with mild than with intense shock (Bitterman & Shoel, 1970) may be explained in terms of the incompatible responses (such as freezing) produced by intense shock.

If the aversive S_2 of the avoidance procedure is replaced with a rewarding S_2, we have the "omission" or "DRO" procedure, the former designation emphasizing the negative contingency and the latter (which stands for "differential reinforcement of other behavior") emphasizing the complementary positive contingency (S_2 produced by responses other than R_i). Very little is known about the effects of omission training, interest in which is of rather recent origin. First-category omission experiments—with salivation in dogs (Sheffield, 1965) and key pecking in pigeons (Williams & Williams, 1969), responses that are classically conditionable—do not seem to produce asymptotic levels of performance comparable to those obtained in first-category avoidance experiments. In omission as well as in avoidance, the contiguity of S_1 and S_2 (when R_i does not occur) strengthens the tendency for S_1 to evoke R_i (Woodard, Ballinger, & Bitterman, 1974), but the nonoccurrence of S_2 seems to have a much more marked suppressing effect on R_i. In fact, the probability of key pecking by pigeons is *less* when an omission contingency is in force than under yoked control conditions in which response is without consequence but the probability of reinforcement is the same (Schwartz & Williams, 1972), a difference directly opposite in direction to that obtained in avoidance. One explanation of these results may be found in the now familiar assumption that competing emotional responses, evoked by the nonoccurrence of anticipated food, are conditioned to proprioceptive feedback from R_i, as well as to exteroceptive stimuli. We may look also to the contiguity with S_2, both of stimuli at localities other than that at which R_i is performed, and of responses other than R_i. According to S–R reward theory, of course, the tendency to make any response incompatible with R_i is strengthened by reward. Whatever the correct interpretation may prove to be, omission training provides a new arena in which the principles of conditioning suggested by results obtained with the older procedures may be tested.

CONCLUSION

Various phenomena have been described which it seems reasonable now to think of as general phenomena of vertebrate learning because they are found in members of rather diverse vertebrate groups. Some ideas about underlying processes, at least in the small number of species studied in any detail, have also been described, with full recognition that what appear to be identical phenomena at the behavioral level may in fact be produced by quite different processes. The broad, functional ideas that have been considered make no real contact with the neural substrate, but they are consistent with what is known about brains; readers familiar with the field will note that the deceptive language of human consciousness that so often serves to beg the question of process has been avoided. Although functional analysis is a precursor of neurophysiological analysis, it is not just the first phase of a task to be completed by the neurophysiologist, but an essential ingredient of a joint, interactive enterprise. Systematic comparative work is expected to yield a patterned array of functional similarities and differences that, together with neurophysiological, biochemical, and even paleontological data, can help to lay bare the mechanisms of learning.

REFERENCES

Amsel, A. The role of frustrative nonreward in noncontinuous reward situations. *Psychological Bulletin*, 1958, **55**, 102–119.

Bindra, D. A unified account of classical conditioning and operant training. In A. H. Black & W. F. Prokasy (Eds.), *Classical conditioning II: Current research and theory*. New York: Appleton-Century-Crofts, 1972.

Bitterman, M. E. Toward a comparative psychology of learning. *American Psychologist*, 1960, **15**, 704–712.

Bitterman, M. E. Phyletic differences in learning. *American Psychologist*, 1965, **20**, 396–410.

Bitterman, M. E., & Shoel, W. M. Instrumental learning in animals: Parameters of reinforcement. *Annual Review of Psychology*, 1970, **21**, 367–436.

Bolles, R. C., & Popp, R. J., Jr. Parameters affecting the acquision of Sidman avoidance. *Journal of the Experimental Analysis of Behavior*, 1964, **7**, 315–321.

Bower, G. H. A contrast effect in differential conditioning. *Journal of Experimental Psychology*, 1961, **62**, 196–199.

Brimer, C. J. Disinhibition of an operant response. In R. A. Boakes & M. S. Halliday (Eds.), *Inhibition and learning*. New York: Academic Press, 1972.

Brown, P. L., & Jenkins, H. M. Autoshaping of the pigeon's key peck. *Journal of the Experimental Analysis of Behavior*, 1968, **11**, 1–8.

Brownlee, A., & Bitterman, M. E. Differential reward conditioning in the pigeon. *Psychonomic Science*, 1968, **12**, 345–346.

Burns, R. A., Woodard, W. T., Henderson, T. B., & Bitterman, M. E. Simultaneous contrast in the goldfish. *Animal Learning and Behavior*, 1974, **2**, 97–100.

Crespi, L. P. Quantitative variation of incentive and performance in the white rat. *American Journal of Psychology*, 1942, **55**, 467–517.

Davenport, D. G., & Olson, R. D. A reinterpretation of extinction in discriminated avoidance. *Psychonomic Science*, 1968, **13**, 5–6.

Doty, R. W., & Giurgea, C. Conditioned reflexes established by coupling electrical excitation of two cortical areas. In J. F. Delafresnaye, A. Fessard, R. W. Gerard, & J. Konorski (Eds.), *Brain mechanisms and learning*. Oxford: Blackwell Scientific, 1961.

Gonzalez, R. C., & Diamond, L. A test of Spence's theory of incentive-motivation. *American Journal of Psychology*, 1960, **73**, 396–403.

Gonzalez, R. C., Gleitman, H., & Bitterman, M. E. Some observations on the depression effect. *Journal of Comparative and Physiological Psychology*, 1962, **55**, 578–581.

Groves, P. M., & Thompson, R. F. Habituation: A dual-process theory. *Psychological Review,* 1970, **77**, 419–450.

Harlow, H. F. Experimental analysis of the role of the original stimulus in conditioned responses in monkeys. *Psychological Record*, 1937, **1**, 62–68.

Hull, C. L. Goal attraction and directing ideas conceived as habit phenomena. *Psychological Review*, 1931, **38**, 487–506.

Hull, C. L. *Principles of behavior*. New York: Appleton-Century-Crofts, 1943.

Hull, C. L. *A behavior system*. New Haven: Yale University Press, 1952.

Kamin, L. J. Predictability, surprise, attention, and conditioning. In B. A. Campbell & R. M. Church (Eds.), *Punishment and aversive behavior*. New York: Appleton-Century-Crofts, 1969.

Lowes, G., & Bitterman, M. E. Reward and learning in the goldfish. *Science*, 1967, **157**, 455–457.

Lubow, R. E., & Moore, A. U. Latent inhibition: The effect of nonreinforced pre-exposure to the conditional stimulus. *Journal of Comparative and Physiological Psychology*, 1959, **52**, 415–419.

Maier, N. R. F., & Schneirla, T. C. Mechanisms in conditioning. *Psychological Review*, 1942, **49**, 117–134.

Morgan, C. L. *An introduction to comparative psychology*. London: Walter Scott, 1894.

Mowrer, O. H. *Learning theory and personality dynamics*. New York: Ronald Press, 1950.

Mowrer, O. H. *Learning theory and behavior*. New York: John Wiley & Sons, 1960.

Pavlov, I. P. *Conditioned reflexes: An investigation of the physiological activity of the cerebral cortex*. Translated by G. V. Anrep. London: Oxford University Press, 1927.

Pert, A., & Bitterman, M. E. Reward and learning in the turtle. *Learning and Motivation*, 1970, **1**, 121–128.

Peterson, G. B., Ackil, J. E., Frommer, G. P., & Hearst, E. S. Conditioned approach and contact behavior toward signals for food or brain-stimulation reinforcement. *Science*, 1972, **177**, 1009–1011.

Prewitt, E. P. Number of preconditioning trials in sensory preconditioning using CER training. *Journal of Comparative and Physiological Psychology*, 1967, **64**, 360–362.

Rescorla, R. A. Effect of US habituation following conditioning. *Journal of Comparative and Physiological Psychology*, 1973, **82**, 137–143.

Rizley, R. C., & Rescorla, R. A. Associations in second–order conditioning and sensory preconditioning. *Journal of Comparative and Physiological Psychology*, 1972, **81**, 1–11.

Schwartz, B., & Williams, D. R. The role of the response-reinforcer contingency in negative automaintenance. *Journal of the Experimental Analysis of Behavior*, 1972, **17**, 351–357.

Sheffield, F. D. Relation between classical conditioning and instrumental learning. In W. F. Prokasy (Ed.), *Classical conditioning: A symposium*. New York: Appleton-Century-Crofts, 1965.

Solomon, R. L., & Wynne, L. C. Traumatic avoidance learning: The principles of anxiety conservation and partial irreversibility. *Psychological Review*, 1954, **61**, 353–385.

Taub, E. Bacon, R. C., & Berman, A. J. Acquisition of a trace-conditioned avoidance response after deafferentation of the responding limb. *Journal of Comparative and Physiological Psychology*, 1965, **59**, 275–279.

Tennant, W. A., & Bitterman, M. E. Blocking and overshadowing in two species of fish. *Journal of Experimental Psychology: Animal Behavior Processes*, 1975, **104**, 22–29.

Thorndike, E. L. *Animal intelligence: Experimental studies*. New York: Macmillan, 1911.

Wagner, A. R. Effects of amount and percentage of reinforcement and number of acquisition trials on conditioning and extinction. *Journal of Experimental Psychology*, 1961, **62**, 234–242.

Wagner, A. R., Thomas, E., & Norton, T. Conditioning with electrical stimulation of motor cortex: Evidence of a possible source of motivation. *Journal of Comparative and Physiological Psychology*, 1967, **64**, 191–199.

Wickens, D. D., & Miles, R. C. Extinction changes during a series of reinforcement-extinction sessions. *Journal of Comparative and Physiological Psychology*, 1954, **47**, 315–317.

Williams, D. R., & Williams, H. Auto-maintenance in the pigeon: Sustained pecking despite contingent non-reinforcement. *Journal of the Experimental Analysis of Behavior*, 1969, **12**, 511–520.

Woodard, W. T., Ballinger, J. C., & Bitterman, M. E. Autoshaping: Further study of "negative automaintenance." *Journal of the Experimental Analysis of Behavior*, 1974, **22**, 47–51.

Woodard, W. T., & Bitterman, M. E. Classical conditioning of goldfish in the shuttlebox. *Behavior Research Methods and Instrumentation*, 1971, **3**, 193–194.

Woodard, W. T., & Bitterman, M. E. Autoshaping in the goldfish. *Behavior Research Methods and Instrumentation*, 1974, **6**, 409–410.

Woodard, W. T., Shoel, W. M., & Bitterman, M. E. Reversal learning with singly presented stimuli in pigeons and goldfish. *Journal of Comparative and Physiological Psychology*, 1971, **76**, 460–467.

10

Vertebrate Learning: Evolutionary Divergences

Kenneth H. Brookshire

Wright State University

The comparative psychology of learning is in a state of such rapid change, with so much disagreement as to strategy and principle, that it is difficult to summarize with any confidence. I can hope here only to represent the direction of movement of the field.

RESEARCH STRATEGY

The Subject of Study

The traditional approach of the comparative psychologist, based on striking similarities in the performance of a wide range of animals studied in the earliest experiments, was to assume a common set of learning processes in all animals (or at least in vertebrates), varying from species to species merely in a quantitative way. On this assumption, it seemed reasonable to try to devise a set of standard tasks designed to measure learning ability and then to determine the relation between performance on these tasks and taxonomic status. Unfortunately, however, performance on learning tasks is affected by a host of variables—sensory, motor, and motivational—that must be equated if meaningful phyletic comparisons are to be made.

Several solutions to the problem have been proposed. Harlow (1951) has been willing at least to compare closely related species (e.g., old world monkeys) under identical testing conditions on the assumption that interspecies differences in sensory, motor, and motivational factors are of little or no consequence, but the solution does not serve the avowed purpose of making comparisons of intelligence among vertebrates as a whole. Moreover, it rests on the dubious premise that natural selection has had no impact on sensory, motor, or motivational processes although (by definition) it has produced substantial morphological changes and is expected to have produced changes in learning. Even more troublesome is the finding that intraspecific variation in performance is so great that interspecies comparisons, even of old world monkeys, become meaningless (Warren, 1973). Another solution that has been proposed (Gossette, 1970) is that, for each nonlearning variable known to affect performance on a given task, the "range" of the effect be charted, providing interspecies comparisons in terms not of single performance scores but of multiple scores. Aside from the impracticality of this approach, intraspecific variation in performance is likely to be so great that it becomes difficult to interpret the performance of the species as a whole. Yet another and very popular solution to the problem has been to ignore it, with the result that a vast literature has accumulated on rates of classical and instrumental conditioning, slopes and asymptotes of learning set and reversal functions, maximal intervals in delayed-response tests, and other scores obtained under conditions in which test parameters have been controlled for individuals within species but varied unsystematically, and often without clear purpose, between species. Notwithstanding the problem of interpreting the differences reported, the scores show no meaningful correlation with taxonomic status (Brookshire, 1970; Warren, 1973).

A second approach to the comparative analysis of learning derives from the ethological study of behavior. The orientation of ethologists traditionally has been toward the study of fixed action patterns, that is, temporally integrated behavior of genetic or epigenetic origin that usually is highly stereotyped and seems to have some selective advantage. Indeed, the presumed function of the behavior in accommodating to the demands of the environment (feeding, reproduction, and the like) forms the basis of the classificational scheme that the ethologists have used (Klopfer & Hailman, 1967), and it is understandable enough for the opinion to have been established that all behavior, including learned behavior, must be studied under natural conditions, for only in this way can its adaptive value presumably be assessed. Furthermore (the argument runs), just as natural selection has produced specialized fixed action patterns that vary greatly among vertebrates, so also it may have produced specialized learning abilities that vary among vertebrates. From this point of view, the direct comparison of learning in different species under what are presumed to be analogous laboratory conditions is considered largely a wasted effort, because an interpretation of learning in any species can only be made in terms of the relationship of the response that is learned to its adaptive value and to the remainder of the animal's behavioral repertoire. Phylogenetic generalization, then, can only take the form of rules about the way in which these special learning abilities emerge through natural selection.

The third major approach to the comparative psychology of learning is that of Bitterman (1960, 1965, 1968, 1973). Its explication requires a little groundwork. Although the comparative psychologist must agree with the ethologist that species-specific learning abilities may exist and that they may well be studied in relation to the animal's total behavioral repertoire, the ethological emphasis on behavioral change in itself rather than on the underlying processes is unacceptable to him. The ethological theory, which has as its core the selective advantage of behavior, tends to ignore the fact that learning is an inferred process rather than an observed instance of a specific behavioral act, and that to understand the manner in which the process operates requires more than merely the statement that a behavioral change has occurred under certain conditions. Furthermore, the emphasis of the ethologists on field conditions does not lend itself as easily to the sort of research necessary to determine process laws as does the laboratory, although laboratory conditions that more nearly mimic or duplicate natural environments may well be used. Whereas it is true that field research on learning has not all been strictly descriptive, it is difficult to imagine the systematic variation of the conditions under which learning takes place that is necessary even for a first approximation to process laws being carried out in the natural environment.

Another concern of the comparative psychologist is that all vertebrate learning, even learning in lower vertebrates, does not appear to be of the specialized type conceived by the ethologists, the specific responses that emerge having little relevance to preconceptions about adaptive significance. Harlow (1958) has remarked that

> the study of animals under laboratory conditions reveals many learning capabilities whose existence is hard to understand in terms of survival value. . . . The observational accounts of these animals make it quite clear that problems of this level of complexity are never solved, indeed, they are never met in the natural environment. It is superficially difficult to see how a trait which was never used gave to an organism some slight selective advantage over another organism which did not use the trait because it did not have it [pp. 273–274].

The argument here is not that the concept of adaptive value has no place in the comparative study of learning but only that current information about animal learning does not fit well into the ethological framework. It seems probable that learning processes of a more generalized and flexible type than that contemplated by the ethologists have evolved in different vertebrates to meet certain not very probable but nonetheless biologically important environmental contingencies and that, because by their nature they are flexible, they can be utilized in the "artificial" learning situations created by psychologists in the laboratory. If so, then it is not quite proper to dismiss this learning as "unnatural" because it occurs in the laboratory. Indeed, the laboratory has provided information about learning that is far less likely to have been obtained in field research. Perhaps in these times it is wise to be wary of theoretical extremes: just as it is clearly wrong for the traditionalists to ignore imprinting, so also is it wrong in the new wave of enthusiasm for adaptive specializations of learning to ignore the basic work on classical and instrumental conditioning.

Bitterman's approach, like that of the ethologists, implies the discontinuity of learning within the animal series, and of course this is quite consistent also with modern concepts of evolution. To quote from one of his recent papers (1973):

> Functional resemblances are to be expected in closely related animals whose brains show rather clear structural resemblances, but it would be remarkable indeed—on the assumption that conditioning is a dominant function of the brain—if, corresponding to the sharp differences evident in the brain structure of, say, fishes and birds, there were not also important differences in conditioning. It would be even more remarkable if the mechanisms of conditioning were fundamentally alike in much more distantly related animals, such as cephalopods and fishes, whose brains seem to have evolved quite independently. Functional similarity in animals of different ancestry is, of course, to be expected in terms of common selective pressures and often carries with it some similarity of mechanism; given the laws of physics and the unique properties of available building materials, any functional problem confronting the organism admits only of a limited number of alternative structural solutions (Pantin, 1951). It may be argued, however, that convergence to the point of identity or even of seriously confusing similarity is extremely unlikely in "elaborately polygenic" behavioral systems (Simpson, 1964), and there is no good reason now to think that the resemblance between the mechanisms of vertebrate and invertebrate conditioning is any more intimate than that between the hand of an ape and the claw of a lobster [pp. 13–14].

Unlike the ethologists, however, Bitterman is not interested in relating learning in any specific case to its adaptive implications and, because of his interest in "process," his work is carried out under rigorous laboratory conditions, indeed with completely automated equipment for determining the order of events in his experiments and for recording responses. Setting out to test the traditional and still widely held view that the principles of learning are the same in all animals, Bitterman has taken the laboratory rat as his point of departure on the ground that it is the animal about the learning of which most is known. For comparison with the rat, he has selected several other vertebrate species (fishes, reptiles, and birds) similar enough to be studied in experiments analogous to those that have shaped our conceptions of learning in the rat, yet "different enough to afford a marked anatomical contrast" (Bitterman, 1965, p. 398). Comparisons are made not in terms of absolute scores but in terms of "functional relations." The first question asked is whether the phenomena of learning are the same in all of these animals—whether their performance in learning situations is "affected in the same way by the same variables"—on the theory that differences in functional relations may provide clues to differences in underlying processes. The results of these experiments are considered in the discussion (below) of divergences in what I call more flexible learning processes. Great attention has been paid to Bitterman's work, which represents an extension of the thinking of Schneirla (1950, 1959), his teacher, who has appreciated that most behavior turns out to be adaptive but has been more interested in the mechanisms underlying it, who has appreciated the importance of field observation but has emphasized the contributions that laboratory work can make.

Historical and Adaptational Analysis

Hodos and Campbell (1969) have distinguished between evolutionary models based on historical analysis and on analysis of adaptation. Together with King and Nichols (1960), they have provided a number of reasons to be skeptical of the fruitfulness of the historical (phylogenetic) approach in behavioral study: (1) behavior cannot be observed in extinct forms; (2) the soft tissue of the CNS (central nervous system) does not usually fossilize, making most structural correlations appropriate for the study of the evolution of learning impossible; and (3) the extinction of many intermediate forms rules out strict historical analysis of behavioral evolution based on observation. There are many possible sources of behavioral similarity among animal forms (such as parallelism and convergence) that make historical inferences based on observation of living species dangerous. Historical inferences about learning certainly cannot be based on the assumption that more complex processes have been evolved from simpler ones because there are many cases of secondary simplification of morphological structures among animals (Klopfer & Hailman, 1967); nor can they be very credibly based on other assumptions about order of emergence, such as that instinctive behavior is ancestral to learned behavior (or vice versa), or that more stereotyped learning processes (special abilities) either precede or succeed the development of more flexible learning processes in any evolutionary line. There seems to be at present, therefore, little hope for a historical analysis of behavior. Because most of the functions of the CNS involve a behavioral output, it may be possible to defend the gross generalization that total behavioral lability is correlated with the evolutionary divergences of encephalization found in fossil remains, but this is hardly a very satisfying conclusion for the student of learning, particularly as it is both inferential and untestable for the reasons mentioned above.

Hodos and Campbell (1969) have suggested that perhaps the historical analysis of learning can be saved by studying more closely related forms, that is, animals with common ancestors. If a choice is made of more "primitive" animals (with characteristics that are little changed from ancestral forms), perhaps learned behavior can be correlated with such quasi-evolutionary descent. Masterton and Skeen (1972) have pointed out, however, that for the comparative psychology of learning this is only half the problem. The other half has to do with determining the "subject of study" (what sort of learning is to be measured and how the results are to be interpreted). Masterton and Skeen present a novel approach to this problem: on the basis of comparative neuroanatomy, they have selected three animals that represent a trend in brain structure (prefrontal system) among mammals thought to be "primitive" in the above sense. Then they selected a learning problem (delayed alternation) for which the available evidence suggests a dependence on an intact prefrontal system—although there is also a dependence on an intact caudate nucleus (Rosvold & Delgado, 1956;

Woodburne, 1971). Their results indicate a relationship between performance with long delay, size of the prefrontal system, and phyletic status. The data are impressive, the approach is novel, and the study represents one of the very few attempts (perhaps the only attempt) to apply the Hodos-Campbell suggestion to research on the historical evolution of learning. It is a shame, therefore, that its prospects are not brighter. Delayed alternation is not a task on which performance can be related with much confidence to the integrity of a single structure in the brain. Nor are there likely to be many such tasks (which means that the behavioral work stemming from such an approach must be quite limited), and, even if there should be, the evidence for localization of function must come from other species. For example, the data on which the involvement of the prefrontal cortex in delayed alternation is based have come from studies of higher mammals. Karamyan (1965) has warned that anatomical localization of behavioral functions may vary with taxonomic status, and therefore the Masterton and Skeen results require confirmation of strict localization in the prefrontal cortex of the species they studied if they wish their research to represent a combined evolutionary analysis of both structure and function. Indeed, delayed response and delayed alternation are found in nonmammalian vertebrates, such as tawny owls (Marcus, 1959), that have no prefrontal cortex and even in such invertebrates as the octopus (Dilley, 1963). Obviously this complicates the problem.

Wilcock (1972) has suggested that behavioral phylogeny can be replaced by psychogenetics. From this point of view, the processes underlying genetic control of learning may be studied, but the actual evolution of learning is ignored, as is the mechanism of natural selection. Although psychogenetics certainly can make an important contribution to behavioral phylogeny, it simply is not an alternative strategy for solving many of the problems and questions that emerge regarding phyletic comparison.

An alternative to a historical analysis of the evolution of learning is the analysis of adaptation. This approach gives no clues to specific sequential patterns in the evolutionary development of learning but may provide a set of principles relating learned behavior to the particular demands of the habitat in which each species has evolved. In an analysis of learning as biological adaptation however, it must be borne in mind that not all evolutionary change is so directly in response to the demands of the environment. During the evolution of a successful group of animals there also is continuous improvement in efficiency of function and, in vertebrates especially, progressively greater independence of the external environment, as, for example, in the regulation of body temperature (Carter, 1967). These changes are adaptive only if the term is used very broadly.

A FRAMEWORK FOR THE COMPARATIVE STUDY OF LEARNING

Implicit in the thinking of many investigators seems to be the assumption that responses to the demands of the environment can be described as falling along a continuum from essentially rigid patterns of activity (highly stereotyped responses elicited by a narrow range of stimuli) to highly flexible patterns of activity (variable responding to more complex and less consistent stimulus situations, with neural mechanisms providing only the "rules" by which particular responses are selectively made to particular stimuli). Three points, or regions, may fairly easily be discerned on this continuum. At one end are the fixed action patterns (FAPs) of the ethologists. These behavioral sequences usually are quite stereotyped and essentially invariant among the members of a species, although the neural processes producing FAPs are quite sensitive to selective pressures, and many instances can be found in which closely related species show considerable behavioral divergence. For example, Huxley (1941) has described two species of pheasant that differ greatly in their foraging behavior, one species scratching for food, the other digging for it with the beak. There is, presumably, little contribution of experience to the development of the processes underlying FAPs other than that they are developed and displayed in an environmental context that sets limits on the expression of the trait. There has long been an argument (Moltz, 1965) as to just what role is played by the environment in the ontogeny of FAPs—whether the environment is "benignly supportive" or "actively implicated" in determining the organization of each system. Probably the role of the environment is no more consistent in its influence from species to species and from process to process than are the FAPs themselves, but few are likely to argue in any case that learning plays a prominant role in the expression of these behavior patterns. FAPs therefore lie generally toward that end of the continuum characterized by nonlability of behavior and minimal effect of experience on the neural processes controlling the behavior.

Somewhere further along this continuum are behaviors in which experience plays a greater role but in which at least some of the most important features have been laid down phylogenetically—behaviors that reflect what I shall call "special learning abilities." Although FAPs usually involve activities of vital importance to individual (or species) survival, it is also true that they are responses made to predictable environmental situations. Examples include consummatory activities, threat displays, courtship and sexual responses, and maternal behavior. In each case, FAPs are controlled by a narrow range of stimuli or "releasers" that are met frequently in the known geographical or ecological range of the species. For animals that have the characteristic of wider geographical range and correspondingly greater likelihood of meeting genetically unanticipated yet critical situations, FAPs may sometimes have been nonadaptive, and new systems may therefore have evolved, more flexible but still highly specialized in the sense of appropriateness to a set of rather well-defined environmental circumstances. For example, an animal may have some preformed defensive reaction to noxious cutaneous stimulation that it can learn to make in response to any neutral stimulus which, in a particular environment, predicts the noxious stimulus. Here there has been a gain in flexibility because the range of eliciting stimuli has been broadened, although the pattern of action remains fixed, and the arrangement is adaptive only to the extent that the response is appropriate to the range of conditions likely to be encountered by the animal. If different reactions are required, now one and now another, an animal that has a strong preference for one of them can adjust readily in some situations but poorly in others. For example, if the task is to flee from a given location when a signal is presented (because the signal predicts electric shock), rats learn very quickly to make the response, whereas if the task is to stay and to press a bar to avoid the same shock, learning proceeds very slowly and indeed some animals may not learn at all (Bolles, 1970). Freeing the form of response to shock improves adjustment in some situations but impairs it in others. By this analysis, then, speed of learning depends on three variables: (1) the number and type of genetically encoded relationships involved in the process; (2) the congruity between those relationships and the requirements of the learning situation; and (3) the nature of the associative processes for any species.

At the other end of the continuum are behaviors primarily under the control of nonspecialized learning processes. Predispositions to respond in particular ways to particular stimuli are minimal. The ability of the animal to deal with environmental contingencies is limited only by the learning processes themselves and by the availability of sensory and motor equipment to process the information from the environment and to mediate appropriate responses to it.

Because most of the special learning abilities that are known seem to involve simple approach and withdrawal responses to stimuli of high biological significance, their adaptive value is easy to appreciate, which may not be true of the more flexible learning processes. As Harlow (1958) has said, the performance of rhesus monkeys on complex problems in the laboratory is difficult to relate to biological adaptation, because the animals never meet such situations in their natural habitat, but it is possible to argue that the learning *processes* used in solving these problems have adaptive value. Their evolution may reflect the fact that in the unpredictable environment of the macaque the number of situations likely to be encountered is too great to be handled adequately by more specialized mechanisms. It is reasonable to think that animals living in a constant ecological niche may develop few nonspecialized processes, whereas animals in variable niches may develop a greater number. Furthermore, in the course of the independent evolution of the classes of vertebrates, different processes may have developed in response to the same environmental problems. In essence, the argument is that here is a case (I assume reasonably rare) in which evolution of nonspecialization is adaptive. (What are the nonbehavioral instances of this?) Animals do learn in incredibly artificial situations, affected by contingencies most unlikely in nature; there have been many decades of laboratory work on animal learning to support this statement. Presumably they can do so only because some of the processes underlying their behavior are not tied to special situations or responses.

Although FAPs are found to vary from species to species, the special learning processes may be the same within all members of a genus or family. "Flexible" or nonspecialized learning processes may vary only among orders or classes for their biological utility is so broadly based that given their introduction in a phylogenetic line it is hard to see how their function can have been replaced by more highly encoded systems (unless there has been a dramatic shrinkage in the size and complexity of the niche). Students of animal behavior have not been inclined to favor attempts to correlate learning processes with taxonomic status (either on the assumption that the laws of learning are the same throughout the animal kingdom or because they believe learning should only be studied in relation to its adaptive value in the natural environment). I have presented the view that the learning processes studied in the laboratory probably are biologically adaptive in a broad sense and that, if so, they must vary with taxonomic status as do other systems in the brain, such as sensory systems (Diamond & Hall, 1969).

DIVERGENCES IN SPECIAL LEARNING ABILITIES

I shall consider here some descriptions of learning in animals that point to the evolution of processes specially adapted to meet environmental contingencies highly critical for survival.

Species-typical responses based on the nature of reinforcement often are found to affect performance in learning situations. The fact that rats learn readily to run to avoid impending electric shock but do not easily learn to press a lever to avoid the same shock (Bolles, 1970) may be understood in terms of the unconditioned response to the shock. Pigeons peck readily at a key the illumination of which is paired with food even though the response does not produce the food; moreover, pecking is maintained even when a peck turns off the key light and prevents the delivery of food (Williams & Williams, 1969). Breland and Breland (1966) have reported a number of cases in which interference with instrumental learning in pigs, raccoons, and chickens may be traced to competition between the response required by the task and the natural response of the animal to food. For example, pigs trained to carry coins over a distance of several feet persist in dropping the coins and rooting them.

Much of the learning connected with species identification shows stimulus selectivity. The responses of herring gulls to their young are preformed, but they must learn to distinguish their own young from others (Tinbergen, 1953) and because they do so very efficiently, there is the suggestion of a predisposition to make such discriminations. Similarly, nidifugous birds rapidly acquire the following response and show a preference for members of the same species (Gottlieb, 1965). Preferences of this sort have been described variously as implying a "search image" (Hinde, 1970) or a "sensory template" (Marler, 1970). Young white-crowned sparrows exposed to the songs of other species develop normal song patterns and when isolated from members of their own species develop normal song patterns, although without local dialects, but birds deafened in their youth develop extremely abnormal songs (Konishi, 1965; Mulligan, 1966). Therefore, a bird must hear its own voice to produce the species song pattern, and yet song development occurs selectively in the presence of multiple song patterns (that is, the conspecific song is learned). In many such cases so-called "critical periods" are involved—which is to say that the learning is largely limited to a narrow age range—but not in all. Critical periods are themselves the result of phylogenesis, and they are likely to be found in learning situations correlated with critical changes in the life cycle, such as those having to do with sexual and parental behavior and with species identification. Situations likely to be encountered throughout the life span cannot be met with such age-limiting processes and remain adaptive.

One special ability that has been studied extensively in recent years has been taste-aversion learning, and this work has been particularly helpful in permitting understanding of some of the characteristics of specialized learning processes. When an animal ingests a specific food and is later poisoned, the food subsequently is rejected. The rejection response itself may be viewed as preformed to accommodate adjustment to noxious substances, but what is surprising is the nature of the associative process. First, contrary to what is found in most other conditioning situations, aversions may develop even with intervals between the conditioned stimulus (CS) and the unconditioned stimulus (US) of several hours (Smith & Roll, 1967). Furthermore, there is some dependence on the novelty of the CS—an aversion is not formed if the substance ingested is one with which the animal has prior familiarity (Revusky & Bedarf, 1967; Wittlin & Brookshire, 1968). Finally, there is the factor of "belongingness"—a selectivity in the development of aversions that seems to be consistent with the manner in which the animals forage. Rats form associations much more readily between the taste of a substance and gastrointestinal upset than between auditory or visual stimuli and gastrointestinal upset (Garcia, McGowan, Ervin, & Koelling, 1968). If, however, electric shock is substituted for illness as the noxious stimulus, an association with visual cues is formed more readily than an association with taste cues. The "belongingness" principle operates differently with bobwhite quail. These animals readily learn to avoid a fluid on the basis of color as well as taste when they are made ill after drinking (Wilcoxon, Dragoin, & Kral, 1971), presumably because visual cues are important for food seeking and food selection in quail, whereas for the nocturnal rat they are not. Unfortunately, there is a paucity of data on interspecific comparisons with respect to the conditioning of food aversions.

Konorski and his associates (Dobrezecka, Szwejkowska, & Konorski, 1966) have reported a type of predisposition in dogs that seems to be different from those already described. In the training of dogs with food as reward, differences in auditory quality (e.g., metronome versus buzzer) are better discriminated than differences in auditory locus (front versus back) when a "go—no-go" procedure is employed (that is, response with the right leg is rewarded if one stimulus is presented but not if the other is presented). Opposite results are obtained, however, when the task of the animal is to respond with the right leg to one stimulus and with the left leg to the other. Seligman (1970) has characterized as "prepared" all instances of learning in which association proceeds on the basis of an innate "tendency" for certain "events" to be associated. This classification ignores the problem of which events are involved—cue, response, or reinforcement. Rozin and Kalat (1971) have treated specialized learning as "diverse" because it is adaptive and tied to ecological considerations. The position here is that there is a sensible middle ground, where adaptive considerations are not ignored but where the data can be ordered in terms of the kinds of "intrusion" on broad associational processes that have taken place in different taxonomic groups. This approach neither preserves traditional learning theory by adding a new correlate nor destroys it by a process of atomization. It is consistent with the succeeding section of this chapter in that learning "principles" are considered to be useful in delineating the methods by which animals adapt to their environments.

DIVERGENCES IN MORE FLEXIBLE LEARNING PROCESSES

Because data useful for the analysis of how learning processes may differ among animals are not in abundance, the description of the present state of knowledge must necessarily be both risky in terms of empirical accuracy and based on rather gross assumptions about the nature of the processes being investigated. Only in the laboratory rat are there enough data to permit reasonably viable descriptions of process—and even here considerable room for error remains.

Processes Involved in Responding for Reward

Evidence of divergence is provided by attempts to demonstrate in other vertebrates three phenomena found in instrumental learning experiments with laboratory rats. The first is the incentive contrast effect of Crespi (1942). If groups of animals are trained to make an instrumental response for high or low reward until they reach asymptotic performance levels, and if the amount of reward for each group then is shifted to the other value, the performance of the animals shifted from high to low reward is disrupted, falling below the level of control animals trained from the outset with low reward ("negative contrast" or "depression"), whereas the performance of animals shifted from low to high reward improves to a level beyond that of control animals trained from the outset with high reward ("positive contrast" or "elation"). The magnitude of the depression effect increases with the extent of the change in amount of reward, rats shifted from high to low reward performing more poorly after the shift than do rats shifted from moderate to low reward (Gonzalez, Gleitman, & Bitterman, 1962). The second phenomenon is an inverse relation between amount of reward in training and resistance to extinction, rats trained with high reward extinguishing more quickly than rats trained with low reward (Hulse, 1958). Here again the animals seem to be influenced by the extent of change in amount of reward, and the inverse relation between amount of reward and resistance to extinction may be regarded as a special case of Crespi's depression effect if extinction is regarded as a shift to zero reward. The third phenomenon is the partial reinforcement effect (PRE): rats that are inconsistently (partially) rewarded in training—say, on a random 50% of trials—extinguish less rapidly than animals that are consistently rewarded in training, even though the total number of rewarded trials is greater for the consistent group (Sheffield, 1949; Weinstock, 1954). Resistance to extinction in partially reinforced rats actually increases with the number of consecutive unrewarded trials in training, the total number of rewarded and unrewarded trials remaining the same (Gonzalez & Bitterman, 1964).

Now it should be perfectly obvious that laboratory rats do not always behave in the manner described in each of these cases. (Indeed, it is in part through the nature of the conditions under which the behavior is not seen that something has been discovered about the processes that control it.) For example, the demonstration of positive contrast in runway situations appears to be hindered by a ceiling effect. If control animals receiving large rewards already are running at the limit of their speed, animals shifted from low to high reward cannot demonstrate better performance (Schrier, 1967). Several investigators (e.g., Shanab, Saunders, & Premack, 1969) have been able to obtain elation by using procedures (such as delay of reward) designed to reduce running speed in the control condition. The depression effect is not found when rats are tested in a low-drive state (Ehrenfreund, 1971) or when a long "retention interval" intervenes between the first and second stages (Gleitman & Steinman, 1964). In spaced trials (that is, with a relatively long intertrial interval), the PRE is not obtained when the amount of reward is small (Sheffield, 1949; Hulse, 1958). The significance of restrictions on the conditions under which various phenomena can be observed is twofold: first, because the behavior does not always occur in rats, a null instance with another species does not necessarily imply the operation of a different sort of learning process. Second, the fact that a given phenomenon appears in two species does not necessarily mean that the underlying processes are the same in both, for different conditional restrictions may point to a difference in process.

Crespi effects have not appeared in experiments with goldfish (Lowes & Bitterman, 1967; Gonzalez, Potts, Pitcoff, & Bitterman, 1972; Mackintosh, 1971). There seems to be some difference of opinion regarding the exact manner in which goldfish adjust to reduction in amount of reward. In Bitterman's experiments (with spaced trials) animals switched from high to low reward do not show any reduction in speed. Mackintosh (with massed trials) has found highly variable performance in a group of goldfish switched from high to low reward, but the general slopes of the curves he presented suggested that the performance might have fallen to the level of the low-reward control group if training had been continued longer. For present purposes, however, this difference is unimportant, because the available evidence is entirely consistent in suggesting the absence of Crespi effects in goldfish. Goldfish also fail to display incentive contrast when groups trained with high and low reward are compared with respect to their resistance to extinction; unlike rats, goldfish show greater resistance to extinction after training with high reward than after training with low (Gonzalez & Bitterman, 1967; Gonzalez, Holmes, & Bitterman, 1967; Mackintosh, 1971; Gonzalez et al., 1972). Painted turtles appear to behave in such experiments much like goldfish (Pert & Bitterman, 1970); they fail to show the depression effect and their resistance to extinction is greater after large than after small reward. Pigeons show the inverse relation between amount of reward and resistance to extinction (Brownlee & Bitterman, 1968) but they have not yet been studied in Crespi experiments. It is unfortunate that there are no data on other vertebrate species, especially on animals representing divergent lines of those described. Such data are necessary before these contrast effects can be placed in an evolutionary context.

Crespi effects often are referred to as "successive contrast effects" because the different amounts of reward are encountered in successive series of training sessions. The inverse relation between resistance to extinction and amount of reward also may be thought of as successive contrast for the same reason. Another negative contrast effect found in rats usually is described as "simultaneous" because the two amounts of reward are encountered concurrently in each session: rats trained with high reward for response to one stimulus and with low reward for response to another stimulus (each stimulus being presented on half the trials) respond less readily to the low-reward stimulus than do rats trained with low reward for response to both stimuli (Bower, 1961). The same effect is found in pigeons (Brownlee & Bitterman, 1968). The fact that goldfish also show the effect (Burns, Woodard, Henderson, & Bitterman, 1974) suggests that despite the intuitive similarity of simultaneous and successive contrast the two phenomena may be mediated by different mechanisms. This is an instance of what Bitterman (1965) terms "phylogenetic filtration." Consider two phenomena, A and B: if, in any animal tested, we find A and B, or neither, then the hypothesis is tenable that the same processes underlie both; if, however, an animal is found that shows A but not B, or B but not A, it may be suspected that the underlying processes are different.

Both goldfish and African mouthbreeders show the PRE in massed trials but, as compared to the rat, under a rather restricted range of conditions (Wodinsky & Bitterman, 1959, 1960; Gonzalez, Eskin, & Bitterman, 1962, 1963; Gonzalez & Bitterman, 1967). A partial reinforcement experiment may be done in either of two ways: trials may be equated (which means that the consistent animals have more reinforcements than the partial) or reinforcements may be equated (which means that the partial animals have more trials than the consistent). If trials are equated (as usually is the case in rat experiments because the equated-reinforcements PRE does not present as difficult an explanatory problem), the fishes studied do not show the PRE when the amount of reward is small and a partial reinforcement schedule is used that involves only short sequences of unreinforced training trials. Bitterman's interpretation is that the weak tendency for inconsistent reinforcement to increase resistance to extinction under these conditions is masked by a greater resistance to extinction in the consistent group that is produced by a greater frequency of reinforcement. Independent tests in fact suggest that goldfish and mouthbreeders are more sensitive than are rats to frequency of reinforcement (Gonzalez, Eskin, & Bitterman, 1961; Gonzalez, Holmes, & Bitterman, 1967). To demonstrate the PRE in fishes, it is necessary only to equate reinforcements or, with trials equated, to heighten the effect of inconsistent reinforcement by using large reward or by incorporating long runs of unreinforced trials in the schedule of partial reinforcement. When, however, trials are widely spaced—when the intertrial interval (ITI) is as long as 24 hr—the fishes studied do not show the PRE at all, even with equated reinforcements, large reward, and long runs of unreinforced training trials (Longo & Bitterman, 1960; Gonzalez, Behrend, & Bitterman, 1965; Schutz & Bitterman, 1969). As do fishes, turtles (*Chrysemys* and *Pseudomys*) show the PRE in massed trials (Murillo, Diercks, & Capaldi, 1961) but not in spaced trials (Eskin & Bitterman, 1961; Gonzalez & Bitterman, 1962; Pert & Bitterman, 1970), and lizards (desert iguana) also have failed to show the PRE in spaced trials (Graf, 1972).

The fact that some animals show the PRE in massed but not in spaced trials has suggested that the spaced-trials PRE, which occurs in pigeons (Roberts, Bullock, & Bitterman, 1963) as well as in rats, may be mediated by mechanisms that are different, at least in part, from those responsible for the massed-trials PRE. Gonzalez and Bitterman (1969) have expressed the view that the spaced-trials PRE is nothing more than negative contrast of the successive type; analysis of the spaced-trials PRE, which occurs in rats only with large reward, indicates that it is caused by decreased resistance to extinction in the consistent group. This interpretation is supported by the fact that fishes and turtles, which fail to show the spaced-trials PRE, also fail to show either Crespi's depression effect or the inverse relation between amount of reward and resistance to extinction, whereas pigeons, which (like rats) do show the spaced-trials PRE, also show the inverse relation between amount of reward and resistance to extinction. According to Gonzalez and Bitterman, the massed-trials PRE in fishes and turtles can be understood in terms of sensory carryover (the partially reinforced animals are frequently reinforced in training for responding to the sensory aftereffects of nonreinforcement in preceding trials—stimuli to be encountered again in extinction—while the consistent group is reinforced only for responding to the aftereffects of reinforcement). The same animals do not show the spaced-trials PRE because stimulus traces decay rapidly and because the processes responsible for successive contrast are absent. In rats and pigeons, the massed-trials PRE is produced both by sensory carryover and by contrast, whereas the spaced-trials PRE is a pure contrast effect.

In summary, then, although all vertebrates investigated show an incentive mechanism in the sense that they perform at a higher level for large as compared with small reward, there is evidence to support the hypothesis that there has been an early divergence making for successive contrast in birds and mammals but not in reptiles or fishes. All vertebrates tested so far also display the PRE, even if some do so under limited circumstances. Whether the processes underlying the PRE are the same in all species is unclear, because it is unclear what those processes are even in the well-studied rat. The rat data are complex and theoretical interpretations vary (Amsel, 1967; Capaldi, 1967). Nevertheless, because the PRE is such a pervasive phenomenon in rats and occurs under a much narrower range of conditions in goldfish, mouthbreeders, and turtles, the possibility of class differences in mechanism cannot be ruled out.

Processes Involved in the Transfer of Discrimination Learning

Suppose that an animal is required to select one of two stimuli (call them A and B), one of the alternatives being consistently rewarded and the other never rewarded. After learning has taken place, its effect on the learning of another discrimination problem or a succession of problems can be studied. When experience with one problem is found to affect performance on a subsequent problem, whether positively or negatively, it can be said that there has been "transfer" of training. The analysis of transfer provides valuable cues to the processes involved in discrimination learning.

A kind of transfer experiment that provides further evidence of evolutionary divergence in learning is the reversal experiment. If rats are trained with A positive and B negative, and if the problem is then reversed (B now positive), the animals show such a strong tendency to continue to respond to A that they may take more trials to learn the reversal than the original problem; that is, they show negative transfer. If the rats then are trained in a series of further reversals, one reversal per day, the negative transfer from each reversal to the next (evidenced by error frequencies greater than chance on the early trials of each reversal) tends to diminish, until finally the animals begin each reversal problem at about the chance level of performance (Bitterman, Wodinsky, & Candland, 1958). This decline in negative transfer in the course of reversal training has been attributed to "proactive interference," the tendency for prior learning to interfere with the retention of subsequently learned material (Gonzalez, Behrend, & Bitterman, 1967). The assumption is that training in prior reversals comes increasingly (as the number of prior reversals increases) to interfere with the retention of the stimulus preference established in Reversal N over the interval between Reversal N and Reversal $N + 1$. There is reason to believe, however, that proactive interference does not appear in goldfish and African mouthbreeders, their errors on the early trials of each reversal problem remaining constant as reversal training continues (Wodinsky & Bitterman, 1957). Setterington & Bishop (1967), working with mouthbreeders, have found modest improvement over reversals in performance on the first five trials of each reversal, but Trial 1 data are not presented, and it should be noted that the finding has not been replicated by Behrend and Bitterman (1967) using a somewhat longer ITI. Gonzalez, Behrend, and Bitterman (1967) trained goldfish on a series of 2-day discrimination reversals (positive and negative stimuli reversed every 2 days), assessing the degree of proactive interference in terms of a retention score based on the difference between performance on the last five trials of Day 1 and the first five trials of Day 2 in each reversal. They have found that the retention of goldfish remains constant throughout the series of reversals, whereas the retention of pigeons trained under analogous conditions declines progressively. Behrend, Powers, and Bitterman (1970) have measured retention directly 1 day and 2 weeks after each of a series of reversal problems. Again, retention in pigeons declines as a function of the amount of prior reversal training, whereas the retention of goldfish does not.

Engelhardt, Woodard, and Bitterman (1973) have reported several new experiments exploring training variables that may improve the efficiency of reversal learning in goldfish. The new data confirm that goldfish can display overall improvement in reversal learning but give little evidence of proactive interference in these animals. Because many earlier studies on discrimination learning in fishes (Bitterman et al., 1958; Warren, 1960; Behrend, Domesick, & Bitterman, 1965) showed an absence of improvement in reversal learning altogether, Bitterman for a time entertained the hypothesis that the processes responsible for improvement in other vertebrates did not operate in fishes. The issue now, however, is not whether progressive improvement occurs in fishes, but how it occurs, although the narrow range of conditions in which it occurs may point to some difference in process. (Improvement in discrimination reversal has been sought in approximately 24 mammalian species, a monotreme, 12 avian species, four reptilian species, and four species of fishes, but to my knowledge there is not a single report of failure to find progressive improvement except in fishes.) Unfortunately, only a few of the many studies of reversal learning in nonhuman vertebrates provide sufficient data for the analysis of the processes underlying reversal learning, their authors usually being concerned merely with the question of whether or not improvement occurs. Bitterman (1972) has suggested that "go–no–go" situations may permit more detailed analysis of the mechanisms of improvement than do the more commonly used choice situations. This is because choice is determined jointly by the properties of the two stimuli, whereas in "go–no–go" experiments the properties of the stimuli are measured in terms of the readiness of the animal to respond to each of them alone (Woodard, Schoel, & Bitterman, 1971; Tennant & Bitterman, 1973a).

Much of the work on (as well as much of the positive evidence for) process differences in nonmammalian learning is owed to Bitterman and his associates. It is a shame that more have not joined the search, for confidence that there are process differences in, say, goldfish is weakened by the fact that the data have been generated primarily in one laboratory. Bitterman also has helped to cloud the issue by (a) emphasizing qualitative differences in behavior (e.g., no improvement in habit reversal); (b) describing his approach at one point in terms of the evolution of intelligence when in fact he can make only meager statements about historical evolution based on his method; (c) describing learned behavior as "ratlike" or "fishlike," which seems to enrage some students of comparative psychology who emphasize that considerable specialization has taken place within an order or a class. There is no need to restrict the range of evolutionary possibilities in this way, although, of course, it is difficult to prevent oneself from generalizing in the absence of fuller knowledge on a broader range of animals. Somehow it seems to be considered that the research strategy of Bitterman is tested when one of his hypotheses is tested by experiment, but that is to be guilty of drawing conclusions at least as unwarranted as those Bitterman is accused of drawing: it is said that a few experiments showing that improvement in reversal has not taken place in, say, a species of frog in no way show that reversal learning can never be found to take place in any species of frog. Although this merely says that it is difficult to generalize from little evidence, it is just as incorrect to deduce that the research strategy is unproductive or incorrect when the results of a few experiments do not conform to a particular hypotheses of the scientist.

Improvement in reversal cannot be attributed to decline in negative transfer alone, because there is also an increase over reversals in the rate of learning within problems as a result of which the animals may make substantially fewer errors at asymptote than they do even in the original problem. According to attention theory (Sutherland & Mackintosh, 1971), animals respond selectively to stimuli, learning to attend to relevant dimensions (defined by differential reinforcement) and to ignore irrelevant dimensions. When the relevant dimension is unchanged from one problem to the next (as is the case in reversal training) performance tends to be better (other things equal) than when the relevant dimension changes. The fact that overtraining on a given problem sometimes is found to produce faster reversal than does mere training to criterion (the so-called "overlearning–reversal effect," or ORE) is explained on the assumption that attention to the relevant dimension continues to be strengthened in the course of overtraining while specific stimulus preferences reach asymptote at criterion. Some of the best evidence for attention theory comes from experiments on intradimensional (ID) versus extradimensional (ED) transfer, in which there is no problem of competing stimulus preferences. Suppose that two groups of animals are trained with stimuli varying in two dimensions (say, form and color), one dimension relevant and the other irrelevant. Then the same animals are trained with entirely new stimuli varying in the same two dimensions, the relevant dimension being the relevant dimension of the first stage for one group (ID shift) and the irrelevant dimension of the first stage for the second group (ED shift). Experiments with rats (Shepp & Eimas, 1964), monkeys (Shepp & Schrier, 1969), and pigeons (Mackintosh, Lord & Little, 1971) consistently show better ID than ED performance.

Sutherland and Mackintosh (1971) have suggested that fishes are not as likely as other vertebrates to show progressive improvement in reversal because their attention to the relevant dimension is strengthened to a lesser extent in the course of reversal training, but the validity of the attentional interpretation of improvement in animals that do consistently show it is by no means established (Schade & Bitterman, 1966; Gonzalez & Bitterman, 1968). Comparative work on attention in fishes is still only in its earliest stages. Two experiments have failed to show the ORE in fishes (Warren, 1960; Mackintosh, Mackintosh, Safriel-Jorne, & Sutherland, 1966), but the meaning of these results is in doubt because the ORE in other vertebrates is the exception rather than the rule (Lukaszewska, 1968). More significant perhaps in view of the results for other vertebrates is the recent finding by Tennant and Bitterman (1973b) that ID transfer is no better than ED transfer in goldfish. The results of a shift experiment with turtles also fail to conform to expectations from attention theory (Graf & Tighe, 1971).

Attention theory does not predict (and certainly cannot account for) all the characteristics of discrimination learning and transfer. Another general process that has been proposed is the formation of "strategies" (Restle, 1958; Levine, 1959). The notion is that animals learn certain "rules" of conduct that are independent of the particular stimuli to be discriminated. For example, in a conventional two-choice discrimination problem, choice on each trial may be based on a "win–stay, lose–shift" rule, and transfer from one problem to another can be determined to some extent by the appropriateness of rules learned in the first problem to solution of the second. Incompatibility of rules is suggested, for example, by the finding of negative transfer in chimpanzees from object-alternation training to conventional object discrimination (Schusterman, 1962), whereas compatibility of rules is suggested by positive transfer in chimpanzees (Schusterman, 1964) and in rhesus monkeys (Warren, 1966) from reversal training to object discrimination. Squirrel monkeys trained in discrimination reversals (Ricciardi & Treichler, 1970), however, are slightly less efficient in subsequent object discrimination. Rumbaugh (1971) has argued that strategy formation may not be a characteristic of all primates. Comparing gorillas with the more primitive talapoins in a series of paired problems, each consisting of a discrimination between two objects followed by a reversal, he found that gorillas are unaffected by substitution of a new stimulus for the previous positive or previous negative stimulus during reversal. In all cases reversal learning occurred in a few trials, suggesting a complicated conditional strategy. Talapoins, in contrast, performed below chance for the first seven trials of regular reversals but performed at chance levels when a new stimulus was substituted for either of the old ones, suggesting simple negative transfer in normal reversal learning and the absence of strategy formation altogether.

Domestic cats do not show positive transfer from position-reversal training to object discrimination under conditions like those in which it is found in monkeys (Warren, 1966). Differences in the strategy formation of cats and rhesus monkeys may be seen in another study by Warren (1965). Both cats and monkeys were given a set of 80 problems, each consisting of an object discrimination followed by reversal, with the initial discrimination learned to a criterion. Half of each group of animals had the initial discrimination training on a white stimulus tray but reversal training on a black stimulus tray, providing a cue for the change in reward. The remaining animals continued on to the reversal with no change in tray color. Performance in the last 20 problems showed that both monkeys and cats utilized the cue provided by change in the color of the tray but apparently in different ways. Monkeys responded to the cue by shifting to the previously unrewarded stimulus (selecting it about 84% of the time in Trial 1), whereas cats responded to the cue by seeming to "begin a new problem" (their choice behavior following a course similar to that displayed on the initial discrimination task). The behavior of the monkeys seemed to conform to a "win–stay, lose–shift" strategy, whereas the behavior of the cats seemed to conform to a "lose–start over again" strategy.

Systematic choice patterns suggestive of rules or strategies have been found in a particular kind of discrimination task that has come to be described as "probability learning." Suppose an animal is given a choice between stimuli A and B, with A rewarded only on a random 70% of the trials and B rewarded on the remaining 30% of trials. If a correction method is used (that is, if the animal is permitted on any trial to respond to the alternative stimulus when the one it chooses first is incorrect, insuring reinforcement on the 70 : 30 ratio regardless of the animal's actual choice behavior), trial-by-trial analysis reveals several systematic response tendencies based on the outcome of the immediately preceding trial. A common finding is that the animals "maximize," that is, respond nearly always to the stimulus with the higher probability of reward, but sometimes they "reward follow," that is, respond much in the manner of a "win–stay, lose–shift" strategy, or they display "negative recency," that is, respond in terms of what seems to be a "win–shift, lose–stay" strategy. Such sequential tendencies have been found in monkeys (Wilson, Oscar, & Bitterman, 1964), cats (Warren & Beck, 1966), rats (Bitterman, 1971), and caimens (Williams & Albiniak, 1972) and often lead to what Bitterman (1965) has called "nonrandom probability matching." The term "probability matching" refers to a close correspondence between choice ratio and reinforcement ratio, that is, in the 70 : 30 problem described above, the animal chooses the 70% stimulus on about 70% of the trials and the 30% stimulus on about 30% of the trials. The matching is said to be "nonrandom" because it is the product of systematic response tendencies. (It should be evident, for example, that strict reward following in a 70 : 30 problem must produce 70% choice of the 70% stimulus.) Usually, however, these strategies interact with a maximizing tendency to produce levels of performance intermediate between matching and maximizing.

Experiments with mouthbreeders and goldfish give little evidence of strategies and under conditions as yet not fully specified produce what Bitterman has called "random matching" because no sequential dependencies are to be found in the data (Bitterman *et al.,* 1958; Behrend & Bitterman, 1961; Behrend, Bauman, & Bitterman, 1965; Behrend & Bitterman, 1966; Marrone & Evans, 1966; Woodard & Bitterman, 1973). Recently, Mackintosh *et al.* (1971) have reported matching in goldfish trained on a 70 : 30 problem but have described it as "nonrandom" on the ground that biases were shown in irrelevant (nondifferentially reinforced) dimensions (position in a color problem, or color in a position problem). This, of course, misses the point. If animals are to respond to the less frequently rewarded "relevant" stimulus, they may do so under a condition that permits them also to choose a stimulus that contains a preferred element in an irrelevant dimension. This is, to be sure, a "response bias," but not one that can account for the distribution of choices in the relevant dimension (Woodard & Bitterman, 1973). In fact, the Mackintosh data support the view that goldfish match randomly in the sense in which Bitterman and his colleagues use the term. Random matching has been found also in pigeons (Bullock & Bitterman, 1962) and in turtles (Kirk & Bitterman, 1965), again only under certain conditions that cannot be fully specified, but it has not been found in mammals. Gonzalez, Roberts, and Bitterman (1964) have presented some surgical evidence that there is a random-matching process in rats that is masked by maximizing and strategic processes.

CONCLUSION

Altogether, there seems to be an impressive array of evidence that evolutionary divergences have been accompanied by the development of functionally disparate learning processes in vertebrates. Some of these processes are tied rather closely to the predictable demands of the environment; others seem to have emerged as the need for flexibility of response has occurred, such as in animals existing in geographically broad habitats and for which the environment is less predictable. The means by which these animals adapt is through the development of flexible learning processes that, although not identical in different classes or orders, are sufficient to meet most contingencies. The evolution of the cerebral cortex in vertebrates has, indeed, been accompanied by greater behavioral plasticity. A problem that remains is to relate the processes involved in different lines to the nature of the demands made on each species.

REFERENCES

Amsel, A. Partial reinforcement effects on vigor and persistence. In K. W. Spence, & J. T. Spence, (Eds.), *The psychology of learning and motivation*. New York: Academic Press, 1967.

Behrend, E. R., Bauman, B. A., & Bitterman, M. E. Probability discrimination in the fish. *American Journal of Psychology*, 1965, 78, 83–89.

Behrend, E. R., & Bitterman, M. E. Probability matching in the fish. *American Journal of Psychology*, 1961, 74, 542–551.

Behrend, E. R., & Bitterman, M. E. Probability matching in the goldfish. *Psychonomic Science*, 1966, 6, 327–328.

Behrend, E. R., & Bitterman, M. E. Further experiments on habit reversal in the fish. *Psychonomic Science*, 1967, 8, 363–364.

Behrend, E. R., Domesick, V. B., & Bitterman, M. E. Habit reversal in the fish. *Journal of Comparative and Physiological Psychology*, 1965, 60, 407–411.

Behrend, E. R., Powers, A. S., & Bitterman, M. E. Interference and forgetting in bird and fish. *Science*, 1970, 167, 389–390.

Bitterman, M. E. Toward a comparative psychology of learning. *American Psychologist*, 1960, 15, 704–712.

Bitterman, M. E. Phyletic differences in learning. *American Psychologist*, 1965, 20, 396–410.

Bitterman, M. E. Comparative studies of learning in the fish. In D. J. Ingle (Ed.), *The central nervous system and fish behavior*. Chicago: University of Chicago Press, 1968.

Bitterman, M. E. Visual probability learning in the rat. *Psychonomic Science*, 1971, 22, 191–192.

Bitterman, M. E. Comparative studies of the role of inhibition in reversal learning. In R. A. Boakes & M. S. Halliday (Eds.), *Inhibition and learning*. London: Academic Press, 1972.

Bitterman, M. E. Conditioning in evolutionary perspective. In V. S. Rusinov, (Ed.) *Mechanisms of formation and inhibition of conditioned reflexes*. Moscow: Nauka, 1973.

Bitterman, M. E., Wodinsky, J., & Candland, D. K. Some comparative psychology. *American Journal of Psychology*, 1958, 71, 94–110.

Bolles, R. C. Species-specific defense reactions and avoidance learning. *Psychological Review*, 1970, 77, 32–48.

Bower, G. H. A contrast effect in differential conditioning. *Journal of Experimental Psychology*, 1961, 62, 196–199.

Breland, K., & Breland, M. *Animal Behavior*. New York: Macmillan, 1966.

Brookshire, K. H. Comparative psychology of learning. In M. H. Marx (Ed.), *Learning: Interactions*. New York: Macmillan, 1970.

Brownlee, A., & Bitterman, M. E. Differential reward conditioning in the pigeon. *Psychonomic Science*, 1968, 12, 345–346.

Bullock, D. H., & Bitterman, M. E. Probability-matching in the pigeon. *American Journal of Psychology*, 1962, 75, 634–639.

Burns, R. A., Woodard, W. T., Henderson, T. B., & Bitterman, M. E. Simultaneous contrast in the goldfish. *Animal Learning and Behavior*, 1974, **2**, 97–100.

Capaldi, E. J. A sequential hypothesis of instrumental learning. In K. W. Spence & J. T. Spence (Eds.). *The psychology of learning and motivation.* New York: Academic Press, 1967.

Carter, G. S. *Structure and habit in vertebrate evolution.* Seattle: University of Washington Press, 1967.

Crespi, L. P. Amount of reinforcement and the level of performance. *Psychological Review,* 1942, **51**, 341–357.

Diamond, I. T., & Hall, W. C. Evolution of neocortex. *Science*, 1969, **164**, 251–262.

Dilley, P. N. Delayed responses in octopus. *Journal of Experimental Biology*, 1963, **40**, 393–401.

Dobrzecka, C., Szwejkowska, G., & Konorski, J. Qualitative versus directional cues in two forms of differentiation. *Science*, 1966, **153**, 87–89.

Ehrenfreund, D. Effect of drive on successive magnitude shift in rats. *Journal of Comparative and Physiological Psychology*, 1971, **76**, 418–423.

Engelhardt, F., Woodard, W. T., & Bitterman, M. E. Discrimination reversal in the goldfish as a function of training conditions. *Journal of Comparative and Physiological Psychology*, 1973, **85**, 144–150.

Eskin, R. M., & Bitterman, M. E. Partial reinforcement in the turtle. *Quarterly Journal of Experimental Psychology*, 1961, **13**, 112–116.

Garcia, J., McGowan, B. K., Ervin, F. R., & Koelling, R. Cues: Their relative effectiveness as a function of the reinforcer. *Science*, 1968, **160**, 794–795.

Gleitman, H., & Steinman, F. Depression effect as a function of retention interval before and after shift in reward magnitude. *Journal of Comparative and Physiological Psychology*, 1964, **57**, 158–160.

Gonzalez, R. C., Behrend, E. R., & Bitterman, M. E. Partial reinforcement in the fish: Experiments with spaced trials and partial delay. *American Journal of Psychology*, 1965, **78**, 198–207.

Gonzalez, R. C., Behrend, E. R., & Bitterman, M. E. Reversal learning and forgetting in bird and fish. *Science*, 1967, **158**, 519–521.

Gonzalez, R. C., & Bitterman, M. E. A further study of partial reinforcement in the turtle. *Quarterly Journal of Experimental Psychology*, 1962, **14**, 109–112.

Gonzalez, R. C., & Bitterman, M. E. Resistance to extinction in the rat as a function of percentage and distribution of reinforcement. *Journal of Comparative and Physiological Psychology*, 1964, **57**, 258–263.

Gonzalez, R. C., & Bitterman, M. E. The partial reinforcement effect in the goldfish as a function of amount of reward. *Journal of Comparative and Physiological Psychology*, 1967, **64**, 163–167.

Gonzalez, R. C., & Bitterman, M. E. Two-dimensional discriminative learning in the pigeon. *Journal of Comparative and Physiological Psychology*, 1968, **65**, 427–432.

Gonzalez, R. C., & Bitterman, M. E. Spaced-trials partial reinforcement effect as a function of contrast. *Journal of Comparative and Physiological Psychology*, 1969, **67**, 94–103.

Gonzalez, R. C., Eskin, R. M., & Bitterman, M. E. Alternating and random partial reinforcement in the fish, with some observations on asymptotic resistance to extinction. *American Journal of Psychology*, 1961, **74**, 561–568.

Gonzalez, R. C., Eskin, R. M., & Bitterman, M. E. Extinction of the fish after partial and consistent reinforcement with number of reinforcement equated. *Journal of Comparative and Physiological Psychology*, 1962, **55**, 381–386.

Gonzalez, R. C., Eskin, R. M., & Bitterman, M. E. Further experiments on partial reinforcement in the fish. *American Journal of Psychology*, 1963, **76**, 366–375.

Gonzalez, R. C., Gleitman, H., & Bitterman, M. E. Some observations on the depression effect. *Journal of Comparative and Physiological Psychology*, 1962, 55, 578–581.

Gonzalez, R. C., Holmes, N. K., & Bitterman, M. E. Resistance to extinction in the goldfish as a function of frequency and amount of reward. *American Journal of Psychology*, 1967, 130, 269–275.

Gonzalez, R. C., Potts, A., Pitcoff, K., & Bitterman, M. E. Runway performance of goldfish as a function of complete and incomplete reduction in amount of reward. *Psychonomic Science,*, 1972, 27, 305–308.

Gonzalez, R. C., Roberts, W. A., & Bitterman, M. E. Learning in adult rats extensively decorticated in infancy. *American Journal of Psychology*, 1964, 77, 547–562.

Gossette, R. L. Note on the calibration of inter-species successive discrimination reversal (SDR) performance differences: Qualitative vs. quantitative scaling. *Perceptual and Motor Skills,* 1970, 31, 95–104.

Gottlieb, G. Imprinting in relation to parental and species identification by avian neonates. *Journal of Comparative and Physiological Psychology*, 1965, 59, 345–356.

Graf, C. L. Spaced-trial partial reward in the lizard. *Psychonomic Science,* 1972, 27, 153–155.

Graf, V., & Tighe, T. Subproblem analysis of discrimination shift learning in the turtle. *Psychonomic Science*, 1971, 25, 257–259.

Harlow, H. F. Primate learning. In C. P. Stone (Ed.), *Comparative psychology*. (3rd ed.) New York: Prentice-Hall, 1951.

Harlow, H. F. The evolution of learning. In A. Roe & G. G. Simpson (Eds.), *Behavior and evolution*. New Haven: Yale University Press, 1958.

Hinde, R. A. *Animal behaviour*. (2nd Ed.) New York: McGraw-Hill, 1970.

Hodos, W., & Campbell, C. B. G. *Scala Naturae:* Why there is no theory in comparative psychology. *Psychological Review*, 1969, 76, 337–350.

Hulse, S. H. Jr. Amount and percentage of reinforcement and duration of goal confinement in conditioning and extinction. *Journal of Experimental Psychology*, 1958, 56, 48–57.

Huxley, J. S. Genetic interaction in a hybrid pheasant. *Zoological Society (London) Proceedings*, 1941, 3, 41–43.

Karamyan, A. I. Evolution of functions in the higher divisions of the central nervous system and their regulating mechanisms. In J. W. S. Pringle (Ed.), *Essays on physiological evolution*. Oxford: Pergamon Press, 1965.

King, J. H., & Nichols. J. W. Problems of classification. In R. H. Walters, D. A. Rethling-shaffer, & W. E. Caldwell (Eds.), *Principles of comparative psychology*. New York: McGraw-Hill, 1960.

Kirk, K. L., & Bitterman, M. E. Probability-learning by the turtle. *Science*, 1965, 148, 1484–1485.

Klopfer, P. H., Hailman, J. P. *An introduction to animal behavior*. Englewood Cliffs, New Jersey: Prentice-Hall, 1967.

Konishi, M. The role of auditory feedback in the control of vocalization in the white-crowned sparrow. *Zeitschrift fur Tierpsychologie*, 1965, 22, 770–783.

Levine, M. A model of hypothesis behavior in discrimination learning set. *Psychological Review*, 1959, 66, 353–366.

Longo, N., & Bitterman, M. E. The effect of partial reinforcement with spaced practice on resistance to extinction in the fish. *Journal of Comparative and Physiological Psychology*, 1960, 53, 169–172.

Lowes, G., & Bitterman, M. E. Reward and learning in the goldfish. *Science,* 1967, 157, 455–457.

Lukaszewska, I. Some further failures to find the visual overlearning reversal effect in rats. *Journal of Comparative and Physiological Psychology*, 1968, 65, 359–361.

Mackintosh, N. J. Reward and aftereffects of reward in the learning of goldfish. *Journal of Comparative and Physiological Psychology*, 1971, 76, 341–348.

Mackintosh, N. J., Lord, J., & Little, L. Visual and spatial probability learning in pigeons and goldfish. *Psychonomic Science*, 1971, 24, 221–223.

Mackintosh, N. J., Mackintosh, J., Safriel-Jorne, O., & Sutherland, N. S. Overtraining, reversal and extinction in the goldfish. *Animal Behaviour*, 1966, 14, 314–318.

Marcus, A. Delayed reactions in the tawney owl *(Strik aluco aluco L.)*. *Folia Biologica*, 1959, 7, 329–348.

Marler, P. A comparative approach to vocal learning: Song development in white crowned sparrows. *Journal of Comparative and Physiological Psychology*, 1970, 71, (monograph supplement), 1–25.

Marrone, R., & Evans, S. Two-choice probability learning in fish. *Psychonomic Science*, 1966, 5, 327–328.

Masterton, B., & Skeen, L. C. Origins of anthropoid intelligence: Prefrontal system and delayed alternation in hedgehog, tree shrew, and bush baby. *Journal of Comparative and Physiological Psychology*, 1972, 81, 423–433.

Moltz, H. Contemporary instinct theory and the fixed action pattern. *Psychological Review*, 1965, 72, 27–47.

Mulligan, J. A. Singing behavior and its development in the song sparrow *Melospiza melodia*. *University of California Publications in Zoology*, 1966, 81, 1–76.

Murillo, N. R., Diercks, J. K., & Capaldi, E. J. Performance of the turtle, *Pseudemys scripta troostii*, in a partial-reinforcement situation. *Journal of Comparative and Physiological Psychology*, 1961, 54, 204–206.

Pantin, C. F. A. Organic design. *Advancement of Science*, 1951, 8, 138–150.

Pert, A., & Bitterman, M. E. Reward and learning in the turtle. *Learning and Motivation*, 1970, 1, 121–128.

Restle, F. Toward a quantitative description of learning set data. *Psychological Review*, 1958, 65, 77–91.

Revusky, S. H., & Bedarf, E. W. Association of illness with prior ingestion of novel foods. *Science*, 1967, 155, 219–220.

Ricciardi, A. M., & Treichler, F. R. Prior training influences on transfer to learning set by squirrel monkeys. *Journal of Comparative and Physiological Psychology*, 1970, 73, 314–319.

Roberts, W. A., Bullock, D. H., & Bitterman, M. E. Resistance to extinction in the pigeon after partially reinforced instrumental training under discrete-trials conditions. *American Journal of Psychology*, 1963, 76, 353–365.

Rosvold, H. E., & Delgado, J. M. R. Effect on delayed alternation test performance of stimulating and destroying electrically structures within the frontal lobes of monkeys' brains. *Journal of Comparative and Physiological Psychology*, 1956, 49, 365.

Rozin, P., & Kalat, J. W. Specific hungers and poison avoidance as adaptive specializations of learning. *Psychological Review*, 1971, 78, 459–486.

Rumbaugh, D. M. Evidence of qualitative differences in learning processes among primates. *Journal of Comparative and Physiological Psychology*, 1971, 76, 250–255.

Schade, A. F., & Bitterman, M. E. Improvement in habit reversal as related to dimensional set. *Journal of Comparative and Physiological Psychology*, 1966, 62, 42–48.

Schneirla, T. C. The relationship between observation and experimentation in the field study of behavior. *Annals of the New York Academy of Sciences*, 1950, 51, 1022–1044.

Schneirla, T. C. An evolutionary theory of biphasic processes underlying approach and withdrawal. In M. R. Jones, (Ed.), *Current theory and research on motivation*. Lincoln, Nebraska: University of Nebraska Press, 1959.

Schrier, A. M. Effects of an upward shift in amount of reinforcer on runway performance of rats. *Journal of Comparative and Physiological Psychology*, 1967, 64, 490–492.

Schusterman, R. J. Transfer effects of successive discrimination reversal training in chimpanzees. *Science*, 1962, 137, 422–423.

Schusterman, R. J. Successive discrimination reversal training and multiple discrimination training in one-trial learning by chimpanzees. *Journal of Comparative and Physiological Psychology*, 1964, 58, 153–156.

Schutz, S. L., & Bitterman, M. E. Spaced-trials partial reinforcement and resistance to extinction in the goldfish. *Journal of Comparative and Physiological Psychology*, 1968, 68, 126–128.

Seligman, M. E. P. On the generality of the laws of learning. *Psychological Review*, 1970, 77, 406–418.

Setterington, R. G., & Bishop, H. E. Habit reversal improvement in the fish. *Psychonomic Science*, 1967, 7, 41–42.

Shanab, M. E., Saunders, R., & Premack, D. Positive contrast in the runway obtained with delay of reward. *Science*, 1969, 164, 724–725.

Sheffield, V. F. Extinction as a function of partial reinforcement and distribution of practice. *Journal of Experimental Psychology*, 1949, 39, 511–526.

Shepp, B. E., & Eimas, P. D. Intradimensional and extradimensional shifts in the rat. *Journal of Comparative and Physiological Psychology*, 1964, 57, 357–361.

Shepp, B. E., & Schrier, A. M. Consecutive intradimensional and extradimensional shifts in monkeys. *Journal of Comparative and Physiological Psychology*, 1969, 67, 199–203.

Simpson, G. G. Organisms and molecules in evolution. *Science*, 1964, 146, 1535–1538.

Smith, J. C., & Roll, D. L. Trace conditioning with X-rays as the aversive stimulus. *Psychonomic Science*, 1967, 9, 11–12.

Sutherland, N. S., & Mackintosh, N. J. *Mechanisms of animal discrimination learning.* New York: Academic Press, 1971.

Tennant, W. A., & Bitterman, M. E. Asymptotic free-operant discrimination reversal in the goldfish. *Journal of Comparative and Physiological Psychology*, 1973a, 82, 130–136.

Tennant, W. A., & Bitterman, M. E. Some comparisons of intradimensional and extradimensional transfer in discriminative learning of goldfish. *Journal of Comparative and Physiological Psychology*, 1973b, 83, 134–139.

Tinbergen, N. *The herring gull's world.* London: Collins, 1953.

Warren, J. M. Reversal learning by paradise fish *(Macropodus opercularis). Journal of Comparative and Physiological Psychology*, 1960, 53, 376–378.

Warren, J. M. Primate learning in comparative perspective. In A. M. Schrier, H. F. Harlow, & F. Stolmitz (Eds.), *Behavior of nonhuman primates: Modern research trends.* New York: Academic Press, 1965.

Warren, J. M. Reversal learning and the formation of learning sets by cats and rhesus monkeys. *Journal of Comparative and Physiological Psychology*, 1966, 61, 421–428.

Warren, J. M. Learning: Vertebrates. In D. Dewsbury (Ed.), *Comparative psychology: A modern survey.* New York: McGraw-Hill, 1973.

Warren, J. M., & Beck, C. H. Visual probability learning by cats. *Journal of Comparative and Physiological Psychology*, 1966, 61, 316–318.

Weinstock, S. Resistance to extinction of a running response following partial reinforcement under widely spaced trials. *Journal of Comparative and Physiological Psychology*, 1954, 47, 318–322.

Wilcock, J. Comparative psychology lives on under an assumed name—psychogenetics. *American Psychologist*, 1972, 27, 531–539.

Wilcoxon, H. C., Dragoin, W. B., & Kral, P. A. Illness-induced aversions in rat and quail: Relative salience of visual and gustatory cues. *Science*, 1971, 171, 826–828.

Williams, D. R., & Williams, H. Auto-maintenance in the pigeon: Sustained pecking despite contingent non-reinforcement. *Journal of the Experimental Analysis of Behavior*, 1969, **12**, 511–520.

Williams, J. T., & Albiniak, B. A. Probability learning in a crocodilian. *Psychonomic Science*, 1972, **27**, 165–166.

Wilson, W. A., Oscar, M., & Bitterman, M. E. Visual probability learning in the monkey. *Psychonomic Science*, 1964, **1**, 71–72.

Wittlin, W. A., & Brookshire, K. H. Apomorphine-induced conditioned aversion to a novel food. *Psychonomic Science*, 1968. **12**, 217–218.

Wodinsky, J., & Bitterman, M. E. Discrimination reversal in the fish. *American Journal of Psychology*, 1957, **70**, 569–576.

Wodinsky, J., & Bitterman, M. E. Partial reinforcement in the fish. *American Journal of Psychology*, 1959, **72**, 184–199.

Wodinsky, J., & Bitterman, M. E. Resistance to extinction in the fish after extensive training with partial reinforcement. *American Journal of Psychology*, 1960, **73**, 429–434.

Woodard, W. T., & Bitterman, M. E. Further experiments on probability learning in goldfish. *Animal Learning and Behavior*, 1973, **1**, 25–28.

Woodard, W. T., Schoel, W. M., & Bitterman, M. E. Reversal learning with singly presented stimuli in pigeons and goldfish. *Journal of Comparative and Physiological Psychology*, 1971, **76**, 460–467.

Woodburne, L. S. Irrelevant tactics, caudate lesions, and delayed response performance in squirrel monkeys. *Physiology and Behavior*, 1971, **7**, 701–704.

11
Issues in the Comparative Psychology of Learning

M. E. Bitterman

University of Hawaii

Two aspects of the foregoing chapter by Brookshire on evolutionary divergences in learning warrant further consideration. I shall comment here first on the strategy of comparative psychology and then on the interpretation of the phenomena described by Brookshire under the heading Divergences in Special Learning Abilities. One of the functions of the conference was to encourage the resolution of disagreements among the participants, and I am confident that Brookshire's treatment of these matters would have been more balanced had he lived to revise his paper, which was written in advance of the conference. Unfortunately, however, he did not, and it remains for me (as his discussant) to supply another view.

STRATEGIC QUESTIONS

One question of strategy is whether what Brookshire has called "emphasizing qualitative differences in behavior (e.g., no improvement in habit reversal)" does indeed serve, as he has said, "to cloud the issue" of divergence. Brookshire has not, of course, recommended an emphasis on quantitative differences in behavior, because he has understood very well that the purely quantitative features of performance in learning situations (such as rate of acquisition or extent of improvement in habit reversal) fluctuate markedly even in the same species as a function of nonlearning variables that there is no way of equating across species. His point is simply that a difference in learning process may be indicated not only by the unqualified failure of a phenomenon discovered in one group of animals to appear in another, but also by a difference in the circumstances under which it appears in the two groups. In my opinion, the distinction is a semantic one.

Consider, for example, that laboratory rats show greater resistance to extinction after partial as compared with consistent reinforcement in spaced as well as in massed trials, whereas goldfish perform analogously only in massed trials. These results may be described by saying that the partial reinforcement effect (PRE) is found under different circumstances in the two animals, or—refining the original definition of the phenomenon by extending the range of defining conditions—it may be said that the massed-trials PRE appears both in rats and in goldfish, whereas the spaced-trials PRE appears in rats but not in goldfish. Although the first formulation may seem preferable on the ground that it does not create the necessity of having to prove the null hypothesis, the advantage is illusory, because both formulations rest on the assumption that goldfish do not show greater resistance to extinction after partial as compared with consistent reinforcement when trials are spaced. Nor need the vagaries of the null hypothesis cause any great anxiety (Bitterman, 1965). After repeated failures to demonstrate some phenomenon or capability in a given species, it becomes reasonable to proceed on the assumption that further efforts must be fruitless. Strategic decisions of this sort are commonly made—indeed, must be made—in science, although often they prove to be wrong.

It may be well to give some careful thought to the distinction between questions about qualitative differences in performance and questions about qualitative differences in learning process. In the first case, it is asked whether there are phenomena of learning that appear in some animals but not in others. In the second case, it is asked whether there are processes of learning that operate in some animals but not in others. It should be noted that qualitative differences in performance do not necessarily mean that there are qualitative differences in process. Consider, for example, the attentional interpretation of the ORE (overlearning—reversal effect) reviewed in Brookshire's paper: a given species may fail to show the ORE not because it lacks an attentional process but because the rate at which attention develops relative to the rate at which specific stimulus preferences develop is different in that species than in one that does show the ORE. (Quantitative differences in process may produce qualitative differences in performance.) Nor, as Brookshire has noted, does the fact that what may seem to be the same phenomenon is found in two groups of animals necessarily mean that the underlying processes are the same in the two groups. In choice situations, for example, both rats and goldfish may show what has been called "probability matching" (correspondence of preference ratio to reinforcement ratio), but detailed analysis reveals certain basic differences in performance that suggest a new behavioral distinction ("random" versus "non-random" matching) and the operation of different processes. Learning processes are, of course, no more than hinted at by learning phenomena, and it may be a long time before it is possible to get at them directly. For the moment, perhaps, one can hope only for a set of functional principles or laws that are arrived at inductively and from which learning phenomena can be deduced. The basic question for the comparative psychologist is whether these laws are the same for all animals (as has long been assumed) or whether they are different. If the laws can be stated as equations, the question about quantitative versus qualitative differences in process reduces to the question of whether the equations required to fit the data for various animals are of the same form, differing only in constants—as g at Hammerfest and Madras—or whether equations of different form are required. It was the opinion of Hull (1945), who first put the problem in this way, that only the constants change.

A second strategic issue has to do with the choice of animals for study. It was Brookshire's contention that comparisons of widely divergent forms permit "only meager statements about historical evolution." This criticism has been made even more sharply by Hodos and Campbell (1969) in a paper that raises the question of "why there is no theory in comparative psychology" and answers it in terms of the inability of comparative psychologists to deal with the "intricacies" of evolutionary history. Comparisons of "teleost fish, turtles, pigeons, rats, and monkeys . . . do not permit generalizations about the evolution of intelligence or any other characteristic of these organisms since they are not representative of a common evolutionary lineage," Hodos and Campbell assert flatly—"no rat," they explain, "was ever an ancestor of any monkey [p. 345] ." In fact, however, the comparative psychology of intelligence has been dominated almost from the outset by a powerful theory that denies divergence, and the discovery of different learning processes in any two species whatever would permit at least the generalization (which, given the history of the subject, I should hardly describe as "meager") that divergence has occurred. In the search both for communalities and for differences in process, furthermore, the only sensible course is to begin with markedly divergent animals. By the same logic which suggests that a process found in two distantly related vertebrate species is more likely to be a process common to vertebrates than one found in two closely related species, it is evident that any differences in process that may exist are more likely to be discovered in comparisons of distantly related species. To compare a fish and a bird may not require much more effort than to compare two fishes or two birds, yet the results are much more apt to be instructive, especially in the early stages of the investigation.

Although I share the interest expressed by Brookshire and others in the suggestion of Hodos and Campbell that it may be profitable to compare animals that comprise "quasi-evolutionary series," I do not believe that only on the basis of such comparisons can an understanding of the phylogeny of learning be hoped for—that "there are no alternatives" (Hodos & Campbell, 1969, p. 343).

In a subsequent discussion of "the evolution of behavioral plasticity," Hodos (1970) himself has shown very nicely how the results of learning experiments with a heterogeneous assortment of animals (among them pigeons, rats, and monkeys) may begin to tell something about the functional properties of common ancestors and about points at which functional divergence may have occurred. Unfortunately, however, the demonstration has no more than formal value, because the performance measure on which Hodos relies—mean percentage of correct choice on the second trial of the hundredth in a series of discrimination problems (for example, 58 for pigeon or 50.0 for tree shrew)—is quite ambiguous. As Brookshire's chapter points out, scores of this sort give us no unequivocal information about learning ability because performance in learning situations is influenced also by sensory, motor, and motivational factors. Hodos is not unaware of the problem of confounding—he recognizes, for example, that discriminative performance must depend in part on sensory capacity—but the solution he proposes (which Brookshire's chapter does not mention) is inadequate. Because "any one of a number of factors can cause an animal to perform at less than its best, but nothing can cause it to exceed its natural limitations" Hodos argues, the technique that results "in the highest performance level for an animal must be taken as giving the best estimate" of the "animal's maximum capability" or level of "attainment" (pp. 36–37). It is not clear, however, just how the problem of confounding is "circumvented" in the comparison of "maximum" performances. Assume for the sake of simplicity that scores are determined by only two factors—sensory capacity and "plasticity"—and that the highest score for Species A exceeds that for Species B. Without an independent measure of sensory capacity, no conclusion can be drawn about plasticity, which may actually be greater in Species B.

Even if it were known how to equate nonlearning variables across species, there would still be doubt as to how to interpret such scores. Available evidence certainly makes it difficult to believe that they reflect but a single dimension of learning (at the very least, the effects both of reinforcement and of nonreinforcement must be involved) or even that the processes underlying improvement are the same in all animals. The remoteness of the relation between learning scores and learning processes should not be underestimated. To get at the processes responsible for the performance of a given animal in some arbitrarily selected learning situation requires systematic experimental analysis; and to provide a comprehensive account of learning in the animal requires analysis of its performance in a great variety of situations. A meaningful comparative psychology can only emerge, not from superficial work with a large number of animals, but from intensive work with a small number of divergent forms. Only when some notion has been gained of the ways in which learning processes differ from animal to animal, and of how best to assess those differences, can it become profitable (following the logic that Hodos makes explicit) to broaden the sample of species studied and to come to grips with the problem of phylogeny.

THE QUESTION OF SPECIAL ABILITIES

Now I turn to Brookshire's interpretation of the phenomena described under the heading Divergences in Special Learning Abilities. The chapter is somewhat ambiguous here, and it is difficult really to say whether Brookshire has assumed the operation of special learning processes (as distinct from those characterized as "more flexible") or of general learning processes under specialized constraints. It is my judgment, in any case, that the phenomena cited by Brookshire provide no good evidence of evolutionary divergence in learning.

The familiar fact that performance in learning situations may be influenced by "species-typical" responses to reinforcement does not necessarily imply the operation of species-typical learning processes. To consider first the Bolles example, suppose that the unconditioned response to shock is jumping in Species A but crouching in Species B, and suppose that the performance of Species A in avoidance training is better when jumping rather than crouching is required to avoid shock and that the opposite is true of Species B. There may be evidence here not of divergence in learning process, but of communality. In both cases, apparently, the pairing of S_1 and S_2 that occurs when the animal fails to avoid produces a tendency for S_1 to elicit a response like that elicited by S_2—a tendency that is advantageous when the same response is required for avoidance but disadvantageous when an incompatible response is required. As to autoshaping, not only pigeons, but rats, goldfish, and monkeys tend to approach, make contact with, or manipulate localized stimuli paired with food or other rewards. To me, these simple instances of classical conditioning suggest communality rather than divergence in learning process. The finding that pigeons continue to peck a key even when pecking prevents the food (although not, of course, with the same probability as when illumination of the key always is followed by food) is perfectly understandable in terms of conditioning principles. The tendency to peck the lighted key, established to begin with by the pairing of key light and food, is weakened by the nonoccurrence of food when pecking occurs but strengthened again by pairing whenever pecking fails to occur. Analogous results have been obtained in salivary conditioning experiments with dogs (Sheffield, 1965). The Breland example can be understood in terms of the same principles as the Bolles example. Tokens associated with food should tend to evoke responses that are like the responses elicited by the food. To the extent that these responses are incompatible with the instrumental response, the animal has difficulty; to the extent that they are compatible, there is much less difficulty. The Konorski example, which Konorski (1967) has explained in terms of "potential" connections in the nervous system, requires further experimental analysis. Even if Konorski's interpretation is correct, however, there is no reason to think that the processes involved in the realization of such potentials are in any way unique. Learning commonly involves the strengthening of existing tendencies rather than their establishment *de novo*.

There is no place here for a comprehensive analysis of the taste-aversion results to which so much attention has been paid in recent years, but I can indicate briefly why I do not find them very illuminating. First, there is the finding that taste aversions may develop even with intervals of several hours between ingestion and illness. In my opinion, the possibility that taste receptors are restimulated at the time of illness by food returned to the mouth from the stomach or by secretion in saliva of components that have been absorbed cannot yet be discounted. In an experiment designed to control for "aftertaste," Rozin (1969) has exposed rats to two concentrations of the same substance, only one of them paired with illness. Unfortunately, however, the presentation of the results is too abbreviated to permit interpretation and the statistical treatment is faulty. One troublesome feature of the data is that there is no absolute decrement in the intake of the concentration associated with illness; instead, intake of the "safe" concentration increases differentially. Another is a striking interaction of change in preference with concentration, an outcome perhaps inherent in such a design but compatible nevertheless with an interpretation in terms of restimulation during illness. Aftertaste is the concern also of an experiment by Garcia, Hankins, Robinson, and Vogt (1972). Their basic idea (taken from Rozin) is a sound one, namely, that differential consumption of wet mash versus dry food plus water after illness paired with one of them cannot easily be explained in terms of aftertaste, but they in fact present no data on differential consumption. A good way to control for aftertaste may be to use a fistula. If the restimulation interpretation is correct, taste aversion may develop even if the food is never taken into the mouth (but placed directly in the stomach), whereas taste aversion may not develop (with long intervals between ingestion and illness) if food is taken into the mouth but prevented from reaching the stomach. This brings to mind the work of Domjan and Wilson (1972) who have simply flushed the oral cavity with a saccharin solution, assuming that the rats do not ingest any of it in the process. Whatever the validity of this assumption, poisoning followed immediately and there was no control for the effect of poisoning per se.

The finding that familiar tastes are not as likely as novel tastes to become aversive in these experiments has the same explanation, I assume, as the finding in conventional experiments that unreinforced presentations of a stimulus to be conditioned retard subsequent conditioning (Lubow & Siebert, 1969), although it should be noted that there are substantial discrepancies in the results of the two novelty experiments cited by Brookshire. The finding that the visual, as distinct from the gustatory, properties of food do not become aversive with long delay is understandable because there is no peripheral mechanism by which the visual stimuli can be reintroduced at the time of illness. The work with quail, to which Brookshire refers in support of the notion that visual properties may in fact become aversive in poisoning experiments when, for the species used, visual cues are "important" in "food seeking and food selection," is quite inadequate. In the main quail study, the food was colored with a blue dye, which (according to the authors) may have had a distinctive taste. Furthermore, there was no control for the possibility of illness-produced aversion to blue apart from the formation of an association between color and illness (for example, a group tested with blue water but poisoned after drinking clear water, or water of another color). With color provided by dye, of course, the possibility of restimulation also must be guarded against. In a supplementary study, blue light rather than dye was used, but there was no control for the effect of illness per se (because only the experimental group was made ill).

Another often-cited result of the experiments on conditioned aversion is that, with electric shock substituted for illness as the noxious unconditioned stimulus, visual aversions are established more readily than taste aversions. It should be noted, however, that when shock is used as the unconditioned stimulus the CS–US interval (the interval between the visual or gustatory stimulus and the shock) usually is short, whereas when illness is used the interval is relatively long; that is, CS–US interval and type of unconditioned stimulus usually are confounded. When shock is used, there may be confounding also of CS–US interval and modality of the conditioned stimulus, as in the experiment by Garcia et al. cited by Brookshire. Because the animals were shocked only on taking the food, and because they must have seen the food before taking it, visual stimuli antedated the shock, whereas taste stimuli were simultaneous with shock or may even have followed it. (Conventional experiments show, however, that conditioning does not occur with strictly simultaneous or backward pairing.) Furthermore, there was a corresponding difference in the testing conditions for the two modalities. The visual animals may have hesitated longer in the test than the taste animals before taking the food (which they previously had been shocked for taking) simply because the stimulus that had been paired with shock was present for the visual group before the food was taken, but for the taste group only after it was taken. Once eating began, in fact, both groups ate readily and both took the same amount of food. The CS–US interval and modality of the CS have not, of course, been confounded in all shock experiments (see, for example, Garcia & Koelling, 1966; Domjan & Wilson, 1972), but I have not been able to find one that is acceptable in every respect. The most common error is failure to control for pseudoconditioning (that is, for the effects of CS and US independently of their temporal relation). Garcia, McGowan, and Green (1972) have expressed the view that whether an aversion is the product of conditioning or pseudoconditioning "doesn't matter." What is important, they say, is that "behavior shows astonishingly organismic properties" (p. 19).

The fact that such work should be so highly esteemed is understandable, I think, in terms of the strong influence of ethology on comparative psychologists, among whom the fashion now is to disparage the traditional psychology of learning and to recite the virtues of work on "natural" problems in "natural" environments. Lockard (1971), for example, admits that the laboratory may have some "limited usefulness" for the analysis of phenomena discovered "in the real world" but warns that, because "animals mismatch the laboratory, severe distortions of behavior are common." The "expenditure of effort on laboratory-born problems" should be avoided, he advises; "today, the interest is not in what an animal can be made to do, but in how it normally functions" (p. 175). Hodos (1970) declares that "wild-reared" subjects are "preferable" to "laboratory-reared" subjects, and that "to obtain the best possible performance from an animal, the testing situation should be designed to reflect its normal interaction with its natural environment" (p. 36). Although the phenomena that have been considered here have been demonstrated in laboratory experiments with common laboratory animals, it is easy to assign adaptive value to them, and all seem to qualify at least for honorary membership in the "real world" on the assumption (however well founded) that they undermine the learning theory of the laboratory world. Emphasis on the natural environment usually is coupled with interest in the adaptive significance of behavior, and speculation about adaptive significance often substitutes for functional analysis (vague teleology masquerading as biological sophistication). The alleged survival value of a rat's ability to associate taste and illness over long intervals may be so impressive, apparently, as to forestall critical examination of the experiments purported to demonstrate the ability.

Even if no more is desired than to understand the behavior of an animal in some restricted environment (such as the one in which the animal is found), it still becomes necessary to alter that environment. It is from "distortions" in behavior correlated with changing circumstances that inferences are made about underlying processes, and there is no way to place a limit in advance on the extent of alteration that may be useful. If the larger task of understanding the animal is set—as it must, I think, if anything is to be said about evolution—only the widest possible variation in circumstances, both of rearing and of testing, can be considered. For the conviction that the capabilities of any species can best be appreciated from tests designed to reflect "normal interaction" with the "natural environment," there is no evidence whatever. As Harlow (1958) has noted, monkeys tested in conventional laboratory settings solve problems far more complex than they ever are likely to meet in the natural environment, and it may be well in this connection to consider also the accomplishments of chimpanzees reared in human homes (Hayes & Hayes, 1952) or linguistically trained in special laboratory programs (Gardner & Gardner, 1969; Premack, 1971). To understand an animal means, of course, to know its limitations as well as its capabilities and to be able to relate both to its structure, which in turn must be explained in terms of its genetic endowment and its life history. The adaptive significance of some capability of the animal in the environment in which it is now found—even if accurately assessed—tells nothing at all about the structural basis of the capability or about the evolutionary history either of the structure or of the capability.

REFERENCES

Bitterman, M. E. Phyletic differences in learning. *American Psychologist*, 1965, **20**, 396–410.

Domjan, M., & Wilson, N. E. Specificity of cue to consequence in aversion learning in the rat. *Psychonomic Science*, 1972, **26**, 143–145.

Garcia, J., Hankins, W. G., Robinson, J. H., & Vogt, J. L. Bait shyness: Tests of CS-US mediation. *Physiology & Behavior*, 1972, **8**, 807–810.

Garcia, J., & Koelling, R. A. Relation of cue to consequence in avoidance learning. *Psychonomic Science*, 1966, **4**, 123–124.

Garcia, J., McGowan, B. K., & Green, K. F. Biological constraints on conditioning. In A. H. Black & W. F. Prokasy (Eds.), *Classical conditioning II: Current research and theory*. New York: Appleton-Century-Crofts, 1972.

Gardner, R. A., & Gardner, B. T. Teaching sign language to a chimpanzee. *Science*, 1969, **165**, 664–672.

Hayes, K. J., & Hayes, C. Imitation in a home-raised chimpanzee. *Journal of Comparative and Physiological Psychology*, 1952, **45**, 450–459.

Harlow, H. F. The evolution of learning. In A. Roe and G. G. Simpson (Eds.), *Behavior and evolution*. New Haven: Yale University Press, 1958.

Hodos, W. Evolutionary interpretation of neural and behavioral studies of living vertebrates. In F. O. Schmidt (Ed.), *The neurosciences: Second study program*. New York: Rockefeller University Press, 1970.

Hodos, W., & Campbell, C. B. G. *Scala naturae:* Why there is no theory in comparative psychology. *Psychological Review*, 1969, **76**, 337–350.

Hull, C. L. The place of innate individual and species differences in a natural-science theory of behavior. *Psychological Review*, 1945, **52**, 55–60.

Konorski, J. *Integrative activity of the brain*. Chicago and London: University of Chicago Press, 1967.

Lockard, R. B. Reflections on the fall of comparative psychology. *American Psychologist*, 1971, **26**, 168–179.

Lubow, R. E., & Siebert, L. Latent inhibition and the CER paradigm. *Journal of Comparative and Physiological Psychology*, 1969, **68**, 136–138.

Premack, D. Language in chimpanzees? *Science*, 1971, **172**, 808–822.

Rozin, P. Central or peripheral mediation of learning with long CS-US intervals in the feeding system. *Journal of Comparative and Physiological Psychology*, 1969, **67**, 421–429.

Sheffield, F. D. Relation between classical conditioning and instrumental learning. In W. F. Prokasy (Ed.), *Classical conditioning: A symposium*. New York: Appleton-Century-Crofts, 1965.

12
Later Mammal Radiations

Leonard B. Radinsky

The University of Chicago

INTRODUCTION

The purpose of this chapter is to survey aspects of the evolutionary history of mammals that are especially relevant for neurologists and behaviorists. From my viewpoint this is information that can aid in selecting experimental animals so as to maximize information gain for a given work input and that can help in interpreting evolutionary changes or differences between species. The most important kinds of data for those purposes are phylogenetic relationships, particularly times of evolutionary divergence of various groups, and knowledge of species that are either unusually primitive or unusually specialized in ways that may be reflected in the central nervous system or behavior or in both. Although the second kind of information, concerning diversity of living forms, is not primarily paleontological, it is included here because it is very relevant and is appropriate in that living species are the current results of past evolutionary radiations.

For many neurological and behavioral studies, particularly those seeking to elucidate anatomical relationships or physiological processes, it may be most efficient to work on the handful of commonly used experimental mammals, owing to the large amount of information already available on them. However, studies that focus on a given anatomical system or on differences between species may well benefit from examination of species other than the common laboratory animals. For example, studies of the auditory system may yield more interesting information if species with unusual auditory specializations, such as bats or kangaroo rats, are studied instead of the usual cat or monkey. Or, comparative studies aimed at exploring the range of diversity of a given anatomical system within a given taxon may be more illuminating if moderately distantly related forms are used rather than closely related ones—e.g., rat versus

227

guinea pig rather than rat versus mouse. For those kinds of studies, knowledge of phylogenetic relationships and of the diversity of species that are currently available and suitable for various experimental work is important.

The above considerations are also important for studies that attempt to reconstruct evolutionary pathways or to assess the significance of differences between species. A common conceptual error leads to the erection of evolutionary series based on living forms, such as rat–cat–macaque–man, whereas, in fact, none of those living forms has given rise to any other one, all have been evolving for the same length of time, and each is specialized for its own particular ecological niche. The assumption that any given species is more primitive than any other and therefore represents an ancestral stage must first be narrowed down to a specific structure or system and, second, serious consideration should be given to the possibility that differences between two species may be the result of both having specialized for different niches rather than of one having remained in a less specialized stage. For example, an erroneous interpretation of how macaque brains have evolved results from assuming that cat brains are representative of an early stage of macaque brain evolution, and the same holds if macaques are taken as representative of human ancestors. Consideration of primate phylogeny suggests that prosimian primates are the best group to examine for features that may have been present in the brains of macaque and human ancestors. This is because modern prosimians may represent in some features stages through which the prosimian ancestors of higher primates have passed. A study of the brains of many different species of living prosimians, monkeys, and hominoids as well as consideration of what information is available from the fossil record of primate brains becomes necessary to assess the probability of any given structure having been present in a common ancestor. Checking with the fossil record is important because even if a structure is present in all living members of a group, such as the central sulcus in higher primate brains, it still may have evolved independently, by parallel evolution. For example, direct evidence from the fossil record indicates that the almost ubiquitous cruciate sulcus has appeared independently several times in the evolution of the brain of carnivores (Radinsky, 1971). Unfortunately, the evolution of most features of mammal brains cannot be checked by reference to the fossil record and can only be interpreted indirectly from distribution in living forms.

NATURE OF THE EVIDENCE

There are two kinds of data from which phylogenetic relationships are reconstructed: direct evidence from the fossil record and indirect evidence from comparisons among surviving species. The fossil evidence is confined to the skeletal system and, for mammals, particularly the teeth. Because enamel is the hardest tissue in the body, teeth are more likely to be preserved as fossils than any other part of the skeleton, and most extinct species of mammals are known primarily from their dentitions. Fortunately, molar teeth of most kinds of mammals have complex morphology and are among the most characteristic parts of the body. The majority of mammal genera may be identified from dentitions alone, and much mammal phylogeny has been based on dental morphology. After the teeth, the next most useful feature for reconstruction of mammal phylogeny has been the morphology of the middle ear region. This region provides a wealth of anatomical detail, and some of the differences between groups, such as the pattern of internal carotid circulation, do not have apparent adaptive significance and are therefore good indicators of phylogenetic relationships.

Evidence of mammalian phylogenetic relationships obtained from living species ranges from comparisons of molecular structure (e.g., Barnicott, 1969; Sarich, 1969; Seal, Phillips, & Erickson, 1970) and of chromosome morphology (e.g., Wurster & Benirschke, 1968) to comparisons of organ systems or whole organisms. Discrepancies in estimated times of divergence derived from the fossil record and from so-called "molecular clocks" have led to the realization that changes in molecular structure (amino acid substitutions) do not occur at the simple regular rate formerly supposed (Uzzell & Pilbeam, 1971; Lovejoy, Burstein, & Heiple, 1972). On the organ system level, dental and middle ear region morphology of living mammals have been an important source of evidence for phylogenetic relationships (e.g., McDowell, 1958; Bugge, 1971), although other parts of the body have been used (e.g., Luckett, 1974), including neuroanatomical features (e.g., Campbell, 1966; Ronnefeld, 1970; Atkins & Dillon, 1971; Radinsky, 1973a).

Specializations of extinct mammals may be inferred from analysis of various parts of the skeletal system and by correlations with skeletal anatomy of living forms for which specializations are known. For example, lengthening of distal limb segments, reduction of lateral digits, and restriction of joint movement to a parasagittal plane are among the features commonly found in recent mammals that run fast, and the functional significance of these features may be inferred from biomechanical analysis. (For examples of this type of analysis, see Schaeffer, 1947; Smith & Savage, 1956; Savage, 1957; Oxnard, 1967; and Butler, 1972.)

External brain morphology is preserved in endocranial casts of fossil mammals and some specializations may be inferred from differential enlargement of various cortical areas. Welker and Campos (1963) have demonstrated enlargement of the hand projection area in primary somatic sensory cortex in raccoons that correlates with their having increased tactile sensitivity of the hand and a behavioral pattern involving palpation of potential food objects. The enlargement of the hand projection area in raccoons is visible on endocasts, for it is delimited by sulci that are reproduced on the endocast. For examples of this type of analysis applied to fossil mammals see Radinsky (1968, 1969).

Finally, paleoecological analysis of the deposits in which a given extinct species occurs may provide information relevant to specialization. Interpretation of sediments and plant fossils may indicate whether the species in question lived in a tropical, temperate, or arctic climate, or whether it was buried in a forest or savannah. Such information is usually too general to be useful for inferring specializations of interest to neuroanatomists and behaviorists. On another level, however, it is becoming increasingly evident that major changes in mammalian evolutionary history are correlated with large-scale changes in climate and plant evolution (Lillegraven, 1972).

EVOLUTIONARY RADIATIONS

The following comments on the evolutionary history of the surviving orders of mammals are of necessity brief and concentrate on those features that seem particularly relevent for neurologists and behaviorists. Good general references on fossil and recent mammals are Romer (1966), Piveteau (1957, 1958, 1961), Anderson and Jones (1967), and Walker (1964). The most recent review of the distribution of mammal families, with good references to the recent primary literature, is that of Lillegraven (1972). Knowledge of fossil endocranial casts up to about 1955 is reviewed by Dechaseaux in the Piveteau references cited above.

Order Insectivora

A major evolutionary radiation about 100 million years ago, with renewed diversification about 76 million years ago, produced several relatively short-lived groups of insectivorans, as well as the ancestors of the surviving families. Living insectivorans include shrews (soricids), moles (talpids), hedgehogs (erinaceids), tenrecs (tenrecids), golden moles (chrysochlorids), and solenodons, as well as the more distantly related tree shrews (tupaiids) and elephant shrews (macroscelidids). Some recent workers classify macroscelidids as a separate order. Most of these families were phylogenetically separated by about 60 million years ago, and for some the time of divergence may go back further than that. Insectivorans include the basal stock from which all placental mammals arose; however, none of the living forms should be considered representative, without qualification, of that common ancestor, for all are specialized in one or more ways as compared to the early stem insectivorans.

Living insectivorans include forms with relatively smaller brains, relatively less neocortex, and relatively larger olfactory bulbs than other living placental mammals (Bauchot & Stephan, 1970; Stephan, Bauchot, & Andy, 1970). Fossil endocasts confirm that the above features are primitive. However, in other features the brains of some living insectivorans are specialized. For example, the visual system is greatly reduced in some moles (Allison & Van Twyver, 1970) and is enlarged in tree shrews (Lende, 1970). Some shrews and some tenrecs are thought to echolocate and therefore elaboration of the auditory system is to be expected, and in some genera of moles the rhinarial area is enlarged into a sensitive tactile organ and therefore elaboration of the somatosensory cortex is to be expected. To approximate what a primitive mammalian brain looked like therefore requires extrapolation from a carefully selected series of modern insectivorans. However, after such a composite is derived, it may be possible to find a living form that closely resembles it.

Tree shrews have been considered primative primates or representatives of ancestral primates and for that reason have received considerable attention from neurologists. However, tree shrews were independently derived from primitive insectivorans, and any similarities to primates are either from retention of primitive features or from independent development of similar specializations, as in the visual system (Van Valen, 1965; Campbell, 1966). Therefore, in the absence of supporting evidence, at present not available, it is incorrect to consider features seen in tree shrews as representative of a stage through which primates passed.

Order Primates

Primates were differentiated from insectivorans by 65 million years ago and deployed in three series of major adaptive radiations. The first radiation, at its height 60 to 55 million years ago, began with forms specialized for insectivorous diets, from which evolved several groups with varying degrees of dental specialization for omnivorous to herbivorous or frugivorous diets. Almost all of these early primates had a pair of enlarged, superficially rodentlike incisors. There are no direct survivors of that radiation. The second radiation, prominent in the fossil record from about 55 to 35 million years ago, produced several lineages that resemble modern prosimian primates in various dental and locomotor specializations (except for tooth combs), and the living prosimians (lorises and galagos, lemurs, and tarsiers), are survivors of that radiation. The times of divergence of those three modern prosimian groups from each other go back at least 35 million years and possibly as far as 60 million years. The third wave of primate radiations appears in the fossil record about 35 million years ago and has produced new world monkeys (ceboids), apes and humans (hominoids), and old world monkeys (cercopithecoids). Paleozoogeographic evidence suggests that ceboids arose independently from prosimians, and it is possible that cercopithecoids and hominoids were also independently derived from prosimian ancestors. The features that distinguish higher primates (anthropoids) from prosimians were therefore probably independently derived at least twice and perhaps three times. (See Simons, this volume, for further information on primate phylogeny).

The fossil record of primate brains reveals that about 55 million years ago, at the base of the second great radiation, primates had reduced olfactory bulbs and expanded visual cortex as compared with most other contemporary mammals (Radinsky, 1970). The oldest anthropoid brains, 25 to 30 million years old and therefore near the base of their respective radiations, show further reduction of olfactory bulbs and expansion of visual cortex, plus the development of the central sulcus, a major cortical-structure that distinguishes anthropoids from most prosimians (Radinsky, 1974a). Frontal lobe evolution in both prosimians and anthropoids lagged behind that of other parts of the brain.

The great diversity of living primate brains (Connolly, 1950; Hershkovitz, 1970; Stephan *et al.*, 1970; Radinsky, 1974b) offers excellent opportunities for studies that may elucidate the functional significance of differences in brain structure among primates. For example, the central sulcus is present in intermediate stages of development in a few living prosimians, and the arcuate sulcus is present in cercopithecoids, absent in hominoids, and present in various stages of development in ceboids.

Order Rodentia

Rodents first appear in the fossil record about 60 million years ago, probably derived from early primates or possibly from unspecialized insectivorans. Their main adaptation, which is evident in the earliest rodents, is modification of the dental apparatus for gnawing (enlarged, ever-growing, self-sharpening incisors) and an omnivorous to herbivorous diet (broad, low-cusped molar teeth). Skull and jaw musculature have been modified to serve those functions. Shortly after their first appearance, rodents underwent a major adaptive radiation, based largely on secondary dental specializations; coincident with that radiation, other groups of small mammals with enlarged incisors and low-cusped cheek teeth (early primates, multituberculates) markedly declined. There appears to have been another wave of adaptive radiations of rodents about 40 million years ago, and most of the approximately 34 surviving families can be traced back to that time. For many rodent families times of divergence go back even further, some back to 50 million years ago. However, one of the most successful groups of rodents, the old world rats and mice (murids), appear in the fossil record only about 10 million years ago and thereafter underwent an explosive radiation.

The great diversity of living rodents is not reflected in their neuroanatomy, which appears relatively stereotyped, although there are differences among major groups (Robertson, 1964). Fossil endocasts of early rodents differ little in external morphology from those of modern forms (Radinsky, personal observation). It therefore appears that evolution of the brain has not been a significant factor in rodent evolution. Some neurological specializations have been noted in rodents, such as hypertrophy of the auditory system in kangaroo rats (Webster, Ackerman, & Longa, 1968) and, to a lesser degree, elaboration of the visual system in squirrels (Kaas, Hall, & Diamond, 1972). Several other groups of rodents besides kangaroo rats have enlarged middle ears and may have similar specialization of the central nervous system. Some rodents, such as mole rats (spalacids and bathyergids), have specialized for subterranean life and presumably may have reduced the visual system (Rees, 1968).

Rodents provide some of the most commonly used laboratory animals, such as rats, mice, guinea pigs and hamsters, and dozens of other species are readily available and easy to raise. The great diversity of rodent families provides a large choice of distantly related forms for comparative study. For example, laboratory rats and mice (*Rattus* and *Mus*) share a common ancestry less than 10 million years old, whereas guinea pigs and chinchillas have evolved in a South American radiation that was isolated over 35 million years ago, and squirrels have been independent from other groups since 40–50 million years ago.

Order Chiroptera

Bats are a very successful order of mammals, second only to rodents in diversity. A complete skeleton of the oldest known bat, 50 million years old, reveals specialization for flight as advanced as in modern bats (Jepsen, 1966). The oldest bat endocranial casts, from about 40 million years ago, have enlarged auditory colliculi, suggesting that echolocation has been developed in bats since at least that long ago (Dechaseaux, 1956). The specializations of flight and echolocation presumably were critical for the origin of bats from early insectivorans, to exploit nocturnal flying insects as food. (There is little competition from insectivorous birds at night.) Most modern bats are insectivorous, but some have specialized to feed on fruit, nectar, fish, or blood.

The fossil record of bats is poor owing to the fragile nature of bat skeletons and to the habitats of bats but there is enough evidence to indicate that several of the modern families were distinct by 35 million years ago. Most bats, including the oldest ones, are in the suborder Microchiroptera, which includes the whole spectrum of dietary types. The much smaller suborder Megachiroptera, composed of the large fruit bats, appears in the fossil record only 25 million years ago.

An important study of the relative sizes of major parts of bat brains (Stephan & Pirlot, 1970) shows correlation of brain specialization with dietary type. For example, olfactory bulbs are most reduced in insectivorous types, and the neocortex is most enlarged in vampire bats. A more detailed but less broadly comparative review of bat neuroanatomy is that of Hensen (1970). Because the auditory system is specialized to varying degrees among the living bats, the order provides an excellent comparative series for study of that system.

Order Carnivora

Carnivorans evolved from early insectivorans about 65 million years ago, with their main adaptation the modification of a pair of cheek teeth for scissorslike shear. Early carnivorans, called "miacoids," underwent a modest evolutionary radiation in competition with other groups of similar-sized archaic carnivores, the creodonts. About 40 million years ago, miacoids underwent a second wave of evolutionary radiations that gave rise to the oldest representatives of most of the modern families of carnivorans: felids, viverrids (civets and mongooses), canids, ursids (bears), procyonids (raccoons, etc.), and mustelids (minks, badgers, otters, etc.). The first two families evolved from a different group of miacoids, the ancestry of which may have diverged from that of other families 50 million years ago. About 20–25 million years ago, hyaenids arose from viverrid stock, phocids (seals) from mustelids, and otariids (sea lions and walruses) from ursids. With the appearance of the modern carnivoran families the creodonts waned and became extinct, apparently unable to compete with the early dogs, cats, etc. The only obvious feature in which carnivorans appeared to have had an advantage over creodonts was in the brain, which increased in relative size and amount of neocortex coincident with the emergence of the modern carnivoran families.

Of the approximately 120 living genera of carnivorans, the brains of only two, the domestic dog and cat, have been seriously studied. There is a fair diversity of external neuroanatomy among living carnivorans, particularly among mustelids, viverrids, and procyonids, and many of these may be good experimental animals. Welker and Campos (1963) have demonstrated unusual somatic sensory specializations in two species of procyonids (raccoon and coati mundi), which has implications for understanding the significance of sulci in cortical localization. From observations of external brain morphology, it appears that somatic sensory specializations have also evolved in several species of mustelids, that the visual cortex is unusually expanded in some mustelids, that the auditory cortex is expanded in some viverrids, and that the new world badger (*Taxidea*), a solitary animal, has an unusually large frontal lobe. Skunks have relatively less neocortex than any other living carnivorans, and some viverrids lack the cruciate sulcus, an otherwise ubiquitous sulcus in carnivorans (Radinsky, 1975). A fairly good fossil record of carnivoran brains has provided information for recent studies of brain evolution in dogs, cats, otters, and carnivorans generally (Radinsky, 1968, 1969, 1971, 1973b).

Order Artiodactyla

Artiodactyls, the even-toed ungulates, are very successful medium-sized to large herbivores, including 25 families of which nine survive: pigs (suids), peccaries (tayassuids), hippos (hippopotamids), camels (camelids), mouse deer (tragulids), deer (cervids), giraffes (giraffids), pronghorns (antilocaprids), and antelope, sheep, cattle, etc. (bovids).

The first artiodactyls arose about 60 million years ago from an archaic group of omnivores, the condylarths, with a modification of the ankle joint that allowed elongation of the foot and enhanced running ability; a modest evolutionary radiation followed. About 40 million years ago, several groups of artiodactyls developed an improved dental apparatus, the selenodont molar, that increased efficiency in shredding plant material, and an explosive adaptive radiation followed. The surviving families with selenodont teeth (camelids through bovids of the above list) also have a ruminating stomach, a specialization that aids in the breakdown of cellulose, and it is possible that the ability to ruminate was also a factor in the second wave of artiodactyl radiation. Each of the last four families on the list has a different type of horn or antler, the development of which was apparently relevant to the radiation of each group.

Pigs and peccaries were distinct families by about 35 million years ago, and their common ancestry probably diverged from that of other artiodactyls 50 million years ago. Hippos and camelids probably also have been distinct groups for about that length of time. The remaining families share a more recent common ancestry, although most were distinct by 25–35 million years ago. The most diverse group, the bovids, underwent an explosive radiation 10 million years ago that resulted in the 44 bovid genera of the present day.

The main specializations of artiodactyls are in the food processing apparatus and locomotor system. However, most if not all of the living ones appear to have expanded snout representation in the primary somatic sensory cortex (Johnson, Rubel, & Hatton, 1974; personal observation), correlated with the expanded use of lips and rhinarial area as a tactile organ, and some, such as antilocaprids, appear to have expanded visual cortex (personal observation). Dechaseaux (1969) has recently reviewed the fossil record of primitive artiodactyl brains, and the relative size and external morphology of the brains of many of the living artiodactyls are described by Kruska (1970) and Oboussier (1972).

Order Perissodactyla

Perissodactyls, the odd-toed ungulates, are medium-sized to large herbivores of which representatives of only three out of 14 families survive: horses, rhinos, and tapirs. Like artiodactyls, perissodactyls evolved from condylarths about 60 million years ago, with a modification of the ankle joint that allowed for lengthening of the foot and more efficient running. Unlike early artiodactyls, they also had a modification of the dental apparatus that provided for more effective mastication of plant material. Immediately after their first appearance, perissodactyls underwent a major evolutionary radiation expressed mainly by secondary specializations of the teeth and differences in size, and by 50 million years ago, 11 families were in existence. Between 40 and 30 million years ago, as the great artiodactyl radiations unfolded, perissodactyls waned, possibly in part because of their inability to compete successfully with the artiodactyls and in part because of the coincident appearance of modern carnivores. Horses underwent a modest secondary radiation correlated with improvements in the cheek teeth for shredding grass and in the feet for running, which began about 20 million years ago and culminated in the one-toed horses about 10 million years ago.

Of living perissodactyls, horses have been phylogenetically separate from rhinos and tapirs for 58 million years. The ancestors of modern tapirs and rhinos probably diverged 50 million years ago.

Little attention has been paid to the limited diversity of living perissodactyl brains. Adrian (1946) has demonstrated the existence of an expanded snout representation in primary somatic sensory cortex in horses, and Edinger's monograph (1948) on 50 million years of horse brain evolution is a classic. Kruska (1973) studied rhino and tapir brains and calculated relative brain sizes for recent perissodactyls.

Order Cetacea

Whales first appeared in the fossil record about 50 million years ago, apparently descended from early condylarths. By the time of their first appearance they were already large and specialized for aquatic locomotion (body lengthened, hind limbs vestigial). The modest evolutionary radiation of early whales, the archaeocetes, gave rise about 40 million years ago to the ancestors of modern whales, the odontocetes, or toothed whales (including porpoises and dolphins), and the mysticetes, or baleen (whalebone) whales. Several modern families can be traced back 25 million years. Both odontocetes and mysticetes show a telescoping of the skull bones, the functional significance of which is unclear. Odontocetes have a fatty organ, the melon, at the front of the head and a marked asymmetry of the external nares that seem to be related to echolocation. The melon is relatively largest in sperm whales, where it apparently has a hydrostatic function (Clarke, 1970). Mysticetes have peculiarly modified oral epithelium and an enlarged tongue as a specialization for straining small organisms from the water.

Noteworthy features of the brains of modern whales include extreme reduction of the olfactory system (olfactory bulbs are absent in adult odontocetes) and hypertrophy of the auditory system, particularly in odontocetes, in relation to echolocation, and the large size of the cerebellum. Neurophysiological experiments on whales involve obvious difficulties, but some interesting work has been done recently (e.g., Lende & Welker, 1972). The most recent review of cetacean brain anatomy is that of Jansen and Jansen (1968). Edinger (1955) and Breathnach (1955) have reviewed fossil cetacean endocranial casts. In the early archaeocetes the olfactory tracts appear to have been unreduced, and the cerebrum was relatively small.

Other Condylarth Derivatives

Elephants, hyraxes, sea cows, and aardvarks are four minor orders of surviving mammals that are generally considered to have evolved from condylarth stock. However, in the absence of an early fossil record or of decisive anatomical evidence, their origins remain problematic.

Two closely related genera of living elephants are the only survivors of the adaptive radiations of the order Proboscidea, which can be traced back to about 45 million years ago. All known members of the order have been large, have had dental modifications for an herbivorous diet, and usually had one or two pairs of enlarged anterior teeth (tusks); most have had various specializations of the upper lip as a prehensile and tactile organ, the proboscis. Brains of living elephants have unusually large temporal lobes, a feature that characterized proboscidean endocasts back to 45 million years ago (Edinger, 1961).

The order Hyracoidea first appeared in the fossil record about 35 million years ago, with its main specialization the modification of the jaws and teeth for a herbivorous diet. Some of the extinct hyraxes were as large as cows, but the three surviving genera are approximately rabbit sized. Their brains, however, are larger than those of rabbits, show a moderate development of the neocortex, and may be interesting to compare with similar-sized brains of other condylarth derivatives, such as small artiodactyls.

Manatees and dugongs are the only living representatives of the order Sirenia, or sea cows, which first appeared in the fossil record about 50 million years ago already quite specialized in locomotor and dental systems for life as large, shallow water browsers. The two living genera are in different families and have been separated phylogenetically for at least 35 million years. Brains of sirenians have reduced olfactory bulbs, as is expected in aquatic mammals, and are unusual in being as large as they are (around 500 cm^3) and yet devoid of neocortical sulci.

The order Tubulidentata contains a single living species, *Orycteropus afer*, the aardvark. It is a medium-sized mammal that is specialized for termite eating, having powerful forelimbs for digging and weak jaws and teeth. The brain of the aardvark is interesting in that it is similar in general proportions and external anatomy to the brains of large condylarths and early artiodactyls and perissodactyls. It therefore might provide insight into what the neuroanatomy of at least those groups of mammals was like about 50 million years ago. Sonntag and Woollard (1925) have provided a good initial description of aardvark neuroanatomy.

Order Edentata

Edentates appeared about 60 million years ago in South America, derived from unknown, presumably insectivoran, ancestors. They underwent major evolutionary radiations during the time that South America was isolated from North America (from about 65 million to about 3 million years ago), that resulted in 11 families of which only three survive: armadillos (Dasypodidae), anteaters (Myrmecophagidae), and tree sloths (Bradypodidae). The known edentate groups, both fossil and living, are so divergently specialized that it is difficult to imagine what were the initial adaptive features responsible for the emergence of the order. The ordinal name derives from the lack of enamel on edentate teeth. Armadillos are small to medium-sized omnivores with relatively weak jaws and teeth, powerful limbs for digging, and a flexible covering of dermal (bony) and epidermal (horny) plates. They first appear in the fossil record about 60 million years ago. Anteaters range from small to medium-sized; have powerful limbs for ripping open termite nests and rotten logs; long, weak, toothless jaws; and a hypertrophied tongue and salivary gland apparatus. Their fossil record goes back about 25 million years. Tree sloths are medium-sized, slow moving arboreal herbivores that have a host of bizarre specializations for hanging upside down from branches. The two living genera, which have no fossil record, apparently were derived from different groups of a large radiation of ground sloths, some of which were larger than the largest bears.

External brain morphology of living edentates has been surveyed by Pohlenz-Kleffner (1969), who noted the characteristics of each of the extant families. With respect to expansion of the neocortex and reduction of the olfactory bulbs, edentate brains appear intermediate between those of insectivorans, on the one hand, and ungulates and carnivorans, on the other. Of the living edentates, brains of armadillos seem least advanced. Sensory projections in a sloth brain have been mapped by Meulders, Gybels, Bergmans, Gerebtzoff, and Goffart (1966).

Order Pholidota

Pholidotans, otherwise known as pangolins or scaly anteaters, are old world termite eaters with cranial and postcranial specializations similar to those seen in the South American anteaters. In addition, the single surviving genus, *Manis*, has a body covering of horny scales. Recent paleontological work has revealed an early (60–30 million years old) radiation of pholidotans in North America. The surviving family can be traced back 35 million years in the old world, with one 35 million-year-old representative in North America. Pholidotan brains superficially resemble those of armadillos.

Order Lagomorpha

Lagomorphs are a small group of mostly small herbivores, including two living families, rabbits and hares (leporids) and pikas (ochotonids). Their main specialization has been the development of enlarged, ever-growing incisors, and cheek teeth modifications for chewing grass and leaves. Any special resemblances to rodents are entirely the consequence of convergent evolution, for lagomorphs and rodents were derived from different groups of insectivorans. Although their origins are obscure, lagomorphs probably arose from a poorly known group of Asiatic insectivorans over 60 million years ago. Ancestors of modern leporids first appear in the fossil record about 40 million years ago, and ochotonids were distinct by 30 million years ago. The locomotor specialization for quadrupedal leaping that reaches an extreme in jack rabbits was a relatively late development.

Order Dermoptera

Dermopterans, or "flying lemurs," consist of a single living genus, *Cynocephalus*, a rabbit-sized herbivore with expanded skin and elongate limbs for gliding and an expanded and notched lower incisor tooth comb that functions in grooming and feeding. Molar tooth patterns and cranial morphology indicate affinity with insectivorans, with which dermopterans are sometimes classified. *Cynocephalus* cannot be traced in the fossil record, but an apparently related group, plagiomenids, existed 60–40 million years ago in North America.

Order Marsupialia

Marsupials and placental mammals diverged about 115 million years ago, with the main difference between them being in the reproductive system. Marsupials never evolved the ability to arrest the estrous cycle, so that young have to finish the equivalent of later fetal development in the maternal pouch rather than in the uterus. There were two main centers of marsupial evolution. In South America, marsupials filled the major insectivore and carnivore niches until the Pleistocene. Only two families of marsupials survived from the South American radiations, didelphids (opossums) and caenolestids. The second center was Australia, where marsupials filled most of the major ecological niches occupied elsewhere by placentals.

Insectivorous and carnivorous dasyurids and thylacinids range from mouse to wolf size and include one form specialized for termite eating; notoryctids are the marsupial equivalent of moles; peramelids, or bandicoots, are insectivorous to omnivorous rat- to rabbit-sized diggers; phalangerids are small to medium-sized omnivores and herbivores, including one nectar feeder; and macropodids (wallabies and kangaroos) and wombats are the medium-sized to large herbivores. Several families have evolved a variety of interesting dental and locomotor specializations. The Australian record of marsupials only goes back to about 30 million years ago, by which time most of the modern families were distinct. The order Marsupialia is broader than placental orders, and most workers now recognize three or four orders of living marsupials.

Brains of marsupials differ from those of placentals in lacking a corpus callosum and in having a neocortical sensorimotor amalgam rather than separable motor and somatic sensory cortical areas (Lende, 1969). Because of their long period of independent evolution apart from placentals, and because of the diversity of brain morphology among marsupials, they may provide important insights into the evolution of mammal brains. References to work on marsupial brains may be found in Reese and Hore (1970) and Walsh and Ebner (1970).

Order Monotremata

Monotremes, the only surviving egg-laying mammals, diverged from other living mammals about 180 million years ago, shortly after mammals evolved from reptiles. Only the duckbilled platypus (*Ornithorhynchus*) and echidna (*Tachyglossus* and *Zaglossus*) survive today, with virtually no fossil record. Both living forms are so bizarrely specialized that they provide little insight into the basic adaptations responsible for the emergence of their order. Echidnas, or spiny anteaters, are adapted for digging and ant and termite feeding (powerful limbs, long, weak edentulous jaws, long tongue). The platypus is a semiaquatic, burrowing insectivore with a highly sensitive expanded rhinarial area.

Because monotremes separated from other mammals so long ago, study of their brains is crucial for insight into the evolution of mammal brains. Echidna brains are unusual in being relatively highly convoluted, with a relatively large frontal lobe, and in having the major neocortical areas skewed around relative to their positions in marsupial and placental mammals (Lende, 1969). Platypus brains are lissencephalic and have only recently been the object of modern experimental studies (Campbell & Hayhow, 1972).

REFERENCES

Adrian, E. D. The somatic receiving area in the brain of the Shetland pony. *Brain*, 1946, **69**, 1–8.

Allison, T., & Van Twyver, H. Sensory representation in the neocortex of the mole, Scalopus aquaticus. *Experimental Neurology*, 1970, **27**, 554–563.

Anderson, S., & Jones, J. K., Jr. (Eds.). *Recent mammals of the world.* New York: Ronald Press, 1967.

Atkins, D. L., & Dillon, L. S. Evolution of the cerebellum in the genus *Canis. Journal of Mammalogy*, 1971, **52**, 96–107.

Barnicott, N. A. Some biochemical and serological aspects of primate evolution. *Science Progress* (Oxford), 1969, **57**, 459–493.

Bauchot, R., & Stephan, H. Morphologie comparee de l'encephal des insectivores Tenrecidae. *Mammalia*, 1970, **34**, 514–541.

Breathnach, A. S. Observations on endocranial casts of recent and fossil cetaceans. *Journal of Anatomy*, 1955, **89**, 532–546.

Bugge, J. The cephalic arterial system in New and Old World hystricomorphs, and in bathyergoids, with special reference to the systematic classification of rodents. *Acta Anatomica*, 1971, **80**, 516–536.

Butler, P. M. Some functional aspects of molar evolution. *Evolution*, 1972, **26**, 474–483.

Campbell, C. B. G. The relationships of tree shrews: the evidence of the nervous system. *Evolution*, 1966, **20**, 276–281.

Campbell, C. B. G., & Hayhow, W. R. Primary optic pathways in the duckbill platypus, *Ornithorhynchus anatinus*: an experimental degeneration study. *Journal of Comparative Neurology*, 1972, **145**, 195–207.

Clarke, M. R. Function of the spermaceti organ of the sperm whale. *Nature (London)*, 1970, **228**, 873–874.

Connolly, C. J. *External morphology of the primate brain.* Springfield, Illinois: Charles C Thomas, 1950.

Dechaseaux, C. Moulages endocraniens naturels de microchiropteres fossiles. *Annales de Paléontologie,* 1956, **42**, 119–137.

Dechaseaux, C. Moulages endocraniens d'artiodactyles primitifs. Essai sur l'histoire du neopallium. *Annales de Paléontologie,* 1969, **55** 195–248.

Edinger, T. Evolution of the horse brain. *Geological Society of America Memoirs*, 1948, **25**, 1–177.

Edinger, T. Hearing and smell in cetacean history. *Monatsschrift für Psychiatrie & Neurologie,* 1955, **129**, 37–58.

Edinger, T. Anthropocentric misconceptions in paleoneurology. *Proceedings of the Rudolf Virchow Medical Society*, 1961, 19(1960), 56–107.

Hensen, O. W., Jr. The central nervous system. In W. A. Wimsatt (Ed.), *Biology of bats.* Vol. 2. Academic Press, New York, 1970.

Hershkovitz, P. Cerebral fissural patterns in platyrrhine monkeys. *Folia Primatologica*, 1970, 13, 213–240.

Jansen, J. & Jansen, J. K. S. The nervous system of cetacea. In H. T. Anderson (Ed.), *The biology of marine mammals.* Academic Press, New York, 1968.

Jepsen, G. L. Early Eocene bat from Wyoming. *Science*, 1966, 154, 1333–1339.

Johnson, J. I., Rubel, E. W., & Hatton, G. I. Mechanosensory projections to cerebral cortex of sheep. *Journal of Comparative Neurology*, 1974, 158, 81–107.

Kaas, J. H., Hall, W. C., & Diamond, I. T. Visual cortex of the grey squirrel (*Sciurus carolinensis*): architectonic subdivisions and connections from the visual thalamus. *Journal of Comparative Neurology*, 1972, 145, 273–305.

Kruska, D. Ueber die Evolution des Gehirns in der Ordnung Artiodactyla Owen, 1848, insbesondere der Teilordnung Suina Gray, 1868. *Zeitschrift für Säugetierkunde*, 1970, 35, 214–238.

Kruska, D. Cerebralisation, Hirnevolution und domestikations bedingte Hirngrössenänderungen innerhalb der Ordnung Perissodactyla Owen, 1848 und ein Vergleich mit der Ordnung Artiodactyla Owen 1848. *Zeitschrift für zoologische Systematiks und Evolutions forschung*, 1973, 11, 81–103.

Lende, R. A. A comparative approach to the neocortex: localization in monotremes, marsupials, and insectivores. *Annals of the New York Academy of Science*, 1969, 167, 262–275.

Lende, R. A. Cortical localization in the tree shrew *(Tupaia)*. Brain Research, 1970, 18, 61–75.

Lende, R. A. & Welker, W. I. An unusual sensory area in the cerebral neocortex of the bottlenose dolphin, *Tursiops truncatus*. *Brain Research*, 1972, 45, 555–560.

Lillegraven, J. A. Ordinal and familial diversity of Cenozoic mammals. *Taxon*, 1972, 21, 261–274.

Lovejoy, C. O., Burstein, A. H., & Heiple, K. G. Primate phylogeny and immunological distance. *Science*, 1972, 176, 803–805.

Luckett, W. P. The phylogenetic relationships of the prosimian primates: evidence from the morphogenesis of the placenta and foetal membranes. In R. Martin (Ed.), *Prosimian biology.* London: Duckworth, 1974.

McDowell, S. B., Jr. The Greater Antillean insectivores. *Bulletin of the American Museum of Natural History,* 1958, 115(3), 113–214.

Meulders, M., Gybels, J., Bergmans, J., Gerebtzoff, M. A., & Goffart, M. Sensory projections of somatic, auditory and visual origin to the cerebral cortex of the sloth (*Choloepus hoffmani* Peters). *Journal of Comparative Neurology*, 1966, 126, 535–546.

Oboussier, H. Morphologische und quantitative Neocortexuntersuchungen bei Boviden, ein Beitrag zur Phylogenie dieser Familie. III. *Mitteilungen aus dem Hamburgischen Zoologischen Museum & Institut*, 1972, 68, 271–292.

Oxnard, C. E. The functional morphology of the primate shoulder as revealed by comparative anatomical, osteometric, and discriminant function techniques. *American Journal of Physical Anthropology*, 1967, 26, 219–240.

Piveteau, J. (Ed.). Primates. Paleontologie humaine. In *Traite de paléontologie*, Tome 7. Paris: Masson, 1957. Pp. 1–675.

Piveteau, J. (Ed.). Mammiferes evolution. In *Traite de paléontologie*, Tome 6, Vol. 2. Paris: Masson, 1958. Pp. 1–962.

Piveteau, J. (Ed.). Mammiferes, origine reptilienne, evolution. In *Traite de paléontologie*, Tome 6, Vol. 1. Paris: Masson, 1961. Pp. 1–1135.

Pohlenz-Kleffner, W. Vergleichende Untersuchungen zur Evolution der Gehirne von Edentaten. II. Form und Furchen. *Zeitschrift für Zoologie Systematiks Evolutionsforschung,* 1969, 7, 180–208.

Radinsky, L. Evolution of somatic sensory specialization in otter brains. *Journal of Comparative Neurology,* 1968, 134, 495–506.

Radinsky, L. Outlines of canid and felid brain evolution. *Annals of the New York Academy of Science,* 1969, 167, 277–288.

Radinsky, L. The fossil evidence of prosimian brain evolution. In C. Noback & W. Montagna (Eds.), *The Primate Brain. Advances in Primatology,* Vol. 1. New York: Appleton-Century-Crofts, 1970.

Radinsky, L. An example of parallelism in carnivore brain evolution. *Evolution,* 1971, 25, 518–522.

Radinsky, L. Are stink badgers skunks? Implications of neuroanatomy for mustelid phylogeny. *Journal of Mammology,* 1973, 54, 585–593. (a)

Radinsky, L. Evolution of the canid brain. *Brain, Behavior, & Evolution,* 1973, 7, 169–202. (b)

Radinsky, L. The fossil evidence of anthropoid brain evolution. *American Journal of Physical Anthropology,* 1974, 41, 15–28. (a)

Radinsky, L. Prosimian brain morphology. Functional and phylogenetic implications. In R. Martin (Ed.), *Prosimian biology.* London: Duckworth, 1974. (b)

Radinsky, L. Viverrid neuroanatomy: phylogenetic and behavioral implications. *Journal of Mammalogy,* 1975, 56, 130–150.

Rees, E. L. A note on the brain of the Cape dune mole rat, *Bathyergus suillus. Acta Anatomica,* 1968, 71, 147–153.

Rees, E. L., & Hore, J. The motor cortex of the brush-tailed possum (*Trichosurus vulpecula*): motor representation, motor function, and the pyramidal tract. *Brain Research,* 1970, 20, 439–456.

Robertson, W. V. The phylogeny of myomorph rodents based on brain morphology. Unpublished doctoral dissertation, Texas A and M University, 1964.

Romer, A. S. *Vertebrate paleontology* (3rd ed.) Chicago: University of Chicago Press, 1966.

Ronnefeld, U. Morphologische und quantitative Neocortexuntersuchungen bei Boviden, ein Beitrag zur Phylogenie dieser Familie. I. *Morphologisches Jahrbuch,* 1970, 115, 163–230.

Sarich, V. M. Pinniped phylogeny. *Systematic Zoology,* 1969, 18, 416–422.

Savage, R. J. G. The anatomy of *Potamotherium,* an Oligocene lutrine. *Proceedings of the Zoological Society of London,* 1957, 129, 151–244.

Schaeffer, B. Notes on the origin and function of the artiodactyl tarsus. *American Museum Novitates,* 1947, 1356, 1–24.

Seal, U. S., Phillips, N. I., & Erickson, A. W. Carnivora systematics: immunological relationships of bear serum albumins. *Comparative Biochemistry and Physiology,* 1970, 32, 33–48.

Smith, J. M., & Savage, R. J. G. Some locomotory adaptations in mammals. *Journal of the Linnean Society Zoology,* 1956, 42, 603–622.

Sonntag, C. F., & Woollard, H. H. A monograph of *Orycteropus afer.* Pt. II. *Proceedings of the Zoological Society of London,* 1925, 1925, 1185–1235.

Stephan, H., Bauchot, R., & Andy, O. J. Data on size of the brain and of various brain parts in insectivores and primates. In C. Noback and W. Montagna (Eds.), *The Primate Brain. Advances in Primatology,* Vol. 1. New York: Appleton-Century-Crofts, 1970.

Stephan, H., & Pirlot, P. Volumetric comparisons of brain structures in bats. *Zeitschrift für Systematiks & Evolutionsforschung,* 1970, 8, 200–236.

Uzzell, T., & Pilbeam, D. Phyletic divergence dates of hominoid primates; a comparison of fossil and molecular data. *Evolution,* 1971, 25, 615–635.

Van Valen, L. Treeshrews, primates, and fossils. *Evolution*, 1965, 19, 137–151.

Walker, E. P. (Ed.). *Mammals of the world.* Baltimore: Johns Hopkins Press, 1964. 3 vols.

Walsh, T. M., & Ebner, F. F. The cytoarchitecture of somatic motor cortex in the opossum (*Didelphis marsupialis virginiana*). *Journal of Anatomy*, 1970, 107, 1–18.

Webster, D. B., Ackerman, R. F., & Longa, G. C. Central auditory system of the kangaroo rat, *Dipodomys merriami. Journal of Comparative Neurology*, 1968, 133, 477–494.

Welker, W. I., & Campos, G. B. Physiological significance of sulci in somatic sensory cerebral cortex in mammals of the family Procyonidae. *Journal of Comparative Neurology*, 1963, 120, 19–36.

Wurster, D. H., & Benirschke, K. Comparative cytogenetic studies in the order Carnivora. *Chromosoma*, 1968, 24, 336–382.

13

Comments on Radinsky's "Later Mammal Radiations"

Malcolm C. McKenna

The American Museum of Natural History

Several years ago I attempted to write a "state of the art" paper summarizing what was then known about the early phylogeny of therian mammals and their differentiation into major taxonomic subdivisions (McKenna, 1969). Most of what is known about early therians concerns their jaws and teeth, which preserve well, and so the first half of that paper deals with the complexities of dental evolution. Some of my braver paleontological colleagues may have read it, but I have an idea that few of the intended audience of neuroanatomists and other nonpaleontologists have plowed through all those details of dental pattern that so delight students of fossil mammals. As a second section of the paper I added some brief comments on Cenozoic eutherian differentiation that were potentially more useful than the first part of the paper, but the second section was too short to be really very helpful. Radinsky's chapter for this volume does a much better job of outlining the major branches of mammalian phylogeny and relating them to neuroanatomical work. I agree with most but not all of his main points and take this opportunity to comment on them lightheartedly.

Radinsky's first important point concerns an unfortunate habit of synecdoche that, although most prevalent among workers who view organisms as bags of chemicals, is also met with all too often among neuroanatomists. A single genus or species becomes "the" bird, "the" rodent, "the" (other) primate. Whole orders and even classes are effectively assigned *en masse* the characteristics of some member species known to be easily maintained in the laboratory, the rest being assumed to follow suit. How much this sort of typological myopia has held back progress is hard to assess, but for anyone with first-hand familiarity with the great diversity of the Mammalia the monotonous litany of "the" mouse, "the" rat, "the" rabbit, "the" cat, "the" dog, "the" macaque sounds about as varied as a one-note concerto. There are plenty of animals living in the world whose last common ancestor with a currently popular "the" animal lived millions of years in the past: take, for example, the genus *Ochotona* (pikas) among the order Lagomorpha. Study of *Ochotona* has been minimal, yet *Ochotona*'s ancestors branched away from other lagomorphs at least 30 million years ago insofar as can be determined from paleontological evidence. Shall Alice's friend the White Rabbit continue to stand for the whole of the Lagomorpha? Perhaps research on *Ochotona* should be increased instead! Possibly not all of the members of the order Lagomorpha share all the character states of the genus *Oryctolagus*! Sampling of available mammalian diversity needs to be broadened; obvious opportunities to do so exist, and few serious obstacles stand in the way of such research.

The remainder of Radinsky's paper deals with the living orders of mammals, eighteen or more in all depending on how they are classified. Paleontological knowledge of ancient extinct members of these orders and of an approximately equal number of wholly extinct archaic orders increases year by year, generally beginning with discoveries of dentitions and later continuing with discoveries of ear regions and foot structure. Complete skeletons are rare. The late Professor A. S. Romer of Harvard was fond of claiming that fossil mammals are mere teeth that lived, gave birth directly to more teeth, and died, but it is a pleasure to report that in truth paleontological ideas are much more balanced than that, as can be seen in Figure 1. As paleontological knowledge increases, iteration toward an accurate depiction of anatomy and phylogeny proceeds apace. That there is still far to go, however, is obvious to all who work actively with the problems of mammalian phylogeny. The following comments apply to some of the orders discussed by Radinsky. No attempt is made here to discuss all of them.

Much work lies ahead before the phylogeny of the heterogeneous order Insectivora can be understood. The affinities of the zalambdodont families are still quite uncertain and the links of all the living insectivoran families to their fossil relatives are weak. Additional paleontological exploration is needed in order to find fossil relatives of the tupaiids (and those of tarsiid primates and dermopterans as well). Most of the known early Tertiary insectivorans resemble tupaiids more than they resemble other living insectivorans, but as collecting proceeds a few primitive relatives of the latter are being found or recognized. A point to remember when comparing living insectivorans with living members of other orders is that for perhaps 70 million years the living insectivorans have been phylogenetically separate from those insectivorans that gave rise to other eutherian orders. The situation is analogous to the relationship of the Reptilia to the Mammalia: living reptiles are only very distantly related to the therapsid "reptiles" that gave rise to mammals. Similarly, hedgehogs are not the "ur-mammals" that Thomas Huxley once imagined. Anatomists' attention should be turned in new directions within the Insectivora, as indeed it has been recently toward the tupaiids, but even in that family the potentially most fruitful genus from an anatomical standpoint, *Ptilocercus*, has been virtually ignored since Le

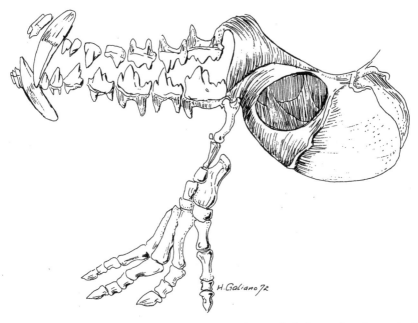

H. Galiano 72

FIG. 1 Romer was wrong! Fossil mammals are not just teeth that reproduced more teeth and then died. It is now known that early mammals possessed a foot and auditory apparatus. A more balanced view of the skeletal anatomy of an early mammal is shown in (a) and a reconstruction, synthesizing all the very latest information, is shown in (b).

Gros Clark's (1926) pioneering effort many years ago. One heartening trend in insectivoran studies, however, is the recent increase in attention given to the enigmatic tenrecoids (Eisenberg & Gould, 1970). Such studies should be expanded and continued.

Regarding primates, it would seem to me that the two greatest problems at present are (1) did the ancestors of South American monkeys reach South America across a water gap from Africa or was it from North America, and (2) which insectivorans give rise to primates? Other contributors to this volume will presumably discuss these problems. The phylogeny of the primates is becoming reasonably well known, thanks to intensive efforts on a broad front by students in many disciplines. Neuroanatomists can contribute to this progress by studying a wide a selection of primates as possible.

Rodent and artiodactyl phylogenies are still in a state of flux, however, and the relationships of living families one to another are still hotly debated. For instance, a new flurry of excitement about the relationships of the hystricognath rodents has recently been triggered by studies of African and South American Mid-Tertiary finds (Hoffstetter, 1972) and by current geological and geophysical studies of the South Atlantic Ocean. Although the South Atlantic was already present as a formidable water barrier long before the existence of rodents, surviving South American hystricognaths ("the" guinea pig and a host of virtually ignored relatives) should be compared to surviving African hystricognaths in greater detail than has been done heretofore in order to gain more accurate knowledge of their evolutionary relationships and history. Evidently the early hystricognaths were once more widely spread on northern continents, including North America (Wood, 1972). South American hystricognaths appear to have arrived on that continent by somehow crossing an oceanic barrier, perhaps on a small "Noah's ark" (McKenna, 1973), but this barrier may have been the Caribbean, not the Atlantic. Living rodent families now restricted to Africa, such as the Ctenodactylidae, Petromuridae, Thryonomyidae, and Pedetidae, as well as South American echimyids and other neglected groups, should become animals of greater interest to neuroanatomists and other students of living rodents.

Turning to another matter, I am inclined to think that there is something wrong with Radinsky's idea of the timing of South American isolation from North America and of its effects on the evolution of edentates. His geological rationale seems to me to be that of stable continents, not plate tectonics, and his implicit concept is that South America became isolated from North America in their present area of connection about 65 million years ago. Perhaps this is so, but they may have been separated there earlier than that, although dispersal of organisms the descendants of which were common to both sides of the gap may have been possible in earlier times via a South America–Africa–Europe–North America route (Reyment & Tait, 1972). In Figure 2 I have tried to summarize recent geophysical and geological estimates of the timing of final separation of

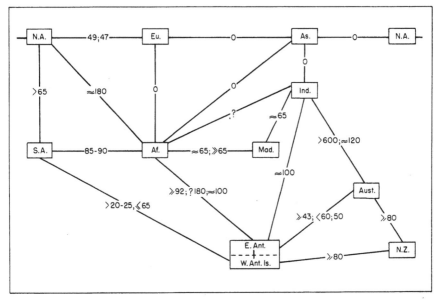

FIG. 2 Various geological and geophysical estimates of approximate times (in millions of years) since the final separation of major landmasses, exclusive of connections caused by new crust (e.g., Panama) and breaks in surface continuity caused by former epicontinental seas now sundered tectonically. Modified from McKenna (1973) Fig. 1. See that paper and its references for discussion and documentation. For the South America–West Antarctic Islands separation date, see Dalziel and Elliot (1971).

the world's major land masses (see McKenna, 1972, 1973, for rationale and references). Edentates might well be survivors of early therians whose genetic continuity with African relatives ceased 85–90 million years or more ago. Both the edentates and their possible sister group the Pholidota deserve renewed study as possible survivors of a very early split from other therians.

Curiously, Radinsky does not come to grips with the problem of marsupial origin and dispersal created by the present revolution in geological thought. See Tedford (1974) for the most up-to-date account of this subject.

I should like to conclude these informal remarks by giving, partly in earnest, a new list of "the" living animals for anatomists to investigate in certain mammalian orders in case they are unwilling or unable to study the broad spectrum I prefer to advocate. It is hoped that they can all be raised in captivity, for I have no wish to cause the extinction of those that are rare by unleashing a new horde of laboratory animal collectors into distant jungles:

The insectivoran *Ptilocercus* (Handle with care!)
The primate   *Tarsius*
The rodent   Any nonmurid except "the" Guinea pig!
The bat   *Pteropus*

The carnivoran   *Nandinia*
The artiodactyl   *Tragulus*
The perissodactyl *Tapirus*
The edentate   *Tamandua*
The lagomorph   *Ochotona*
The marsupials   *Marmosa, Caenolestes,* and *Antechinus*

REFERENCES

Dalziel, I. W. D., & Elliot, D. H. Evolution of the Scotia Arc. *Nature* (London), 1971, **233**, 246–252.

Eisenberg, J. F., & Gould, E. The tenrecs: a study in mammalian behavior and evolution. *Smithsonian Contributions to Zoology*, 1970, (27) i–vi, 1–137.

Hoffstetter, R. Origine et dispersion des rongeurs hystricognathes. *Comptes Rendus des séances des Academie des Sciences (Paris)*, Serie D, 1972, 274, 2867–2870.

Le Gros Clark, W. E. On the anatomy of the pen-tailed tree-shrew (*Ptilocercus lowii*). *Proceedings of the Zoological Society of London*, 1926, **2**, 1179–1309.

McKenna, M. C. The origin and early differentiation of therian mammals. *Annals of the New York Academy of Sciences*, 1969, **167**, 217–240.

McKenna, M. C. Eocene final separation of the Eurasian and Greenland-North American landmasses. *International Geological Congress, 24th Session*, section 7, Paleontology, pp. 275–281. Montreal, Canada, 1972.

McKenna, M. C. Sweepstakes, filters, corridors, Noah's arks, and beached Viking funeral ships in palaeogeography. In D. H. Tarling & S. K. Runcorn (Eds), *Continental drift, sea floor spreading and plate tectonics: Implications to the earth sciences. Proceedings of the NATO Advanced Study Institute, Newcastle upon Tyne, April 10–14, 1972*. London: Academic Press, 1973. Pp. 291–304.

Reyment, R. A., & Tait, E. A. Biostratigraphical dating of the early history of the South Atlantic Ocean. *Philosophical Transactions of the Royal Society of London, Series B, Biological Sciences*, 1972, **264**, 55–95.

Tedford, R. H. Marsupials and the new paleogeography. *In* C. A. Ross (Ed.), *Paleogeographic Provinces and Provinciality. Special Publication of the Society of Economic Paleontologists and Mineralogists*, 1974, **21**, 109–126.

Wood, A. E. An Eocene hystricognathous rodent from Texas: its significance in interpretations of continental drift. *Science*, 1972, **175**, 1250–1251.

14

Brain Evolution in Mammals: A Review of Concepts, Problems, and Methods

Wally Welker

University of Wisconsin

INTRODUCTION

Brains have evolved to be the most complicated of all living tissues. Nervous systems assess the relevance of myriad physical energies in the environment and plan and execute adaptive reactions to them. The paths along which mammalian brains have evolved may be estimated from study of the nervous systems of living mammals. Although the brains of all mammals have most of the same basic neural circuits, there is considerable interspecific diversity and variation in relative size, connectivity, and complexity of their components. In addition, there may be novel circuits in some groups. Because inherited differences in neural circuitry are expected to result in variations in perception, judgment, and reactivity to the environment, behavioral differences in living mammals are expected to be important indicators of circuit differences. Studies of species diversity in neuroanatomical, neuroelectric, and neurochemical aspects of neural circuits are important because they provide direct evidence of species differences in mechanisms underlying heritable behavioral variations. However, many structures in skin, muscles, skeleton, and other nonneural tissues also vary in different animals and contribute in specific ways to adaptation. Because neural and nonneural tissues are so intimately interrelated, developmentally and functionally, it is essential that they be considered together in the search for understanding of the course and causes of brain evolution.

Because most major neural circuits present in mammals also exist in other vertebrates, it appears that the basic adaptive neural mechanisms had been worked out during early vertebrate evolution. These mechanisms are revealed by the existence in all vertebrates of the perception of five general types of physical energy, organized postural and locomotor activities, reflexes and fixed action patterns related to procreation and protection of young, ingestion and elimination, escape and defense, maintenance of homeostatic equilibria, selective attention and orientation toward environmental stimuli, learning and forgetting, and timed ontogenetic development of the behavioral repertoire. The latest stages of neurobiological evolution, especially in some mammals, have produced refined and elaborate neural circuits that provide for greater perceptual, conceptual, and reactive accuracy and complexity in transactions with the environment. I believe that these later evolutionary advances consist of improved abilities to perceive, symbolize, and selectively react to specific aspects of the physical environment. The refinements in sensory, conceptual, and motor systems that have resulted are expressed in greatly expanded and diversified behavioral and cognitive repertoires of transactions with the environment.

Although the major advances in vertebrate brain evolution have occurred during premammalian reptilian radiations, the present review is restricted to examination of neurobiological features of living mammals. My reasons for this are that (a) a great deal more must be learned than is now known about reptilian brain circuitry before appropriate comparisons with mammals can be made; and (b) I believe that studies of the multiple interrelated neurobiological structures in living mammals have provided such a great wealth of information regarding the kinds of factors at work during evolutionary radiation of the mammalian brain that it seems important and fruitful to summarize them here.

Consequently, in the present paper I shall try to identify and catalog features of neural circuits and of associated body structures that vary in different living mammals. Particular attention is given to those specialized sensory and motor structures that play prominent roles in perception of, attention to, and choice of specific environmental niches. How these structural specializations result in specializations in both neuroelectric function and in behavioral and cognitive repertoires is also explored, and the role of environment in the selection of heritable variations of neural circuits is discussed. Neural mechanisms responsible for an animal's invasion of specific ecological and ethological niches are postulated. The role of nonneural tissues in such evolutionary changes is also reviewed. Genetic determinants of the formation of neural and nonneural structural assemblies during ontogeny are briefly mentioned, as is the role of natural selection in the evolution of neural diversity. All these issues pertain to suggestions for promoting continued and improved multidisciplinary studies of the brain. The importance is emphasized of using specific kinds of experimental and observational methods in the search for a more comprehensive understanding of brain evolution.

I take the view that insufficient comparative neurobiological data are available to permit specific conclusions to be drawn regarding the course or causes of evolution of neural circuits within the class Mammalia. First of all, a sufficient body of factual details regarding existing diversity of neural circuits and of their functions is not yet available. In addition, the great diversity of neurobiologically interrelated phenomena pertinent to adaptive function of neural circuits is poorly understood. Moreover, the number of different living mammals that have been carefully studied is still too small to permit appropriate inferences regarding phylogenetic sequences that have produced the great variety of existing forms. Finally, the methods and experimental designs used to study the important problems are in many cases still rudimentary, or they do not appropriately address themselves to the questions asked. In general, it seems to me that the strategies often adopted in searches for understanding of brain evolution are poorly conceived.

Consequently, I attempt in this review to outline a broad, neurobiologically based, conceptual framework that, as a guide to the organization, collection, and analysis of data, seems necessary to improve understanding of brain evolution. Most of these conceptions are borrowed from researchers in this field and modified to reflect a mixture of my own biases, beliefs, hypotheses, intuition, and just plain faith. A great deal of what follows is devoted to developing the general conceptual framework and defining concepts. Because a primary mission of this chapter is to provide general orientation to readers who are not very familiar with the technical and conceptual tools in this neurobiological field, I have decided to simplify the presentation wherever I feel I can. However, the list of references provides many sources for those who wish to probe further into these matters. The complexity of structures and processes revealed even by this rather superficial review may at first glance seem overwhelming and tend to discourage further interest. However, I wish to encourage the reader to simply appreciate and realize the truths implied in this marvelous complexity and to open him- or herself to the excitement of its challenge, mystery, and beauty.

ROLE OF ENVIRONMENT IN BRAIN EVOLUTION

Neural tissue evolved from other tissues so as to enable organisms to react to features of the environment and to control and regulate the animal and its internal processes in such transactions. Specific features of environment played crucial roles in selecting and shaping specific aspects of brain evolution (Drake, 1968; Levins, 1968). The total physical environment consists of a large number of diversified physical energy sources. From every and any place in space and on our planet there emanates a mosaic of physical energies with potential to influence evolving receptor systems. (Figure 1; Barnothy, 1964; Clayton, 1970, 1971; Hoff & Riedesel, 1969). There are also numerous mechanical, thermal, chemical, endocrine, and electrical energy sources within an animal's body that provide additional microenvironments potentially able to affect neural evolution.

All stimulus energies have trajectories that extend outward from their sources. Each type and intensity of energy has its own velocity, effective range, path of

NEURAL REPRESENTATION OF REALITY

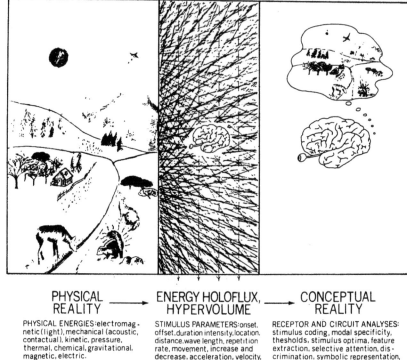

PHYSICAL REALITY →	ENERGY HOLOFLUX, HYPERVOLUME →	CONCEPTUAL REALITY
PHYSICAL ENERGIES:electromag-netic(light), mechanical (acoustic, contactual), kinetic, pressure, thermal, chemical, gravitational, magnetic, electric.	STIMULUS PARAMETERS:onset, offset,duration intensity,location. distance.wave length, repetition rate, movement, increase and decrease, acceleration, velocity, spatial pattern, temporal pat-tern, number, chemical con-figuration and concentration.	RECEPTOR AND CIRCUIT ANALYSES: stimulus coding, modal specificity, thesholds, stimulus optima, feature extraction, selective attention, dis-crimination, symbolic representation, learning and memory of spatiotem-poral patterns, estimation, adaptive exploration, flexible reasoning, creativity, judgement, planning, pro-grammed activity.

FIG. 1 Schematic drawing of the brain's conceptualization of the physical universe by analysis and abstraction from the energy holoflux. Only the visual mode is depicted.

travel, and optimal modes and media for transmission. Energies radiate through a vacuum and/or matter, are conducted, convected, diffused, and/or mechanically transmitted. Each specific locus in space is transected by a confluent mixture of some of these energy patterns emanating from their real physical sources. Therefore, any locus in space is characterized by polymodal, multivectorial fluxes of energies (energy holoflux, Figure 1). Moreover, each locus has a unique configuration of such energy patterns. The physical composition of this flux varies widely from place to place on our planet.

Hutchinson (1965) has proposed the term "hyperspace" or "hypervolume" to refer to those myriad physical energies in the biosphere capable of influencing life. The axes of the hypervolume are the energy variables that are relevant to the life of a species. The hypervolume is a polyvectorial space. The concept of "niche" refers to those particular sets of environmental features with which a specific animal is capable of interacting (Shugart & Patten, 1972). In this chapter, the term "niche transaction" is used to refer to specific response transactions with particular aspects of niche. My definition of a niche transaction therefore involves specifying not only a particular set of stimulus patterns but also an animal's reactions to that set. In this review, I am concerned with only those kinds of niche transactions mediated by the brain. Hutchinson uses the term "realized niche" to refer to that portion of the hypervolume actually inhabited by an organism.

Actually, the concept of "microniche" or microenvironment seems useful in referring to a highly specific set of stimulus patterns to which an animal is receptive and toward which special kinds of action patterns commonly are directed (Figure 2). From the point of view of the brain, the total realized niche consists of a set of specific microniches, each being spatiotemporally unique and having a specific spatial pattern and a specific occurrence in time. For example, a

FIG. 2 Specialized niche transactions. Depicted are locomotor, postural, perceptual, orientational, exploratory, social, defense, and ingestive patterns. Each behavioral pattern deals with a portion of the energy holoflux. Each specific behavioral sequence or expression is intimately transactional with a specific stimulus pattern in the available niche. Each such behavioral sequence is tied to a microniche feature. The total repertoire of microniche transactions is exceedingly large for each mammal. Each species has its own characteristic repertoire.

wolf turning its head toward a rustling sound in the leaves is realizing a microniche. The sudden visual appearance of a tiny rodent from beneath a blanket of leaves provides an additional microniche to which a different response transaction is appropriate. No other creature can occupy that same microniche. The microniche transactions of the mouse at that same instant have different features. Viewed in this way, there are vast numbers of potential microniches for each animal. No two are alike. Some are more complex than others, involving patterns of microniches. Some animals have simple-niche habits and others are capable of inhabiting complex niches. An animal as a whole is designed to respond overtly to only one microniche pattern at a time. However, body parts individually may respond to those portions of the environment to which they are uniquely exposed (e.g., as the left hindfoot in a scratch reflex).

The energy sources within an animal also provide an assortment of energies (carbon dioxide and blood sugar levels, temperature, endocrine concentrations, etc.) potentially available as stimuli to an organism. Moreover, because the nervous system only knows its inputs, such internal stimulus patterns also may be viewed as microniches. Correspondingly, neural circuits that sense and react to these internal states have evolved. Intraorganismic microniches usually can be transacted simultaneously with external ones.

Using these definitions, daily life for an animal is conceived as consisting of fluctuating sequences of microniche transactions. In this review the concept of niche is used to represent a general class of transactions of nervous systems with environments.

The neural circuits of each animal type are designed to detect, differentiate, organize, reconstruct, abstract, and translate some of the available energy patterns into the spatiotemporal symbolic language of nerve cells. For example, animals have evolved exquisite mechanisms for representing and recording within themselves adaptively important features of the physical world. Each species, and each individual animal within a species, has its own specialized neurobiological apparatus, equipping it for specific unique sets of microniche transactions (Beidler, Fishman, & Hardiman, 1955; Mrosovsky, 1971; Norris, 1966, 1969; Schmidt-Nielsen, 1964; Van Gelder, 1969). Different species have deployed to exploit the environmental resources of available habitats in different ways. The degree to which the members of a taxon have evolved specialized transactions with environments has depended on (a) the complexity of the available environment itself during any time period, (b) competitive pressures to exploit the available environments, and (c) the degree of differentiation of sensory, cognitive, and motor mechanisms already attained at that time.

There are certain optimal ranges of available energies that are reacted to and explored in niche invasion and occupation (Carpenter, 1956; Duncan, 1962). Stimuli may be too intense or too mild, too large or too small, too bright or too dark, too loud or too quiet, too hard or too soft, too sweet or too sour, too moist or too dry, etc. Such energies are avoided or ignored, whereas others of intermediate character are just right and are preferred. Competition for optimal ranges within each energy domain has led to gradual selection, from among naturally occurring variants, of those neural circuits that allow a creature to transact with less dangerous, more advantageous, and less competitive extremes of these ranges. Therefore, preferences for new optima have arisen and new transactional capabilities have been realized. Catastrophe and sudden change in energy domains have likewise contributed to selection of those circuit variants more appropriate to success in the new domains (Levins, 1968). The resultant diversified perceptual, reactive, and locomotor capacities in different mammalian lines have biased the activities of different taxa toward various combinations of aerial, arboreal, fossorial, aquatic (marine, estuarine, lacustrine), diurnal, crepuscular, nocturnal, seasonal, frugivorous, carnivorous, herbivorous, browsing, grazing, scavenging, predatory, social, nonsocial, solitary, familial, etc., life styles (Aschoff, 1963; Bourliére, 1956; Cloudsley-Thompson, 1961; Klopfer, 1962; Lofts, 1970; Van Gelder, 1969).

However, inherited, preprogrammed niche transactions may be shaped and modified by individual experience and learning. The degree to which animals are able to learn, remember, and forget stimulus patterns affects the manner in which they deal with the present environment in terms of past ones. Such learning capabilities predispose each animal to prefer, or in some cases to avoid, those environmental features in which its parents have existed. Yet learning capabilities also provide some flexibility in subsequent niche selection and in adaptability to changes in niche. Learning, memory, and forgetting, combined with ability to generalize, often allow for the establishment of reliable and predictable transactions with relative features of stimulus patterns, regardless of many of their absolute energy values. Evolution of neural circuits responsible for such flexibility has been especially prominent in some animal groups.

Changes in environment produced by increasingly numerous living animal and plant groups increased the number and variety of available niches potentially able to influence selection of evolving structures and functions. Thus, food preferences and specialized eating habits were provoked by evolution of new varieties of flora and fauna. Some animals created environmental features, as in nest building, or they maintained and restored eroding features, as in the dam building of beavers. In addition, environmental features were destroyed by overutilization. Thus, living creatures produced a multitude of changes in environments.

The physical appearance of other animals, as well as their behaviors, are also facets of environment, and body structures and neural circuits that produce or perceive such patterns has eventually led to the evolution of social repertoires. Evolution of the brain's ability to discriminate different behavioral patterns often appears to parallel the evolution of the motor ability to produce such patterns in the same or different species. Circuits for sensory perception and motor expression may thereby parallel one another in interactive ways during phylogeny (see section on Neuroanatomical Evolution).

Trends such as these increased the number and diversity of both nervous systems and environments. Each heritable line tended to differentiate further by evolving neurobiological structures and functions that gradually took advantage of greater numbers and varieties of the potentially infinite energy patterns available in the hypervolume. The accumulation of multiple variations in nervous systems and associated body structures accelerated and slowed down from time to time. Periodically, populations of highly specialized forms were eliminated by catastrophe, by competition with other forms more adequately adapted to portions of a niche pattern, by accession of new environments to which a form was not adapted, or by gradual modification or elimination of specific features of the environment that had become crucial for survival (Simpson, 1953, 1961).

Distinct types of neural mechanisms have evolved to deal with specific types of environmental patterns. The great diversity of niche transactions that have resulted is truly amazing. The search for detailed understanding of these mechanisms and how they have evolved is the subject of the remainder of this review.

BEHAVIORAL AND MENTAL EVOLUTION

As indicated above, much of brain evolution has had to do with development and improvement of circuits capable of detecting, discriminating, abstracting, and registering stimulus patterns accurately. Because the operations of such circuits are not directly revealed to an observer, they are referred to here as "covert," "mental," or "cognitive" processes. The brain has also evolved circuits capable of producing a great variety of specific overt action patterns that move an organism in specific organized ways that result in adaptive niche transactions. Species differences in both overt behavior and in mental processes reflect underlying differences in neural circuitry (Caspari, 1958). Consequently, a study of these phenomena, the domain of the behavioral sciences, is an important aspect of any search for understanding of brain evolution.

Overt Behavior

The full repertoire of overt behaviors in most adult mammals (see Table 1; Welker, 1971) is relatively large when compared with that of other vertebrates (Welker, 1971). Almost all items of the behavioral repertoire are specialized to deal with particular classes of transactions with stimulus patterns. Overt behaviors can be graded roughly into several different hierarchical levels that differ from one another in spatiotemporal complexity, duration, and directness of relationship of the responses to the stimuli. The more complex behaviors involve transactions with more complex stimulus patterns.

Simple reflexes consist of specific responses to specific microenvironments. Many are of short duration and are concerned with achievement, maintenance, and restoration of stable postures. They allow an animal to keep itself "together" with its skeletal members spatiotemporally coordinated and integrated, or they establish special relationships to kinetic energy changes involved in body movements induced by gravity or by bodily movements themselves. Other simple reflexes are nociceptive or aversive, involving protective reactions or reactions tending to reduce or avoid high intensities of stimulus energy or to tissue stresses associated with intestinal cramps or bowel and bladder tensions. However, many simple reflexes tend to increase stimulation, such as sniffing, pupillary dilation, head orientation to acoustic or light stimuli, grasp reflexes, supporting reactions and oral "rooting." Still others are instigated by excessive alterations of blood pressure or of blood and tissue levels of oxygen, carbon dioxide, pH, sugar, urea, and poisons. Other simple reflexes are incorporated within simple fixed action patterns relating to ingestion, elimination, sexual and maternal behavior, escape, defense, and social behavior.

Postures and postural changes are adjustive reactions to kinetic energy associated with either bodily movement within gravitational fields or with changes or disturbances in motion of the whole body. They may involve specific sequences of actions that are often repetitive. Others are simple protective reactions to sustained forces associated with gravity, as they produce strains on joints, tendons, and other tissues during maintained postures. However, some are designed to prepare an animal selectively to "tune in" or "tune out" external stimuli (as in awakening and sleeping), to alert to stimuli (sitting up, crouching down, "freezing"), or to facilitate simple fixed action patterns, such as elimination of body wastes and copulation.

Locomotor sequences move an animal into or out of specific environments, permit it to approach or avoid specific stimulus patterns, and are instrumental in making possible the great bulk of the more complex behavioral transactions with the environment mentioned below. Locomotor sequences typically involve repetitive actions and in most mammals are highly stylized and invariant in pattern. All mammals have several locomotor modes, each adapted to effect a specific kind of result (Gray, 1968).

The simple fixed action patterns of ingestive, eliminative, sexual, maternal, escape, defensive, and social behaviors are more or less rigorously organized in ways that bring the animal into, or away from, contact with special kinds of stimuli. Many of these involve specialized (social) reactions to stimulus patterns provided by other animals (Cairns, 1966; van Hooff, 1967; Johnson, 1972). These aggregates of simple and complex response patterns are rather reliably induced by specific sets of external stimuli, especially when the intraorganismic

TABLE 1
The Overt Behavioral Repertoire[a]

1. *Simple reflexes* include muscle twitches and quivers; myotactic (stretch) reflexes; reciprocal "inhibition" of antagonists; lengthening reactions (inverse myotactic reflexes); positive supporting (magnet) extension reactions; negative supporting reactions; ipsilateral extensor reflexes; flexion reflexes; crossed extension reflexes; bilateral intersegmental (crawling) reflex "figures", startle reflexes; scratch reflexes; mouth wiping (by tongue head, hand, or foot) reflexes; head and body shaking reflexes; skin-flick reflexes; urination and defecation reflexes; panting, breathing (inspiration, expiration), gasping, choking, coughing and sneezing, and vomiting reflexes; rooting, mouthing, and biting reflexes; licking, and swallowing reflexes; clasping and grasping (hand or foot) reflexes; pinna, tympanic, and eyeblink and squinting reflexes; lip and other facial reflexes; lordosis; pelvic thrust reflexes; nystagmus and saccadic and drift eye movements; and narial dilatation reflexes. The simple responses listed above primarily involve striated muscle but the effects of nonstriated muscle responses may also be observed as the following list illustrates: milk letdown reflexes; bladder and anal reflexes (retentive and releasing); piloerection, gastric and intestinal reflexes; pupillary dilation; lens accomodation; uterine and vaginal reflexes; and penile erection and ejaculation reflexes.

2. *Postures and postural changes* include immobile sitting and standing, leaning, lying (supine or prone), sprawling, crouching, curling up, squatting, balancing, hanging, floating, urination and defecation postures, rearing, hopping, placing, stretching, yawning, tonic head-neck-body reflexes, righting (free-fall and supine) reflexes (optic, labyrinthine), labyrinthine acceleratory reflexes, and labyrinthine positional reflexes, "freezing" immobilization (e.g., of neonatal kittens when carried by head or scruff of neck). The adjustive recovery from a particular assumed posture (such as stretching) must also be considered as a postural change.

3. *Locomotor sequences* change body locus either in space or usually more specifically with respect to certain types of environmental stimuli and in such cases may also involve orientation sequences (see below). Locomotor patterns include: walking (quadrupedal, bipedal, bimanual), running (trotting, pacing, cantering, galloping), lunging, charging, creeping, crawling, stalking, dragging, struggling, wriggling, climbing (up or down), rolling, somersaulting, sliding, circling, backing up, skipping, flying, gliding, hopping, sideling, swinging, brachiating, dropping, leaping, jumping, rearing, bucking, diving, surfacing, swimming, and floating.

4. *Simple receptor orientation sequences* include head turning, nodding, lifting, and fixation; eye opening; eye fixation (visual "grasp"); ear turning; sniffing (polypnea); tasting; licking; biting; touching; poking; grasping; holding; batting; dropping; releasing; rubbing; tapping; pushing; pulling; slapping; scratching; twisting; throwing (by proboscis, mouth, hands, feet, or tail).

5. *Simple "fixed" action patterns.*

    (a) *Ingestion (eating and drinking) related components* include mouthing, biting, lip and tooth grasping, grazing, browsing, nibbling, tasting, chewing, cudding, lapping, sipping, drinking, licking, sucking, swallowing, "chop" licking, and lip smacking.

    (b) *Elimination-related components* include site selection, scratching, digging, squatting, leg lifting, immobilized straining postures, urinating, defecating, parturition, postural recovery, urine and fecal covering, and "scenting" or gland secretion "marking".

    (c) *Sexual-related components* include (1) *Courtship sequences*, such as partner orientation and selection, approach, "teasing", chasing and retreating, dancing, vocalization, body and facial displays or gestures, strutting, specific postering and "presenting", orogenital and oroanal (nose, lip, and mouth) contacts, biting, sniffing, blowing, sucking, kissing, licking, caressing, nuzzling, and embracing; and (2) *Copulatory and orgastic sequences* include partial and complete mounting (male), immobilization and lordosis (female), clasping and grasping of mate, pelvic thrusting, masturbation, orgasm and ejaculation, dismounting, separation, and afterreactions (rolling, running, cavorting, vocalizing).

(continued)

Table 1 (*continued*)

(d) *Maternal-related components* include licking, nosing, nuzzling, handling, holding, carrying, retrieving and piling, nipple presenting and nursing, huddling and covering, cradling, corraling of offspring.

(e) *Escape-, defense-, and attack-related components* include immobilization (freezing), cringing or submission postures and gestures (e.g., rolling over on back), cowering or running away; crouching, stalking, threst gestures and displays, hissing, tail erecting, twitching or slapping, abortive charging, full charge and chasing, bunting, biting, tearing, holding and pinioning, grasping and shaking, hitting, clawing, kicking, sneak attacking, leaping upon and wrestling, growling, barking, grunting, squealing, and screaming.

(f) *Other instinct sequences* include grooming, washing, preening, licking, picking and scratching specific body zones, shaking, awakening, going to sleep, sleeping, "dreaming", nest material retrieval and nest construction, burrowing, digging, excavating, retrieving, holding, carrying, packing, shredding, arranging, weaving, piling, and territorial marking (with urine, feces, or glandular secretions), chest beating, various vocalization patterns (cooing, babbling), spitting, blowing, bucking, bathing, basking, cooling, painting, building, wrecking and testing, purring, and claw protraction and retraction.

(g) *Miscellaneous gestures, displays, and social components* include grimaces, tail twitching, ear retraction, eye squinting, lip retraction and tooth baring, protective gestures and postures, smiling, kissing, nuzzling, hugging, huddling, tickling, grabbing, grapling, taking, stealing, staring, gaze avoiding, scratching, specific muscle tensions or relaxations, imitation, encouragement, beckoning, begging, avoidance, separation, mouth sounds, vocalizations (whining, purring, hissing, squealing, etc.), pacing, exhileration, excitement, restlessness and agitation, rocking, clasping, and self-biting and autisms.

6. *Complex organized response sequence aggregates* include exploration, searching or hunting for food, shelter, escape, a mate, novelty, etc.), building, territorial demarcation, identification and maintenance, games (hide and seek), housekeeping, and various social "structuring" and interactive behaviors.

---

[a]Most of these behavioral sequences are seen only in adult animals or at least in those old enough to have developed complete or matured patterns.

Reprinted with permission from Moltz (1971).

---

milieu of the nervous system is biased by certain concentrations of specific metabolic agents or hormones. These fixed action patterns consist of multiple individual act sequences, each patterned in a characteristic way to a specific kind of stimulus situation.

Finally, complex, organized response sequences include longer term activity patterns the finer details of which are individually stylized for each animal by learning. Such activities involve multiple, complex, and variable responses in time and space. They are guided by specific conceptual "plans," "sets" or "maps" that are fashioned by the individual's experience. They are deliberate, exploratory, and changeably adaptive to a highly complex physical environment that is forever changing.

Each of these kinds of overt behavior consists of transactions with limited aspects of the total available environment: with a microniche or microenvironment (Figure 2). Overt behaviors bias each animal to inhabit particular sets of niche and habitat that promote adaptive survival. As an animal matures its behaviors gradually become "chained," organized, and integrated in time and space into patterns and life styles that are relatively consistent and reliably embedded in regular daily events.

Although most behaviors either are initiated by stimuli or occur because they produce sensory stimulation, some items of an animal's behavioral repertoire (e.g., some intersegmental reflexes) are not "stimulus bound" but are expressions of efferent, central programs not requiring antecedent or subsequent stimuli. Such behaviors function as associated, supportive, auxiliary, or guiding actions for those movements that are aimed at specific stimuli.

Mental Processes

Although the movement patterns of overt behavior are direct expressions of central neural activities, they represent only a small portion of the brain's operations. For example, movement sequences are isofunctional only with the firing patterns of motor horn cells. That is, the patterns of discharge of only motor neurons exactly coincide with the patterns of contraction and relaxation of the muscles they innervate. In contrast, the operations of most neural circuits are not directly observable, simply because they do not innervate muscles. Such circuits may influence behavior, but their firing patterns are not isofunctional with patterns of overt behavior. In this sense they are mental, being part of abstracting, discriminating, symbolizing, judging, planning, etc., machinery. As such, they may regulate, control, and determine certain specific aspects of movement patterns, such as onset, offset, amplitude, temporal and spatial patterning, duration, force, timing, velocity, acceleration, and repetition rate (Figure 3). Consequently, it is important that each of these finer features of overt behavior be studied carefully and in detail in order to appreciate the roles that central circuits play in overt behavior (see Discussion). The complexity of movement control is further emphasized when the nicety is considered of the spatiotemporal patterning demanded of contractions of the several muscles that usually participate in even the simplest discriminative action sequences (Figure 4; Bernstein, 1967; Compton, 1973; Lashley, 1951).

These facts make it clear that analyses of overt behavior patterns themselves directly reveal the operation of only a small subset of neurons in the central nervous system. Consequently, inferences from behavioral evidence regarding functions of most central circuits must depend on rather tricky and elaborate logical games the strategies of which are based on introspective, intuitive, and hypotheticodeductive methods. Inferences and hypotheses regarding mental processes are tested by observation of details of overt behavior in experimentally controlled environments. That is, the covert functions of neural circuits are defined operationally by identification of several specific characteristics of behavior that are believed to depend on these functions. This is done in empirical investigations of such topics as learning, forgetting, motivation, rein-

CIRCUIT CONTROL OF A SLOW DISCRIMINATIVE MUSCLE MOVEMENT

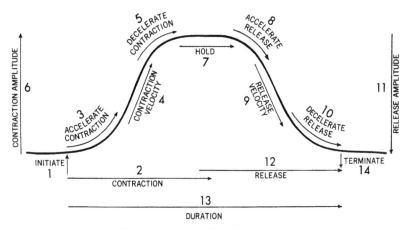

FIG. 3 Schematic diagram illustrating 14 possible different features of a simple movement requiring control to produce a careful, timed, accurate discriminative movement.

forcement, perception, fixed action patterns, problem solving, preferences, aversions, habituation, extinction, and discrimination.

Several different types of central states and processes are postulated (see Table 2; Welker, 1971). General responsiveness or unresponsiveness to stimulation requires postulation of inactive states or active states, and the change from one type to another implies arousal processes. However, awake animals may exhibit

PROFILE OF A SIMPLE COORDINATED ACTION SEQUENCE

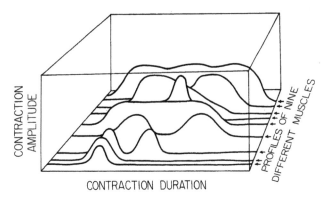

FIG. 4 Schematic diagram illustrating a profile of muscle contractions in a hypothetical movement in which nine different muscles are coordinated in a particular temporal pattern to produce a relatively simple coordinated discriminative movement sequence.

TABLE 2
Repertoire of Covert Mental Processes[a]

Conceptual category	Specific concepts
Inactive states	Sleeping, unaware, unconscious, inattentive
Arousal processes	Activation, arousal, alerting
Awake states	Aware, conscious, alert, vigilant
Attentive states	Alerting, attending, expectancy, scanning, focusing, detection, vigilance, sensitivity, orientation
Specific-reactivity processes	Mobilization, threshold, set, preference, aversion, differentiation, image, expectancy, fixed action pattern, scanning, focusing, attitude, perception, detection, hallucination, goal orientation, sensitivity, excitation, orientation, discrimination, tendency, illusion, displacement, identification
Cognitive states and processes	Perception, thinking, planning, purpose, judgement, guessing, trying, will, wish, hypothesis, evaluation, imitation, cognitive content, expectancy, set, decision, insight, optimizing, competence, self-actualizing, recognition, reasoning, understanding, concept formation, abstraction, symbol formation, cognitive map, cognitive model, ideation, aim, creativity, innovation, volition, plasticity, confidence, certainty, effectance, choice, purpose, assumption, conception, goal orientation, seeking
Integrative processes	Generalization, consolidation, judgement, introspection, deduction, homeostasis, programming, mediation, repression, inhibition, facilitation, insight, fixation, plasticity, closure, abstraction, assimilation, feedback, planning, ideation, learning, transaction, creativity, conditioning, symbol formation, association, integration, summation, irradiation, reafference, displacement, incubation, regulation
Experiential processes	Perception, detection, insight, confidence, introspection, perception, discrimination, surprise, confusion, competence, knowing, symbolizing, feeling, empathy
Motivational processes	Energy, attitude, compulsion, interest, homeostasis, optimizing, thirst, love, aspiration, hope, perseverence, craving, disposition, appetite, preference, aversion, fear, hate, joy, will, wish, drive, need, habit, strength, volition, urge, curiosity, hunger, anger, anxiety, value
Affective states	Tension, boredom, sensitivity, joy, emotion, anxiety, conflict, surprise, satiation, anxiety, love, preference, aversion, impact, desire, passion, amusement, sentiment, longing
Learning processes	Discrimination, familiarization, symbolization, consolidation, incubation, learning, insight, fixation, abstraction, imprinting, conditioning, generalization, deduction
Reinforcement processes	Impact, meaning, significance, reward, reinstatement, reinforcement, inhibition, supression, repression, trace, engram, facilitation, closure, feedback

(continued)

TABLE 2 (*continued*)

Conceptual category	Specific concepts
Other change-type processes and states	Adaptation, adaptation level, satiation, suppression, feedback, forgetting, creativity, innovation, plasticity, inhibition, recovery, displacement, habituation, dishabituation, accommodation, disinhibition
Fixation states and processes	Persistence, fixation, instinct, habit, generalization, fixed action pattern, sterotype, consistency
Memory processes	Memory (immediate, delayed), recall, habit, recognition, amnesia, forgetting, retention, trace, engram, storage, retrieval
Ability states	Discrimination, learning set, perception, achievement, adaptability, habit, acuity, capability, capacity
Maturational processes and states	Critical period, growth, differentiation, readiness, histogenesis, neurogenesis, regionalization, induction, morphogenesis, pattern formation, organization

[a] Behaviorally and phenomenologically derived concepts, constructs and intervening variables referring to hypothesized, internal, central and covert phenomena, functions, states, contents and processes. Reproduced with permission from Moltz (1971).

various types of attentive states (Mackintosh, 1965) as well as a wide variety of specific-reactivity states to specific microniches. A large number of terms representing cognitive, integrative, and experiential processes has been inspired by the abilities of animals to perceive, evaluate, analyze, seek, discover, recognize, judge, and plan their interactions with microniches. Motivational processes and affective states are postulated to account for a broad, diversified class of internal prompting or "driving" forces or conditions that activate the organism to invade or retreat from different types of stimulus patterns. Behavioral changes resulting from environmental transactions result from learning, reinforcement, and other change-type processes. Memory or recall processes retrieve the resultant fixated and ability states. Concepts referring to maturational processes are needed to account for behavioral differences at different ontogenetic phases of life (see section on Construction of Neural Circuits during Ontogeny).

Outcomes of Behavioral and Mental Evolution

The behavioral and mental repertoire of an animal reflects its repertoire of niche transactions. Clearly, mammals possess a relatively large and diversified repertoire of such transactions. Although all mammals exhibit all the phenomena listed in Tables 1 and 2, it appears that during evolution the more complex mammals have developed increased capabilities of dealing accurately and adaptively with an ever increasing diversity of stimulus patterns in the environment. It seems that speciation has involved the development of genetically inspired neurobiological inventions for the exploitation of increasingly numerous features of the universe (Teilhard de Chardin, 1971).

Overt behavior has been an instrument of this "complexification" and diversification of organisms because it constitutes an active interface between the internal systems of an organism and its external environment. As such, behavior is one means that an organism uses to settle the fate of its genotype. Overt behavior brings home the goods. It gets the nourishment, avoids injury, finds a mate, cares for the young, etc. It is the means for resolving all kinds of practical problems posed by the confrontation with physiological needs and with the external world.

Translation of Bodily Needs into Behavior

All behavior survives in evolution because it plays a part in maintaining viability of the organism's internal biochemical machinery, which constitutes the essence of life (Adám, 1967; Dell, 1958; Kare & Maller, 1967). Although the bulk of this review deals with neural circuits responsive to stimulus energies in the external environment, mention must be made of the mechanisms that have evolved for detecting different kinds of internal biochemical states and translating these into adaptively appropriate patterns of behavior (Black, 1970). These mechanisms are referred to as "homeostatic" because they are responsive to features of the internal biochemical milieu the activities of which they regulate and the optimal levels of which they maintain (Bajusz, 1969).

Biochemical changes affect neural circuits by altering firing thresholds of some of their neurons. Such neurons are especially responsive to coursing hormones, metabolites, hydrogen ions, oxygen or carbon dioxide tensions, etc. (Cohen, 1970; Cross, 1964). Each type of biochemical alteration affects specific types of neurons, synapses, or receptor junctions. Changes in body temperature likewise affect certain neurons (Hellon, 1967). Circuits and systems containing chemosensitive neurons are activated or deactivated, depending on the type and level of circulating chemical. They typically energize circuits that prompt specific types of behavioral sequences in response to certain types of external stimulus patterns. Such behaviors are commonly rigorously programmed as relatively fixed action patterns. The type of action sequence elicited varies with the type of biochemical alteration involved and is sufficiently appropriate and persistent to enhance the probability that the relevant environmental stimuli can be attained.

This direct effect of internal chemicals on behavior toward external stimuli is assisted by additional mechanisms that have evolved to integrate the effects of internal and external determinants. For example, not only may biochemical changes alter sensory thresholds to external stimuli, but the latter may be able to alter biochemical levels reflexly by activating or deactivating secretory circuits.

Comparative Studies of Behavioral and Mental Phenomena

Observational and experimental studies of behavior have generated a voluminous literature documenting species differences in a wide variety of behaviors (Denny & Ratner, 1970; Eisenberg, 1967; Kaufman & Rosenblum, 1966; Kellogg, 1961;

King 1968; Lorenz, 1965; Roe & Simpson, 1958). Many similarities in behavior exist because all mammals have evolved exposed to certain common environmental problems. All are committed to eating and drinking through their mouths, breathing through their noses, eliminating wastes via anus and urethra, locomoting with four limbs, copulating with their genitals, seeking and attracting a mate, giving birth and carrying for their young, defending themselves and attacking or escaping enemies, socializing, etc. All mammals look with their eyes; feel with their skin; listen with their ears; sniff with their nostrils; taste with their tongue; grasp food, objects, or other animals with lips or appendages; and so on. Yet, the exact ways in which these general activities are carried out differ widely. The numerous beasts within each mammalian order are therefore differentially committed, by anatomical or physiological differences, to behave in certain ways that are special for that order. For example, the hooves, long legs, and skeleton of artiodactyls and perissodactyls predispose them to certain types of posture and locomotor modes. Likewise, their eating and digestive structures predispose certain behaviors and preclude or render others improbable.

Because there are only a few major types of habitat situations available, animals in several phylogenetically divergent orders and families have evolved certain specializations that are similar but that are considered convergent. For example, there are arctic carnivores, rodents, lagomorphs, pinnipeds, cetaceans, and artiodactyls. Creatures from several different orders have specialized to invade terrestrial, arboreal, aquatic (Andersen, 1969), or aerial habitats. Similar specializations have also evolved in different animals in the ways that they respond to visual, auditory, somatic sensory, or olfactory stimuli. For example, whales and bats echolocate, and creatures in several orders have prominent nocturnal vision. Convergent similarities also occur for types of locomotor behaviors or for body parts used in touching and grasping. Because these specializations are convergent, they do differ in certain ways.

Although such behavioral specializations are influenced by obvious structural specializations of eyes, ears, hands, mouth, lips, feet, and tail, much behavioral diversity cannot be clearly assigned to such bodily specializations. Instead, there often appear to be covert, central neural, or mental specializations that confer on some animal types expertise in dealing with problems of the physical world by intellectual effort; by utilizing thinking, problem solving, evaluating, abstracting, memorizing, and estimating skills. Little is understood about species differences in such mental phenomena. Documentation of such differences among mammals is hard, being fraught with problems of procedure and interpretation (see Discussion). Evolution of such mental phenomena is discussed more extensively in the section on Cognitive Evolution.

Significance of Comparative Behavioral Studies for Understanding Brain Evolution

Studies of overt behavior in different animal types in their natural habitats reveal species differences in niche transactions. Such studies indicate which stimulus energies can be detected, which are preferred or avoided, and which elicit and guide specialized organized behaviors. In addition, the spatiotemporal features of overt behaviors (onset, offset, velocity, acceleration, strength, and braking of contraction, repetitive actions, cocontractions, reciprocal actions, etc.) are direct clues to certain spatial and temporal properties of the neuroelectric processes of some central neural circuits. Species differences in spatiotemporal features of homologous behaviors, such as vocalizations, gait, postures, or fixed action patterns, reflect differences in firing frequencies, timing, and in some instances even spatial characteristics of the underlying neural circuits. Comparative studies of all such behavioral details can indicate which features of movement sequences are heritable. Finally, the number and hierarchical complexity of the items of the behavioral repertoire also provide clues to the hierarchical complexity of the underlying neural circuits and systems (Eisenberg, 1967).

Earlier studies of mental processes attempted to compare animals on generalized learning abilities, memory, and problem solving (Bitterman, 1960; Doty, Jones, & Doty, 1967). These general approaches placed primary emphasis on hypothetical drive states, reinforcement, attention, and fixation processes. Because of their generality, they were not able to shed much light on species differences in mentation. Subsequently, studies of behavioral differences among animals have become increasingly sophisticated. Not only are details of behavioral sequences being more carefully examined, but greater care is also being employed in identifying the exact stimulus patterns to which an animal may respond selectively (Shinkman, 1962). In addition, increasing emphasis is being placed on the search for the neurobiological mechanisms that underlie behavior. Studies are now possible in which the behavior of animals is carefully measured while electrophysiological activity is simultaneously recorded from neural circuits that are presumably performing certain mental operations (Phillips, 1973). Such studies may provide new specific insights into species differences in mental processes underlying adaptive behaviors.

However, most of these studies of behavioral and mental phenomena have not utilized truly comparative methods (Beach, 1960; Hodos & Campbell, 1969; Lockard, 1971; see Discussion), so few conclusions can yet be made regarding taxonomy of the mammalian repertoire of behavioral and mental processes.

COGNITIVE EVOLUTION

The incredibly vast array of behavioral and mental abilities that have evolved in mammals allows them to make precise and delicate adjustments to a wide variety of complex, changing stimulus patterns. To these more sophisticated brain functions the term "cognition" is applied. However, in newborn mammals cognitive processes are rudimentary and most of the niche transactions are relatively simple or generalized. Therefore, stimulus patterns are poorly perceived and reactions to them are relatively crude. However, with experience and motivation, behavioral and mental repertoires proliferate. The discriminable stimulus patterns increase in number, variety, and spatiotemporal complexity, and the responses to specific stimuli become more discrete, precise, and adaptively effective. The potential or capacity for acquiring these more diversified, elaborate, accurate, and flexible niche transactions has increased in more advanced mammals. The concepts of learning, memory, intelligence, language and culture refer to different aspects of these more complex manifestations of neural evolution (Alland, 1967).

Learning and Memory

It is tempting to emphasize learning and memory as primary features in evolution (Adám, 1971; Honig & James, 1971). Perhaps this is so because of some of man's outstanding adaptive features, and because of his heritage whereby he teaches, stresses learning and achievement, and accumulates and stores knowledge (Pilbeam, 1970). Indeed, learning is an important feature in brain evolution. Learning consists of the restructuring of neural circuits in subtle yet influential ways. It allows each individual animal to shape its repertoire of responses adaptively to the unique environments that it experiences.

Like most brain functions, learning and memory are inferred from changes in overt behavior subsequent to exposure to, or experience with, environmental events (Richter, 1966; Sheldon, 1968). That is, learning and memory are defined operationally. Several different types of learning have been postulated: classical conditioning, operant learning of several types, imitation, imprinting, habituation, adaptation, extinction, relearning, etc. Different types of neural changes may underlie the enduring behavioral changes known to occur for each of these. The assumption that there may be a single mechanism or neural locus for all learning changes has never been supported. Indeed, learning may involve multiple changes in several neural circuits. Certain circuit components may be especially important in producing the reinforcing aspects of learning; or certain types of neurons or small neural assemblies within many nuclear components may play important roles in memory fixation. Selective growth of certain pre- and post-synaptic structures may be essential for some aspects of learning to occur, as may biochemical alterations within synaptic microstructures. Learning may require unique spatiotemporal patterns of electrical activities within certain synaptic assemblies. Very likely, the interaction of several circuits, such as those presumed to be involved in attentive, motivational, and reinforcement processes, is required for adaptive learning to occur. Glial cells have been viewed as playing a role in learning processes (Altman & Das, 1964), as have conformational changes in neuronal proteins or alterations in molecular mechanisms (Albert, 1966; Ansell & Bradley, 1973; Bogoch, 1968; Byrne, 1970; Coleman, Pfingst, Wilson, & Glassman, 1971; Fjerdingstad, 1971; Gurowitz, 1969; John, 1967;

Ungar, 1970). Different phases of learning have been postulated and it is possible that different processes are responsible for the consolidation and storage phases of learning (Barondes & Cohen, 1966; Mackworth, 1962). The fact that learning changes can be attributed to circuits within the spinal cord, brainstem, or forebrain suggests that learning processes, perhaps of different sorts or different levels of complexity, can occur in different parts of the brain (Kimble, 1965).

Valid conceptions of these hypothetical phenomena are not at hand (Gerard, 1961). Current understanding of what takes place in the brain during learning and forgetting is still largely speculative. However, new sorts of physiological and anatomical facts are steadily accumulating. They extend understanding beyond that of two decades ago, when published reports of experimental studies of purely behavioral approaches to animal learning were especially prominent.

Remembering and memory are thought of as involving different processes than learning. For example, recognition or recall phenomena may employ specialized neural circuits to retrieve or select those assemblies that have been modified by learning. The conditions under which memory and recall can be evoked have been extensively studied, but little can be said about possible neural mechanisms involved in such recall. Until more can be learned about the exact neural substrates for learning and memory, comparative studies of these phenomena are difficult to conceive.

Intelligent Behavior

Concepts of intelligence have also been prominent in discussions of evolution of adaptive capabilities (Bitterman, 1965; Cunningham, 1972; Viaud, 1960). Difficulties in defining intelligence may be viewed as caused by the inherent complexity of the multiple, interacting processes that contribute to overall adaptive performance. The numerous concepts listed in Table 3 are reminders of the variety of specialized adaptive activities that may be considered components of intelligent activity. Because such a large number of capabilities apparently underlie most adaptive activities, the broad concept of intelligence has little value in analytical attempts to conceive of the underlying neurobiological processes.

Intelligent behavior is always manifest in performance as successful, orderly, adaptive, problem-oriented transactions with the environment. Problem-oriented behavior that is unsuccessful, inappropriate, confused, or maladaptive is stupid, foolish, idiotic, or insane. All adaptive performance involves behavioral units at many levels of hierarchical complexity. For example, simple reflexes, postures, locomotor patterns, fixed action patterns, and complex sequences are usually all integrated into even the simplest discriminative actions. Numerous mental and behavioral processes have coevolved to produce the many familiar intelligent cognitive activities. For example, abilities to attend, abstract, simplify, create, transpose, symbolize, explore, change, generalize, ponder, and abstain are all cognitive phenomena that contribute to adaptive intelligent activities. Consciousness and free will may also prove to have evolved as specialized processes to provide flexible attentive and directed concentration on specific aspects or abstractions of the physical world (Sperry, 1970). There must be many such central processes that, acting together, have resulted in man's unusually expanded and flexible repertoire of thinking and doing. There is no reason to believe a priori that any one of these interrelated functions has been preeminent in phylogeny. They have probably coevolved from relatively simple into more elaborate and complicated functions. The more complex phenomena of tool making, tool using, fabrication and construction, teaching, plant and animal tending, and similar activities in man have certainly evolved as progressive expansions and extensions of the simpler creative cognitive repertoire of exploratory and play behaviors that are seen in many mammals (Kortlandt & Kooij, 1963; Kurth, 1968). However, as with curiosity, cognitive capacities are most prominent and efficient under certain balanced conditions of mild stress and safety, when homeostatic and emergency circuits are not operating. Consequently, under stress of some instinctual needs, cognitive functions are held in check and their more subtle, creative features are transcended and displaced.

Species differences in learning ability and adaptive performances are common. Comparative studies of animals in natural situations, or in experimental test situations, have led to suggestions as to why animals in different taxa learn and solve problems differently (Beach, 1960; Bitterman, 1965; Lashley, 1949). However, there have been few hypotheses formulated that attempt to explain such differences in terms of validated neurobiological constructs. The generalizations offered suggest evolutionary increases in ability to abstract, generalize, learn, respond on the basis of reduced cues, attend to multiple cues or complex spatiotemporal stimulus patterns, employ a large number of responses in trial and error fashion, inhibit or interrupt fixed action patterns, explore, invent, fabricate, etc.

Caution is required here because there are several special methodological and interpretational problems inherent in comparative studies of learning, memory, and intelligence (Bitterman, 1960, 1965; Hodos & Campbell, 1969; Lockard, 1971; see Discussion). Understanding of neural mechanisms underlying these manifold phenomena requires that current rather generalized intuitive notions regarding these phenomena be fractionated and validated. For the explanatory power of such conceptions to be improved, their formulation should be guided by knowledge of underlying neurobiological structures and processes (Corning & Balaban, 1968).

Language

The unusual language abilities of humans allow them to symbolize the phenomena of physical reality with remarkable efficiency, speed, and effectiveness (Lenneberg, 1967). Speech and gestural languages are used in highly discriminative ways to signal, teach, punish and otherwise control, manipulate, and

TABLE 3
Substrates of Intelligent Behavior[a]

A. *Perception and abstraction of physical energies*
 1. Multimodal perception of wide ranges of physical energies (light, sound, mechanical, kinetic, thermal, chemical, gravitational)
 2. Accurate perception and discrimination of specific details of environmental niches: onset and offset, location, number, intensity, duration, size, shape, distance, temporal order and pattern, spatial pattern, movement, acceleration, velocity, change, novelty, increases or decreases, and integrals of many of these; especially in visual, auditory, and somatosensory modalities
 3. Time perception and estimation.
 4. Selective feature-specific attention or disattention

B. *Learning and memory*
 1. Selective learning and memory (short and long term) of any of these features
 2. Reinforcement
 3. Simultaneous and sequential registration of different stimulus features

C. *Cognitive abilities*
 1. Abstraction, generalization, representation and symbolization; ability to identify any of these features, or combinations of them
 2. Multiple-cue abstraction and integration of cues into specialized concepts
 3. Synchronous, integrated, multimodal perception and learning; sensory fusion
 4. Reconstruction of percepts and concepts from reduced cues
 5. Concepts of reality; universals and absolutes
 6. Misperceptions, misconceptions, illusions
 7. Maps, hypotheses, sets, plans, strategies, goals: all reflecting multiple-niche patterns
 8. Flexible rearrangement of concepts, insight, association, crossmodal transfer

D. *Thinking, problem solving, judging*
 1. Rational thinking; logical, scientific, and mathematical reasoning
 2. Consciousness, introspection
 3. Judgements of probability, truth, morality, and conscience
 4. Calculating, estimating, inferring, guessing, speculating, predicting; relativistic and absolutist thinking, imagination
 5. Reality testing, cross-checking, validation, sampling, comparing
 6. Curiosity, questioning, searching, discovery, invention, creativity
 7. Imitation
 8. Flexible, adaptive, and variable selection of percepts and behavioral patterns in optimal combinations: "optimizing"
 9. Decision and control over choices of action; volition, ego, will power, subjective certainty, conviction, faith, belief, self-control, discipline
 10. High-speed sorting, rearranging, judging, and decision processes
 11. Comparison of expectations with perceptions
 12. Error detection and correction
 13. Holding and maintaining in readiness or in imagery; waiting

E. *Programming for action*
 1. Simultaneous programming of separate action sequences
 2. Smooth, integrated programming of multiple, individuated movements in organized spatiotemporal patterns

(continued)

TABLE 3 (continued)

    3. Ability to break up, abort, interrupt, and rearrange behavioral units and fixed-action patterns or other learned sequences
    4. Restoration and reinstatement of sequences after interruption
    5. Coordination of separate action sequences (e.g., hand-eye coordination)
    6. Hierarchical organization and integration of discrete units of the behavioral repertoire into complex, long-term sequences and patterns
    7. Temporal ordering

F. *Behavioral niche transactions*
    1. Behavioral repertoire differentiated into hundreds of units at each of several hierarchically distinct levels of movement pattern
    2. Behavioral units intimately transactional with features of environment
    3. Niche-specific action patterns of great accuracy and delicacy
    4. Exploration; trial and error, niche testing and searching, playing
    5. Imitation
    6. Communicating; gesturing, speaking, crying; language, syntax, writing, naming, signaling, labeling, showing, displaying, teaching, story telling
    7. Tool using, manufacturing, producing, constructing, disassembling, cultivating, domesticating
    8. Special social and interpersonal action patterns: family, group and cultural transactions; role playing, organized interanimal organizations and associations

G. *Adaptive results and effects on the environment*
    1. Flexible utilization of all the above abilities and skills in service of fundamental hunger, aggressive, territorial, sexual, security, affectional, maternal, and mastery motives, emotions and instinctive tendencies
    2. Organized, adaptive practices and institutions that are guides to the utilization, control, and regulation of environmental features and of animal and human transactions
    3. Tools, machines, objects, materials, structures, buildings, etc.
    4. Libraries of laws, codes, beliefs, practices, and systems of thought

[a]These terms and phrases have been gleaned from various sources in which the components of intelligent behavior and thought have been discussed.

influence others. Advanced language performance has required sophisticated improvement in ability to integrate labial, oral, buccal, glossal, glottal, nasal, pharyngeal, and respiratory muscle action sequences to produce highly individuated patterns of sound. The enormous vocabulary of phonemes, syllables, words, and sentences and the complex grammatical and syntactical structure of the human linguistic repertoire is marvelous indeed. Man's story-telling capabilities are unique in the animal kingdom. Evolution of abilities to recognize, discriminate, and perceive these acoustic speech patterns undoubtedly paralleled the evolution of vocal patterns (Andrew, 1962, 1963a,b; Erulkar, 1972). In addition, evolution of capabilities to symbolize such patterns in writing and to perceive written language signs visually developed along with the increasingly complex verbal expressive and auditory perceptive capabilities. By means of speech and writing and reading, each individual can quickly abstract and summarize for her- or himself, refresh his or her memory, and communicate to other humans certain adaptively important essences of the environment. These abstractions are objectified and made a concrete part of reality in written and spoken words. Accurate communication regarding the environment that such symbols represent requires only that other individuals share a reasonably similar experiential history with the environment and that they share similar learned repertoires of expressive, acoustic, and visual symbols.

Studies of the production and perception of vocal behavior and its role in communication in different mammals suggest that there are multiple interacting behavioral and mental determinants of this specialized class of behaviors that deal with acoustic physical energies. In humans, the improved integration of these abilities with others, such as those in visual perception and writing dexterity, parallels the increased complexity of our many other transactions with the environment (Alland, 1967; Asimov, 1963; Pfeiffer, 1955).

Culture

The concept of culture refers to the great variety of teachable and learnable abstractions, thought, and habit patterns that are particularly prominent in human societies. "Culture" refers to man's tool-using, handicraft, technological, institutional, ideological, familial, and social habits, beliefs, and customs and to the concrete structures he creates to represent them. However, cultural habits and transactions are not merely abstract. In the life of each individual they incorporate things seen and touched and of sounds heard. The things seen and heard are the people around one and their actions, their articulations, their patterns of behavior, and the concrete objects that they create, build, and utilize. Society and its laws and beliefs are comprised of abstractions of such things and events that are perceived directly. Social life consists of those specific transactions each has with specific people at specific times and places. Society, institutions, knowledge, laws, practices, beliefs, and many other such conceptions that refer to cultural habits and structures can be thought and talked about abstractly. However, these concepts, like such others as universe, atoms, atmosphere, love, and life, are abstractions or generalizations that derive from stimulus patterns that are experienced directly. They are all equally abstract or real because they all consist of neural representations or codes of environmental events. The role that cultural institutions have played in human adaptation is to provide concrete guides to perception, utilization, regulation, and control of the complexities of the physical world to which humans are exposed and of which they conceive. Such institutions as government, education, corporate business, science, marriage, religion, and specialized fraternities exist because they provide organized structure and stability to the interpretation of worldly phenomena, just as do words and drawings. Just as language and writing have grown out of increasingly complex perceptual, speaking, thinking, and manual skills, so too, architecture, tool and equipment manufacture, and acoustic and visual art forms have emerged from improved abilities to assign thoughts into the space–time coordinates of the visual, auditory, and somatic sensory world (Napier, 1970; Wallace, 1971).

From a deterministic neurobiological perspective, it seems to me inappropriate to conceive that the latest human evolutionary developments have been determined by culture and society rather than by further advances in neurobiological complexity. That latest advances have been caused by cultural rather than biological evolution is claimed to be supported primarily by evidence that the brain size of *Homo* has failed to increase throughout his cultural history (Tobias, 1971). This view holds that no new abilities have developed to respond to the increasingly complex patterns of the environment that comprise culture, those that have resulted from man's own increased accumulation of cultural objects and learned habits. However, there is no real validation for this view. It seems equally likely that a great deal of evolution has occurred, and is continuing to occur, throughout man's cultural history, and the greater niche complexity accompanying cultural advance is the "stimulus" for such changes (Wilson, 1975). There is no reason to believe that refined improvement of neural circuits has not continued to keep pace with increased complexity of the cultural environment. Such reorganization conceivably could occur without increases in total brain size. Indeed, there are examples of differences in behavioral complexity in closely related species of mammals of similar relative brain size (e.g., different races of dogs). The few available studies of patterns of neural organization in such animals support the notion that fine-structural differences can occur without noticeable changes in brain size (see the section on Neuroanatomical Evolution).

Reality and Truth

More advanced mammals gradually developed increasingly numerous and complex, subtle and variable, repertoires of thought and action (Schneirla, 1950). In order to maintain stability and organization of transactions with the environment in the face of such greater complexities, improvements and refinements in the accuracy of perception, abstraction, judgement, and action were required. Circuits gradually evolved that were capable of higher order abstractions, which allowed the animal to maintain a literal on-going assessment of environmental changes (Beritoff, 1965). Basic exploratory capabilities became more finely tuned to respond to and search for basic physical truths and relativities and constancies in the multiple changing stimulus patterns of the physical environment (Berlyne, 1960). Out of these interelated advancements in feature abstracting and symbolizing capabilities there emerged new abilities to seek and formulate abstractions and generalizations (absolutes) that were "better Gestalts" or "best fits" of the variable shifting patterns of energies in the environment.

Humankind's unusual ability to classify and cognize absolutes and invariants from broad experience with a great diversity of environmental events has led to a variety of "higher order" abstractions (Dubos, 1968, 1972). One of these that is paramount is the conception of "reality." Another is "truth" or ultimate reality. Others refer to subtle regularities of physical phenomena of which humans are only able to dimly conceive because of inherent limitations of their sensory, cognitive, and motor structures. Science deals with many of these conceptions, but philosophy, mysticism, and religion also deal with the more elusive of these abstractions, ideas, ideals, and idealogies. Conceptions of beauty, goodness, God, freedom, equality, etc., as well as of truth, refer to some of these experiential phenomena. Symbols (flags, statues, emblems, icons, signs) assigned to abstractions give them perceptual permanence and provide for easy reference to and recall of the original experiences. Such abstractions often tend to persist as representations more of the concrete symbols than of the perceptions for which such symbols stand. As has been seen, language, inventions and materialistic constructions (art objects, buildings, etc) also objectify or externalize such conceptions. So too do the organized portrayals of concepts that characterize stories, legends, narratives and epics. Populations capable of achieving more "realistic" and "truthful" abstractions of the great mass of learning and environmental experience to which they have been exposed have tended to survive. For some lineages of mammals, this has been a direction along which adaptive speciation has progressed.

The overall aim of the comments in this section and in the last is to rationalize my belief that behavioral, mental, and cognitive phenomena (a) have increased in complexity as animal brains have evolved; (b) can be assigned to the operations of neural circuits; and (c) exist in such diversity in living mammals to encourage search for understanding of both their neurobiological mechanisms and their probable evolutionary course.

NEUROANATOMICAL EVOLUTION

Evolution of specific features of neural circuits of the brain is largely responsible for the evolution of species differences in behavior and in responsiveness to the environment. The brains of all vertebrates have a common ground plan of their five major embryological subdivisions, but the telencephalon and diencephalon in mammals have diverged from those of the other vertebrates in several ways. Although recent work on reptiles and birds reveals that there are several forebrain circuits that are probably humologous with those known in mammals (Ebbesson, 1970), many neuron populations in mammals are quite differently deployed and interconnected than are those of lower vertebrates.

Among mammals there is considerable variation in brain size and shape and in relative sizes of specific subdivisions and in certain features of external morphology. Many of these features, such as fissural patterns, are stable for the representatives of a given taxon and can be used to characterize the group. Of special importance is the fact that species differences in functional specialization can sometimes be directly assessed and compared merely by inspection of the brain, because some external morphological features reflect aspects of underlying brain circuitry (Welker & Campos, 1963). Consequently, such features prove useful not only in establishing taxonomic relationships but also in estimating the degree of development of specific sensory or motor areas (Radinsky, 1968, 1971, 1972, 1973). Indeed, study of convolutional and sulcal impressions of endocranial casts of extinct mammals provides the only source of data regarding brain phylogeny (Edinger, 1948). Paleoneurological studies of endocasts and of cranial foramina are therefore the only direct means of evaluating the course and characteristics of brain evolution (Jerison, 1973). Paleontological and stratigraphic studies also provide valuable information regarding evolution of some of the habits of extinct animals (LeGros Clark, 1964; Colbert, 1961; Romer, 1968). For example, postures, modes of locomotion, specialized behaviors (e.g., digging), eating habits, etc., can be reasonably estimated from studies of skeletons (e.g., of dentition, dimensions of bones, joints), fossils, flora, fauna, and artifacts found in the same geological strata. However, because the morphological features of brains or of habits determined by these methods are relatively gross, nothing can be learned about important specific details of brain circuitry of extinct mammals. Fine details of function and structure can be studied only in living animals. It is my aim in this section to survey those neuroanatomical characteristics of living mammals that I believe can help in understanding which factors may have operated during evolution to produce the known diversity in brain structure and function.

Basic Neuroanatomical Units

Before the features that may have characterized mammalian brain evolution are reviewed, it may be useful to establish a nomenclature of basic neuroanatomical units. Because all living mammals possess similar basic types of neuroanatomical structures, mammalian brains must have diversified during evolution by addition, deletion, and rearrangement of these same basic units in the construction of their neural circuits. These structures differ with respect to the hierarchical complexity of their organization (Figure 5). They are of ten basic types, and there are many variants of each type, depending on its specific location within the numerous circuits of the brain (Bodian, 1962, 1972; Friend & Gilula, 1972; Pappas & Purpura, 1972; Phillis, 1970; Poláček, 1965; Reese & Shepherd, 1972; Shkol'nik-Yarros, 1971; Straile, 1969). The morphological details of a type of structure are unique at each location.

The basic types are as follows.

1. Neurohumoral transmitter structures include receptor junctions, synapses, and effector junctions (motor end plates). These three are the elemental structures responsible for the basic transduction and coding characteristics of nerve

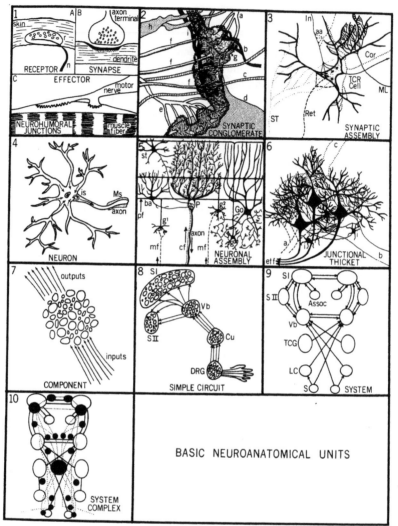

FIG. 5

FIG. 5 Basic neuroanatomical units.

1. Neurohumoral Transmitter Structures. (A) Receptor Junctions: Skin Mechanoreceptor. r = Merkel cell, n = innervating neurite. Similar neurochemicals in receptor–neurite complexes of all other mechanoreceptors, as well as in thermoreceptors, nociceptor, and in olfactory, gustatory, acoustic vestibular, and visual receptors. (B) Synapses. axodendritic synapse shown but other varieties also occur: axoaxonic, axosomatic and dendrodendritic. Types of biochemical transmitters and their structural substrates vary with the type of synapse that occurs. Electrical synapses are an additional synaptic type that occur in certain regions. They are structurally different and do not utilize biochemical transmitters for impulse transmission. (C) Effector Junctions. Neuromuscular junction shown of motor neurite with striated muscle. Smooth muscle is similarly innervated. Some receptors are also innervated by efferent neurohumoral junctions. Neuroglandular and neurosecretory junctions are specialized to liberate secretory products into glandular tissues (e.g., in neurohypophysis, adrenals).

2. Synaptic Conglomerate. A specialized configuration of synapses on a single cell that originate from neurons of the same component or from sources in other components in the

278

same or different circuit. Abbreviations: (a and i) two axons giving off multiple terminals and forming axodendritic synapses with dendritic spines. (b) Axoaxonic synapses upon axon terminals of another axon. (c) Passing axon making an axodendritic synapse of passage. (d) Nerve cell soma. (e) Terminal ramifications of axon making cluster of axosomatic synapses. (f) Three axons making single synaptic contacts with dendrite. (g) Glial cell envelope. (h) Dendrite of another neuron making dendrodendritic synapses.

3. Synaptic Assembly. The total aggregate of several types of synapses on a single neuron that derive from all afferent sources. TCR cell: Thalamocortical relay cell of ventrobasal thalamus. Afferent sources from spinothalamic tract (ST), reticular formation (Ret), medial lemniscus (ML), corticothalamic (Cor), and interneurons (In). Each afferent terminates in specialized ways on distal and/or proximal dendrites or cell body or presynaptically as axoaxonal contacts (aa).

4. The Neuron. The next higher integrative level of neuroanatomical structure. The integral of the electrical activity of all input synapses upon its dendrites and soma is determined by the neuron. Its output, expressed as digital pulses propagated down the axon, is determined by the bioelectric integral established at the initial segment (is).

5. The Neuronal Assembly. This consists of two or more types of neurons, some sending axons to distant components, others with axons distributed internally. The types of inputs, intrasynaptic component connections, and outputs are the same for all adjacent assemblies in a given component. All the basic operations of the component are performed by each assembly. The assembly depicted is that known for the cerebellum. Abbreviations: st, stellate cell; ba, basket cell; P, Purkinje cell; Go, Golgi cell; g^1, g^2, g^3, granule cells with parallel fibers (pf) at different levels from the cerebellar surface (at top); mf, mossy fiber; cf, climbing fiber.

6. Junctional Thicket. This concept refers to the overlapping aggregate of neuropile that results from the closely interlaced and overlapping character of densely packed neuronal assemblies. The concept seems required by the likelihood that the spatially distributed, overlapping of inputs to (a, b, c), and interconnections between adjacent neuronal assemblies (via recurrent collaterals, not shown, of efferents and axons from interneurons, i), are functionally important in the normal operation of a component. The intimate, spatially distributed, overlapping character of synaptic and ephaptic junctions provides a substrate whereby patterns of graded responses produce a slow waveform.

7. The Component. This consists of many assemblies. Each component deals with sets of inputs and outputs peculiar to itself and performs a particular set of transformations of inputs to outputs.

8. A Simple Circuit. This consists of several components in a series or with some components in parallel. Each component in a circuit performs particular sets of transforms of inputs to outputs. The functions of a circuit are reflected in the functions available at the components situated at its most forward level. The circuit illustrated is somatic sensory, involving neurons at dorsal root ganglia (DRG) and medullary cuneate (Cu), thalamic ventrobasal (Vb), and cerebral neocortical S1 and S11 areas. Outputs to S1 and S11 from Vb are in parallel.

9. A System. This consists of two or more simple circuits that are interconnected. A somatic sensory system is illustrated, showing the circuits of both sides of the brain and their interconnections. Abbreviations: s, spinal cord; LC, lateral cervical; TCG, trigeminal–cuneate–gracile complex; Vb, ventrobasal; S1 and S11, the two cortical areas; Assoc, association cortex.

10. A System Complex. This is an interconnected set of different systems, in this case a somatic sensory (open circles) and a motor system (solid circles), a sensory-motor system complex.

The basic units proposed are abstractions. Except for Drawing 5, all these illustrations are schematic and oversimplified. Each drawing portrays only one conceivable (not necessarily actual) structural configuration. Variants of each "type" are probably innumerable. Drawing 5 is redrawn by permission from Fig. 1 of Eccles *et al.* (1967).

terminals (Auerbach, 1972; Brightman & Reese, 1969; Cauna, 1961; Davis, 1961; Triggle, 1971; Zacks, 1964). "Electrical" synapses (not shown in Figure 5) are also found in neural circuits but do not utilize neurohumoral mechanisms.

2. The concept of synaptic conglomerate refers to a more complex kind of morphological aggregate in the neuropile. It consists of a network of interconnected synaptic junctions the input sources, types (axoaxonic, axodendritic, axosomatic, dendrodendritic), and spatial arrangements of which result in a subset of functionally important inputs to specific parts of the postsynaptic neuron. In most places glial elements encapsulate special aggregates of such synapses (Hydén & Lange, 1962). Each type of neuron within a neuron population supports one or more synaptic conglomerate that is specialized and peculiar to itself.

3. A synaptic assembly consists of the total aggregate of pre- and postsynaptic junctions that impinge on a single neuron. The sources, numbers, sizes, types, arrangement, and locations of the synapses of this entire assembly predetermine the kinds and characteristics of neuroelectric influences that can reach the postsynaptic neuron.

4. The neuron (including dendrites, soma, and axon) is the simplest basic structural unit capable of integrating the multiple neuroelectric effects of all its individual synapses, synaptic conglomerates, and synaptic assemblies and then coding this integral in the form of spike discharges initiated at the axon hillock and transmitted by the axon. Many morphologically distinguishable types of neurons have evolved (Brazier, 1969; Ramón y Cajal, 1909–1911; Horridge, 1968; Hydén, 1967).

5. A neuronal assembly is a hierarchically more complex kind of structural unit. The assembly is composed of all those basic cell types that occur in a component (e.g., relay neurons, internuncials), together with their interconnections and special sets of afferents and efferents. Patterns of connectivity exhibited by each type of assembly determine the kinds of relay, transfer, coding, and integrative functions of the component of which it is a part (Burns, 1958; Caianiello, 1968; Eccles, Ito, & Szentágothai, 1967; Eccles & Schadé, 1964; von Euler, Skoglund, & Söderberg, 1968).

6. However, the concept of junctional thicket seems necessary to refer to certain important structural and functional microfeatures not usually depicted in "telephone exchange" notions of circuitry (Pribram, 1971). A junctional thicket is defined as the aggregate of neuropile that results from the overlapping features of inputs to, outputs from, and interconnnections within adjacent neuronal assemblies. This feltwork of densely packed dendrites, axon terminals, and synapses is relatively denuded of insulating myelin so that neuroelectric potentials are capable of nonsynaptic transmembrane effects, simply because they are in close contact (an ephapse). The entire thicket is enmeshed by elaborate envelopes of glial membranes. The fact that the acreage of "naked" unmyelinated pre- and postjunctional neurites packed tightly in small volumes of

neuropile is enormous suggests that the thicket is capable of computing integrals of the graded electrical activities within its neuropile. Consequently, the junctional thicket provides a substrate for the production of slow waves, which are believed to play an important role in transmission, coding, and integration within all components of the nervous system (Pribram, 1971).

7. The circuit component consists of multiple replications of assemblies and thickets. The major difference among the numerous assemblies within a component is the spatial locations of their afferent sources and efferent destinations. A component is the basic processing or integrating unit of a circuit. It transforms and integrates patterns of discharge arriving from its several inputs and delivers particular kinds of patterns to its outputs. Components also differ in significant ways in different parts of the brain. They are usually visibly distinct to the naked eye or in stained sections viewed by the light microscope. It has been a major mission of early neuroanatomists to identify and name the basic brain components or nuclei (see Figure 6).

8. A component is part of an even greater functional unit, the simple circuit, which consists of two or more components arranged in series, with some components often distributed in parallel. It is axiomatic that, except for bilateral symmetry, no two circuits of the brain are alike. There is no structural–functional redundancy among circuits, although there may be similarities of some general capabilities of two different circuits. Likewise, no circuit can assume the functions of another if one is damaged.

9. Higher level structural–functional units may be designated as "systems," which consist of branched or interconnnected circuits (e.g., somatic sensory or visual systems).

10. Finally, system complexes consist of interconnected systems (e.g., a sensory-motor system couplex). Using this nomenclature, the brain is more than a nervous system. It is a nervous system complex.

Although this terminology is clearly provisional in some respects, the brains of different mammals can be compared in quantitative terms with respect to these idealized types of structural unit.

Parameters of Neuroanatomical Variation

In an individual living mammal, there are many structural variations in each of these types of basic unit in different parts of the brain (Lemkey-Johnston & Larramendi, 1968; Lenn & Reese, 1966; LeVay, 1973; Ramón-Moliner, 1968; Winkelmann, 1963). Although different individuals of the same species are very similar in all homologous neuroanatomical features, members of different species may differ considerably when homologous structures are compared. Because each of the basic types of structural unit can be constructed in several ways, the potential diversity for heritable variations is large.

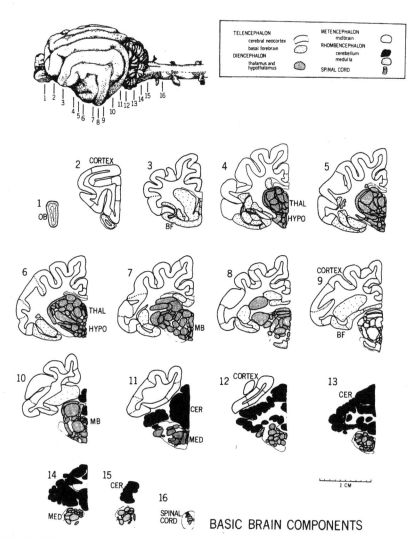

FIG. 6 Major neuronal components. Diagram of adult cat brain at top left. Numbers 1–16 indicate levels of transverse sections shown below. Areas outlined on these sections by solid lines indicate major components or groups of components, of indicated shape, size, and position, identified in major embryonic subdivisions (key at upper right) for mammals. Subcomponents are not demarcated. OB, olfactory bulb; BF, basal forebrain; THAL, thalamus; HYPO, hypothalamus; MB, midbrain; CER, cerebellum (basal nuclei included); MED, medulla. Dotted lines in cortex demarcate different cortical components.

Documentation of species differences in neural organization must ultimately be quantitative. Neuroanatomical structures can vary from one neural component to another and from one animal species to another along the following major parameters: size and shape of cell body, dendritic tree, axon, and synaptic terminal; length of dendrites and axons; number of cells, dendrites, dendritic spines, axon collaterals, synaptic terminals, afferent sources, and efferent destinations; spatial distribution of cell types, axon collaterals, synaptic terminals, and dendritic branches; complexity of dendritic arborization and synaptic architecture; and diversity of types of cells, dendrites, synapses, afferent sources, and efferent destinations. All of these structural features may be under genetic control and thus subject to natural selection. Unfortunately, documentation of species differences in any one of these features in itself tells little about how such a change may influence overall function of a unit. The brains of different mammals typically differ in multiple ways, and the role of any one of these differences (such as cell size or number) in adaptive function is difficult to assess (see Discussion). To be most useful, interspecies comparisons should be made with respect to several interrelated microfeatures of these basic neuroanatomical units. Nevertheless, the neuroanatomical literature contains abundant instances of quantitative neuroanatomical studies (Shariff, 1953). The question remains as to what use this data can be put in attempting to understand brain evolution (see section on Outcomes of Brain Evolution).

Major Neuronal Components

The relative location of distinguishable neuronal populations or nuclei (referred to as components in the vocabulary of this review) that have been described at the several levels of the brain for mammals are schematically illustrated in Figure 6. The five general embryonic divisions of the cephalic neuraxis do not have functional significance in the operational nervous system. They probably have physiological importance only in the early embryonic phases of neural ontogeny. Many neural components are easily distinguished architectonically from one another. In some nuclear regions (e.g., cerebral neocortex) functionally different neural components are not always easily distinguished from one another by anatomical criteria alone. Subdivision of such regions often relies on differences in afferent source or efferent destination of their connections or on differences in fine structure of their neuronal assemblies and synaptic glomeruli.

Depending on the criteria used to distinguish components, there may be 200–400 anatomically distinct components in mammalian brains. They have been given names that reflect their location, shape, color, architectural appearance, or presumed function. Many are named after their discoverer. Catalogs of these components can be found in atlases, textbooks, and research reports (e.g., Berman, 1968; Ariens-Kappers, Huber & Crosby, 1936; Rose, 1942). However, a descriptive list of neuronal components is of little value for understanding function. They are components in a vast web of interconnnected circuits and systems.

Major Neural Circuits

No nuclear component has functional independence but contributes only its own type of influence to the overall transactions of the circuit. Figure 7 illustrates the three major sensory circuits and two motor circuits the connections of which are relatively well known. Visual, auditory, somatic sensory, and motor circuits are shown because they are major ones used by most mammals in dealing with highly specific stimulus patterns in environmental niches. Space prohibits a thorough review of details of sensory and motor circuits. The point to be made here is that such circuits are apparently complex anatomically and are interconnnected in ways sufficiently elaborate and intricate to be responsible for the subtle, organized behavioral and mental phenomena that bind an animal adaptively to its environment.

This general review of basic neuroanatomical units and of brain components and circuits leads now to consideration of species differences in these morpho-

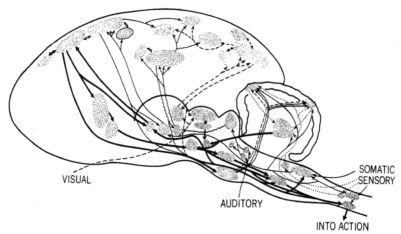

FIG. 7 Major sensory and motor circuits. Dotted ellipses stand for major neural components. Components for each circuit are located at several levels of the neuraxis and are interconnected in a variety of ways. Character of lines (consistent for each modality) interconnect components within each circuit. Visual, dashed lines; auditory, wavy lines; somatic sensory muscle sense, dotted lines; somatic sensory touch and joint position sense, thin solid lines; motor, heavy lines. The prime purpose of this drawing is to give the impression of spatial dispersion of major circuits through several levels of the neuraxis as well as of the interactive complexity, yet high specificity, of connections between components of each circuit. Although the relative locations and numbers of components involved in these mammalian circuits reflects their actual arrangement, accuracy in detail is neither claimed nor intended.

logical entities, of the results or outcomes of neuroanatomical evolution as revealed in living mammals.

Outcomes of Neuroanatomical Evolution

Understanding brain evolution requires understanding of its causes, its course, and its outcomes. The genetic causes of heritable brain changes and of their differential survival among populations can be assessed from experimental studies on living forms. Because data of sufficient detail come primarily from living animals, a starting point in identifying the course of brain evolution is in the study of the outcomes of evolution as revealed by neuroanatomical variations among living mammals.

Figure 8 depicts 28 different kinds of neuroanatomical variations revealed in the brains of living mammals. I have attempted to conceptualize and illustrate these effects so that they may be viewed as hypotheses capable of verification. Many of these have been explicitly proposed in the literature and are based on careful quantitative comparative studies. Others are less well documented and are either suggested or implied by data. Still others are presented because they seem logically possible. In most instances, the existence of variations of a particular feature (e.g., brain size) among living mammals suggests the nature (e.g., cell number, cell size) and direction (e.g., increase or decrease) of that brain feature. No attempt has been made to be exhaustive or specific in this review. The data relevant to the outcomes depicted in Figure 8 are scattered in an enormous literature. The following publications, not specifically referred to later, provided much data and most of the ideas that I have utilized in preparing these interpretive illustrations: Bishop and Smith (1964), Blinkov and Glezer (1968), Bodian (1972), Brazier (1969), Bullock and Horridge (1965), Burns (1958), Ramón y Cajal (1960), Campbell (1972), Diamond and Chow (1962), Ebner and Myers (1965), Eccles and Schadé (1964), Edinger (1948), Harman (1957), Harting, Hall, and Diamond (1972), Hassler (1967), Holloway (1969), Ariëns-Kappers et al. (1936), Kruger (1966, 1970), Lane, Allman, Kaas, and Miezen (1973), Lane, Allman, and Kaas (1971), Mehler (1957, 1969), Llinás (1969), Morest (1965), Nauta and Ebbesson (1970), Pappas and Purpura (1972), Peters, Palay, and Webster (1970), Petras (1969), Petras and Noback (1969), Polyak (1957), Smith (1902), Tigges (1970), Tower and Schadé (1960). This list, of course, is not thorough or comprehensive.

Mammals with larger bodies have larger brains (Cobb, 1965; Dubois, 1920; Jerison, 1963). This generalization holds better among more closely related taxa (e.g., among primates or rodents). More advanced or complex creatures have larger brains. The available data on brain–body ratios (by weight or volume) is best summarized by·Jerison (1955, 1961, 1963, 1973). The ratio of brain size to body size (cephalization quotient) is greater in the brains of mammals with more complex life styles (Holloway, 1968; Jerison, 1973). The progressive increase in

286

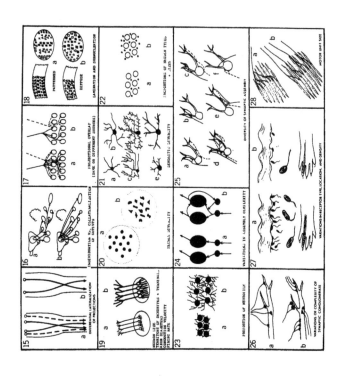

OUTCOMES OF BRAIN EVOLUTION

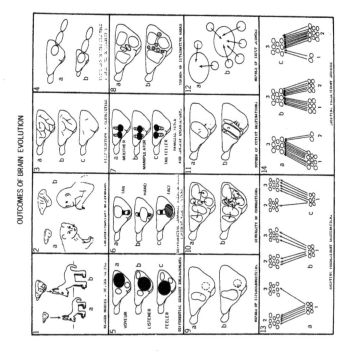

FIG. 8.

relative brain size in these more "advanced" forms is also documented in phylogeny from paleoneurological evidence (Edinger, 1948; Harman, 1957; Jerison, 1973; Mettler, 1956).

Larger and more complex brains are more fissured and convoluted. This generalization refers specifically to the cerebral neocortex and holds better among more closely related taxa (Connolly, 1950). Brains of similar size in different taxa may not be equally fissured or convoluted. For example, the relatively large brain of *Castor canadensis* (beaver) is relatively smooth, and that of the other giant rodent, the capybara (*Hydrochoerus*), is relatively fissured. Likewise, the tiny brain of the least weasel (*Rixosa*) is relatively fissured, whereas the larger brain of the manatee (*Trichecus*) is said to have a relatively smooth cortical surface (Smith, 1902). In addition, the echidna (*Tachyglossus*), a relatively small-bodied monotreme that is primitive in some respects has a relatively large, fissured cerebral cortex (Lende, 1964). Determinants of such variations in fissuration in brains of similar size may be clarified when the multiple causes of fissural dynamics are better understood (Connolly, 1950; Welker & Campos, 1963).

Not all parts of the brain are enlarged to equal degrees. Among mammals of related taxa, the more cephalic regions (i.e., forebrain or cerebral neocortex) are relatively larger in those animals with larger brains. "Differential corticalization" refers to the finding of a relative increase in cerebral or cerebellar cortex as a systematic trend in mammals of larger size and/or of more advanced levels of complexity (Campos & Welker, 1960; Diamond & Hall, 1969; Jerison, 1973). Differential enlargements of specific portions of neural circuits is the most obvious and well-documented source of differences among living mammals (Holloway, 1969; Ives, 1971; Ariëns-Kappers *et al.*, 1936; Towe, 1973; Welker, 1973; Welker & Campos, 1963). The finer details of these differential increases or decreases are of several kinds and are mentioned more specifically in the paragraphs that follow.

Differential enlargement of major sensory systems is well documented. Such specializations appear to bias an organism differentially to employ vision, audition, somatic sensory, or olfactory modalities in its transactions with the environment. Differential enlargement of subregions of a specific circuit component occur in motor, sensory, and central integrative neural components. It is noteworthy that these two types of differential development occur in homologous portions of all ascending components of sensory circuits (Welker, Johnson, & Pubols, 1964).

Parallel differential enlargements occur in sensory and motor circuits (Hardin, Arumugasamy, & Jameson, 1968). For example, spider monkeys with sensitive

---

FIG. 8 Outcomes of brain evolution. Schematic drawings illustrating major conceptions and hypotheses regarding evolutionary changes in neuroanatomical structures that have resulted in known types of variations in brains of living mammals.

tail pads can move their tails in numerous feeling and grasping ways (Pubols, 1966; Pubols & Pubols, 1972). The tail portions of the motor circuit controlling these movements and the sensory circuit from the sensory skin doing the touching are both relatively enlarged. Enlargements of sensory and motor circuits are typically expressed in specialized behavioral transactions with the environment (Welker, 1973). Several such brain-behavior correlations have been demonstrated.

Associated with such differential enlargements, other circuit changes may also occur. For example, the number of integrative components may differ in different mammals. This is reflected primarily in cerebral neocortex where the relative amount of "nonsensory" cortex varies considerably from one species to another (Diamond & Hall, 1969; Erulkar, Nelson, & Bryan, 1968; Hubel & Wiesel, 1965). It is from such differential developments, together with those occurring in the sensory and motor circuits, that many species differences in behavioral and mental processes must result. However, there is little direct evidence on this point. Likewise, the number of output destinations may differ, such as in the multiplication of projections presumed to occur from thalamic somatic sensory, auditory, or visual components to cortical subareas (Allman & Kaas, 1971; Paul, Merzenich, & Goodman, 1972; Welker, 1973). The number of input sources to or output destinations from a particular component may also change (Spatz & Tigges, 1972). There may have been shifting sources of inputs to or shifting destinations of outputs from a particular component in evolution (Creel & Giolli, 1972). Variations in degree of collateralization of outputs from a component may occur. Therefore, outputs from a component may reach several destinations via separate "lines" or as collaterals from a single "line." There is differential lateralization of projections of some circuits in different animals (Lund, 1965). For example, certain ipsilateral projections are relatively large in some species (Cabral & Johnson, 1971; Towe, 1973).

Overlap of inputs to a circuit component may vary so that in some species the degree of overlap or specificity of adjacent inputs from different sources is more discrete than in others. Species differences in lamination and subnucleation are common for somatic sensory, auditory, and visual circuits (Campos-Ortega & Hayhow, 1970; Guillery, 1970; Johnson, Welker, & Pubols, 1968; Welker & Johnson, 1965). For example, the neurons within a component may become grouped into clusters and lamina. These more structurally organized components reflect specialized, highly organized kinds of input, output, and intracomponent connections that probably confer greater functional specificity on such neuronal assemblies (Colonnier, 1968; Killackey & Ebner, 1972).

Neuron size is a neuroanatomical feature that may vary in homologous components of different species. Cell size confers on neurons several interrelated structural and functional features (Henneman & Olson, 1965; Henneman, Somjen, & Carpenter, 1965). For example, larger cell bodies also have larger diameter axons, which possess faster conduction velocities of propagated impulses. Larger cells also have larger dendritic trees capable of supporting more synapses, a greater membrane time constant with a corresponding decrease in maximum firing rates, and greater territorial dispersion of their axon terminals. The importance of this size principle for function has been stressed by several authors (cf. Bishop, 1959; Bishop & Smith, 1964). Diversity of cell type or diversity of dendritic specialization are characteristics of neural populations that also may vary in different species (Morest, 1965; Ramón-Moliner, 1968). Proportion of specific neuron types and sizes in a neuronal assembly may also be an interspecies variable. Each of these kinds of differences can be expected to affect the functional capacities of neuronal assemblies and components.

The proportion of neuropile in a component (e.g., in the cerebral neocortex) varies in different mammals (Ariëns-Kappers *et al.*, 1936). It has been suggested that the functional significance of this factor (measured as cell density or gray-cell coefficient) lies in the possibility that greater proportions of neuropile allow for more interactive and integrative effects to be produced by greater numbers and varieties of synaptic assemblies and afferent inputs. The validity of this general suggestion has not been tested. More specific hypotheses are required for it to become meaningful.

Variations in connectivity of neuronal assemblies, in synaptic assemblies and in complexity of synaptic conglomerates are other features of neural components, that vary in major ways from one nuclear region to another in individual animals (Bodian, 1972; Colonnier, 1968; White, 1973). They may also account for species differences in homologous components.

Variations in receptor density, type, and location occur for homologous sensory organs in different species (Cauna, 1961; Grundfest 1963a; Pringle, 1963; Pubols, Welker, & Johnson, 1965; Zollman & Winkelmann, 1962). Finally, size of motor unit in the efferent innervation of striated muscles may vary from one muscle to another in an individual animal, as well as in homologous muscles in different mammals (Wray, 1969).

Problems of Interpretation

Only a few of the comparative differences listed above are adequately documented. Most are merely indicated as possible evolutionary differences in a few scattered studies. Few have received quantitative analysis, and none have been subject to appropriate comparative surveys (see Discussion). Possible adaptive importance of any of the suggested differences can be inferred in some cases but validation of such hypotheses is essential.

Because the brains of different species of mammals usually differ in many ways (Blinkov & Glezer, 1968), the role played by any one changed feature in adaptive performance is difficult to analyze. Caution is therefore required in interpreting the evolutionary significance of any particular neuroanatomical difference. There is always the possibility that an observed difference is related to some other, even incidental, factor, such as body size or brain size. In such cases the observed difference (axon size : conduction velocity) may not contribute to a particular functional difference (reaction time) between two animals. For example, although larger axons in larger species may provide faster conduction velocities, they may not necessarily result in faster reaction times because impulses have to travel longer distances in the larger bodies. Moreover, additional structural–functional differences (e.g., synaptic delays) in central components of such circuits in different animals may also have contrasting effects.

In summary, although many species differences in a wide variety of neuroanatomical structures are known, very little systematic data have yet been gathered that clarify either the course of brain evolution or even the main features of its differential outcomes. This review has been presented merely as a sketch of the kinds of available conceptions that may serve as a guide for more adequate study of living mammals.

A major question still requiring answer pertains to how existing neural circuits manage to change during phylogeny. Especially important to know is how novel circuits arise. Structural novelties must add new functional possibilities. These issues are reviewed in the sections on the Construction of Neural Circuits During Ontogeny, and Genetic Control of Neurobiological Evolution, below. Before they are considered, the kinds of functional neuroelectric changes that are conferred on evolving neural circuits must be examined.

NEUROELECTRIC EVOLUTION

In order to evaluate how neuroelectric activities may have evolved among mammals, it is essential to review what these activities are, which structural features underlie them, and how heritable variations in them may influence adaptive success (Grundfest, 1963a; Pringle, 1963, 1965).

The Languages of the Brain

The waking brain is incredibly busy. Spike discharges shuttle back and forth, in and out, around and through the highways and byways of neural circuits. Their brisk spatiotemporal patterns of movement through a complex maze of interconnected neural circuits contain multiplex messages with specific meanings. The multitude of diversified messages are intercepted at hundreds of specialized synaptic stations (components) every second. Each component of a circuit decodes the messages it receives according to its own rules. And, depending on integrals of the melange of steady, slow, and fast potential "information" prevalent within the neuropile at the moment, each component discharges its obligate functions and generates new modulated messages that are sent on their way along predetermined routes. Certain regions of the brain may become especially active or inactive at any given moment and these multiple foci of action or inaction are constantly shifting and switching among different assemblies, components, circuits, and systems. These permutable patterns take shape swiftly, persist briefly as do standing waves, or are swept through or away either by new waves of action arriving from other sources or by organized patterns of quiescence (Pribram, 1971). These shifting montages of impulses, slow potentials, and synaptic activities must leave slight, as yet obscure, traces in passing, and they disappear along paths often never to be retraced in the same ways. These intricate, hierarchically organized, complex transactions can often be envisaged introspectively and subjectively in conscious thought. Because of their speed, their delicate evanescence, their uniqueness and their transient nature, they never may be objectively recorded and measured in their full richness, except occasionally as vivid splashes in memory. Many of these dynamic neuroelectric sequences become transfigured into behavior by means of complex high-speed spatiotemporal action sequences of muscle contraction. It is because of this correlation of neuroelectric and behavioral events that analysis of behavioral sequences can provide sensitive cues as to some spatiotemporal aspects of neuroelectric activities in the underlying determinant neural circuits.

Structural Determinants of Neuroelectric Capabilities

Structures dictate function (Bullock & Horridge, 1965; Eccles, 1964, 1969, 1970; Grenell & Mullins, 1956; Ungar, 1963). Knowledge of the fine details of neuroanatomical features (e.g., fiber diameters, type of synapse, degree of arborization of axon terminals) often suggests some aspect of the associated neuroelectric activities (e.g., conduction velocity, inhibitory or excitatory action at synapses, degree of spatial specificity of synaptic interactions). It is true that the kinds of neuroelectric activities that dendrites, cell bodies, synapses, and axons can generate are determined by highly specific neuroanatomical and associated chemical and physical features of these structures (Figure 9; Adelman, 1971; Aidley, 1971; Brazier, 1968; Grundfest, 1967). However, structural differences cannot reveal the dynamic action patterns that may occur in working neural circuits. Only neuroelectric studies can illuminate such adaptively important circuit transactions.

There is not much evidence that homologous neurons in different mammals have different neuroelectric capabilities. More likely, species-specific neuroelectric differences are to be revealed in the input—output coding profiles—differences caused by unique patterns of variation in fine structure of conglomerates, assemblies, components, and circuits. Because species differences in these neuroanatomical features are known, comparative neurophysiological studies should prove fruitful in the search for such functional differences. Table 4 lists some of the neuroelectric phenomena that can be measurably compared.

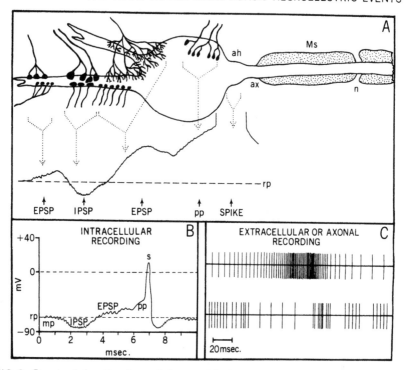

FIG. 9 Structural–functional correlations and basic neuroelectric events. (A) Diagram of neuron with dendrites (and impinging neurites) at left, cell body at center, and myelinated axon at right. Beneath the neuron is depicted the intracellular electrical events induced by electrical activity in synapses of different types and locations. The exact location of the different endings are drawn in arbitrary locations. Thus the prepotential is not necessarily produced by an ending on the soma in the location shown. In addition, the types of synaptic endings depicted are schematic only. The dashed line indicates resting potential (rp) of the neuron. Depolarization is upward and hyperpolarization is downward. Time reads from left to right. Dotted arrows with brackets point to the neuroelectric events induced by activity in the synaptic structures included above the brackets. Symbols: ah, axon hillock; Ms, myelin sheath; n, node; EPSP, excitatory postsynaptic potential; IPSP, inhibitory postsynaptic potential; pp, prepotential. (B) Drawing showing amplitude and temporal features of typical intracellular events. Symbols the same as in (A); s, spike potential. (C) Drawing of spike discharges of single neuron recorded by an extracellular electrode. These traces illustrate that spike frequency (interspike interval) and temporal pattern are the major parameters available to a neuron for transmitting information to distant neurons.

TABLE 4
Measurable Neuroelectric Phenomena

1. Transduction at:	Receptors	Motor endplates
	a. generator potentials b. adaptation c. relationships of spike frequency to stimulus parameters	a. endplate potentials

2. Spike discharges
   a. frequency
   b. rate of onset and offset
   c. duration
   d. amplitude
   e. conduction velocity
   f. accommodation
   g. latency
   h. temporal pattern
   i. thresholds

3. Synaptic transfer
   a. excitatory postsynaptic potentials (EPSP's)
   b. inhibitory postsynaptic potentials (IPSP's)
   c. summation
   d. synaptic delay
   e. modes of coding input frequency and temporal pattern: replication, selective gating (on, off, window), pattern extraction, addition, subtraction, transformation (abstraction)
   f. following frequency to afferent spikes
   g. after-discharge capability
   h. duration
   i. refractive period

4. Postsynaptic activities of soma, dendrite
   a. spatiotemporal summation and integration of individual synaptic effects
   b. soma spike, dendritic spikes and fast prepotentials, resting and slow changing membrane potentials, depolarizing currents, field currents
   c. relationship of PSP's, membrane potentials, and spike generation
   d. postspike hyperpolarization
   e. glial-neuronal neuroelectric relationships

5. Assembly and component transactions
   a. recurrent collateral excitation and inhibition
   b. presynaptic and postsynaptic inhibition
   c. summation
   d. surround inhibition and facilitation
   e. cyclic discharge, rebound
   f. conditional gating
   g. input-output coding: relationship of input patterns to output patterns

6. Circuit transactions
   a. code transformation or alteration: relationships of temporal patterns of initial codes to output codes at successive and final components of the circuit; coding of input spatial patterns to temporal or spatial outputs
   b. distribution of inputs to several outputs
   c. amplification of outputs over inputs
   d. feedback and feedforth operations
   e. abstraction
   f. conditional effects

7. System transactions
   a. relationships of output codes to spatial and temporal patterns of inputs from participating circuits
   b. facilitation, excitation, summation, inhibition
   c. feedback and feedforth operations
   d. integration of multiple different temporal patterns of inputs
   e. output codes of spatial interactions of inputs to each component

Basic Neuroelectric Activities

There are four basic types of neuroanatomical locations where functionally important neuroelectric transactions occur (Table 4): (1) within the neuropile, where decoding, encoding, and integration take place; (2) in the dendrites and soma where the postsynaptic integration and spike initiation are generated; (3) in the axons where spike discharges carry the coded messages from one component to another; and (4) in receptors and motor endplates that receive information from and result in reactions to features of the environment.

Neuropile transactions. These transactions occur between and among axon terminals and dendrites (Grundfest, 1967; Hubbard, Llinás, & Quastel, 1969; Pappas & Purpura, 1972; Pribram, 1971). They are usually situated close to cell bodies but they are in some instances deployed in specialized zones at some distance from the cell body. Known variations in complexity and arrangement of synaptic conglomerates, assemblies, and thickets are responsible for the unique ways in which neuroelectric messages are decoded, integrated, and transformed in different components (Eccles *et al.*, 1967; Gerard & Duyff, 1962; Leibovic, 1969). The synaptic neurohumoral and neuroelectric activities in neuropile are spatiotemporally complex. Their major neuroelectric features are prescribed by two sets of factors: (a) location, type (inhibitory, excitatory, axoaxonal, etc.), number, composition, and architecture of the synaptic assemblies that contact the dendrites and soma of the postsynaptic neuron (Pappas & Purpura, 1972); and (b) the spatiotemporal pattern of discharges arriving at their inputs. It should be noted that electrical synapses exhibit some features different from chemical synapses (Bennett, 1972; Pappas & Waxman, 1972). With each discharge a neurohumoral synapse is capable of producing a particulate quantity of neurohumoral transmitter at the pre- and postsynaptic junction (Katz, 1969). This results in either depolarizing or hyperpolarizing current flow. The amplitude and time course of the postsynaptic effects are determined by type and location of the synaptic structures and by the frequency and spatiotemporal pattern of arriving discharges. If the impulse traffic within the terminal arbors and within dendrites is sufficiently frequent and regular, steady potentials or slow potentials may be recorded. It is believed that these slower neuroelectric effects exist in relatively large domains (junctional thickets, Figure 5; Pribram, 1971) of neuropile at sufficient current amplitudes to provide a neuroelectric bias within the synaptic territory of many neurons. Such generalized potentials presumably are functionally important, and some circuits may be particularly concerned with producing and regulating such regional neuroelectric states. The configurations of the glial envelopes and of myelin sheaths within the neuropile play an important role in promoting or preventing "volume conduction" spread of current among adjacent neuropile elements.

Integration in dendrites and soma; spike initiations. Integration of the myriad postsynaptic neuroelectric activities and spike initiation are the basic functions of the dendrites and soma of the neuron. Dendrites and soma conduct graded slow potentials and, in some cases, fast spike potentials. The length, shape, size, architecture, and location of dendrites (Purpura, 1967); the size and shape of the soma; and the location of the initial segment of the soma in relation to the proximal axon are all geometric determinants of the spike patterns generated by the neuron (see Figure 9). The nature of temporal patterns in the influx of synaptic neuroelectric activities is the other major determinant of spike output of the neuron (Purpura, 1972).

Axonal transmission. The spike discharge conducted along axons is the basic signal that transmits information from one component to another within circuits (Grundfest, 1959, 1963b; Hodgkin, 1964). It is a fast, all or none electrical event (Bishop, 1956). The major measurable parameters of such impulses that are functionally important for transmission are discharge frequency (interspike interval), temporal pattern, and conduction velocity. Spike amplitude is considered invariant under normal conditions. Each axon delivers its discharges into its terminals. The number of these terminals, their diameter, the number and sizes of synaptic contacts they make, and their location and spatial dispersion determine the axons' overall potency to influence postsynaptic neuroelectric activity. Discharge frequency and temporal patterns are set up at the initial segment of the soma. Apparently, the axon is capable of faithfully following the fastest impulse frequency that the initial segment can generate. Conduction velocity is the only major parameter that is determined by axonal and axonal sheath characteristics (cross-sectional area, degree of myelinization, internodal distance).

Transductions at receptors and motor endplates. These are the fourth major type of structural–functional unit. They lie at the receptive and reactive interfaces of the animal with aspects of its internal and external environment (Granit, 1955; Katz, 1962; Matthews, 1972). They are the basic transducers without which there can be no niche transactions. As do synapses, receptors use neurohumoral and neuroelectric activities for transduction of stimulus features to the spike discharges that constitute the signaling language of the nervous system (Cauna, 1961; Davis, 1961; Grundfest, 1963a). Motor endplate potentials, as do synapses, translate the output decisions of the nervous system into contraction patterns of muscles (Katz, 1969; Zacks, 1964).

Each type of structural unit in the brain potentially plays a particular type of role that is important in improving the overall smooth and orderly adaptive transactions of an animal with its environment. It is likely that each type of synaptic complex or cell assembly has optimal patterns of excitatory and inhibitory input for its most effective performance. It is also likely that different types of input pattern to any synaptic complex, cell assembly, or component yield different output patterns. Such things are not well understood. "Feedback" circuits are found in many systems (Escobar, 1964; Machin, 1964). Spatiotemporal patterns are the preeminent features used by the brain to achieve adaptively appropriate coding and integration patterns among its circuits and systems. The spatial parameters of such patterns are genetically determined in the main, although the finer details of such patterns must be altered and shifted by afferent activities associated with learning and experience.

Comparative Studies

The existence of all these basic structural–functional phenomena in invertebrates as well as in all vertebrates (Bullock & Horridge, 1965; Grundfest, 1959, 1963a, b; Horridge, 1968; Wiersma, 1967) makes it clear that the major evolutionary developments in brain function that are seen in mammals have probably occurred by adding, subtracting, changing, and rearranging synapses, synaptic conglomerates, and interconnections within assemblies, components, circuits, and systems. It is possible that some evolutionary changes have occurred among mammals in some of the transmembrane neuroelectric phenomena mentioned above. However, there have been no truly comparative studies designed to examine species differences in such phenomena.

BIOCHEMICAL EVOLUTION

Biochemical transactions are the essences of life. The chemical fires raging in and among living structures are fierce, powerful, and all pervasive. These processes are restless, merciless, and timeless and throughout evolution have undergone cycles of creation and destruction. However, prebiological chemical evolution has also consisted of such processes in the fabrication of increasingly complex physicochemical reactions of compounds of C, H, N, and O in a primeval milieu of gases, minerals, acids, bases, and water.

Evolutionary processes in early living cells and organisms have perfected structures capable of increasingly complex and reliable metabolic, synthetic, respiratory, oxidative, secretory, and excretory machinery (Barry, 1964; Bernhard, 1968; Caspar, 1966; Florkin, 1966; Kenyon & Steinman, 1969; Mazia & Tyler, 1963; Ingram, 1965; Kalmus, 1966; Margulis, 1970). The development of reliable chemical means for inheritance of these biochemical factories set the stage for their further adaptive diversification in the tissues and organs of organisms (Bryson & Vogel, 1965; Wolstenholme & Knight, 1970; Wolstenholme & O'Connor, 1965). The major proportion of expended energy in cells has been devoted to growth, maintenance, and replication. The cellular biochemical mechanisms that have been evolved early have been remarkably persistent and constant in vertebrates (Anfinsen, 1959; Pattee, 1968; Pringle, 1965; Willmer, 1970). Biochemical structures capable of simple detection, discrimination, selection or rejection, ingestion, excretion, utilization, and metabolism of water, vitamins, minerals, carbohydrates, and proteins are similar in all animals.

With increased size and complexity, simple organisms evolved specialized cells to perceive, assess, and respond to those environmental energies that were required by the essential biochemical reactions (Duncan, 1967; Eakin, 1963; Pringle, 1963; Reiner, 1968). Neurons were among such specializations. Evolution of hierarchically more complex specializations of cells and tissues within each organ resulted in nervous systems capable of interactions with endocrine systems (Adolph, 1968). Neuroendocrine controls became more complex and sophisticated (Frieden & Lipner, 1971). Special sets of circuits and neurohumoral transactions were concocted and became progressively specialized and spatially distributed to deal with particular relationships of the organism with its environment, as well as with specific internal body processes. The portions of evolving nervous systems dealing with the external environment developed novel sensory, integrative, and motor processes that progressively became less directly related to the fundamental biochemical processes themselves and more concerned with judgement and estimation of probabilities of energy patterns in the holoflux (Kare & Maller, 1967; Ohloff & Thomas, 1971; Tamar, 1972). Actually, neurons evolved their own specialized biochemical features, which were refinements of the processes found in all living cells (Albers, Siegal, Katzman, & Agranoff, 1972; Schneider, 1973). Certain membrane, transport, trophic, biophysical, electrical, metabolic, biochemical, and neurohumoral features therefore differentiated further in neurons and were employed in specialized ways (Bullough, 1967).

Biochemical homeostasis was only the incidental concern of many of these new neurobiological structures and processes (Cort, 1965). In their more complex manifestations, many of these new structures became autonomous or only indirectly concerned with the total body's biochemical problems. The higher order operations that evolved were increasingly concerned with adaptive perception, judgement, estimation, choice, and reactivity to specific changing and changeable physical and chemical features of the environment. Such functions worked because certain environmental features tended to be highly associated, probabilistically, with important chemical features of the environment.

However, biochemical homeostasis continued to be the concern of several other progressively specialized neural circuits. Specialized sensory systems evolved to sense chemical features of the external world directly (Kare & Maller, 1967; Pfaffmann, 1969; Zotterman, 1963). Neurons in some circuits became modified to detect alterations in circulating hormones, metabolites, hydrogen ions, oxygen or carbon dioxide tensions, etc. Changes in body temperature likewise selectively affected certain neurons (Bligh, 1966; Hellon, 1967; Mrosovsky, 1971; Whittow, 1971). Other neurons in these circuits came to regulate heart or breathing rates, secretions of endocrine glands, and the functions of other body organs (Coupland, 1958) or activated behavior that might lead the organism to obtain food, water, warmth, etc.

These issues have been raised not only to recall that the most fundamental concerns of organisms derive from the chemical necessities of life. They also serve to illustrate that most biochemical mechanisms have long been available to evolving animals. My particular concern here is with the specialized biochemical activities and reactivities of neural circuits. Here too, all mammals utilize almost identical mechanisms.

As far as nervous systems are concerned, biochemical processes underlie all neuroanatomical structures and neuroelectric events. Even seemingly stable anatomical structures contain a mosaic of seething foci of chemical activities. All neurons derive their capability of communicating with one another primarily from their neurohumoral activities. In each mammal there are billions of micro-secretory pre- and postsynaptic structures at receptors, synapses (Pappas & Purpura, 1972; DeRobertis, 1964), motor endplates on muscles, or endings in endocrine organs (Namba, 1971). The several chemical transformations that occur at all these loci are of similar kinds, differing only in specific details at different receptors or at different types of synapse (inhibitory, excitatory). Neurons in certain populations are also capable of neuroendocrine secretory activities and are especially receptive to either circulating hormones secreted from distant endocrine glands or circulating metabolites (Bern, 1963; De-Robertis, 1964; Ford, 1971; Martini & Meites, 1970; Martini, Motta, & Fraschini, 1970). Many of these glands are, in turn, innervated by specialized neural circuits (Itoh, 1968). These reciprocally interacting neural and endocrine systems, among others, help coordinate basic life processes and adaptive activities of the thyroid, parathyroid, stomach, intestine, pancreas, testis or ovary, adrenal, neurophypophysis, adenohypophysis, pineal and subcommissural organs (Barrington, 1964; Lissák & Endröczi, 1965; Sawin, 1969; Scharrer, 1959; Scharrer & Scharrer, 1963). Each of these nonneural glandular structures is adapted to control or regulate specific aspects of biochemical functions.

With respect to neural circuits themselves, the general biochemical constitution of cellular populations or fiber tracts differs in different parts of the brain in all mammals. Several neural components exhibit relatively high concentrations of proteins, such as noradrenalin, adrenaline, dopamine, GABA, 5-HT, succinode-hydrogenase, glutamine, and glutamic and aspartic acids. All mammals probably exhibit similar regional specializations in biochemical constitution but species differences probably exist (Kety & Elkes, 1961; Okada, Nitch-Hassler, Kin, Bak, & Hassler, 1971; Palladin, 1964; Roberts & Sano, 1963). Variations in regional biochemistry within an animal are reflected in the underlying differentially specialized neurons and glia, but the functional significance of such differences in different animals is not yet clear (Hoffman & Sladek, 1973; Richter, 1964; Torre, 1972; Waehneldt & Shooter, 1973).

Although the basic biochemical building blocks do not differ much among the different living vertebrates, and certainly differ little among mammals, considerable variability in fine details of molecular construction and composition of the neurohumoral complexes in neurons, axons, and dendrites and at synaptic complexes is expected to exist. Because many species differences in biochemical processes are evident only during early ontogeny when timing of appropriate construction events is critical, the role of biochemical evolution is probably more likely to be revealed by developmental studies. Such studies may disclose specialized heritable biochemical transactions during the selective maturation, growth, and maintenance of specialized synaptic complexes, neuronal types, neurohumoral synapses, neuronal—glial relationships, and neuroendocrine specializations as well as during learning processes (Barondes, 1965). It is through variations in these processes and in their timetables of appearance during development that comparative studies can elucidate subtle details regarding the causes and outcomes of biochemical evolution. Were the character of such heritable ontogenetic variations known, so could be the biochemical sequences which constitute the basic processes capable of being selected for during brain evolution.

Although study of specialized biochemical events during development can provide the bulk of data relevant to evolution of biochemical transactions, several such processes continue to function after the nervous system has been constructed and during the remainder of the organism's life. These are (1) maintenance of transmitter sensitivity and of transmitter production at synapses; (2) operation of chemical systems that result in membrane polarization, depolarization, and hyperpolarization; (3) biochemical interactions between glia and neurons; (4) maintenance of chemical composition of blood and cerebrospinal fluid (CSF) by the blood—brain barrier; (5) generation of neurosecretory activity of specialized neurons and of those especially receptive to endocrines and metabolites; (6) secretion of several types of hormones by endocrine glands (Gorbman & Bern, 1962); (7) neurochemical changes at synapses associated with electrical activity related to learning; (8) processes associated with regeneration of axons and their terminals; and (9) specialized neuroendocrine and metabolic transactions of neurons during drive states and sleep.

These diverse potentialities for specialized neurobiochemical transactions are spatially distributed in particular ways among the neuron populations of the brain. The patterns of distribution of these numerous biochemical specialties are as intricately and highly organized as are the patterns of connectivity of neuroanatomical conglomerates, assemblies, circuits, and systems themselves.

In summary, certain structural features of biochemicals associated with receptors, synapses, and motor endplates are similar in most mammals. However, species differences are likely to be revealed when appropriate, more detailed, comparative studies are carried out. Some myelin proteins, for example, differ quantitatively in different species (Morell *et al.*, 1973). Fine differences in programming of biochemical sequences by the genotype during ontogeny are expected and may be identified with development of more refined means of studying such things. However, major species differences in normal functioning of the brain must be attributed to differences in ways that neural circuits are constructed.

EVOLUTION OF OTHER ORGAN SYSTEMS

The brain has evolved in a context of skin and bones, muscle, viscera, and blood. Indeed, all these tissues and their organ systems have evolved together and interdependently. All organ systems are coadapted to one another, each specialized in ways that result in sets of interrelated effects that are successful for the population of organisms. The nervous system establishes special physicochemical contacts with these nonneural tissues. Motor innervation of muscles attached to skin, cartilage, and bone permits the animal to move itself, its parts, and its sense organs. Sensory innervation of specialized structures, such as cornified, papillary glabrous skin; vascular, collagenous fatty dermal skin; and hair, and of accessory supporting structures, such as joints, muscles, and other deep tissues, allows the brain to better perceive objects and surfaces touched, skeletal movement and position, and lengths and tensions of muscles (Kenshalo, 1968; Matthews, 1972; Montagna, 1962; Sinclair, 1967). Visual perception is facilitated by such special ocular structures as eyelids, cornea, lens, ciliary muscles that focus the lens, iris diaphragm with sphincter and dilator muscles, transparent aqueous and vitreous humors, choroid and sclera, pigments, and extraocular muscles capable of moving the eyeball (Duke-Elder, 1958; Gregory, 1966; Polyak, 1957; Walls, 1942). Auditory perception is promoted by the external ear, tympanum, bony ossicles, osicular muscles, evacuated middle ear cavity with eustachian tube, round and oval windows opening on the scalae vestibuli and tympani, Reissner's membrane, stria vascularis, spiral ligament, limbus, lymph fluids, tentorial and basilar membranes, and the petrous bone (Busnel, 1963; De Reuck & Knight, 1968; Rasmussen & Windle, 1960). Taste perception is facilitated by specializations of lips, tongue, buccal cavity, and associated musculature (Pfaffman, 1969; Zotterman, 1963). Olfaction is enhanced by special muscular, skeletal, and endodermal structures in the rhinarium, epithelium, cartilage, and nasal cranium. Finally, perception of changes in bodily orientation in three-dimensional space has resulted from the coevolution (with vestibular receptors) of the bony labyrinth, membranous semicircular canals oriented in three rectangular planes, utricle, sacculus, endolymph, ampulla, crista, and cupula (Busnel, 1963; Rasmussen & Windle, 1960; Richards,

1971). These various accessory nonneural specializations participate in selectively filtering the stimulus holoflux, allowing transduction and reception by sensory organs of only certain ranges and intensities of available environmental energies (Walls, 1962). These accessory structures have evolved differentially in different mammals to predispose them to perceive and react to the physical world in ways that are adaptively relevant. I believe it remiss to attempt developing hypotheses regarding features of brain evolution that do not encompass the coevolution of nonneural structures.

Nonneural specializations can influence behavioral expression in ways other than by filtering stimulus inputs. For example, the range and amplitude of movement possible at a joint are determined by the structural features of the articulation of its two bones (Alexander, 1968; Cornwall, 1956; Hall, 1965). The locations of the origins and insertions of the tendon attachments on bones also influence characteristics of the contracting muscles in moving the joint. Homologous muscle groups in different animals often differ in relative size, shape, and number of bellies, thereby resulting in species differences in movement patterns (Compton, 1973; Huber, 1930). Prehensile hands and bipedal postures are structural features of importance in the evolution of adaptive behavior in primates (Napier, 1970). The snipping, shearing, or grinding capabilities of teeth in certain animals predispose that they be used in these ways. Likewise, the presence of horny ischial callosities may influence the sitting postures assumed. A long prehensile tongue affects the kinds and locations of food objects taken; buccal pouches play a role in food gathering habits; and claws tend to facilitate fighting, grasping, and climbing. Variations in the elaborate peripheral vocal apparatus of the lips, mouth, tongue, pharynx, and larynx enable the complex and varied complex communicative skills of many mammals (Andrew, 1963a, b; DuBrul, 1958; Norris, 1969). Behavioral differences among orders, families, and lower taxa are associated with such nonneural, nonbehavioral traits and are documented in detail in several instances (Geist, 1966; Goffart, 1971; LeGros Clark, 1964, 1970; Napier, 1960, 1970; Van Gelder, 1969; Washburn & Jay, 1968). Indeed, many habits and habitats of creatures long extinct are reliably inferred from skeletal details. All these structural features selectively influence the parameters of physical energies that reach the receptor and organs. The receptors themselves, of course, also have accessory nonneural structures (merkel cells, supporting cells, hairs, etc.) that exercise the final enhancement, focusing, filtering, gating, and localizing effects on those stimulus energies actually transduced and coded.

Other tissues and tissue products play a role in the general structuring and maintenance of neural tissue. For example, the cranium, vertebrae, and meninges conceal and protect the brain and spinal cord. Cerebrospinal fluid and the vasculature make available essential metabolites and nutrients to appropriate populations of neurons. Endocrine organs, digestive organs, and other tissues, interacting with the nervous system, continue to carry out the essential physiological functions required to maintain the biochemical transactions that are the essences of life of all tissues (Burn, 1963; Carlson, 1968).

All of these enabling nonneural structures develop and mature during ontogeny in the course of intimate transactions with simultaneously developing neurons and their processes. This association of neural and nonneural tissues during ontogeny is essential for normal development of the nervous system and is programmed by the genotype. These complex, timed interplays among tissues are the major events that influence the early creation and establishment of neural circuits. To the extent that there is maldevelopment in any important structural or functional feature of these nonneural tissues, the development of some aspects of neural ontogeny is likewise altered (Kalter, 1968; Purpura, 1974).

CONSTRUCTION OF NEURAL CIRCUITS DURING ONTOGENY

In trying to comprehend how the nervous systems of different species have come to differ from one another, much can be understood by observing exactly how neural circuits are constructed during ontogeny (Sperry, 1964). At early embryonic stages, the developing neural tube appears indistinguishable in different mammals. However, as the neuroblasts migrate, differentiate, develop dendrites, send out axons, and establish synapses, the neural circuits of different mammals become progressively more distinct from one another. Careful examination of structural and biochemical details of specific circuits at these points of divergence should reveal the exact ways that the heritable genotype utilizes to produce the species differences found in mature animals. Moreover, microenvironmental determinants in developing brains may also be revealed. Therefore, documentation of species differences in developmental sequences may help identify the kinds of determinants that have been subject to change in evolution.

Development of Specificity of Integrated Neurobiological Structures

Previous sections have attempted to illustrate the marvelous complexity of neurobiological processes and structures in mammalian brains. The superbly precise and elaborately regulated assembly of interconnections and interactions of the brain develops that way during ontogeny because the mosaic of developing neurobiological components interact according to rigorously timed and regulated programs (De Beer, 1958; Ebert, 1965). Conceptions of some of the determinants of these programs are reviewed in the following paragraphs (cf. Barth, 1964; Ramón y Cajal, 1960; Eayrs & Goodhead, 1959; Gaze, 1967; Goss, 1964, 1972; Himwich, 1970; Hughes, 1968; Jacobson, 1970; Locke, 1968; Mugniani, 1971; Podolski, 1971; Sperry, 1965, 1971; Weber, 1967; Weiss, 1955, 1968).

The great structural complexity and highly patterned specificity of neural circuits is established progressively and sequentially according to a logic that is incredibly complex. Basically, the nervous system builds itself under guidance of the genotype; yet at all locations microenvironmental determinants play important roles. The relatively stable milieu of the embryo and fetus permits the orderly development of an increasingly diversified and spatially extended mosaic of neurobiological microenvironments. This increasingly complex distributed mosaic of microenvironments determines the resultant great complexity of neural circuits. Neurons, synaptic conglomerates and assemblies, neuronal assemblies, components, circuits, and systems all differentiate and establish their major connections during early ontogeny. The construction of fine features of circuits are regulated in spatial patterns and according to timetables that are highly specific and orderly. Most of the microenvironmental determinants of developmental sequences in neurobiological ontogeny are understood fairly well in general outline but poorly in detail.

In the following paragraphs, I review some of these developmental determinants of circuit formation. As an aid to readers, the drawings in Figure 10 have been prepared to provide additional imagery regarding these determining factors. The italic numbers in the following paragraphs refer to the numbered diagrams in Figure 10. In reviewing these determinants the reader is asked to keep in mind the search for the genetically inspired causes of species differences in brain circuitry. It then becomes clear that major bases for anatomical differences lie in the realm of early biochemical transactions (Davison, 1971).

As conceived by Willmer (1970), the major determinants of differentiation (Figure 10), the genome (*1*; in Figure 10), or all the genes carried by an individual, provides the necessary library of biochemical transactions; but the econome (the component parts and inclusions of a given cell) and agoranome (the cell's external environment) as well as the genome of each developing neuron determine the particular portions of the genome (hegemon) that are active in a particular cell at any given time (see also Curtis, 1967; Wright, 1973). Specialized microenvironments that are established at the cellular level are essential for the further differentiation of the series of ever more complex microstructures that characterize mature neural circuits (Pease, 1971).

Initially the neural crest and specialized ganglia separate from the neural tube and their processes interact with ectoderm, mesoderm, and endoderm (*2*). Several specialized ectodermal plaques are formed in interactive contiguity with differentiating neurites deriving from olfactory, gustatory, optic, acoustic, vestibular, and somatic sensory ganglion cell populations (Weiss, 1955). Other neurons in ganglia send their peripheral neurites into endodermal and mesodermal tissues. Independent differentiation of the neural tube (*2a*), of endodermal and mesodermal tissues, and of ectodermal plaques and ridges (*2a*) occurs early in the embryo. These structural specializations, and the subsequent independent formation of a mosaic of different adjacent cell populations (mosaic self-deter-

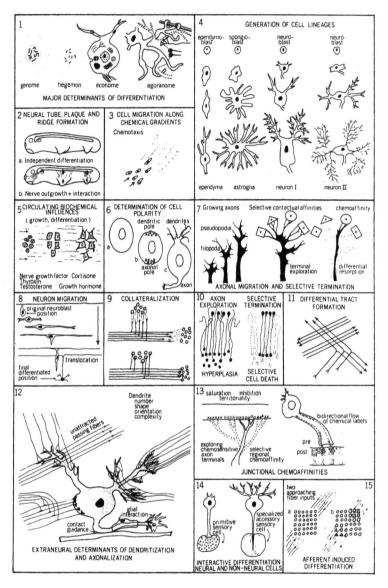

FIG. 10A

FIG. 10 Determinants of circuit construction during ontogeny. These illustrations and their associated descriptive words and phrases are intended to symbolize my interpretations of concepts and results obtained from studies of neural ontogeny. Each numbered drawing illustrates a known developmental feature, a factor affecting such a feature, or a set of interrelated structures and processes. These drawings are arranged only in a general sequential order of early to later stages of ontogeny. In most drawings, the structures and

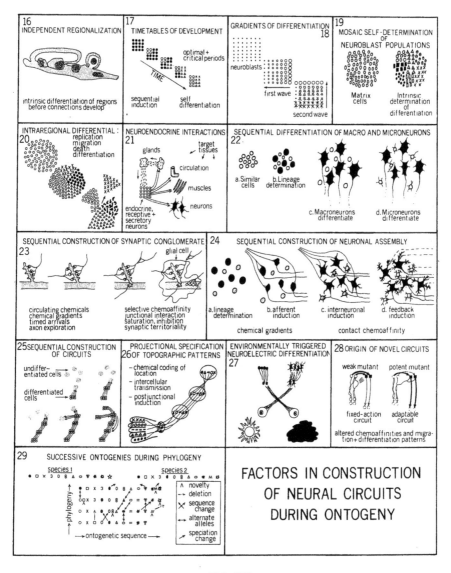

16 INDEPENDENT REGIONALIZATION	17 TIMETABLES OF DEVELOPMENT	GRADIENTS OF DIFFERENTIATION 18	19 MOSAIC SELF-DETERMINATION OF NEUROBLAST POPULATIONS

FIG. 10B

cellular arrangements shown are illustrative only, depicting only a general class of structure or function. They are not to be interpreted as representing specific structures, sequences, or functions at a particular location in the brain of any one kind of animal. Concepts in (*1*) above are those of Willmer (1970), and those in (*8*) are redrawn from Ramón y Cajal (1909–1911).

mination, *19*; regionalization, *16, 20*; Minkowski, 1967), appear to be determined by factors intrinsic to some developing regions. However, specific inductive interactions between neural and nonneural tissues (Angeletti & Vigneti, 1971) and between the separate populations within each level of the neuraxis soon follow (*14*). These intercellular transactions are of several kinds and are fundamental to circuit formation (Silvestri, 1972; Trinkhaus, 1969).

Specific physicochemical interactions of neurons or neuroblasts with each other and with adjacent nonneural (ectodermal, mesodermal, and endodermal) tissues are fundamental to circuit formation (Lowenstein, 1968, 1973; Michalke & Lowenstein, 1971; Pitts, 1971). Biochemical influences reach developing neurons via the circulation (*5*) (Balázs, 1971; Hamburgh, Mendoza, Burkart, & Weil, 1971; Harris & Levine, 1965), by diffusion through adjacent tissues, or by direct intercellular contacts (*13*). Blast cells grow (*5*), multiply (*10*) (Langman, Shimada, & Haden, 1971; Newton, 1964), and differentiate along different cell lineages to result in ependymal cells, astroglia, or neurons of different types (*4, 22*; Morest, 1969; Saxén & Toivonen, 1962). Neuron polarity may be intrinsically determined (*6*) but extraneural determinants effect the subsequent development of dendrite and axonal poles (*12*). Entire neurons undergo migration or cell body translocation (*8*) under external influences (*3*), but axonal exploration (*7, 10*) comes to play a major role in making selective contacts in circuit building (*7, 10, 13;* Herndon, 1971). Glial cells (Vaughn & Peters, 1971) and mechanical factors as well as localized chemical influences play specific roles in the migratory direction of axons and their collaterals (*9, 11;* Causey, 1960; Guillery, Sobkowicz, & Scott, 1970; Sidman & Rakic, 1973). Glial cells also wrap some axons in myelin (Martinez & Friede, 1970; Yakovlev & Lecours, 1967), a feature that alters the axon's conduction velocity. Maturation of receptors, motor endplates, and synaptic conglomerates is dependent on several specific localized mechanical and biochemical contacts (*23*; Bloom, 1972; Curtis, 1967; Herndon, 1971). In such contacts, biochemical labels pass in both directions (*13*), so that subsequent development of connections between neural and between neural and nonneural structures is interdependent, involving specific determinants (*13, 14, 15*; Gilula, Reeves, & Steinbach, 1972; Sperry, 1963). One or more gradients of differentiation occur in some neuroblast populations (*18*; Rose, 1942). Within any region, different neuroblasts may selectively undergo differentiation, migration, multiplication, or cell death (*20*). Large neurons differentiate first in most regions of the brain (*22*; Jacobson, 1970). These differential developments proceed in sequential pattern (*25*) according to specific timetables (*17*; Altman, 1969; Caviness & Sidman, 1973; Hamburgh *et al.*, 1971; Sperry, 1971), as does the construction of many features of circuit connectivity (*23, 24, 25*). As circuits become established, neuroelectric processes, especially those triggered via environmental stimulus energies impinging on receptors (*27*), affect fine features of maturation (Coleman & Riesen, 1968; Schapiro & Vukovich, 1970; Walsh, Budtz-Olson, Penny, & Cummins, 1969).

Early interaction of peripheral neural and nonneural tissues is essential for the development of precise, orderly interconnections of specific sense organs with appropriate central circuits (26). All major sensory systems have distinctly different end organs. Each of these differentiates in ways determined in part by the distinctive characteristics of structures in its local milieu. In each type of end organ, there are variations in the submodal type of receptor. Each of these is also associated with specific characteristics of the several structures they innervate in the local milieu. For example, somatic and visceral tissues, glabrous skin, scaly skin, hairy skin, deep dermal, corneal, mucocutaneous, sexual, tendinous, muscular, joint, fascial, gut, and bladder tissues all contain receptor types peculiar to themselves (De Reuck & Knight, 1966; Kenshalo, 1968; Sinclair, 1967; Matthews, 1972). Motor efferents from spinal cord neurons make specialized contacts with striated and nonstriated muscle tissues. These contacts are followed by neuroelectric and biochemical transactions that are essential for continued differentiation and maintenance of both pre-and postjunctional structures (Bernstein & Goodman, 1973; Close, 1971; Guth, 1971). The presence of specific microstructural and chemical features in each of these tissues is important for continued and proper development of neural–nonneural assemblies. First-order peripheral neurons transmit these modality and topographical locus codes (26) by axoplasmic flow into the cell body and out through efferent neurites to contact second-order populations of primitive central neurons. These neurons, thereby chemically tagged, likewise establish specific physiochemical connections with other third-order neural populations or with peripheral muscles or secretory tissues (13, 15, 21, 25). Specialized neurons with heightened secretory functions and endocrine organs develop and come to interact via the circulation or via innervation of the endocrine organs (Bleier, 1971).

Central sensory, motor, and associational circuits are built up sequentially in this general way (25). The developmental program is additive; each new circuit connection derives from axons and dendrites of neurons with contacts within already established circuits (25). Specializations of structural and biochemical features in nonneural tissues are also promoted and maintained by contacts from neurons with them. Little is known about the physicochemical causes of many specific details of neural differentiation, or about which microstructural characteristics of specific neurons are under direct intrinsic control of the genes of those neurons themselves.

This brief review of determinants of circuit construction is intended to point out that the genotypic and microenvironmental factors involved are manifold, transient, subtle, and complex. The data also indicate that there is a rigorous, orderly program of simultaneous and sequential development of a vast mosaic of interacting neurobiological structures. These organized events result in an increasingly complicated hierarchy of neural circuits and systems during ontogeny. There is currently fragmentary understanding of the exact details regarding either the resultant brain circuits or the timetable of their formation. The overall picture is only dimly conceived. However, enough details are known to encourage a search for those species differences in these phenomena that may indicate the kinds of genetic determinants associated with speciation of mammalian nervous systems.

Phylogeny Viewed as a Succession of Changing Ontogenies

It is misleading to view phylogeny as consisting only of a succession of adults forms (De Beer, 1958). More correctly, phylogeny is conceived as a succession of progressively altered complete ontogenies. In any animal, the different portions of the heritable genotype are expressed in an organized program as an animal constructs itself during development. The architectural plans contained in the genome determine the particular building blocks used at any locus as well as the sequences of their assembly. Evolutionary processes may select for new genes, delete old ones, and rearrange others in ways that alter the timetables of their expression (Figure 10, *29*). Fine details of phenotypic expression of the genotype are most fully revealed by examination of development of the entire assemblage during ontogeny. The adult form is only one phase of expression of the genotype in a steadily changing, ongoing sequence of developments during an individual's life span. It seems appropriate, therefore, that comparative studies of species differences in neurobiological structures benefit from examination of the full range of these features as they are laid down during ontogeny. Using the comparative ontogenetic approach, it should be possible to trace species differences in neural circuits of adults to specific differences in the ontogenetic sequences of the forms compared.

Viewed in this way it is not useful to conceive of ontogenetic sequences as recapitulating phylogenetic sequences (De Beer, 1958). Much evidence reveals that, except for a few general similarities, specific details of construction of normal embryos and fetuses do not pass through stages that resemble those that characterize stages of vertebrate evolution. For example, in more advanced forms, evolutionary modifications may be inserted at all phases of development, not just near the end (De Beer, 1958). Also, some early ontogenetic sequences, present in fish, for example, may be deleted or drastically modified in mammalian embryos. The organism must be adapted to its environment at all stages of its life history.

Origin of Novelty during Ontogeny

Modifications or changes in circuit construction are necessary for brains to evolve. Such genetic novelties are expected to reveal themselves as altered fine details of structures, connections, and processes during the building of the nervous system. Brain evolution results from the steady and progressive accumulation of mildly novel sequences and the deletion or suppression of outmoded ones. All such novelties begin as trivial changes and are selected for overall fitness to a way of life. Sudden, large novelties do not ordinarily survive. To last, even minor novelties must be biochemically and structurally compatible with the vast array of existing neurobiological structures and processes. It follows that major observable heritable changes consist of changed developmental sequences and cumulating batches of minor alterations.

The kinds of structural changes that may be observed during ontogeny are changes in cell size, number, axonal destinations, and any of a great number of variations in architecture of neurons, dendrites, axonal terminals, and synapses, such as have been reviewed in the section on Neuroanatomical Evolution. Because species differences in biochemical sequences underlie the development of species-typical microstructures, studies of biochemical ontogeny should also prove fruitful. Potentially, such studies can catalog species differences in time-table, sensitivity, and topographical locus of key biochemical transactions (e.g., hormones) that affect developing circuits. Such information can also point to features of the genotype that are responsible for these biochemical sequences.

Role of Learning in Circuit Building

Learning involves structural and biochemical alterations in those neural structures that "experience" transsynaptic electrical activities. Literally, such changes are incarnations of features of the external world; and the brain incorporates the environment as coded residues within itself. Neural circuits are potentially able to be restructured and shaped in minor ways by merely being active during the experiencing that inevitably proceeds throughout an individual's lifespan (Coleman & Riesen, 1968; Globus & Scheibel, 1967). Atrophy of such learned effects also occurs with disuse or lack of experience. This experiential moulding and nourishing of neural circuits can be viewed as the means that brains have evolved to keep their circuits tuned to the classes of environmental phenomena that are current for the organism at a particular span in time and locus in space (Denenberg, 1966; Newton, 1968). Yet, many of these "traces" are not readily erased, especially if they have occurred early in ontogeny.

These considerations lead to the conclusion that no two individuals of a species, even "identical" twins with identical genotype, can be expected to have identical neural circuits. This is expected for two reasons. First of all, during construction of circuits in embryonic and fetal periods, most structural and functional details are determined by microenvironmental factors (econome and agoranome) rather than directly by the hegemon or genome (Figure 10, *1*; Willmer, 1970). This being so, variations in circuitry are likely to occur because of numerous differential fluctuations in microenvironment within each developing embryo. Such variations are also likely to be differentially affected by fluctuations of maternal biochemistry, nutrition, stress factors, etc. (Galli, Jacini, & Pecile, 1973; Jacobson, 1970; Kalter, 1968; Scrimshaw & Gordon, 1968). Such determinants are epigenetic (Caspari, 1958). Second, in postnatal life, the experiential shaping of circuits is likely to vary because no two individuals can occupy identical spatial and temporal locations in the hypervolume or holoflux. The differential shaping effects on neural circuits resulting from different experiences with the environment, with time, progressively impels each twin into separate, although largely similar, niches.

There is constant interplay of generative, degenerative, and regenerative processes always in progress in many biological tissues, including muscle (DeHaan & Ursprung, 1965; Kalter, 1968; Peter, 1971; Thornton & Bromley, 1973; Willier, Weiss, & Hamburger, 1955), epidermis, blood, endodermis and bone, as well as neural tissue. These processes are affected by stimulation, its lack, trauma, nourishment, aging, etc. (Bernstein & Goodman, 1973; Birren, Imus, & Windle, 1959; Rockstein, 1972).

Neuroscientists are becoming increasingly aware of the importance of such processes in the continued normal maturation and maintenance of brains throughout the life span, (Newton, 1968). The precise nature of these changes and how they have evolved are not well understood.

Relevance of Behavioral Ontogeny to Neural Ontogeny

As neurons differentiate, develop neuroelectric capabilities, and make functional sensory, integrative, and motor connections, behavioral sequences begin to emerge. Study of the developing behavioral repertoire can provide clues to how relatively simple behavioral units are progressively added, altered, and integrated into the more complex, patterned behavioral sequences of mature animals (Barnett, 1972; Fox, 1971; Hailman, 1967; Hamburger, 1963; Kuo, 1967; Moltz, 1971; Schneirla, 1966; Schneirla & Rosenblatt, 1963; Scott, 1962; Welker, 1971). The diversity, complexity, and subtlety seen in behavioral sequences reflect corresponding functional features of underlying circuits. Spatiotemporal patterns of gross behavior have homologs in the detailed patterns of functioning neural circuits (Compton, 1973). For example, behavior may externalize certain features of the neuroelectric patterns. Consequently, studies of behavioral maturation, coupled with those of underlying neuroanatomical, neurophysiological, and neurochemical maturation, can contribute to understanding which growing neural mechanisms underly the expanding behavioral repertoire. Study of this developing repertoire in its transactions with the environment also may reveal ways that these neurobiological structures and processes have contributed to adaptive evolution.

GENIC CONTROL OF
NEUROBIOLOGICAL EVOLUTION

The ways that neural circuits are constructed reveal that genic controls operate on the constituent neurobiological aggregates throughout the lifespan of an animal. It is to the phenotypic, rather than genotypic, aspects of these aggregates that most attention is usually given, and this is especially important because of the pervasive role that microenvironmental factors of the agoranome and econome (Willmer, 1970) play in circuit building. However, in attempting to assess phylogenetic relations in any search for understanding of neurobiological evolution, comparative studies of genic material (chromosomes, DNA, proteins) as well as of morphological and physiological characters can also be helpful, (Caspari, 1958).

Developmental studies have revealed the structural and functional complexity of neurobiological aggregates and the spatiotemporal diversity of their construction during ontogeny. Such data signify that the genic influences on these structures and their functions are multiple, spatially dispersed, and interactive. Complex neurobiological characters have polygenic determinants. Rarely do single genes operate in isolation. Typically, the effect of a particular gene depends on those of other genes. To survive, genes must comprise coadaptive complexes (Dobzhansky, 1962). Genic effects must be biochemically compatible, at least up to reproductive age.

Genes regulate enzyme syntheses (Barry, 1964; Bernhard, 1968; Garber, 1972). However, for each cell the program of expression and suppression of their biochemical activities in the spatial and temporal domains is rigorously prescribed by the specific features of econome and agoranome (Willmer, 1970), as well as by the constitution of coadaptive gene complexes themselves (Markert & Ursprung, 1971).

Minor variations in genic material, or in its arrangement, constitute random genetic noise (Murray, 1972; Ohno, 1970; Mettler & Gregg, 1969; Stahl, 1964; Stebbins, 1966; Willmer, 1970). Mutational pressures are omnipresent, and sufficiently numerous and diversified that they continually press the organism's phenotypic characters to saturate the available niches within the energy holoflux. Many of these genic pressures are cut short by their own excesses. However, coadaptive complexes occasionally emerge above the "noise" level to improve adaptive potential by providing the organism with novel means of conducting transactions with some sector of the available holoflux.

Studies of inherited developmental abnormalities (Guillery & Kaas, 1973; Sidman, Green, & Appel, 1965; Yoon & Frouhar, 1973) and of behavioral genetics (Fuller & Thompson, 1960; Hirsch, 1967; Manosevitz, Lindzey, & Thiessen, 1969; Thiessen, 1972) have provided much information regarding genetic determinants of a great variety of phenotypic neural and behavioral phenomena.

These studies so far have examined only the most obvious or gross genic expressions. These strong effects are commonly broadcast both among neural and nonneural tissues, and their behavioral effects are manifest broadly in behavior (Sidman, Green, & Appel 1965). Most of these widespread effects indicate polygenic inheritance. Subtle effects require study of fine morphological and behavior features, especially in congenic strains (Hirsch, 1967). Because of the frequent occurrence of gene linkage, inversions, and crossing over in chromosomes, such genetic studies are potentially able to provide much greater resolving power regarding genes and their locations on chromosomes.

Greater care is required in the analysis of the fine phenotypic details of neural, nonneural, and behavioral expression in order that a level of analysis be reached commensurate with that at which particulate gene expression is clearly manifest.

SELECTION, ADAPTIVE RADIATION, AND PHYLOGENY

Environments participate in selection of genic characters (Drake, 1968; Levins, 1968; Mayr, 1963). Niche features are constantly changing and place selective pressures on the varying, mutating gene pool via the expressed phenotype. From the point of view of a population of animals, changes in milieu may be brought about by migration and discovery of new environments or by environmental change itself (Mettler & Gregg, 1969; Orr, 1970). Changes in environment may be sufficiently small that genetic variation impels easy exploitation of the new niches. If environmental changes are sufficiently sudden or drastic, extinction, removal, or geographical segregation and isolation occur (Colbert, 1961; Olson, 1965; Simpson, 1965, 1969; Udvardy, 1969).

Any animal type is specialized for certain optimal kinds and features of niche, being less effective, maladaptive, or defective for survival in microenvironments at the extremes of its optimal ranges (Bajusz, 1969; Hoff & Riedesel, 1969; Reiner, 1968; Williams, 1966). Evolution of novel features in neural development may improve or lessen an animal's adaptive capabilities in the specific microenvironments in which it finds itself (De Beer, 1958). Such alterations, if not too drastic, may merely shift in small ways the animal's optimal range of transactions with specific environmental features.

Mutations that result in novel neural structures are selected if they improve the adaptive transactions of the population with specific features of available niche. Each genetically affected neurobiological aggregate, however small, must gain for the population's gene pool some potent piecemeal advantage for adaptive success to persist in successive populations. Hundreds and thousands of minute, genetically regulated features of developing neural circuits are probably affected, with the result that the overall aggregate of these individually trivial features is optimized (De Beer, 1958; Rosen, 1967). All such structures and functions that are selected must be mutually compatible to be sustained (Adolph, 1968; Goss, 1964). Selective pressures in an environment operate on the total aggregate of neural and nonneural structures, although it is by means of the functions of neural circuits themselves, and especially by means of their behavioral and homeostatic manifestations, that the organism accomplishes the specialized and specific adaptive transactions with fine features of the holoflux (Hinde, 1970; Marler & Hamilton, 1966).

Adaptive radiation, speciation, and the evolution of higher order taxa eventually occur by accumulation of neurobiological aggregates exhibiting new and different hierarchical levels of organization and function (Rensch, 1960, 1971;

Whyte *et al.*, 1969). This occurs very gradually and only over very long periods of time among similar groups that differ, at first unrecognizably, in small features. Different mammalian orders have thereby evolved and differentiated as basic new adaptive types (Mayr, 1963, 1969, 1970; Olson, 1965).

DISCUSSION

Multiplicity of Neurobiological Determinants in Evolution

In this review, I have tried to illustrate that both the causes and outcomes of brain evolution are multifold and complex. I believe that this emphasis on complexity and multiplicity is important in attempts to reach realistic conceptions of possible determinants of phylogenetic sequences. Awareness of neurobiological multiplicity opens new conceptual channels. It fosters search for more adequate and comprehensive explanations of neurobiological evolution (Dobzhansky, Hecht, & Steere, 1967). Indeed, answers to questions about the causes, course, and outcomes of brain evolution are revealed in many of the multiple microdetails of neurobiological aggregates. It seems especially important to realize that causes of neurobiological complexity are pluralistic. Because of the many different ways that neurobiological structures and processes interact and are interdependent both spatially and temporally, it is essential that explanatory constructs be developed that bear witness to the several hierarchical levels of interactive complexity that can be described.

Importance of Multidisciplinary Studies of Neurobiological Aggregates

Because of the inherent spatial complexity of neural circuits, the incredible spatiotemporal complexity of their discharge patterns and resultant overt behavioral sequences, and the intimate interdependence of neural and nonneural structures and functions, the comparative neural sciences have been obliged to utilize analytical tools from many scientific disciplines in the search for understanding of brain evolution (Minckler, 1972; Quarton, Melnechuk, & Schmitt, 1967). Studies of single-variable events are of limited value for understanding how brains work or how they have evolved. As in other branches of comparative biology, analysis of multivariate phenomena is best achieved if many measurable characters are studied by many methods. In the sections that follow, some special problems and values associated with such neurobiological methods are discussed.

Comparative Methods

This review has attempted to show that the paths by which neural circuits have evolved to result in the brains of living mammals are unknown. The value of the comparative approach to the understanding of brain evolution has only been partially realized. Living mammals represent phylogenetic experiments with variations on a common theme of neural organization. A more thorough, systematic, and comprehensive examination of these evolutionary experiments is necessary before conclusions can be reached regarding determinants and trends in brain evolution. More detailed comparative neuroanatomical, neurophysiological, and behavioral studies are also required in order that hypotheses regarding factors in brain evolution be adequately tested.

However, the choice of animal types used in comparative studies is important. Comparative statements cannot necessarily be considered to be valid if the animal series used consists only of mouse, mongrel, monkey, and man, although comparisons of such creatures are valuable in exploratory searches for differences. The concept of a representative animal is suspect. To be meaningful, systematic assays of several closely related species are required (Chiarelli, 1968; Kaufman & Rosenblum, 1966; King, 1968; Schrier, Harlow, & Stollnitz, 1965; Tuttle, 1972; Welker & Campos, 1963). Or, if comparisons among higher order taxa are desired, several species from each taxon should be examined. In any case, choice of animals studied should be guided by all available information regarding the degree of phylogenetic affinity of the species chosen (Crowson, 1970; Cullen, 1959; Hennig, 1966; Napier & Napier, 1970; Simpson, 1945, 1965). It is particularly useful to choose several closely related animals that are similar in most respects but that differ in their degree of specialization in some particular type of neural structure, circuit, overt behavior, niche occupation, etc. The greater the number of such differences among the species to be compared, the less specific and definitive can be the conclusions regarding the neural mechanisms underlying the behavioral or other adaptive effects.

A difficulty in all comparative studies relates to the use of hypotheticodeductive methods in assessing either whether a particular character is adaptive or how mutation, variation, and natural selection have permitted the evolution of neurobiological aggregates that have new adaptive roles (Williams, 1966). Speculation in this area is entertaining but not particularly useful if the hypotheses are not testable. In any study of species differences in neurobiological structures and functions it is necessary to demonstrate that an observed difference (e.g., an increase in axon diameter in a particular circuit or an increased number of corticospinal fibers making direct synaptic contact with motor horn cells) is indeed responsible for some specific adaptive effect. It is important to reemphasize that no two species differ in a single character. Differences are always multivariate. Species that are more distantly related differ in greater numbers of characters. Such differences are probably never to be found only within a single circuit, system, or type of tissue. This is so because of the interdependent nature of neurobiological aggregates.

The question of homology is important (Campbell & Hodos, 1970). In comparing neurons, components, circuits, and systems, it is essential, if the comparisons are to be meaningful, that only homologous characters be compared (Dillon

& Brauer, 1970). To compare visual synapses in the superior colliculus in one animal with auditory synapses at the inferior colliculus in another is trivial. Fortunately, among mammals, homologous cellular components are usually identifiable in almost all species studied. Some neural circuits are relatively large and robust, are located at similar positions in the brains of different mammals, and are therefore easily homologized. However, other circuits are delicate, relatively small, difficult to identify by neuroanatomical and neurophysiological methods, or have variable locations in different animals, and as a result are not yet easily homologized.

It is essential to remember that living species are end points of phylogenetic sequences. Ancestral features cannot easily be reckoned from living forms. Only paleoneurological studies can provide direct evidence of some general features of brain evolution (Edinger, 1948; Jerison, 1973; Radinsky, 1972). The courses and causes of fine details of neurobiological evolution of extinct creatures must be estimated from the brains of living animals by the direct means at our disposal.

Behavioral Methods

Differences in overt behavior are often the first clues that suggest species differences in neural circuitry. In attempting to diagnose brain differences from behavioral data it is important to realize that behavior typically is guided, either simultaneously or sequentially, by auditory, visual, gustatory, olfactory, kinesthetic, and nociceptive stimuli as well as by tactile, thermal, and proprioceptive stimuli. The potential spatiotemporal complexity of interactions of these inputs suggests the level of complexity of interaction of the several systems subserving these different perceptual systems. The character of naturally occurring action patterns requires accurately timed integration between motor and sensory systems (Granit, 1970; Eccles *et al.,* 1967; Roberts, 1967). It is a long-term goal to learn how these multiple systems interact to produce the organized actions that are known and to learn what circuit differences occur that produce species differences in transactions with the environment (Hebb, 1949; Lashley, 1949; Paillard, 1960).

Overall performance on many tasks is determined by multiple sensory cues, by complex chains of responses (Gibson, 1966; Montagu, 1971), and by multilevel central integrating mechanisms. The integral of all these transactions is what is expressed in overt behavior. Most mammals have evolved multimodal systems to deal with the polymodal stimulus features normally encountered in the environmental holoflux. These systems typically work in concert.

Determination of exactly what aspects of the environment an animal is responding to is a difficult task. Special tests are essential to delineate the specific cues or stimulus patterns that are related to each type of response pattern. Assessment of such stimulus determinants is difficult, requires much time and care, and consequently is rarely accomplished. However, such informa-

tion is essential in the search for valid species differences in behavioral repertoires. In studying an animal's responsiveness to stimuli, concern must be given
to all parameters of each stimulus so that the ranges and optima of adequate
stimuli can be specified. Most comparative studies of function of sensory circuits
have not employed appropriate discriminative tasks that require the subject to
attend to the subtle or complex features that single-neuron recording studies
suggest can be detected by a circuit (Welker, 1973). The test stimuli used are
often so gross or strong as to preclude differentiation between the possible
exquisite sensitivity of one sensory circuit and the more general sensitivity of
other different, but related, sensory circuits. Tests of absolute and relative
thresholds can reveal an animal's ability to discriminate specific aspects of
several parameters of stimulation in each of several stimulus modalities (Blough,
1961, 1966). However, stimulation procedures have been standardized in only a
few instances and consequently the results from different studies have not been
comparable. A more serious problem involves failure to test for thresholds and
for responsiveness to the full range of values for each of these stimulus parameters. It usually has been assumed that the particular behavioral responses used
as indicators of sensory thresholds, acuity, or discriminative ability are not very
important as long as they are relatively easy to perform (Bitterman, 1960, 1965;
Denny & Ratner, 1970; Warren, 1965). However, different species may not find
such tasks equally easy. More appropriate would be to test each type under
conditions that are optimal for that species. To examine different animals on the
same tests implicitly assumes that they are capable of occupying, and competing
for, the same niche, which is not true for species that are as different as are
squirrel monkeys and slow lorises, or raccoons and cats.

Initially, stimulus patterns that an animal is naturally exposed to can be used
to study its response repertoire (Ploog, 1971). However, stimulus patterns not
commonly encountered should also be explored because sensory circuits also
may be able to respond to them and therefore may suggest prospective capabilities for responding to niches for which the species is "preadapted" or that have
not yet been encountered (Williams, 1966). Ethological studies are especially
designed to assess such transactions (Bourliére, 1956; Breland & Breland, 1966;
Eisenberg, 1967; Ewer, 1968; Hinde, 1970; Kummer, 1970). These studies, more
than any others, are concerned with the entire behavioral repertoire of a species
and with exactly how various species differ with respect to their behavioral
transactions with specific aspects of the environment.

Most formal tests of behavior do not involve behavioral patterns that have a
high natural probability, or they restrict an animal's response capabilities to a
limited (often arbitrary) subset of its available repertoire of environmental
transactions. In testing for motor accuracy or strength, the details of which
specific actions are performed and in which sequence they occur are usually
recorded with much greater accuracy. Several motion analysis techniques are
used to make such measurements (Alexander, 1968; Bernstein, 1967; Doty &

Bosma, 1956). However, the delicate, precisely timed spatiotemporal response patterns that characterize most discriminative behaviors are rarely noted.

It is important to keep in mind that overt behavioral sequences are isofunctional only with firing patterns of motor neurons. Consequently, only certain aspects of the several available parameters of movement can provide clues as to the operation of central circuits (see Figures 3 and 4 and section on Behavioral Evolution). However, many central neural operations probably are purely covert, having no measureable manifestation in any parameter of behavior. Such central processes can only be hypothesized from studies of long-term sequences or patterns of overt behavior in specified situations.

Different behavioral tests are used for different purposes. Attempting to assess species differences in behavior by testing several different animals on the same or similar tasks is often a fruitless venture. This is especially true of tests that record only an animal's success or failure scores. The difficulty lies in the possibility that different animals (even of the same species) may solve a problem in different ways using different combinations of neural circuits. Such differences, especially between species, may be subtle yet, from the viewpoint of phylogeny, may reflect important adaptive changes in circuit function that are not manifest in crude composite scores. Indeed, if animals have different repertoires or optimal modes of dealing with the environment, they cannot be expected to react to the same situation in the same way. For a long time the bulk of behavior study has been devoted to phenomena associated with learning, motivation, and reinforcement. The time seems ripe to examine species differences in some of the more subtle, yet rapid and complex, response transactions with the environment. Such behavioral sequences are the normal modes of doing business.

Methods for Circuit Analysis

Accurate identification and description of the structural and functional properties of synapses, neurons, conglomerates, assemblies, thickets, components, circuits, and systems are central goals of the neural sciences. All neuroscience disciplines are hard at work tracing neural circuits. Assessing how neural circuits function cannot proceed far if their structural and biochemical features have not been fully delineated. The enormous complexity of connectivity of mammalian neural circuits, their extensive distribution in the brain, and the relatively small size and compactness of their functional elements are real impediments to study of their details. The relatively inelegant methods of the past are being replaced steadily by more refined micromethods capable of assessing finer structures and functions of neural circuits.

Neuroanatomical mapping methods, revealing the fine structure of circuits, have great resolving power and are needed to complement detailed microelectrode recording methods (Nauta & Ebbesson, 1970; Peters et al., 1970; Ralston,

1969). Lesion-degeneration methods trace circuit connections (Szentágothai, 1965), but in most studies the lesions are large and destroy afferents, interneurons, and efferents, typically by the tens or hundreds of thousands. Large cortical lesions may destroy millions of neurons and cannot provide detailed information about microcircuits, especially of those within the damaged component. Tiny lesions produced by various methods are essential if specificity of circuit details is to be revealed. For accuracy, such lesions should be placed in cellular components rather than in fiber tracts. Lesions of fiber tracts are out of the question when attempting to determine fine details of circuit connectivity because fiber tracts usually contain admixtures of axons from several different circuits, or of several different functional types. Autoradiographic tract-tracing methods are especially useful (Cowan, Gottlieb, Hendrickson, Price, & Woolsey, 1972) because only orthodromic pathways seem to be labeled, but they also may be limited by the large number of neurons typically affected. Antidromic paths can be delineated by means of the method involving retrograde transport of horseradish peroxidase (LaVail, Winston, & Tish, 1973). Electron microscopic (EM) methods are indispensible for determining the kinds and locations of synaptic, dendritic, axonal, and perikaryal structures in normal tissue (Pappas & Purpura, 1972; Peters *et al.,* 1970). To specify circuits in fine-structural detail, the exact locations within a neural component from which the EM sample is taken must be known. Special staining methods, such as those involving selective impregnation of entire neurons, reveal sizes, shapes, locations, and orientations of cell bodies and dendritic arborizations. Together with fine-lesion methods and special staining of selected inputs, EM studies permit assessment of which structures derive from each known input source. They also suggest the types and extent of possible synaptic interactions. Tissue culture methods can also provide valuable information regarding cell movements, sequences of differentiation, intercellular interactions, and the role of chemicals in metabolism, growth, and differentiation (Ruščák & Ruščáková, 1971; Sobkowicz, Guillery, & Bornstein, 1968). The several types of neuroanatomical studies have led to valuable hypotheses regarding circuit functions within the nervous system. All these neuroanatomical data suggest hypotheses regarding the class of neuroelectric actions produced within synaptic conglomerates and assemblies. Such data, combined with neurophysiological mapping of the same circuit elements, can validate structural–functional hypotheses.

Comparative neuroanatomical studies increasingly require that quantitative methods be employed in all phases (Dubin, 1970; Mendell & Henneman, 1971). Several procedures have been developed to determine allometric relationships of various brain components and the absolute and relative number of neurons within a component (Abercrombie, 1946; Dornfeld, Slater, & Scheffé, 1942; Gould, 1966; Jerison, 1973). As it becomes more and more important to count and measure synaptic profiles, axon diameters, proportions of different types of synaptic structures, cell types, etc. (Ralston, 1969), several more precise quantitative methods have been developed. These include computer procedures for scanning populations of neural structures, computation of statistics of the major measureable parameters, and display of the results of these analyses. In addition, it is now relevant and important to determine the functional significance of these parameters of quantitative neuroanatomy. As mentioned above, only comparison of homologous portions of circuits is of value in comparative studies.

Biochemical mapping of locations within the brain of enzymes, proteins, lipids, endocrines, neurohumors, transmitter substances, and other species of molecules is an active feature of neurochemical research. The methods and applications of numerous chemical techniques for labeling, staining, and analyzing chemicals associated with even the smallest neural structures are increasing in number and sophistication (Albers *et al.*, 1972; Bogoch, 1968; Carlson, 1968; DeRobertis, 1964; Frieden & Lipner, 1971; Gorbman & Bern, 1962; Hanly, 1970; Itoh, 1968; Kety & Elkes, 1961; Martini & Meites, 1970; Minkowski, 1967; Palladin, 1964; Phillis, 1970; Richter, 1964; Ruščák & Ruščáková, 1971; Scharrer, 1959; Torre, 1972; Triggle, 1971). Knowledge of the varieties and exact distribution of chemical functions within the brain is essential for an adequate comprehension of how the multifold circuits of the brain develop and diversify in highly specific ways during growth, maturation, regeneration, degeneration, and aging (Birren *et al.*, 1959; Kohn, 1971). Neurochemical studies increasingly are able to identify exact locations of the many different kinds of biochemical activities involved in synaptic activity, in neurohumoral production and receptivity, and in regional specialized neurochemical differentiation of neural components. Systematic comparative studies of these biochemical features are now possible but are still rare.

Neurophysiological mapping of sensory circuits consists of the rational utilization of the following set of methods: (a) microelectrode recording or stimulation; (b) systematic sampling of the population of neurons in the component being examined; (c) controlled, quantified, threshold natural stimulation of receptor surfaces, or electrical stimulation (using microelectrodes) of neurons or groups of neurons in specific circuit components; (d) marking the recording loci; and (e) identifying electrode locations histologically. Thoroughness of sampling is required for both recording and stimulating locations (Johnson *et al.*, 1968; Welker, 1973). Such mapping is essential in any attempt to (a) delineate accurately and thoroughly the connections between components, (b) specify exact locations of inputs and outputs from a component, and (c) elucidate details of the microcircuitry or of excitatory and inhibitory neuroelectric events within neuronal assemblies, components, and circuits.

Analysis of neuroelectric data is a major problem in neurophysiological studies. Neural spikes, slow waves, and other particulate electrical events can be quickly and easily recorded. It is essential that the great mass of this neuroelectric data be counted, measured, and analyzed by quantitative methods. Because of the high frequency and complex spatiotemporal sequences of neuroelectric patterns, the use of computer techniques has been crucial, and modeling methods are being used increasingly to test hypotheses regarding neuroelectric functions of neural structures (Bogoch, 1968; Deutsch, 1967; Fogel *et al.*, 1966; Gelfand, Gurfinkel, Fomin, & Tsetlin, 1971; Harmon & Lewis, 1966; Leibovic, 1969; Machin, 1964; Ramsey, 1967; Reiner, 1968; Reiss, 1964; Sayre & Crosson, 1963; Stark, 1968). Sampling problems are enormous. It seems that the statistical properties of relatively large populations of the neural elements within each type of neuroanatomical structure (see Figure 5) are likely to constitute the functionally most relevant data (Burns, 1968; Caianiello, 1968; Pribram, 1971). Consequently, it is essential that statistically adequate samples of neuroelectric phenomena be taken from each population of neuroanatomical structures being studied (Griffith, 1971). The application of pattern analysis (Grusser & Klinke, 1971), systems theory (Mesarović, 1968), and optimizing theory (Rosen, 1967) is becoming increasingly necessary to improving any conceptual grasp of the real spatiotemporal domains of neuroelectric events within the tangled circuits of mammalian brains.

Electrical stimulation methods are appropriate for identifying and analyzing certain aspects of circuit functions. Comparative studies using electrical stimulation of cortex have determined the location of several motor areas in different animals (Woolsey, 1958, 1964). Electrical stimulation of sensory areas has shown that efferents from them also activate circuits that produce movements. Such studies reveal that animals exhibiting specialized behaviors have unusually large motor areas that, when stimulated electrically, activate movements used in those behaviors (Hardin *et al.*, 1968; Woolsey, 1958).

Electrical stimulation of receptors or of central and peripheral components or fiber tracts also provides accurate timing to the impulses projected to target neurons being studied. By using appropriate electrical stimulus programs, conduction velocity, latency, utilization time, following frequency, excitatory and inhibitory interactions, and directionality of conduction can be determined. Systematic mapping of circuit components and fiber tracts with the stimulating electrode can locate neurons capable of activating those from which recordings are being taken.

However, electrical stimulation has a few serious drawbacks. When suprathreshold stimulation through macroelectrodes is used (as is common), axons, cell bodies, dendrites and synapses may be activated simultaneously and unselectively and probably in unnatural sequences, patterns and directions of conduction. Moreover, the exact volume of tissue excited is unknown and, with the macroelectrodes usually used, a desired effect may not be produced until several cubic millimeters of tissue, containing tens or hundreds of thousands of neurons, have been activated. Such a barrage may "jam" the normal functional capabilities of a circuit. When either peripheral nerves or central fiber tracts are stimulated by electrical pulses, fibers with many different functional connections or involved in several distinctly different circuits all may be activated (Welker, 1973). In addition, they are activated synchronously and, for low stimulation rates, in a temporal pattern like that of the electrical stimulus, which differs from that of most naturally occurring stimuli. As a result, central interactions may be set up or synapses traversed that are ordinarily improbable or rare.

Electrical stimulation via either extracellular or intracellular microelectrodes, or dissection of nerves so that only one functionally coherent group is stimulated, may avoid some of these difficulties. However, because of the high specificity of most circuit connections, locating distant cells that respond to such focal stimuli becomes difficult if the number of neurons activated is too small. Yet, if specificity of functional connections exists, microelectrode stimulation should be used whenever possible. Many studies have utilized rather gross macroelectrode electrical stimulation methods and consequently some of the conclusions reached must be viewed with caution. Also, electrical stimulation of receptors or nerves, no matter how precisely applied, cannot provide definitive answers as to how the nervous system detects and processes information from the environment.

Diagnosing Circuit Functions by Brain Lesion Methods

Brain lesions have been used for a long time to assess both brain–behavioral mechanisms and circuit connectivity. As more has been learned about the high degree of specificity of neural circuits, lesions have been made smaller and confined to more specific, anatomically or functionally distinct regions. There is a definite limit, however, to the degree of specificity of connectivity or function that can be assessed by brain lesions, however small. This is so because any lesion of a neural component destroys at once the entire subcomponent circuitry. It obliterates together the several cell types and their differentiated connections. Also, it usually destroys adjacent cells that may have different inputs and outputs. In tracing axonal degeneration, either retrograde or anterograde, resulting from even small lesions, such finer details of connectivity are obliterated.

From a functional point of view, the ablation of a component produces an animal divested not only of that component and its outputs, but probably also of the normal transactions of other components that it normally influences by its outputs. In addition, the components projecting their outputs into the destroyed component can no longer exert their normal effects. As a consequence, their internal operations, as well as their outputs to other circuits yet intact, also may be abnormal. Thus, the destruction of a single component likely has far-reaching effects, not only on the other components in its circuit but also on those other circuits that normally interact with the tampered circuit.

Other important effects of lesions also are likely to occur. For example, because the effects of many brain lesions become manifest as defects in motor control or perceptual ability, an animal usually tends to compensate for the loss in several ways, by spared mechanisms. For example, locomotor patterns and postures may be readjusted merely to prevent falling or loss of balance. Likewise, an animal may look, listen to, and touch the environment in different ways than before so as to utilize those remaining stimulus cues that activate circuits not directly affected by the lesion. Such changes, if subtle, may escape notice.

Clearly, functional disturbances produced by simple, even relatively small, lesions can have far-reaching effects. Any observed motor, perceptual, cognitive, or learning defect cannot be assumed to indicate directly the function of the component that has been destroyed. Such functions are distributed throughout many circuits of the brain. The problem of interpretation is even worsened if a lesion destroys more than a single functional subdivision.

This is not to say that lesion studies are to no avail. A carefully prepared series of selective partial ablations of different sorts may provide information of value, especially if the ablations result in distinctly different effects. However, the assumptions and interpretations that are made regarding the neural mechanisms underlying these defects must also consider several other sets of neurobiological data regarding the circuits and systems affected.

Chronic Neurophysiological Recording Methods

How neural circuits function during thinking and behaving can be directly assessed only by recording from single neurons or clusters of neurons in unanesthetized animals engaged in these activities (Brooks, Adrien, & Dykes, 1972; DeLong, 1971; Evarts, 1968; Funkenstein, Nelson, Winter, Wollberg, & Newman, 1972; Luschei, Garthwaite, & Armstrong, 1971; Schiller & Koerner, 1971; Thach, 1968). Although the sampling problem is enormous and the data analysis required is complex and difficult, this method, coupled with appropriate behavioral testing and neuroanatomical methods, is the only way to obtain real-time data regarding the spatiotemporal action patterns of interacting neural circuits. As technology in this area advances and experimental designs become more sophisticated, chronic recording studies potentially become more able to provide direct tests of hypotheses regarding central mechanisms of thought and action that have long been of interest to psychologists and behaviorists. Comparison of homologous circuits in different animals faced with different problems and environments should clarify the different ways that they organize their perceptual–behavioral transactions with the environment.

The Search for Neurobiological Mechanisms

The search for explanations of perceptual, cognitive, and behavioral phenomena leads to conceptions of mechanisms in terms of the spatiotemporal features of active neural circuits. There have been many reviews of the literature and symposia on brain mechanisms: (Altman, 1966; Glassman, 1967; Harlow & Woolsey, 1958; Hinde, 1970; Horn & Hinde, 1970; Karczmar & Eccles, 1972; Marler & Hamilton, 1966; Morgane, 1969; Rasmussen & Windle, 1960; Rupp, 1968; Smith, 1970; Whalen, Thompson, Verzeano, & Weinberger, 1970; Yahr & Purpura, 1967; Zotterman, 1967). Neurobiological mechanisms have neuro-anatomical, neuroelectric, neurobiochemical, and nonneural anatomical and physiological characteristics; but they also exhibit behavioral and mental manifestations. Information regarding any of these features sheds light on neurobiological mechanisms of an animal's transactions with the holoflux of environmental energies. The importance, from a comparative point of view, of seeking to understand underlying mechanisms is that their differential operation in different animals determines species differences in niche selection, occupation, and adaptability to environmental changes.

From neuroanatomical, neurophysiological, and brain–behavioral studies, it is necessary to conclude that circuits that deal with perception of, and behavior toward, specific features of an environment are distributed throughout many levels of the entire neuraxis not in the diffuse manner of a network, or in overlapping fashion, but in specific, highly organized, articulated ways (McIlwain & Fields, 1971; Sprague, 1966). The fact that different circuits typically are closely packed, or that their interconnections cross or intermingle with one another, is probably one reason that relatively large brain lesions, or lesions across the boundaries of two or more different circuits, have led to the misleading conception of "mass action" in brain function (Lashley, 1949, 1951). Because most neural circuits and systems are distributed throughout many levels of the brain, any function, such as vision, pattern perception, learning, movement control, learning, or memory, cannot be localized to a single component or nuclear mass. It is difficult, at the present state of knowledge, to know accurately which processing functions a single component performs in a simple sensory or motor circuit. Interacting neural mechanisms are unbelievably complex and conceptions of their operational features can only be roughed out at present.

Role of the Brain in Evolution

The brain is the grand instrument for detection, abstraction, exploration, and utilization of internal and external realities. It monitors the energy holoflux and guides behavior, thought, and internal physiological processes that have evolved to carry out transactions with the holoflux. Neural mechanisms are incredibly numerous and diversified. They are so integrated that the total organism functions as a smoothly working whole. However, our conceptions of neural mechanisms are certainly overly simplified. In adult animals, neural circuits may seem distinct from other tissues, but they exist as they do only because they have been built, phylogenetically and ontogenetically, together with all the other bodily tissues by which they are sustained and to which they are always inextricably linked. No one portion of an organism easily can be shown to have played a greater role in evolution than another. All parts have coevolved and are coadapted. The most wonderfully complex brains have been built only with permission and collaboration of all the other tissues. Natural selection operates on the population of individual organisms at all phases of their ontogeny. The particular populations that survive are those in which all tissues are optimally or best suited for conducting survival tactics with the holoflux.

CONCLUSIONS

It has been the preoccupation of this essay to outline and review several rational conceptions regarding factors in brain evolution. I have emphasized the organized diversity and multifold patterned complexity of neurobiological structures and processes. The known phenomena exhibit numerous facets when illumined by a broad battery of research methods and strategies. Much ground has been covered in this review. The narrative has pursued many ramifications of this vast subject. It probably has rambled excessively in some places and may seem too cryptic and condensed in others.

In almost all respects, I have attempted to show that the truth regarding brain evolution is still beyond the horizons of our vision and lies yet veiled by the limitations of our rational verbal concepts and exploratory tools. Unconscious inference and intuition are still preeminent modes of providing understanding of basic truths. However, these truths are often provisional and ephemeral figments limited by the stylistic logic of rational descriptions, operational definitions, and explanations.

Perhaps an insurmountable problem is that the incredibly vast and diversified network of interrelated, perpetually changing neurobiological structures and funtions are of such a nature as to be beyond the ken of the thinking, perceiving, analytical, and explaining machinery of human brains. The logical operations by which human neural circuits are constrained to function surely do not provide adequate or faithful analogs of the spatiotemporal flux of phenomena that we so pretentiously wish to detect and divine. For example, only a few of the features of these natural phenomena may be sufficiently discrete or optimally suited to be transmuted accurately into our verbal and symbolic language. Consequently, the bulk of reality regarding the dynamic flow of life processes may forever remain hidden behind a veil of ignorance imposed by these intrinsic limitations. If new structures were to evolve to conceive more accurately what now lies still hidden, even then additional impalpable mysteries regarding the functions of these newest structures would concurrently have been created.

It is a manifestation of our biological heritage that we have the vanity, pride, and subjective certainty to believe that we can know all, or that our ability to explain and comprehend is boundless, requiring only further technological improvements, money, numbers of investigators, and time. We are, perhaps, incurably enthralled by the magnitude of our wit, wisdom, and cognitive development. We often tend to forget that as individuals we are all born in essential ignorance of the great wealth of accumulated knowledge; that we are in many ways constrained from ascending those pinnacles of wisdom of our dreams by a powerful endowment with ageless and insistent instincts for adaptive survival; and that for each one of us the channel toward the simple basic truths is clogged with an endless succession of mistakes, misconceptions, wrong turns, oversights, self-deceptions, and blunders, as well as old age, senility and natural death.

We are constrained by imponderable realities to guess what courses brain evolution has taken, and we cannot know in which directions it will go. Future directions in understanding and explaining comparative neurobiological phenomena are probably unknown. As this review has suggested, we are even at pains to understand life adequately in the living present. New directions of search commonly follow unexpected paths, perceived in advance only dimly or not at all. Fads of methodology and conception are always dominant forces, and the vagaries of human emotional and social motives to pursue certain lines of scientific investigation, are strong influences. However, questions regarding where all this search and accumulated wisdom is to lead need not be of major concern, for the phenomena themselves constantly reveal novel directions of inquiry. The only possible step toward new wisdom must always be the next one, leading directly from the one just taken. Indeed, the best guides are in the data themselves. Luckily, the excitement engendered by man's natural curiosity and by its simple realization can be personal fulfillment enough and provides sufficient reward for continued and renewed exploration of the myriad untold mysteries that lie ahead.

ACKNOWLEDGMENTS

This project has been supported by U.S. Public Health Service Grants M-2786, NS-3249, and NS-6225. I wish to thank Drs. R. W. Guillery, S. E. Kornguth, K. H. Pribram, E. N. Willmer, and especially W. G. Reeder for their helpful comments on the manuscript; Shirley Hunsaker for her invaluable assistance in photography and especially her patience and careful and attentive proofreading of the manuscript; Terry Stewart for preparing photographic materials; Betty Bablitch, Gloria Postel, Maree Giese, and Barbara Caddock for artful typing of the manuscript; and Dee Urban for her beautiful artistic renditions of the neural structures and functions reviewed in this chapter. I apologize to those authors whose relevant work is not directly cited because it was included in an edited volume that also contained many other relevant papers. This was often necessary to keep the bibliography to an acceptable length, or because persistent sloth, coupled with approaching deadlines, necessitated a too hasty preparation of the final draft.

REFERENCES

Abercrombie, M. Estimation of nuclear population from microtome sections, *Journal of Anatomy*, 1946, **94**, 239–247.

Adám, G. *Interception and behavior: An experimental study*. Budapest: Publishing House of the Hungarian Academy of Sciences, 1967.

Adám, G. (Ed.). *Biology of memory*. New York: Plenum, 1971.

Adelman, W. J., Jr. (Ed.). *Biophysics and physiology of excitable membranes*. Princeton, New Jersey: Van Nostrand-Reinhold, 1971.

Adolph, E. F. *Origins of physiological regulations*. New York: Academic Press, 1968.

Aidley, D. J. *The physiology of excitable cells*. Cambridge: Cambridge University Press, 1971.

Albers, R. W., Siegal, G. J., Katzman, R., & Agranoff, B. W. (Eds.), *Basic neurochemistry*. Boston: Little, Brown, 1972.

Albert, D. J. Memory in mammals: evidence for a system involving nuclear ribonucleic acid. *Neuropsychologia*, 1966, **4**, 79–92.

Alexander, R. McN. *Animal mechanics*. Seattle: University of Washington Press, 1968.

Alland, A., Jr. *Evolution and human behavior*. Garden City, N. Y.: The Natural History Press, 1967.

Allman, J. M., & Kaas, J. H. A representation of the visual field in the caudal third of the middle temporal gyrus of the owl monkey (*Aotus Trivirgatus*). *Brain Research*, 1971, **31**, 85–105.

Altman, J. *Organic foundations of animal behavior*. New York: Holt, Rinehart & Winston, 1966.

Altman, J. Autoradiographic and histological studies of postnatal neurogenesis. III Dating the time of production and onset of differentiation of cerebellar microneurons in rats. *Journal of Comparative Neurology*, 1969, **136**, 269–294.

Altman, J., & Das, G. D. Autoradiographic examination of the effects of enriched environment on the rate of glial multiplication in the adult rat brain. *Nature (London)*, 1964, **204**, 1161–1163.

Andersen, H. T. (Ed.). *The biology of marine mammals*. New York: Academic Press, 1969.

Andrew, R. J. The situations that evoke vocalization in primates. *Annals of the New York Academy of Science*, 1962, **102**, 296–315.

Andrew, R. J. The origin and evolution of the calls and facial expressions of the primates. *Behavior*, 1963, **20**, 1–109.   (a)

Andrew, R. J. Trends apparent in the evolution of vocalization in the old world monkeys and apes. *Symposium of the Zoological Society of London*, 1963, **10**, 89–101.   (b)

Anfinsen, C. B. *The molecular basis of evolution*. New York: Wiley, 1959.

Angeletti, P., & Vigneti, E. Assay of nerve growth factor (NGF) in subcellular fractions of peripheral tissues by micro complement fixation. *Brain Research*, 1971, **33**, 601–604.

Ansell, G. B., & Bradley, P. B. (Eds.). *Macromolecules and behaviour*. London: Macmillan, 1973.

Ariëns-Kappers, C. U., Huber, G. C., & Crosby, E. C. *The comparative anatomy of the nervous system of vertebrates, including man*. Reprint of original 1936 edition, 3 vol. New York: Hafner Publishing Co., 1960.

Aschoff, J. Diurnal rhythms. *Annual Review of Physiology*, 1963, **25**, 581–600.

Asimov, I. *The Human Brain.* Boston: Houghton-Mifflin, 1963.

Auerbach, A. A. Transmitter release at chemical synapses. In G. D. Pappas & D. P. Purpura (Eds.), *Structure and function of synapses.* New York: Raven Press, 1972.

Bajusz, E. (Ed.). *Physiology and pathology of adaptation mechanisms.* Oxford: Pergamon, 1969.

Balázs, R. Biochemical effects of thyroid hormones in the developing brain. In D. C. Pease (Ed.), *Cellular aspects of neural growth and differentiation.* Berkeley: University of California Press, 1971.

Barnett, S. A. The ontogeny of behavior and the concept of instinct. In A. G. Karczmar & J. C. Eccles (Eds.), *Brain and human behavior.* Berlin: Springer-Verlag, 1972.

Barnothy, M. F. (Ed.). *Biological effects of magnetic fields.* New York: Plenum, 1964.

Barondes, S. H. Relationship of biological regulatory mechanisms to learning and memory. *Nature* (London), 1965, **205**, 18–21.

Barondes, S. H., & Cohen, H. D. Puromycin effect on successive phases of memory storage. *Science*, 1966, **151**, 594–595.

Barrington, E. J. W. *Hormones and evolution.* Princeton, New Jersey: D. Van Nostrand, 1964.

Barry, J. M. *Molecular biology: Genes and the chemical control of living cells.* Englewood Cliffs, New Jersey: Prentice-Hall, 1964.

Barth, L. J. *Development, selected topics.* Reading, Massachusetts: Addison-Wesley, 1964.

Beach, F. A. Experimental investigations of species-specific behavior. *American Psychologist*, 1960, **15**, 1–18.

Beidler, L. M., Fishman, I. Y., & Hardiman, C. W. Species differences in taste responses. *American Journal of Physiology*, 1955, **181**, 235–239.

Bennett, M. V. L. A comparison of electrically and chemically mediated synaptic transmission. In G. D. Pappas & D. P. Purpua (Eds.), *Structure and function of synapses.* New York: Raven Press, 1972. Pp. 221–256.

Beritoff, J. S. *Neural mechanisms of higher vertebrate behavior.* Boston: Little, Brown, 1965.

Berlyne, D. E. *Conflict, arousal and curiosity.* New York: McGraw-Hill, 1960.

Berman, A. L. *The brain stem of the cat.* Madison, Wisc.: The University of Wisconsin Press, 1968.

Bern, H. A. The secretory neuron as a doubly specialized cell. In D. Mazia & A. Tyler (Eds.), *General physiology of cell specializations.* New York: McGraw-Hill, 1963.

Bernhard, S. *The structure and function of enzymes.* New York: Benjamin, 1968.

Bernstein, J. J., & Goodman, D. C. (Eds.) Neuromorphological plasticity. Symposium at the 1972 Cajal Club Meeting. *Brain, Behavior, Evolution,* 1973, **8**, 5–164.

Bernstein, N. *The coordination and regulation of movements.* London: Pergamon, 1967.

Birren, J. E., Imus, H. A., & Windle, W. F. (Eds.). *The process of aging in the nervous system.* Oxford: Blackwell Scientific Publications, 1959.

Bishop, G. H. Natural history of the nerve impulse. *Physiological Reviews*, 1956, **36**, 376–399.

Bishop, G. H. The relation between fiber size and sensory modality: phylogenetic implications of the afferent innervations of cortex. *Journal of Nervous & Mental Diseases*, 1959, **128**, 89–114.

Bishop, G. H. & Smith, J. M. The sizes of nerve fibers supplying cerebral cortex. *Experimental Neurology*, 1964, **9**, 483–501.

Bitterman, M. E. Toward a comparative psychology of learning. *American Psychologist*, 1960, **15**, 704–712.

Bitterman, M. E. The evolution of intelligence. *Scientific American*, 1965, **212**(1), 92–100.

Black, P. (Ed.). *Physiological correlates of emotion.* New York: Academic Press, 1970.

Bleier, R. The relations of ependyma to neurons and capillaries in the hypothalamus: a Golgi-Cox study. *Journal of Comparative Neurology*, 1971, **142**, 439–464.

Bligh, J. The thermosensitivity of the hypothalamus and thermoregulation in mammals. *Biological Reviews,* 1966, **41,** 317–367.

Blinkov, S. M., & Glezer, I. I. *The human brain in figures and tables.* New York: Basic Books and Plenum Press, 1968.

Bloom, F. E. The formation of synaptic junctions in developing rat brain. In G. D. Pappas & D. P. Purpura (Eds.), *Structure and function of synapses.* New York: Raven Press, 1972.

Blough, D. S. Experiments in animal psychophysics. *Scientific American,* 1961, **205,** 113–122.

Blough, D. S. The study of animal sensory processes by operant methods. In W. K. Honig (Ed.), *Operant behavior.* New York: Appleton-Century-Crofts, 1966.

Bodian, D. The generalized vertebrate neuron. *Science,* 1962, **137,** 323–326.

Bodian, D. Synaptic diversity and characterization by electron microscopy. In G. D. Pappas & D. P. Purpura (Eds.), *Structure and function of synapses.* New York: Raven Press, 1972.

Bogoch, S. *The biochemistry of memory: with an inquiry into the function of the brain mucoids.* New York: Oxford University Press, 1968.

Bourlière, F. *The natural history of mammals.* (3rd ed.) New York: Knopf, 1956.

Brazier, M. A. B. *The electrical activity of the nervous system.* Baltimore, Williams & Wilkins, 1968.

Brazier, M. A. B. (Ed.). *The interneuron.* Berkeley: University of California Press, 1969.

Breland, K., & Breland, M. *Animal behavior.* New York: The Macmillan Co., 1966.

Brightman, M. W., & Reese, T. S. Junctions between intimately apposed cell membranes in the vertebrate brain. *Journal of Cell Biology,* 1969, **40,** 648–677.

Brooks, V. B., Adrien, J., & Dykes, R. W. Task-related discharge of neurons in motor cortex and effects of dentate cooling. *Brain Research,* 1972, **40,** 85–88.

Bryson, V., & Vogel, H. J. (Eds.). *Evolving genes and proteins.* New York: Academic Press, 1965.

Bullock, T. H., & Horridge, G. A. *Structure and function in the nervous systems of invertebrates.* San Francisco: Freeman, 1965.

Bullough, W. S. *The evolution of differentiation.* New York: Academic Press, 1967.

Burn, J. H. *The autonomic nervous system.* Oxford: Blackwell Scientific Publ., 1963.

Burns, B. D. *The mammalian cerebral cortex.* London: Edward Arnold, 1958.

Burns, B. D. *The uncertain nervous system.* London: Edward Arnold, 1968.

Busnel, R.-G. (Ed.). *Acoustic behavior of animals.* Amsterdam: Elsevier, 1963.

Byrne, W. L. (Ed.). *Molecular approaches to learning and memory.* New York: Academic Press, 1970.

Cabral, R. J., & Johnson, J. I., Jr. The organization of mechanoreceptive projections in the ventrobasal thalamus of sheep. *Journal of Comparative Neurology,* 1971, **141,** 17–35.

Caianiello, E. R. (Ed.). *Neural networks.* New York: Springer-Verlag, 1968.

Cairns, R. B. Attachment behavior of mammals. *Psychological Reviews,* 1966, **73,** 409–426.

Campbell, C. B. G. Evolutionary patterns in mammalian diencephalic visual nuclei and their fiber connections. *Brain, Behavior & Evolution,* 1972, **6,** 218–236.

Campbell, C. B. G., & Hodos, W. The concept of homology and the evolution of the nervous system. *Brain, Behavior, & Evolution,* 1970, **3,** 353–367.

Campos, G. B., & Welker, W. I. Physiological and anatomical comparisons between brains of Capybara (*Hydrochoerus*) and guinea pig (*Cavia*). *The Physiologist,* 1960, 3(3), 35.

Campos-Ortega, J. A., & Hayhow, W. R. A new lamination pattern in the lateral geniculate nucleus of primates. *Brain Research,* 1970, **20,** 335–339.

Carlson, F. D. (Ed.). *Physiological and biochemical aspects of nervous integration.* Englewood Cliffs, N. J.: Prentice-Hall, 1968.

Carpenter, J. A. Species differences in taste preferences. *Journal of Comparative Physiology & Psychology,* 1956, **49,** 139–144.

Caspar, D. L. D. Design and assembly of organized biological structures. In T. Hayashi & A. G. Szent-Györgyi (Eds.), *Molecular architecture in cell physiology.* Englewood Cliffs, N. J.: Prentice-Hall, 1966.

Caspari, E. Genetic basis of behavior. In A. Roe and G. G. Simpson (Eds.), *Behavior and evolution.* New Haven, Connecticut: Yale University Press, 1958.

Cauna, N. Cholinesterase activity in cutaneous receptors of man and of some quadrupeds (Histochemistry of Cholinesterase Symposium, Basel). *Bibliotheca Anatomica (Karger),* 1961, **2**, 128–138.

Causey, G. *The cell of schwann.* Edinburgh: E. and S. Livingstone, 1960.

Caviness, V. S., Jr., & Sidman, R. L. Time of origin of corresponding cell classes in the cerebral cortex of normal and reeler mutant mice: an autoradiographic analysis. *Journal of Comparative Neurology,* 1973, **148**, 141–152.

Chiarelli, B. *Taxonomy and phylogeny of old world primates with references to the origin of man.* Torino, Italy: Rosenberg and Sellier, 1968.

Clayton, R. K. *Light and living matter: A guide to the study of photobiology, Vol. 1. The physical part.* New York: McGraw-Hill, 1970.

Clayton, R. K. *Light and living matter: A guide to the study of photobiology. Vol. 2: The biological part.* New York: McGraw-Hill, 1971.

Close, R. Neural influences on physiological properties of fast and slow limb muscles. In R. J. Podolsky (Ed.), *Contractility of muscle cells and related processes.* Englewood Cliffs, N. J.: Prentice-Hall, 1971.

Cloudsley-Thompson, J. L. *Rhythmic activity in animal physiology and behavior.* New York: Academic Press, 1961.

Cobb, S. Brain size. *Archives of Neurology* (Chicago), 1965, **12**, 555–561.

Cohen, M. I. How respiratory rhythm originates: evidence from discharge patterns of brainstem respiratory neurones. In R. Porter (Ed.), *Breathing: Hering-Breuer centenary symposium.* London: Churchill, 1970.

Colbert, E. H. *Evolution of the vertebrates.* New York: Science Editions, 1961.

Coleman, M. S., Pfingst, B., Wilson, J. E., & Glassman, E. Brain function and macromolecules. VIII. Uridine incorporation into brain polysomes of hypophysectomized rats and ovariectomized mice during avoidance conditioning. *Brain Research,* 1971, **26**, 349–360.

Coleman, P. D., & Riesen, A. H. Environmental effects on cortical dendritic fields. *Journal of Anatomy,* 1968, **102**, 363–374.

Colonnier, M. Synaptic patterns on different cell types in the different laminae of the cat visual cortex. An electron microscopic study. *Brain Research,* 1968, **9**, 268–287.

Compton, R. W. Morphological, physiological and behavioral studies of the facial musculature of the Coati *(Nasua). Brain, Behavior, & Evolution,* 1973, **7**, 85–126.

Connolly, C. J. *External morphology of the primate brain.* Springfield, Ill.: Charles C Thomas, 1950.

Corning, W. C., & Balaban, M. (Eds.). *The mind: Biological approaches to its functions.* New York: John Wiley and Sons, 1968.

Cornwall, I. W. *Bones for the archaeologist.* London: Phoenix House, 1956.

Cort, J. H. *Electrolytes, fluid dynamics and the nervous system.* New York: Academic Press, 1965.

Coupland, R. L. The innervation of pancreas of the rat, cat, and rabbit as revealed by the cholinesterase technique. *Journal of Anatomy,* 1958, **92**, 143–149.

Cowan, W. M., Gottlieb, D. I., Hendrickson, A. E., Price, J. L., & Woolsey, T. A. The autoradiographic demonstration of axonal connections in the central nervous system. *Brain Research,* 1972, **37**, 21–51.

Creel, D. J., & Giolli, R. A. Retinogeniculostriate projections in guinea pigs: albino and pigmented strains compared. *Experimental Neurology,* 1972, **36**, 411–425.

Cross, B. A. The hypothalamus in mammalian homeostasis. *Symposia of the Society of Experimental Biology*, 1964, **18**, 157–194.

Crowson, R. A. *Classification and biology*. London: Heinemann Educational Books, 1970.

Cullen, J. M. Behavior as a help in taxonomy. *Systematics Association Publication No. 3*, 1959, 131–140.

Cunningham, M. *Intelligence: Its organization and development*. New York: Academic Press, 1972.

Curtis, A. S. G. *The cell surface: Its molecular role in morphogenesis*, New York: Logos Press (Academic), 1967.

Davis, H. Some principles of sensory receptor action. *Physiological Reviews*, 1961, **41**, 391–416.

Davison, A. N. Lipids and brain development. In D. C. Pease (Ed.), *Cellular aspects of neural growth and differentiation*. Berkeley: University of California Press, 1971.

De Beer, G. *Embryos and ancestors* (3rd ed.). Oxford: Clarendon, 1958.

DeHaan, R. L., & Ursprung, H. (Eds.). *Organogenesis*. New York: Holt, Rinehart and Winston, 1965.

Dell, P. C. Some basic mechanisms of the translation of bodily needs into behaviour. In G. E. W. Wolstenholme & C. M. O'Connor (Eds.), *Symposium on the Neurological Basis of Behaviour*. Boston: Ciba Foundation, 1958.

DeLong, M. R. Activity of Pallidal neurons during movement. *Journal of Neurophysiology*, 1971, **34**, 414–427.

Denenberg, V. H. Animals studies on developmental determinants of behavioral adaptability. In O. J. Harvey (Ed.), *Experience, structure and adaptability*. New York: Springer-Verlag, 1966.

Denny, M. R., & Ratner, S. C. *Comparative psychology: Research in animal behavior*. (Rev. ed.) Homewood, Ill.: Dorsey Press, 1970.

De Reuck, A. V. S., & Knight, J. (Eds.). *Touch, heat and pain*. Boston: Little, Brown, 1966.

De Reuck, A. V. S., & Knight, J. (Eds.) *Hearing mechanisms in vertebrates*. London: Churchill, 1968.

DeRobertis, E. D. P. *Histophysiology of synapses and neurosecretion*. New York: Macmillan (Pergamon Press), 1964.

Deutsch, S. *Models of the nervous system*. New York: John Wiley and Sons, 1967.

Diamond, I. T., & Chow, K. L. Biological psychology. In S. Koch (Ed.), *Psychology, a study of science*. New York: McGraw-Hill, 1962.

Diamond, I. T., & Hall, W. C. Evolution of neocortex. *Science*, 1969, **164**, 251–262.

Dillon, L. S., & Brauer, K. A proposed method for establishing homologies among the lobules of the anterior lobe of the mammalian cerebellum. *Jahrbuch für Hirnforschung*, 1970, **12**, 217–232.

Dobzhansky, T. *Mankind Evolving: The Evolution of the Human Species*. New Haven, Connecticut: Yale University Press, 1962.

Dobzhansky, T., Hecht, M. K., & Steere, W. C. *Evolutionary biology*, Vol. 1. New York: Appleton-Century-Crofts, 1967.

Dornfeld, E. J., Slater, W., & Scheffé, H. A method for accurate determination of volume and cell numbers in small organs. *Anatomical Record*, 1942, **82**, 255–259.

Doty, B. A., Jones, C. N., & Doty, L. A. Learning-set formation by mink, ferrets, skunks and cats. *Science*, 1967, **155**, 1579–1580.

Doty, R. W., & Bosma, J. F. An electromyographic analysis of reflex deglutition. *Journal of Neurophysiology*, 1956, **19**, 44–60.

Drake, E. T. (Ed.). *Evolution and environment*. New Haven: Yale University Press, 1968.

Dubin, M. W. The inner plexiform layer of the vertebrate retina: a quantitative and comparative electron microscopic analysis. *Journal of Comparative Neurology*, 1970, **140**, 479–506.

Dubois, E. The qualitative relations of the nervous system determined by the mechanism of the neurone. *Koninklijke Nederlandse Akademie van Wetenschappen*, 1920, **22**, 665–680.

Dubos, R. *So human an animal*. New York: Scribner and Sons, 1968.

Dubos, R. *A god within*. New York: Scribner and Sons, 1972.

DuBrul, E. L. *Evolution of the speech apparatus*. Springfield, Illinois: Charles C Thomas, 1958.

Duke-Elder, S. *System of ophthalmology. Vol. 1. The eye in evolution*. St. Louis, Mo.: C. V. Mosby, 1958.

Duncan, C. J. Salt preferences of birds and mammals. *Physiological Zoology*, 1962, **35**, 120–132.

Duncan, C. J. *The molecular properties and evolution of excitable cells*. Oxford: Pergamon Press, 1967.

Eakin, R. M. Lines of evolution of photoreceptors. In D. Mazia & A. Tyler (Eds.), *General physiology of cell specialization*. New York: McGraw-Hill, 1963.

Eayrs, J. T., & Goodhead, B. Postnatal development of the cerebral cortex in the rat. *Journal of Anatomy*, 1959, **93**, 385–402.

Ebbesson, S. O. E. On the organization of central visual pathways in vertebrates. *Brain, Behavior, & Evolution*, 1970, **3**, 178–194.

Ebert, J. D. *Interacting systems in development*. New York: Holt, Rinehart & Winston, 1965.

Ebner, F. F., & Myers, R. E. Distribution of corpus callosum and anterior commissure in cat and raccoon. *Journal of Comparative Neurology*, 1965, **124**, 353–365.

Eccles, J. C. *The physiology of synapses*. Berlin and New York: Springer-Verlag, 1964.

Eccles, J. C. *The inhibitory pathways of the central nervous system*. Springfield, Ill.: Charles C Thomas, 1969.

Eccles, J. C. *Facing reality*. Berlin and New York: Springer-Verlag, 1970.

Eccles, J. C., Ito, M., & Szentágothai, J. *The Cerebellum as a neuronal machine*. New York: Springer-Verlag, 1967.

Eccles, J. C., & Schadé, J. P. (Eds.). *Progress in Brain Research, Organization of the spinal cord*. Vol. 11, Amsterdam: Elsevier, 1964.

Edinger, T. *Evolution of the horse brain*. Memoir 25 of the Geological Society of America. Baltimore: Waverly Press, 1948.

Eisenberg, J. F. A comparative study in rodent ethology with emphasis on evolution of social behavior. *International Proceedings of the United States National Museum*, 1967, **122**, 1–51.

Erulkar, S. D. Comparative aspects of the localization of sound. In. M. B. Sachs (Ed.), *Physiology of the auditory system*. Baltimore: National Educational Consultants, Inc. 1972.

Erulkar, S. D., Nelson, P. G., & Bryan, J. S. Experimental and theoretical approaches to neural processing in the central auditory pathway. *Contributions to Sensory Physiology*, 1968, **3**, 149–189.

Escobar, A. (Ed.). *Feedback systems controlling nervous activity*. Villahermosa, Tabasco, Mexico: Sociedad Mexicana de Ciencias Fisiologicas, A. C., 1964.

Evarts, E. V. Relation of pyramidal tract activity to force exerted during voluntary movement. *Journal of Neurophysiology*, 1968, **31**, 14–27.

Ewer, R. F. *Ethology of mammals*. New York: Plenum Press, 1968.

Fjerdingstad, E. J. (Ed.). *Chemical transfer of learned information*. Amsterdam: North-Holland, 1971.

Florkin, M. *A molecular approach to phylogeny*. Amsterdam: Elsevier, 1966.

Fogel, L. J., Owens, A. J., & Walsh, M. J. *Artificial intelligence through simulated evolution*. New York: John Wiley and Sons, 1966.

Ford, D. H. (Ed.). *Influence of hormones on the nervous system*. Basel: S. Karger, 1971.

Fox, M. W. *Integrative development of brain and behavior in the dog.* Chicago: University of Chicago Press, 1971.

Frieden, E., & Lipner, H. *Biochemical endocrinology of the vertebrates.* Englewood Cliffs, N. J.: Prentice-Hall, 1971.

Friend, D. S., & Gilula, N. B. Variations in tight and gap junctions in mammalian tissues. *Journal of Cell Biology,* 1972, 53, 758–776.

Fuller, J. L., & Thompson, W. R. *Behavior genetics.* New York: John Wiley and Sons 1960.

Funkenstein, H. H., Nelson, P. G., Winter, P., Wollberg, Z., & Newman, J. D. Unit responses in auditory cortex of awake squirrel monkeys to vocal stimulation. In M. B. Sachs (Ed.), *Physiology of the auditory system.* Baltimore: National Educational Consultants, Inc., 1972.

Galli, C., Jacini, G., & Pecile, A. *Dietary lipids and postnatal development.* New York: Raven Press, 1973.

Garber, E. D. *Cytogenetics: An introduction.* New York: McGraw-Hill, 1972.

Gaze, R. M. Growth and differentiation. *Annual Review of Physiology,* 1967, 29, 59–86.

Geist, V. The evolution of horn-like organs. *Behaviour,* 1966, 27, 175–214.

Gelfand, I. M., Gurfinkel, V. S., Fomin, S. V., & Tsetlin, M. L. (Eds.). *Models of the structural-functional organization of certain biological systems.* Cambridge, Mass.: MIT Press, 1971.

Gerard, R. W. The fixation of experience. In J. F. Delafresnaye (Ed.), *Brain mechanisms and learning.* Council for International Organizations of Medical Sciences Symposium. London: Oxford University Press, 1961.

Gerard, R. W., & Duyff, J. W. (Eds.). *Information processing in the nervous system. Vol. III, Proceedings of the International Union of Physiological Sciences.* Amsterdam: Excerpta Medica Foundation, 1962.

Gibson, J. J. *The senses considered as perceptual systems.* Boston: Houghton, 1966.

Gilula, N. B., Reeves, R., & Steinbach, A. Metabolic coupling, ionic coupling and cell contacts. *Nature (London),* 1972, 235, 262–265.

Glassman, E. *Molecular approaches to psychobiology.* Belmont, California: Dickenson Publishing Co., 1967.

Globus, A., & Scheibel, A. B. The effect of visual deprivation on cortical neurons: A golgi study. *Experimental Neurology,* 1967, 19, 331–345.

Goffart, M. *Function and form in the sloth.* Oxford: Pergamon Press, 1971.

Gorbman, A., & Bern, H. A. *A textbook of comparative endocrinology.* New York: Wiley, 1962.

Goss, R. J. *Adaptive growth.* London: Logos Press, 1964.

Goss, R. J. (Ed.). *Regulation of organ and tissue growth.* New York: Academic Press, 1972.

Gould, S. J. Allometry and size in ontogeny and phylogeny. *Biological Reviews of the Cambridge Philosophical Society,* 1966, 41, 587–640.

Granit, R. *Receptors and sensory perception.* New Haven: Yale University Press, 1955.

Granit, R. *The basis of motor control.* New York: Academic Press, 1970.

Gray, J. *Animal locomotion.* New York: Norton, 1968.

Gregory, R. L. *Eye and brain, the psychology of seeing.* New York: McGraw-Hill, 1966.

Grenell, R. G., & Mullins, L. J. (Eds.). *Molecular structure and functional activity of nerve cells.* Washington, D.C.: American Institute of Biological Sciences, 1956.

Griffith, J. S. *Mathematical neurobiology.* New York: Academic Press, 1971.

Grundfest, H. Evolution of conduction in the nervous system. In A. D. Bass (Ed.), *Evolution of nervous control.* Washington, D.C.: American Association for the Advancement of Science, Publ. 52, 1959.

Grundfest, H. Evolution of electrophysiological varieties among sensory receptor systems. In J. W. S. Pringel (Ed.), *Problems of the evolution of function and enzymochemistry of excitation processes.* New York: Pergamon Press, 1963a.

Grundfest, H. Impulse conducting properties of cells. In D. Mazia & A. Tyler (Eds.), *General physiology of cell specialization.* New York: McGraw-Hill, 1963b.

Grundfest, H. Synaptic and ephaptic transmission. In G. C. Quarton, T. Melnechuk, & F. O. Schmitt (Eds.), *The neurosciences.* New York: The Rockefeller University Press 1967.

Grusser, O.-J., & Klinke, R. (Eds.). *Pattern recognition in biological and technical systems.* Berlin: Springer-Verlag, 1971.

Guillery, R. W. The laminar distribution of retinal fibers in the dorsal lateral geniculate nucleus of the cat: a new interpretation. *Journal of Comparative Neurology*, 1970, **138**, 339–368.

Guillery, R. W., & Kaas, J. H. Genetic abnormality of the visual pathways in a "white" tiger. *Science*, 1973, **180**, 1287–1289.

Guillery, R. W., Sobkowicz, H. M., & Scott, G. L. Relationships between glial and neuronal elements in the development of long term cultures of the spinal cord of the fetal mouse. *Journal of Comparative Neurology*, 1970, **140**, 1–34.

Gurowitz, E. M. *The molecular basis of memory.* Englewood Cliffs, New Jersey: Prentice-Hall, 1969.

Guth, L. A review of the evidence for the neural regulation of gene expression in muscle. In R. J. Podolsky (Ed.), *Contractility of muscle cells and related processes.* Englewood Cliffs, New Jersey: Prentice-Hall, 1971.

Hailman, J. P. The ontogeny of an instinct. *Behaviour Supplement*, 1967, **15**, 1–159 pp. xi-viii.

Hall, M. C. *The locomotor system: Functional anatomy.* Springfield, Illinois: Charles C Thomas, 1965.

Hamburger, V. Some aspects of the embryology of behavior. *Quarterly Reviews in Biology*, 1963, **38**, 342–365.

Hamburgh, M., Mendoza, L. A., Burkart, J. F., & Weil, F. The thyroid as a time clock in the developing nervous system. In D. C. Pease (Ed.), *Cellular aspects of neural growth and differentiation.* Berkeley: University of California Press, 1971.

Hanly, E. W. *Problems in biology: RNA in development.* Salt Lake City: University of Utah Press, 1970.

Hardin, W. B., Arumugasamy, N., & Jameson, H. D. Pattern of localization in 'precentral' motor cortex of raccoon. *Brain Research*, 1968, **11**, 611–627.

Harlow, H. F., & Woolsey, C. N. (Eds.). *Biological and biochemical bases of behavior.* Madison, Wisc.: University of Wisconsin Press, 1958.

Harman, P. J. *Paleoneurologic, neoneurologic, and ontogenetic aspects of brain phylogeny,* James Arthur Lecture on the Evolution of the Human Brain. New York: American Museum of Natural History, 1957.

Harmon, L. D., & Lewis, E. R. Neural modelling. *Physiological Reviews*, 1966, **46**, 513–591.

Harris, G. W., & Levine, S. Sexual differentiation of the brain and its experimental control. *Journal of Physiology*, 1965, **181**, 379–400.

Harting, J. K., Hall, W. C., & Diamond, I. T. Evolution of the Pulvinar. *Brain, Behavior, & Evolution*, 1972, **6**, 424–452.

Hassler, R. Comparative anatomy of the central visual system in day-and night-active primates. In R. Hassler & H. Stephan (Eds.), *Evolution of the forebrain. Phylogenesis & Ontogenesis of the Forebrain.* New York: Plenum Press, 1967.

Hassler, R., & Stephan, H. (Eds.). *Evolution of the forebrain. phylogenesis and ontogenesis of the forebrain.* New York: Plenum Press, 1967.

Hebb, D. O. *The organization of behavior: A neuropsychological theory.* New York: John Wiley and Sons, 1949.

Hellon, R. F. Hypothalamic neurones responding to temperature in conscious rabbits. *Journal of Physiology*, 1967, **191**, 37.

Henneman, E., & Olson, C. B. Relations between structure and function in the design of skeletal muscles. *Journal of Neurophysiology*, 1965, **28**, 560–580.

Henneman, E., Somjen, G., & Carpenter, D. O. Functional significance of cell size in spinal motoneurons. *Journal of Neurophysiology*, 1965, **28**, 117–131.

Hennig, W. *Phylogenetic systematics.* Urbana, Illinois: University of Illinois Press, 1966.

Herndon, R. M. The interaction of axonal and dendritic elements in the developing and the mature synapse. In D. C. Pease (Ed.), *Cellular aspects of neural growth and differentation.* Berkeley: University of California Press, 1971.

Himwich, W. A. (Ed.). *Developmental neurobiology.* Springfield, Illinois: Charles C Thomas, 1970.

Hinde, R. A. *Animal behaviour: A Synthesis of ethology and comparative psychology.* (2nd ed.) New York: McGraw-Hill, 1970.

Hirsch, J. (Ed.). *Behavior–Genetic analysis.* New York: McGraw-Hill, 1967.

Hodgkin, A. L. *The conduction of the nervous impulse.* Liverpool: University Press, 1964.

Hodos, W., & Campbell, C. B. G. *Scalae Naturae:* Why there is no theory in comparative psychology. *Psychological Reviews*, 1969, **76**, 337–350.

Hoff, C. C., & Riedesel, M. L. (Eds.). *Physiological systems in semiarid environments.* Albuquerque: University of New Mexico Press, 1969.

Hoffman, D. L., & Sladek, J. R., Jr. The distribution of catecholamines within the inferior olivary complex of the gerbil and rabbit. *Journal of Comparative Neurology*, 1973, **151**, 101–112.

Holloway, R. L. The evolution of the primate brain: some aspects of quantitative relations. *Brain Research*, 1968, **7**, 121–172.

Holloway, R. L. Some questions on parameters of neural evolution in primates. *Annals of the New York Academy of Science*, 1969, **167**, 332–340.

Honig, W. K., & James, P. H. R. (Eds.) *Animal memory.* New York: Academic Press, 1971.

Horn, G., & Hinde, R. A. (Eds.). *Short-term changes in neural activity and behavior.* Cambridge: Cambridge University Press, 1970.

Horridge, G. A. *Interneurons. Their origin, action specificity, growth and plasticity.* San Francisco: W. H. Freeman, 1968.

Hubbard, J. I., Llinás, R., & Quastel, D. M. J. *Electrophysiological analysis of synaptic transmission.* London: Edward Arnold, 1969.

Hubel, D. H., & Wiesel, T. N. Receptive fields and functional architecture in two non-striate visual areas (18 and 19) of the cat. *Journal of Neurophysiology*, 1965, **28**, 229–289.

Huber, E. Evolution of facial musculature and cutaneous field of trigeminus. *Quarterly Reviews in Biology*, 1930, **5**, 133–188.

Hughes, A. F. W. *Aspects of neural ontogeny.* London: Logos Press, 1968.

Hutchinson, G. E. *The ecological theater and evolutionary play.* New Haven: Yale University Press, 1965.

Hydén, H. (Ed.). *The neuron.* Amsterdam: Elsevier, 1967.

Hydén, H., & Lange, P. A. Kinetic study of the neuron-glia relationship. *Journal of Cell Biology*, 1962, **13**, 233–237.

Ingram, V. M. *The biosynthesis of macromolecules.* New York: W. A. Benjamin, 1965.

Itoh, S. (Ed.). *Integrative mechanism of neuroendocrine system.* Hokkaido: Hokkaido University School of Medicine, 1968.

Ives, W. R. The interpeduncular nuclear complex of selected rodents. *Journal of Comparative Neurology*, 1971, **141**, 77–94.

Jacobson, M. *Developmental neurobiology.* New York: Holt, Rinehart and Winston, 1970.

Jerison, H. J. Brain to body ratios and the evolution of intelligence. *Science*, 1955, **121**, 447–449.

Jerison, H. J. Quantitative analysis of evolution of the brain in mammals. *Science*, 1961, **133**, 1012–1014.

Jerison, H. J. Interpreting the evolution of the brain. *Human biology*, 1963, **35**, 263–291.

Jerison, H. J. *Evolution of the brain and intelligence*. New York: Academic Press, 1973.

John, E. R. *Mechanisms of memory*. New York: Academic Press, 1967.

Johnson, J. I., Jr., Welker, W. I., & Pubols, B. H. Jr. Somatotopic organization of raccoon dorsal column nuclei. *Journal of Comparative Neurology*, 1968, **132**, 1–44.

Johnson, R. N. *Aggression in man and animals*. Philadelphia: W. B. Saunders, 1972.

Kalmus, H. *Regulation and control in living systems*. New York: Wiley, 1966.

Kalter, H. *Teratology of the central nervous system*. Chicago: University of Chicago Press, 1968.

Karczmar, A. G. & Eccles J. C. (Eds.). *Brain and human behavior*. New York: Springer-Verlag, 1972.

Kare, M. R., & Maller, O. (Eds.). *The chemical senses and nutrition*. Baltimore: The Johns Hopkins Press, 1967.

Katz, B. The transmission of impulses from nerve to muscle, and the subcellular unit of synaptic action. *Proceedings of the Royal Society of London, Series B*, 1962, **155**, 455–477.

Katz, B. *The release of neural transmitter substances*. Springfield, Illinois: Charles C Thomas, 1969.

Kaufman, I. C., & Rosenblum, L. A. A behavioral taxonomy for *Macaca nemestrima* and *Macaca radiata*: based on longitudinal observation of family groups in the laboratory. *Primates*, 1966, 7, 205–258.

Kellogg, W. N. *Porpoises and sonar*. Chicago: University of Chicago Press, 1961.

Kenshalo, D. R. (Ed.). *The skin senses*. Springfield, Illinois: Charles C Thomas, 1968.

Kenyon, D. H., & Steinman, G. *Biochemical predestination*. New York: McGraw-Hill, 1969.

Kety, S. S., & Elkes, J. (Eds.). *Regional neurochemistry*. New York: Pergamon Press, 1961.

Killackey, H. P., & Ebner, F. F. Two different types of thalamocortical projections to a single cortical area in mammals. *Brain, Behavior, & Evolution*, 1972, 6, 141–169.

Kimble, D. P. (Ed.). *The anatomy of memory*. Vol. 1. Palo Alto, Calif.: Science and Behavior Books, 1965.

King, J. A. Species specificity and early experience. In G. Newton & S. Levine (Eds.), *Early experience and behavior. The psychobiology of development*. Springfield, Illinois: Charles C Thomas, 1968.

Klopfer, P. H. *Behavioral aspects of ecology*. Engelwood Cliffs, New Jersey: Prentice-Hall, 1962.

Kohn, R. R. *Principles of mammalian aging*. Englewood Cliffs, New Jersey: Prentice-Hall, 1971.

Kortlandt, A., & Kooij, M. Protohominid behaviour in primates. *Symposia of the Zoological Society of London*, 1963, **10**, 61–87.

Kruger, L. Specialized features of the cetacean brain. In K. S. Norris (Ed.), *Whales, dolphins and porpoises*. Berkeley and Los Angeles: University of California Press, 1966.

Kruger, L. The topography of the visual projection to the mesencephalon: a comparative survey. *Brain, Behavior & Evolution*, 1970, 3, 169–177.

Kummer, H. Behavioral characters in primate taxonomy. In J. R. Napier & P. H. Napier (Eds.), *Old world monkeys, evolution, systematics and behavior*. New York: Academic Press, 1970.

Kuo, Z. Y. *The dynamics of behaviour development*. New York: Random House, 1967.

Kurth, G. (Ed.). *Evolution and hominisation*. Stuttgart: Gustav Fischer, 1968.

Lane, R. H., Allman, J. M., Kaas, J. H., & Miezin, F. M. The visuotopic organization of the superior colliculus of the owl monkey (*Aotus trivirgatus*) and the bush baby (*Galago senegalensis*). *Brain Research*, 1973, **60**, 335–349.

Lane, R. H., Allman, J. M., & Kaas, J. H. Representation of the visual field in the superior colliculus of the grey squirrel (*Sciurus carolinensis*) and the tree shrew *(Tupaia glis)*. *Brain Research*, 1971, **26**, 277–292.

Langman, J., Shimada, M., & Haden, C. Formation and migration of neuroblasts. In D. C. Pease (Ed.), *Cellular aspects of neural growth and differentiation*. Berkeley: University of California Press, 1971.

Lashley, K. S. Persistent problems in the evolution of mind. *Quarterly Reviews in Biology*, 1949, **24**, 28–42.

Lashley, K. S. The problem of serial order in behavior. In L. A. Jeffress (Ed.), *Cerebral mechanisms in behavior*. New York: Wiley, 1951.

LaVail, J. H., Winston, K. R., & Tish, A. A method based on retrograde intraaxonal transport of protein for identification of cell bodies of origin of axons terminating within the CNS. *Brain Research*, 1973, **58**, 470–477.

LeGros Clark, W. E. *The fossil evidence for human evolution*. (2nd ed.) Chicago: University of Chicago Press, 1964.

LeGros Clark, W. E. *History of the primates: An introduction to the study of fossil man*. London: Trustees of the British Museum of Natural History, 1970.

Leibovic, K. N. (Ed.). *Information processing in the nervous system*. New York: Springer-Verlag, 1969.

Lemkey-Johnston, N., & Larramendi, L. M. H. Types and distribution of synapses upon basket and stellate cells of the mouse molecular layer: an electron microscopic study. *Journal of Comparative Neurology*, 1968, **34**, 73–113.

Lende, R. A. Representation in the cerebral cortex of a primitive mammal; sensorimotor, visual, and auditory fields in the echidna *(Tachyglossus aculeatus)*. *Journal of Neurophysiology*, 1964, **27**, 37–48.

Lenn, N. J., & Reese, T. S. The fine structure of nerve endings in the nucleus of the trapezoid body and the ventral cochlear nucleus. *American Journal of Anatomy*, 1966, **118**, 375–389.

Lenneberg, E. H. *Biological foundations of language*. New York: Wiley, 1967.

LeVay, S. Synaptic patterns in the visual cortex of the cat and monkey. Electron microscopy of Golgi preparations. *Journal of Comparative Neurology*, 1973, **150**, 53–86.

Levins, R. *Evolution in changing environments*. Princeton, New Jersey: Princeton University Press, 1968.

Lissák, K., & Endröczi, E. *The neuroendocrine control of adaptation*. Oxford: Pergamon Press, 1965.

Llinás, R. (Ed.). *Neurobiology of cerebellar evolution and development*. Chicago: AMA Education Research Foundation, 1969.

Locke, M. (Ed.). *The emergence of order in developing systems*. New York: Academic Press, 1968.

Lockard, R. B. Reflections on the fall of comparative psychology: Is there a message for us all? *American Psychologist*, 1971, **26**, 168–179.

Loewenstein, W. R. Communication through cell junctions. Implications in growth and differentiation. In M. Locke (Ed.), *The emergence of order in developing systems*. New York: Academic Press, 1968.

Loewenstein, W. R. Membrane junctions in growth and differentiation. *Federation Proceedings*, 1973, **32**, 60–64.

Lofts, B. *Animal photoperiodism*. London: Edward Arnold, 1970.

Lorenz, K. *Evolution and modification of behavior*. Chicago: University of Chicago Press, 1965.

Lund, R. D. Uncrossed visual pathways of hooded and albino rats. *Science,* 1965, **149,** 1506–1507.

Luschei, E. S., Garthwaite, C. R., & Armstrong, M. E. Relationship of firing patterns of units in face area of monkey precentral cortex to conditioned jaw movements. *Journal of Neurophysiology,* 1971, **34,** 552–561.

Machin, K. E. Feedback theory and its application to biological systems. *Symposium of the Society of Experimental Biology,* 1964, **18,** 421–426.

Mackintosh, N. J. Selective attention in animal discrimination learning. *Psychological Bulletin,* 1965, **64,** 124–150.

Mackworth, J. F. The visual image and the memory trace. *Canadian Journal of Psychology,* 1962, **16,** 55–59.

Manosevitz, M., Lindzey, G., & Thiessen, D. D. *Behavioral genetics method and research.* New York: Appleton-Century-Crofts, 1969.

Margulis, L. (Ed.). *Origins of life.* New York: Gordon and Breach, 1970.

Markert, C. L., & Ursprung, H. *Developmental genetics.* Englewood Cliffs, New Jersey: Prentice-Hall, 1971.

Marler, P., & Hamilton, W. J. III. *Mechanisms of animal behavior.* New York: Wiley, 1966.

Martinez, A. J., & Friede, R. L. Changes in nerve cell bodies during the myelination of their axons. *Journal of Comparative Neurology,* 1970, **138,** 329–338.

Martini, L., & Meites, J. (Eds.). *Neurochemical aspects of hypothalamic function.* New York: Academic Press, 1970.

Martini, L., Motta, M., & Fraschini, F. (Eds.). *The hypothalamus.* New York: Academic Press, 1970.

Matthews, P. B. C. *Mammalian muscle receptors and their central actions.* London: Edward Arnold, 1972.

Mayr, E. *Animal species and evolution.* Cambridge, Massachusetts: Harvard University Press, 1963.

Mayr, E. *Principles of systematic zoology.* New York: McGraw-Hill, 1969.

Mayr, E. *Populations, species, and evolution. An abridgment of animal species and evolution.* Cambridge, Mass.: Harvard University Press, 1970.

Mazia, D., & Tyler, A. (Eds.). *General physiology of cell specializations.* New York: McGraw-Hill, 1963.

McIlwain, J. T., & Fields, H. L. Interactions of cortical and retinal projections on single neurons of the cat's superior colliculus. *Journal of Neurophysiology,* 1971, **34,** 763–772.

Mehler, W. R. The mammalian "pain" tract in phylogeny. *Anatomical Record,* 1957, **127,** 332.

Mehler, W. R. Some neurological species differences—a posteriori. *Annals of the New York Academy of Science,* 1969, **167,** 424–468.

Mendell, L. M., & Henneman, E. Terminals of single Ia fibers: location, density and distribution within a pool of 300 homonymous motoneurons. *Journal of Neurophysiology,* 1971, **34,** 171–187.

Mesarović, M. D. (Ed.). *Systems theory and biology.* New York: Springer-Verlag, 1968.

Mettler, F. A. *Culture and the structural evolution of the neural system,* James Arthur Lecture on the Evolution of the Human Brain, 1955. New York: American Museum of Natural History, 1956.

Mettler, L. E., & Gregg, T. G. *Population genetics and evolution.* Englewood Cliffs, New Jersey: Prentice-Hall, 1969.

Michalke, W., & Loewenstein, W. R. Communication between cells of different types. *Nature (London),* 1971, **232,** 121–122.

Minckler, J. (Ed.) *Introduction to neuroscience.* St. Louis: C. V. Mosby, 1972.

Minkowski, A. (Ed.). *Regional development of the brain in early life.* Oxford: Blackwell, 1967.

Moltz, H. *The ontogeny of vertebrate behavior.* New York: Academic Press, 1971.

Montagna, W. *The structure and function of skin.* (2nd ed.) New York: Academic Press, 1962.

Montagu, A. *Touching: The human significance of the skin.* New York: Columbia University Press, 1971.

Morell, P., Lipkind, R., & Greenfield, S. Protein composition of myelin from brain and spinal cord of several species. *Brain Research,* 1973, 58, 510–514.

Morest, D. K. Identification of homologous neurons in the posterolateral thalamus of cat and Virginia opossum. *Anatomical Record,* 1965, 151, 390.

Morest, D. K. The growth of dendrites in the mammalian brain. *Zeitschrift für Anatomical & Entwicklungsgeschichte,* 1969, 128, 290–317.

Morgane, P. J. (Ed.). Neural regulation of food and water intake. *Annals of the New York Academy of Science,* 1969, 157, 531–1216.

Mrosovsky, N. *Hibernation and the hypothalamus.* New York: Appleton-Century-Crofts, 1971.

Mugnaini, E. Developmental aspects of synaptology with special emphasis upon the cerebellar cortex In D. C. Pease (Ed.), *Cellular aspects of neural growth and differentiation.* Berkeley: University of California Press, 1971.

Murray, J. *Genetic diversity and natural selection.* Edinburgh: Oliver & Boyd, 1972.

Namba, T. Cholinesterase activity of muscle fibers and motor end plates: comparative studies. *Experimental Neurology,* 1971, 33, 322–328.

Napier, J. R. Studies of the hands of living primates. *Proceedings of the Zoological Society of London,* 1960, 1134, 647–657.

Napier, J. R. *The roots of mankind.* Washington, D.C., Smithsonian Institution Press, 1970.

Napier, J. R. and Napier, P. H. (Eds.) *Old world monkeys, evolution, systematics and behavior.* New York: Academic Press, 1970.

Nauta, W. J. H., & Ebbesson, S. O. E. (Eds.) *Contemporary research methods in neuroanatomy.* New York: Springer-Verlag, 1970.

Newton, A. A. Synchronous division of animal cells in culture. In E. Zeuthen (Ed.), *Synchrony in cell division and growth.* New York: Interscience Publishers, 1964.

Newton, G. (Ed.). *Early experience and behavior.* Springfield, Illinois: Charles C Thomas, 1968.

Norris, K. S. (Ed.) *Whales, dolphins and porpoises.* Berkeley: University of California Press, 1966.

Norris, K. S. The echolocation of marine mammals. In H. T. Andersen (Ed.), *The biology of marine mammals.* New York: Academic Press, 1969.

Ohloff, G., & Thomas, A. F. (Eds.) *Gustation and olfaction.* London: Academic Press, 1971.

Ohno, S. *Evolution by gene duplication.* New York: Springer-Verlag, 1970.

Okada, Y., Nitsch-Hassler, C., Kin, J. S., Bak, I. J., & Hassler, R. Role of *b*-aminobutyric acid (GABA) in the extrapyramidal motor system. I. Regional distribution of GABA in rabbit, rat, guinea pig and baboon CNS. *Experimental Brain Research,* 1971, 13, 514–518.

Olson, E. C. *The evolution of life.* New York: New American Library, 1965.

Orr, R. T. *Animals in migration.* New York: Macmillan, 1970.

Paillard, J. The patterning of skilled movements. In J. Field, H. W. Magoun, & V. E. Hall (Eds.), *Handbook of physiology,* Sec. 1, Vol. III. Washington, D.C.: American Physiological Society, 1960.

Palladin, A. V. (Ed.) *Problems of the biochemistry of the nervous system.* New York: Pergamon Press, 1964.

Pappas, G. D., & Purpura, D. P. (Eds.) *Structure and function of synapses.* New York: Raven Press, 1972.

Pappas, G. D., & Waxman, S. G. Synaptic fine structure: morphological correlates of chemical and electrotonic transmission. In G. D. Pappas & D. P. Purpura (Eds.), *Structure and function of synapses*. New York: Raven Press, 1972.

Pattee, H. H. The physical basis of coding and reliability in biological evolution. In C. H. Waddington (Ed.), *Towards a theoretical biology 1. Prolegomena*. Chicago: Aldine Publishing Co., 1968.

Paul, R. L., Merzenich, M., & Goodman, H. Representation of slowly and rapidly adapting cutaneous mechanoreceptors of the hand in Brodmann's areas 3 and 1 of *Macaca mulatta*. *Brain Research*, 1972, **36**, 229–249.

Pease D. C. (Ed.) *Cellular aspects of neural growth and differentiation*. Berkeley: University of California Press, 1971.

Peter, J. B. Histochemical, biochemical, and physiological studies of skeletal muscle and its adaptation to exercise. In R. J. Podolsky (Ed.), *Contractility of muscle cells and related processes*. Englewood Cliffs, New Jersey: Prentice-Hall, 1971.

Peters, A., Palay, S. L., & Webster, H. DeF. *The fine structures of the nervous system: The cells and their processes*. New York: Harper and Row, 1970.

Petras, J. M. Some efferent connections of the motor and somatosensory cortex of simian primates and felid, canid and procyonid carnivores. *Annals of the New York Academy of Science*, 1969, **167**, 469–505.

Petras, J. M., & Noback, C. R. (Eds.) Comparative and evolutionary aspects of the vertebrate central nervous system. *Annals of the New York Academy of Science*, 1969, **167**, 1–513.

Pfaffmann, C. (Ed.) *Olfaction and taste*. New York: The Rockefeller University Press, 1969.

Pfeiffer, J. *The human brain*. New York: Pyramid Publications, 1955.

Phillips, M. I. (Ed.) *Brain unit activity during behavior*. Springfield, Illinois: Charles C Thomas, 1973.

Phillis, J. W. *The pharmacology of synapses*. Oxford: Pergamon Press, 1970.

Pilbeam, D. *The evolution of man*. New York: Funk & Wagnalls, 1970.

Pitts, T. D. *Direct interaction between animal cells*. Third Lepetit Colloquium. Amsterdam; North Holland, 1971.

Ploog, D. The relevance of natural stimulus patterns for sensory information processes. *Brain Research*, 1971, **31**, 353–359.

Podolsky, R. J. (Ed.) *Contractility of muscle cells and related processes*. Englewood Cliffs, N. J.: Prentice-Hall, 1971.

Poláček, P. Differences in the structure and variability of spray-like nerve endings in the joints of some mammals. *Acta Anatomica*, 1965, **62**, 568–583.

Polyak, S. In H. Klüver (Ed.), *The vertebrate visual system*. Chicago: University of Chicago Press, 1957.

Pribram, K. H. *Languages of the brain: Experimental paradoxes and principles of neuropsychology*. Englewood Cliffs, N. J.: Prentice-Hall, 1971.

Pringle, J. W. S. (Ed.) *Problems of the evolution of function and enzymochemistry of excitation processes*. New York: Pergamon Press, 1963.

Pringle, J. W. S. (Ed.) *Essays on physiological evolution*. Oxford: Pergamon Press 1965.

Pubols, B. H., Jr., & Pubols, L. M. Neural organization of somatic sensory representation in the spider monkey. *Brain, Behavior & Evolution*, 1972, **5**, 342–366.

Pubols, B. H., Jr., Welker, W. I., & Johnson, J. I., Jr. Somatic sensory representation of forelimb in dorsal root fibers of raccoon, coatimundi and cat. *Journal of Neurophysiology*, 1965, **28**, 312–341.

Pubols, L. M. Some behavioral, physiological and anatomical aspects of the somatic sensory nervous system of the spider monkey (*Ateles*). Unpublished doctoral dissertation, University of Wisconsin, Madison, 1966.

Purpura, D. P. Comparative physiology of dendrites. In G. Quarton, T. Melnechuk, & F. O. Schmitt (Eds.), *The neurosciences. A study program.* New York: The Rockefeller University Pres, 1967.

Purpura, D. P. Intracellular studies of synaptic organizations in the mammalian brain. In G. D. Pappas & D. P. Purpura (Eds.), *Structure and function of synapses.* New York: Raven Press, 1972.

Purpura, D. P. (Ed.) *Methodological approaches to the study of brain maturation and abnormalities.* Baltimore: University Park Press, 1974.

Quarton, G. C., Melnechuk, T., & Schmitt, F. O. (Eds.) *The neurosciences. A study program.* New York: Rockefeller University Press, 1967.

Radinsky, L. A new approach to mammalian cranial analysis, illustrated by examples of prosimian primates. *Journal of Morphology*, 1968, **124**, 167–180.

Radinsky, L. An example of parellelism in carnivore brain evolution. *Evolution*, 1971, **25**, 518–522.

Radinsky, L. Endocasts and studies of primate brain evolution. In R. Tuttle (Ed.), *The functional and evolutionary biology of primates.* Chicago: Aldine-Atherton, 1972.

Radinsky, L. Evolution of the canid brain. *Brain, Behavior & Evolution*, 1973, 7, 169–202.

Ralston, H. J. III. The synaptic organization of lemniscal projections to the ventrobasal thalamus of the cat. *Brain Research*, 1969, **14**, 99–115.

Ramón-Moliner, E. The morphology of dendrites. In G. H. Bourne (Ed.), *Structure and function of nervous tissue.* New York: Academic Press, 1968.

Ramón y Cajal, S. *Studies on Vertebrate Neurogenesis.* Transl. by L. Guth, Charles C Thomas, Springfield, Illinois, 1960.

Ramón y Cajal, S. *Histologie du Systeme Nerveux de l'Homme et des Vertébres*, 2 vols. L. Azoulay, Translator (Reprinted by Instituto Ramón y Cajal el Consejo Superior de Investigacions Cientisicas Madrid, 1952–1955), 1909–1911.

Ramsey, D. M. (Ed.) *Information and control processes in living systems.* New York: New York Academy of Science, 1967.

Rasmussen, G. L., & Windle, W. F. (Eds.) *Neural mechanisms of the auditory and vestibular systems.* Springfield, Ill.: Charles C Thomas, 1960.

Reese, T. S., & Shepherd, G. M. Dendo-dendritic synapses in the central nervous system. In G. D. Pappas & D. P. Purpura (Eds.), *Structure and function of synapses.* New York: Raven Press, 1972.

Reiner, J. M. *The organism as an adaptive control system.* Englewood Cliffs, N.J.: Prentice-Hall, 1968.

Reiss, R. F. (Ed.) *Neural theory and modeling.* Stanford, California: Stanford University Press, 1964.

Rensch, B. *Evolution above the species level.* New York: Columbia University Press, 1960.

Rensch, B. *Biophilosophy.* New York: Columbia University Press, 1971.

Richards, W. Motion detection in man and other animals. *Brain, Behavior, & Evolution*, 1971, 4, 162–181.

Richter, D. (Ed.) *Comparative neurochemistry.* New York: Macmillan, 1964.

Richter, D. *Aspects of learning and memory.* New York: Basic Books, 1966.

Roberts, E., & Sano, K. The specific occurrence of the γ-aminobutyric acid system in nervous tissue of animals. In D. Mazia & A. Tyler (Eds.), *General physiology of cell specialization.* New York: McGraw-Hill, 1963.

Roberts, T. D. M. *Neurophysiology of postural mechanisms.* London: Butterworths, 1967.

Rockstein, M. (Ed.) *Development and aging in the nervous system.* New York: Academic, 1972.

Roe, A., & Simpson, G. G. (Eds.) *Behavior and evolution.* New Haven, Connecticut: Yale University Press, 1958.

Romer, A. S. *Notes and comments on vertebrate paleontology.* Chicago: University of Chicago Press, 1968.

Rose, J. E. The ontogenetic development of the rabbit's diencephalon. *Journal of Comparative Neurology*, 1942, 77, 61–129.

Rosen, R. *Optimality principles in biology.* New York: Plenum, 1967.

Rupp, C. (Ed.) *Mind as a tissue.* New York: Harper and Row, 1968.

Ruščák, M., & Ruščáková, D. *Metabolism of the nerve tissue in relation to ion movements in vitro and in situ.* Baltimore, Maryland: University Park Press, 1971.

Sawin, C. T. *The hormones: Endocrine physiology.* Boston: Little, Brown and Co., 1969.

Saxén, L., & Toivonen, S. *Primary embryonic induction.* London: Logos Press, 1962.

Sayre, K. M., & Crosson, F. J. (Eds.) *The modeling of mind. Computers and intelligence.* New York: Clarion, 1963.

Schapiro, S., & Vukovich, K. R. Early experience effects upon cortical dendrites: a proposed model for development. *Science*, 1970, **167**, 292–294.

Scharrer, E. General and phylogenetic interpretations of neuroendocrine interrelations. In A. Gorbman (Ed.), *Comparative endocrinology*, New York: Wiley, 1959.

Scharrer, E., & Scharrer, B. *Neuroendocrinology.* New York: Columbia University Press, 1963.

Schiller, P. H., & Koerner, F. Discharge characteristics of single units in superior colliculus of the alert rhesus monkey. *Journal of Neurophysiology*, 1971, **34**, 920–936.

Schmidt-Nielsen, K. *Desert animals: Physiological problems of heat and water.* Oxford: Clarendon Press, 1964.

Schneider, D. J. (Ed.). *Proteins of the nervous system.* New York: Raven Press, 1973.

Schneirla, T. C. Levels in the psychological capacities of animals. In R. W. Sellers, V. J. McGill, & M. Farber (Eds.), *Philosophy for the future.* New York: Macmillan, 1950.

Schneirla, T. C. Behavioural development and comparative psychology. *Quarterly Reviews in Biology*, 1966, **41**, 283–302.

Schneirla, T. C., & Rosenblatt, J. S. "Critical periods" in the development of behavior. *Science*, 1963, **139**, 1110–1115.

Schrier, A. M., Harlow, H. F., & Stollnitz, F. (Ed.) *Behavior of non-human primates.* Vols. I & II. New York: Academic Press, 1965.

Scott, J. P. Critical periods in behavior development. *Science*, 1962, **138**, 949–958.

Scrimshaw, N. S., & Gordon, J. E. *Malnutrition, learning and behavior.* Cambridge, Mass.: The M.I.T. Press, 1968.

Shariff, G. A. Cell counts in the primate cerebral cortex. *Journal of Comparative Neurology*, 1953, **98**, 381–400.

Sheldon, M. H. Learning. In L. Weiskrantz (Ed.), *Analysis of behavioral change.* New York: Harper and Row, 1968.

Shinkman, P. G. Visual depth discrimination in animals. *Psychological Bulletin*, 1962, **59**, 489–501.

Shkol'nik-Yarros, E. G. *Neurons and interneuronal connections of the central visual system.* New York: Plenum Press, 1971.

Shugart, H. H. Jr., & Patten, B. C. Niche quantification and the concept of niche pattern. In B. C. Patten (Ed.), *Systems analysis and simulation in ecology.* New York: Academic Press, 1972.

Sidman, R. L., Green, M. C. & Appel, S. H. *Catalog of the neurological mutants of the mouse.* Cambridge, Massachusetts: Harvard University Press, 1965.

Sidman, R. L., & Rakic, P.. Neuronal migration, with special reference to developing human brain: a review. *Brain Research*, 1973, **62**, 1–35.

Silvestri, L. G. *Cell interactions.* Amsterdam: North-Holland, 1972.

Simpson, G. G. The principles of classification and a classification of mammals. *Bulletin of the American Museum of Natural History*, 1945, **85**, i–xvi, 1–350.

Simpson, G. G. *Life of the past. An introduction to paleontology.* New Haven: Yale University Press, 1953.

Simpson, G. G. *Principles of animal taxonomy.* New York: Columbia University Press, 1961.

Simpson, G. G. *The geography of evolution. Collected essays.* New York: Capricorn Books, 1965.

Simpson, G. G. *Biology and man.* New York: Harcourt-Brace-Jovanovich, 1969.

Sinclair, D. *Cutaneous sensation.* London: Oxford University Press, 1967.

Smith, C. U. M. *The brain; Towards an understanding.* New York: Putnam and Sons, 1970.

Smith, G. E. *The physiological series of comparative anatomy* (Royal College of Surgeons of England). London: Taylor and Francis, 1902.

Sobkowicz, H. M., Guillery, R. W., & Bornstein, M. B. The neuronal organization in long term cultures of the spinal cord of the fetal mouse. *Journal of Comparative Neurology*, 1968, **132**, 365–396.

Spatz, W. B., & Tigges, J. Species difference between Old World and New World monkeys in the organization of the striate-prestriate association. *Brain Research*, 1972, **43**, 591–594.

Sperry, R. W. Chemoaffinity in the orderly growth of nerve fibre patterns and connections. *Proceedings of the National Academy of Sciences, U.S.*, 1963, **50**, 703–710.

Sperry, R. W. *Problems outstanding in the evolution of brain function*, James Arthur Lecture on the Evolution of the Human Brain. New York: American Museum of Natural History, 1964.

Sperry, R. W. Embryogenesis of behavioral nerve nets. In. R. L. DeHaan & H. Urspring (Eds.), *Organogenesis.* New York: Holt, 1965.

Sperry, R. W. An objective approach to subjective experience: further explanation of a hypothesis. *Psychological Review*, 1970, 77, 585–590.

Sperry, R. W. How a developing brain gets itself properly wired for adaptive function. In E. Tobach, L. R. Aronson, & E. Shaw (Eds.), *The biopsychology of development.* New York: Academic Press, 1971.

Sprague, J. M. Interaction of cortex and superior colliculus in mediation of visually guided behavior in the cat. *Science*, 1966, **153**, 1544–1547.

Stahl, F. W. *The mechanics of inheritance.* Englewood Cliffs, N.J.: Prentice-Hall, 1964.

Stark, L. *Neurological control systems: Studies in bioengineering.* New York: Plenum Press, 1968.

Stebbins, G. L. *Processes of organic evolution.* Englewood Cliffs, N.J.: Prentice-Hall, 1966.

Straile, W. E. Encapsulated nerve end-organs in the rabbit, mouse, sheep and man. *Journal of Comparative Neurology*, 1969, **136**, 317–336.

Szentágothai, J. The use of degeneration methods in the investigation of short neuronal connexions. *Progress in Brain Research*, 1965, **14**, 1–32.

Tamar, H. *Principles of sensory physiology.* Springfield, Ill.: Charles C Thomas, 1972.

Teilhard de Chardin, P. *Man's place in nature.* R. Hague (transl.). London: Wm. Collins Sons and Co. (Fontana Books), 1971.

Thach, W. T. Discharge of purkinje and cerebellar nuclear neurons during rapidly alternating arm movements in the monkey. *Journal of Neurophysiology*, 1968, **31**, 785–797.

Thiessen, D. D. *Gene organization and behavior.* New York: Random House, 1972.

Thornton, C. S., & Bromley, S. C. (Eds.) *Vertebrate regeneration.* Stroudsburg, Pa.: Dowden, Hutchinson and Ross, 1973.

Tigges, J. Retinal projections to subcortical optic nuclei in diurnal and nocturnal squirrels. *Brain, Behavior, & Evolution*, 1970, **3**, 121–134.

Tobias, P. V. *The brain in hominid evolution.* New York: Columbia University Press, 1971.

Torre, J. C. *Dynamics of brain monoamines.* New York: Academic Press, 1972.

Towe, A. L. Relative number of pyramidal tract neurons in mammals of different sizes. *Brain, Behavior, & Evolution*, 1973, 7, 1–17.

Tower, D. B., & Schadé, J. P. (Eds.) *Structure and function of the cerebral cortex.* Amsterdam: Elsevier, 1960.

Triggle, D. J. *Neurotransmitter-receptor interactions.* New York: Academic Press, 1971.

Trinkaus, J. P. *Cells into organs. The forces that shape the embryo.* Englewood Cliffs, N.J.: Prentice-Hall, 1969.

Tuttle, R. (Ed.) *The functional and evolutionary biology of primates.* Chicago: Aldine-Atherton, 1972.

Udvardy, M. D. F. *Dynamic zoogeography, with special reference to land animals.* New York: Van Nostrand Reinhold, 1969.

Ungar, G. *Excitation.* Springfield, Ill.: Charles C Thomas, 1963.

Ungar, G. (Ed.) *Molecular mechanisms in memory and learning.* New York: Plenum Press, 1970.

Van Gelder, R. G. *Biology of mammals.* New York: Scribner and Sons, 1969.

van Hooff, J. A. R. A. M. The facial displays of the catarrhine monkeys and apes. In D. Morris (Ed.), *Primate ethology.* London: Weidenfeld and Nicolson, 1967.

Vaughn, J. E., & Peters, A. The morphology and development of neuroglial cells. In D. C. Pease (Ed.), *Cellular aspects of neural growth and differentiation.* Berkeley: University of California Press, 1971.

Viaud, G. *Intelligence. Its evolution and forms.* (transl.) by A. J. Pomerans London: Arrow Books, 1960.

von Euler, C., Skoglund, S., & Söderberg, U. (Eds.). *Structure and function of inhibitory neuronal mechanisms.* Oxford: Pergamon Press, 1968.

Waehneldt, T. V., & Shooter, E. M. A comparison of the protein composition of the brains of four rodents. *Brain Research*, 1973, 57, 361–371.

Wallace, A. F. C. *Culture and personality.* (2nd ed.) New York: Random House, 1971.

Walls, G. L. *The vertebrate eye and its adaptive radiation.* Bloomfield Hills, Mich.: Cranbrook Institute of Science, 1942.

Walls, G. L. The evolutionary history of eye movements. *Vision Research*, 1962, 2, 69–80.

Walsh, R. N., Budtz-Olson, O. E., Penny, J. E., & Cummins, R. A. The effects of environmental complexity on the histology of the rat hippocampus. *Journal of Comparative Neurology*, 1969, 137, 361–366.

Warren, J. M. The comparative psychology of learning. *Annual Review of Psychology*, 1965, 16, 95–118.

Washburn, S. L., & Jay, P. C. (Eds.) *Perspectives on human evolution.* New York: Holt, Rinehart and Winston, 1968.

Weber, R. (Ed.) *The biochemistry of animal development.* New York: Academic Press, 1967.

Weiss, P. A. Nervous system (neurogenesis). In B. H. Willier, P. A. Hamburger, & V. Hamburger (Eds.), *Analysis of development.* Section VII: Special Vertebrate Organogenesis. Philadelphia: W. B. Saunders, 1955.

Weiss, P. A. *Dynamics of development: Experiments and inferences.* New York: Academic Press, 1968.

Welker, W. I. Ontogeny of play and exploratory behaviors: a definition of problems and a search for new conceptual solutions. In H. Moltz (Ed.), *The ontogeny of vertebrate behavior.* New York: Academic Press, 1971.

Welker, W. I. Principles of organization of the ventrobasal complex in mammals. *Brain, Behavior, & Evolution*, 1973, 7, 253–336.

Welker, W. I. & Campos, G. B. Physiological significance of sulci in somatic sensory cerebral cortex in mammals of the family Procyonidae. *Journal of Comparative Neurology*, 1963, 120, 19–36.

Welker, W. I., & Johnson, J. I., Jr. Correlation between nuclear morphology and somato-topic organization in ventro-basal complex of the raccoon's thalamus. *Journal of Anatomy*, 1965, **99**, 761–790.

Welker, W. I., Johnson, J. I., Jr., & Pubols, B. H. Jr. Some morphological and physiological characteristics of the somatic sensory system in raccoons. *American Zoologist*, 1964, **4**, 75–94.

Whalen, R. E., Thompson, R. F., Verzeano, M., & Weinberger, N. M. (Eds.) *The neural control of behavior*. New York: Academic Press, 1970.

White, E. L. Synaptic organization of the mammalian olfactory glomerulus: new findings including an intraspecific variation. *Brain Research*. 1973, **60**, 299–313.

Whittow, G. C. (ed.) *Comparative physiology of thermoregulation, Vol. II, Mammals*. New York: Academic Press, 1971.

Whyte, L. L., Wilson, A. C., & Wilson, D. (Eds.) *Hierarchical structures*. New York: Elsevier, 1969.

Wiersma, C. A. G. (Ed.) *Invertebrate nervous systems. Their significance for mammalian neurophysiology*. Chicago: University of Chicago Press, 1967.

Williams, G. C. *Adaptation and natural selection. A critique of some current evolutionary thought*. Princeton: Princeton University Press, 1966.

Willier, B. H., Weiss, P. A., & Hamburger, V. (Eds.) *Analysis of development*. Philadelphia: W. B. Saunders Co., 1955.

Willmer, E. N. *Cytology and evolution*. New York: Academic Press, 1970.

Wilson, E. O. *Sociobiology: A new synthesis*. Cambridge, Massachusetts: Harvard University Press, 1975.

Winkelmann, R. K. Nerve endings in the skin of primates. In J. Buettner-Janusch (Ed.), *Evolutionary and genetic biology of primates*. New York: Academic Press, 1963.

Wolstenholme, G. E. W., & Knight, J. (Eds.) *Control processes in multicellular organisms*. London: J. and A. Churchill, 1970.

Wolstenholme, G. E. W., & O'Connor, M. *Principles of biomolecular organization*. Ciba Foundation Symposium. London: J. and A. Churchill, 1965.

Woolsey, C. N. Organization of somatic sensory and motor areas of the cerebral cortex. In H. F. Harlow & C. N. Woolsey (Eds.), *Biological & biochemical bases of behavior*. Madison: University of Wisconsin Press, 1958.

Woolsey, C. N. Cortical localization as defined by evoked potential and electrical stimulation studies. In G. Schaltenbrand & C. N. Woolsey (Eds.), *Cerebral localization and organization*. Madison: University of Wisconsin Press, 1964.

Wray, S. H. Innervation ratios for large and small limb muscles in the baboon. *Journal of Comparative Neurology*, 1969, **137**, 227–250.

Wright, B. E. *Critical variables in differentiation*. Englewood Cliffs, N. J.: Prentice-Hall, 1973.

Yahr, M. D., & Purpura, D. P. (Eds.), *Neorphysiological basis of normal and abnormal motor activities*. New York: Raven Press, 1967.

Yakovlev, P. I., & Lecours, A.-R. The myelogenetic cycles of regional maturation of the brain. In A. Minkowski (Ed.), *Regional development of the brain in early life*. Oxford: Blackwell, 1967.

Yoon, C. H., & Frouhar, Z. R. Interaction of cerebellar mutant genes I. Mice doubly affected by "staggerer" and "weaver" conditions. *Journal of Comparative Neurology*, 1973, **150**, 137–146.

Zacks, S. I. *The motor endplate*. Philadelphia: W. B. Saunders, 1964.

Zollman, P. E., & Winkelmann, R. K. The sensory innervation of the common North American raccoon *(Procyon lotor)*. *Journal of Comparative Neurology*, 1962, **119**, 149–157.

Zotterman, Y. (Ed.) *Olfaction and taste*. New York: Macmillan, 1963.

Zotterman, Y. (Ed.) *Sensory mechanisms*. Vol. 23 in *Progress in Brain Research*. Amsterdam: Elsevier, 1967.

15

Comparative Anatomy of the Tetrapod Spinal Cord: Dorsal Root Connections

J. M. Petras

Walter Reed Army Institute of Research

INTRODUCTION

The attempt is made in this chapter to examine the comparative anatomy of the tetrapod spinal cord from the viewpoints of structure and evolution. Only the anatomy of dorsal root connections is considered; this is to evaluate whether morphological questions with implied evolutionary significance can be formulated and then experimentally tested. The further purpose of this exercise is to attempt to weigh whether the data of comparative morphology can provide a theoretical description and explanation of the history of the central nervous system in terms of the adaptations of species as a functioning population group modified over time to new environmental challenges.

MAMMALS

Mammals are classified into four subclasses: Prototheria, Allotheria, Theria, and a fourth group of Triassic and Jurassic docodont and triconodont mammals of uncertain subclass status. The Prototheria include the monotremes of presumed triconodont origins (Hopson, 1969). Contemporary monotremes are included in two families: the Ornithorhynchidae, the duckbilled platypus *Ornithorhynchus* of Australia, and the Tachyglossidae, the spiny anteater or echnida *Tachyglossus* of Australia and New Guinea. The allotheres are the multituberculates of the Mesozoic and Late Paleocene. They have not given rise to subsequent higher taxa.

There are three infraclasses of therian mammals: †Trituberculata,[1] the symmetrodonts, and the pantotheres considered ancestral to metatherians and placental (eutherian) mammals (Simpson, 1945; Patterson, 1956; McKenna, 1969; Romer, 1966).

The metatherians include recent marsupials of the Australian continent and the New World oppossum *Didelphis*. The largest and most successful living or extinct group of terrestial mammals is the placentals. Thirty-two therian orders have been identified, half of which are extinct. Approximately 44% of identified eutherians are extinct also. The Insectivora are considered primitive placentals. The order is divided by Romer (1966) into five suborders: Proteutheria, Macroscelidae, Dermoptera, Lipotyphla, and Zalambdodonta. Other authorities consider the Dermoptera as a separate mammalian order. The proteutherian insectivores may be considered to represent the basal eutherian stock from which the modern insectivores, such as the macroscelidids (elephant shrews), lipotyphlans (erinaceoids and soricoids), and zalambdodonts (tenrecs and chrysochlorids), arose. From Cretaceous insectivore stock the leptictids and deltatheridians appear as central groups ancestral to all derived eutherian mammals. The deltatheridians have perhaps led to creodont carnivores (Romer, 1966), whereas the leptictids may be potential ancestors for other progressive placentals. The major insectivoran radiations include the following derived placentals: (1) modern insectivores; (2) Primates, Rodentia, Lagomorpha, Chiroptera, Dermoptera (Simpson, 1945; McKenna, 1969), and Edentata [the Pholidota are more recently derived from palaenodont edentates (McKenna, 1969)] ; and (3) arctocyonid insectivores, which have formed a primitive and central family group of condylarths from which approximately seven lineages have arisen (McKenna, 1969). Three of these families have given origin separately perhaps to the †Dinocerata, †Pantodonta, and †Tillodontia. The didolodontids are ancestral to a large assemblage of South American Tertiary ungulates (†Litopterna, †Notoungulata, †Astrapotheria, †Trigonostylopoidea, †Pyrotheria, and †Xenoungulata). The phenacodont arctocyonids may be ancestral to Proboscidea, Sirenia, †Demostylia, and Perissodactyla. Cetaceans may have arisen from mesonychid condylarths, and hypsodontids may have established the Artiodactyla. Authorities differ about the number of major taxa derived from basal condylarths. There are between 15 and 19 orders (see Simpson, 1945; Romer, 1966; McKenna, 1969). Utilizing McKenna's classification, 61% of the condylarth radiation is extinct.

Eutherians occupy a very extraordinary range of habitats throughout the globe. Besides their continental conquests they have invaded marine and aerial niches. Interpretation of the neuroanatomy of mammals necessitates a careful consideration of the taxon's phylogentic history and its current adaptive niche before anatomically significant historical correlations can be made. The con-

---

[1] A † placed before a taxonomic category denotes that the taxon is extinct.

temporary neuroanatomical data are limited, unfortunately, to a small number of so-called "common laboratory species" of distant lineage.

The Mammalian Spinal Cord

The central connections of dorsal root fibers have been most thoroughly studied in the domestic cat and rhesus monkey (Sprague, 1958; Sprague & Ha, 1964; Petras, 1965; Carpenter, Stein, & Shriver, 1968; Shriver, Stein, & Carpenter, 1968). The general pattern of central connections as seen in these two species is currently the primary basis for studying other vertebrates.

Dorsal root transection at all levels results in a typical pattern of central fiber degeneration that can be identified in all major regions of the spinal cord: cervical, thoracic, lumbar, sacral, and coccygeal. Incoming dorsal root fibers travel longitudinally in three spinal areas: Lissauer's tract, the dorsal funiculi, and the gray matter. The Golgi studies of the late nineteenth and early twentieth centuries have demonstrated small-caliber axons entering and dividing in Lissauer's tract. After running short distances cranially and caudally, collateral and terminal branches enter the substantia gelatinosa and the nucleus magnocellularis pericornualis. Larger caliber dorsal root axons, designated as the medial division, occupy the dorsal funiculus. Their cranial branch continues in the dorsal column to eventually end, depending on their segmental level of origin, in the nucleus gracilis or nucleus cuneatus. Both cranial and caudal branches furnish numerous collaterals to cells of the spinal gray matter, particularly the nucleus proprius cornus dorsalis, column of Clarke, nucleus intermediomedialis, and trunkal and limb somatic motor neurons. The most massive terminal degeneration occurs at the level of entry of each dorsal root. This degeneration extends for various segments above and below the level of entry, sometimes as much as six segments.

Figures 1–7 illustrate how the populations of various spinal nerve cells are organized in the rhesus monkey. This organization is similar in the cat. A very brief description of each pertinent cell group and its relationship to incoming dorsal root fibers is presented together with some comments regarding the functional properties of these cell groups. Currently there are two prevailing theories regarding the organization of cell groups in the spinal cord: nuclear column and laminar organization (for example, see Massazza, 1922, 1923, 1924; Rexed, 1952, 1954, 1964). Both classifications face the same basic difficulty, namely, that not all neuronal cell bodies of the spinal cord are grouped into circumscribed populations. Furthermore, the dendrites of some neurons extend beyond nuclear or laminar boundaries. The cytoarchitectonic anatomical scheme described below is that preferred by myself. Although some degree of lamination is present in the head of the dorsal horn, namely, the nucleus posteromarginalis and the substantia gelatinosa, in my experience many areas of the spinal gray matter do not lend themselves to a laminar parcellation (see Figure 1). The cytoarchitectonic scheme is summarized pictorally in Figures 2–7. A fuller

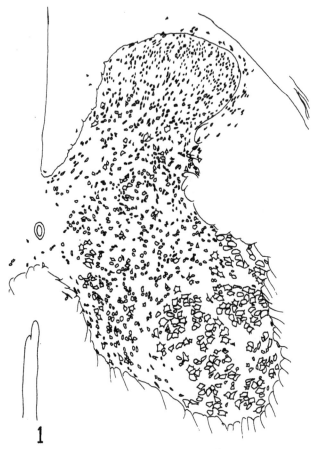

1

FIG. 1 Drawing of a Nissl-stained section taken through the cervical enlargement of the rhesus monkey spinal cord. The locations of various nerve cell bodies were confirmed microscopically. See text for a description of the findings.

descriptive account is omitted because it lies beyond the scope and present intent of this chapter. The figures are presented in order to identify those nuclear groups that have direct bearing on a consideration of afferent dorsal root connections.

The *nucleus posteromarginalis* of Waldeyer (W in Figure 4; nucleus magnocellularis pericornualis, marginal cells) is formed by a thin band of neurons interspersed among the fine-caliber axons running longitudinally atop the dorsal horn. Some large multipolar and deeply chromatophilic neurons are present and are characteristic of this cell group. The nucleus occurs in all spinal segments.

Terminal dorsal root fibers are present in the nucleus, but compared with the terminal degeneration occurring in other nuclei of the dorsal horn, terminal fibers are not abundant. Many of the terminal fibers in the nucleus postermarginalis are small-caliber fibers originating as "collaterals" from dorsal root axons in the zone of Lissauer (Ramón y Cajal, 1952). The functional properties of this nucleus are unknown.

The *substantia gelatinosa* of Rolando (R in Figure 4; nucleus sensibilis proprius) is a prominent and large nuclear group present in the head of the dorsal horn and found in all spinal segments. In myelin-stained sections the neuropil is pale and almost totally translucent except for the passage of small bundles of myelinated dorsal root fibers. This delicate neuropil is composed of very many small dendritic stems and branches and axon collaterals that are best visualized with the electron microscope. The ventral border of the nucleus presents a distinctive dentate contour. Many small neurons populate the nucleus and appear to be arranged in two cellular layers. The cells of the dorsal layer are densely packed. The neurons of the ventral layer are oriented dorsoventrally, in vertical arrays (Rexed, 1952, 1954; Petras, 1968).

The cells of both layers receive terminal dorsal root connections (Ramón y Cajal, 1952; Heimer & Wall, 1968; Ralston, 1968; Sprague & Ha, 1964; Petras, 1968). From the dorsal surface of the nucleus incoming axons are small-caliber fibers, whereas those entering the nucleus from the medial and ventral sides are large-caliber myelinated recurrent axons (Ramón y Cajal, 1952; Petras, 1968). Golgi-stained sections have provided the first brilliantly vivid demonstration of these afferent connections (Ramón y Cajal, 1952); the modern anatomical methods initially have given more varied results (Petras, 1968).

A substantia gelatinosa is present in the spinal portion of the trigeminal nuclear complex (Olszewski, 1950). Trigeminal rhizotomy or medullary tractotomy are employed to alleviate neuralgia. Neurosurgical experience suggests that the integrity of the trigeminal portion of the substantia gelatinosa is necessary for the phenomenon of trigeminal neuralgia to occur. More recent studies by Melzac and Wall (1965) have explored further the role of the substantia gelatinosa in this aspect of sensory discrimination and the reader is referred to this article for a discussion of pain.

The *nucleus proprius cornus dorsalis* (PD in Figure 4; nucleus centrodorsalis [CD], nucleus magnocellularis centralis) occupies the largest area in the head of the dorsal horn. It is an extremely cell-rich nuclear group that is very easily distinguished by the presence of a wide variety of cell types. Included within this population are some of the largest cells of the spinal cord. Besides the heterogeneity of its cellular makeup, the nucleus assumes a deeply black or blue-black color after staining with the methods of Weigert or Weil. This staining property is indicative of the great abundance of myelinated axons in the nucleus. The ventrally adjacent basolateral and basomedial areas of the dorsal horn are not as darkly stained. Following dorsal rhizotomy, the PD is easily recognized in sections stained with the Nauta method. This is because of the heavy impregnation of numerous degenerated fibers and their terminals ending in the nucleus

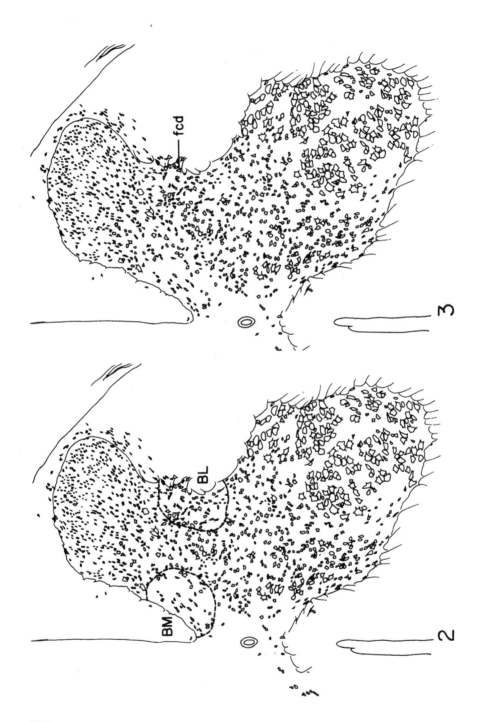

(Ramón y Cajal, 1952; Sprague, 1958; Sprague & Ha, 1964; Petras, 1965, 1966; Ralston, 1968; Carpenter *et al.* 1968; Shriver *et al.* 1968).

All segments of the spinal cord exhibit massive terminal degeneration in the nucleus proprius cornus dorsalis following transections of their appropriate dorsal roots. The nucleus is a major terminus for incoming cutaneous and visceral sensory inputs to the spinal cord (Wall, 1967; Selzer & Spencer, 1969). These afferent fibers innervate the integument, ligaments, tendons, and intrafusal muscle fibers as well as the epithelial and smooth muscle receptors of the viscera. Some cells of the PD are sensitive to one type of incoming stimulus, whereas other cells may be excited by multiple stimuli. For example, Wall (1967) presents evidence for a segregation of cell types according to their excitability by one or more types of afferent nerve stimulation: cutaneous, skeletal muscle, joint, or visceral nerve. The neurons of the PD may also participate in the reflex phenomena of the spinal cord by contributing to the formation of polysynaptic circuits, such as crossed extension and nociceptive responses.

For many generations the belief has persisted that this nucleus is the source of fibers for the ventral and lateral spinothalamic tracts. This concept has been reexamined recently by Trevino, Coulter, and Willis (1973). They find that the cells of origin for the spinothalamic tract in rhesus and squirrel monkeys are distributed over wide areas of the spinal gray matter: nucleus pericornualis, substantia gelatinosa, nucleus proprius cornus dorsalis, basolateral and basomedial dorsal horn regions, nucleus centrobasalis, and nucleus proprius cornus ventralis. An abundant concentration of cells appears in the nucleus proprius cornus dorsalis and the basolateral region of the dorsal horn. Curiously, this basolateral region, based on my experience, is not a principal site for dorsal root terminals. It is a major target of rubrospinal and "motor" corticospinal fibers, however. This differential input suggests an important segregation in the afferent connections to spinothalamic neurons and, furthermore, corresponds very well with the cytoarchitectonic findings reviewed here.

FIGS. 2–7 Architectonic division of spinal cell groups in the rhesus monkey. The nuclei named in the drawings are comparable to those in the domestic cat. A number of the major nuclei are outlined in this series of drawings and each nucleus is described in the text. The borders defining each group are approximate and are based on a large body of histological information: cytoarchitecture, myeloarchitecture, and afferent connections in the cat and rhesus monkey. Abbreviations: BL, basolateral region of the dorsal horn; BM, basomedial region of the dorsal horn; CB, nucleus centrobasalis; CCd, nucleus cornucommissuralis dorsalis; CCv, nucleus cornucommissuralis ventralis; fcd, bed nucleus of fasciculus cornus dorsalis (the nucleus reticularis spinalis of earlier authors); Ml, nucleus motorius lateralis; Mm, nucleus motorius medialis; PD, nucleus proprius cornus dorsalis; PVc, nucleus proprius cornus ventralis, pars centralis; PVdl, nucleus proprius cornus ventralis, pars dorsolateralis; PVdm, nucleus proprius cornus ventralis, pars dorsomedialis; PVvm, nucleus proprius cornus ventralis, pars ventromedialis; R, nucleus substantia gelatinosa of Rolando; W, nucleus posteromarginals. (Figures 4 and 5 on page 352; Figures 6 and 7 on page 353.)

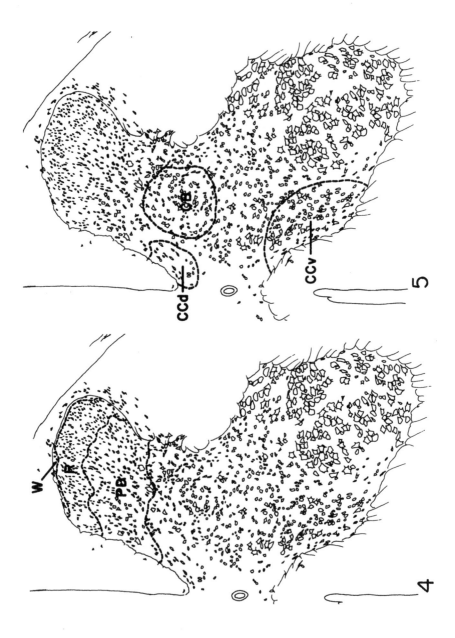

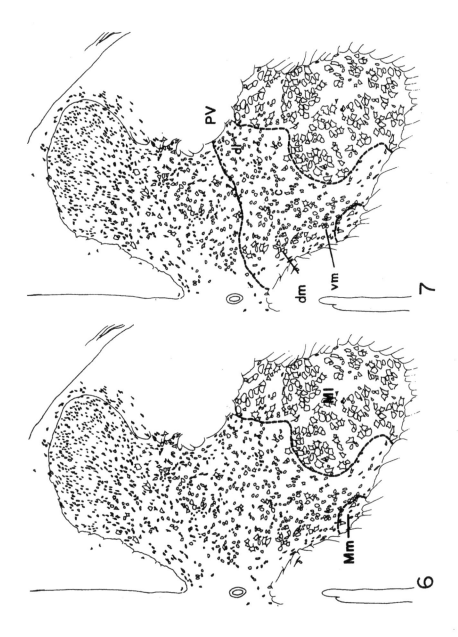

The basolateral dorsal horn region (BL in Figure 2) does not appear as a well-defined cell cluster when viewed cytoarchitectonically. It is difficult to separate the neurons of this region from a larger field that is coextensive ventrally with the zona intermedia. The results of experimental studies, however, enable a clearer distinction to be made between the BL and a more medial neuronal field, the *nucleus centrobasalis* (CB in Figure 5; the intermediate nucleus of Ramón y Cajal). The BL is the target of massive corticospinal fibers from the precentral cortex in many simian primates and apes and the equivalent cortical (motor) field in carnivores (Petras, 1969).

The basomedial dorsal horn region (BM in Figure 2) is located between the PD dorsally, the dorsal white and gray spinal commissures and the central spinal canal ventromedially, and the nucleus centrobasalis (CB) laterally. Encompassed within this geographic region is the *nucleus cornucommissuralis dorsalis* (CCd in Figure 5). The neurons of this nucleus are small and their long axes are oriented ventromedially. The cells form a longitudinal cell column in the ventromedial angle of the dorsal horn with the dorsal commissure. The nucleus cornucommissuralis dorsalis (CCd) appears to be continuous rostrally with a very prominent lens-shaped nucleus in the upper cervical region (tentatively labeled "nucleus cervicalis lenticularis" but not illustrated in this chapter). This nucleus is the site of abundant numbers of degenerated terminal corticospinal (Petras, 1967; Figure 20) and rubrospinal fibers (Petras, 1967; Figure 16). The remaining neurons of the basomedial region are not organized as compactly; consequently, no further label is currently applied to them.

Neither the basolateral (BL) nor the basomedial (BM) dorsal horn regions appear to receive dorsal root connections. The number of degenerated fibers in these regions is small and they are scattered. Compared with other dorsal horn or intermediate zone nuclei, the fiber degeneration in the BL region suggests fibers of passage; the available histological data are not dramatically in favor of an interpretation of terminal connections on cells of this region (Petras, 1965, 1966, 1968, and unpublished observations).

Another distinctive nucleus of the upper cervical spinal cord is the *nucleus cervicalis centralis* (Rexed, 1954). It is the site of massive terminal dorsal root fiber connections (Petras, 1965, 1966). This nucleus occupies a medial position in the zona intermedia of upper cervical (C_1-C_4) segments. The functional significance of the nucleus cervicalis centralis has not been established.

The *nucleus centrobasalis* (CB in Figure 5) is located, as its name suggests, in the center and basal region of the dorsal horn. It is a very prominent nucleus in both the cervical and lumbosacral enlargements. Studies of the connections of the spinal cord performed in my laboratory have not lead to an identification of an equivalent cell group in thoracic or upper lumbar segments. It is interesting to observe that the CB appears to be absent from exactly those spinal segments that have a *nucleus dorsalis* (nucleus magnocellularis dorsalis; the column of Clarke). A role in the kinesthetic physiology of the body may be ascribed to the nucleus centrobasalis. This anatomical speculation is strengthened by the finding that neurons in this spinal region are excited by afferent signals traveling over joint nerves (Wall, 1967). The presence of the nucleus centrobasalis at levels serving the sensory and motor innervation of the limbs suggests an important role in the coordinated kinesthetic physiology of the limbs. In this way it "supplements" perhaps the kinesthetic information from limb receptors that reach the column of Clarke and the *nucleus cuneatus externus* of the medulla oblongata. Neurons in the two latter cell groups are electrically excited by Group IA and Group IB afferent fibers.

The zona intermedia is usually identified by drawing arbitrary lines across the basal areas of the dorsal and ventral horns to separate these two great districts. The intervening region of gray matter is called the zona intermedia. This technique is an unsatisfactory and unfortunate method for subdividing the gray matter for it does not directly address the problem of cell group organization. There are no less than four discrete nuclei in this region. Also present in this region are nerve cell populations that remain unclassified. These cells appear continuous with dorsal and ventral horn cell groups. The better cytoarchitecturally defined nuclei present in this region include the *nucleus intermediomedialis* (IM), *nucleus intercalatus* (IC), *nucleus intermediolateralis thoracolumbalis* (ILp), *nucleus intermediolateralis sacralis* (ILS) (Petras & Cummings, 1972), and a portion of the *nucleus centrobasalis* (CB). The term "zona intermedia" in this chapter is intended to signify neurons not clearly belonging to the CB, IM, IC, ILp, and the ILS cell groups. It is also used in cases where there are no easily defined cytoarchitectural borders, as is the case between the BL and the parvocellular portion of the nucleus proprius cornus ventralis, pars dorsolateralis (PVdl).

The ventral horn is classically subdivided into two great territories: central cells and root cells. The root cells are divided into two groups (Figures 6 and 7): (1) a longitudinal medial (or ventromedial) group, the *nucleus motorius medialis*, for the innervation of the muscles of the neck, thorax, back, abdominal wall and tail; and (2) a lateral (or ventrolateral) longitudinal population, the *nucleus motorius lateralis* which is in part responsible for the formation of the cervical and lumbosacral enlargements of the cord. The cells of this nucleus innervate the muscles of the pectoral and pelvic girdles and limbs. The more numerous central cells constitute a more heterogeneous population, contribute fiber systems that form complex intraspinal pathways, and may contribute to the formation of the long ascending tracts that project to the cerebellum, caudal brainstem, and thalamus.

Except for the nuclei mentioned above, the remaining neurons of the zona intermedia are continuous ventrally with a very large neuronal field that includes all of the intrinsic (central) cell population of the ventral horn, the *nucleus proprius cornus ventralis* (PV in Figure 7). This nucleus is a highly mixed cellular population. Figure 1 illustrates the extent of the nucleus and the cellular diversity of this community. Figure 7 illustrates the general cytoarchitectonic

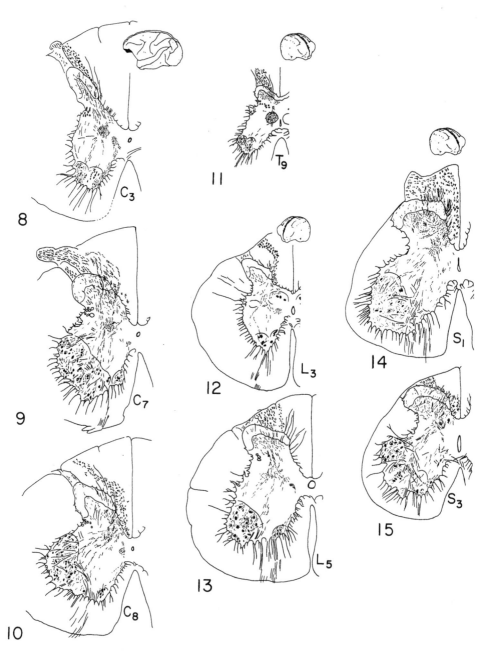

FIGS. 8–15 The distribution of degenerated dorsal root connections are illustrated in this series of chartings through different segments of the rhesus monkey (Figs. 8–10) and cat (Figs. 11–15) spinal cords. Dorsal root degeneration is abundant in the nuclei PD, CB, IM, column of Clarke, and nucleus cervicalis centralis and among trunkal and limb motor neurons of both species. See text for fuller explanation of the findings.

356

parcellation of this nucleus into four subnuclear units: PVdl, the nucleus proprius cornus ventralis, pars dorsolateralis; PVc, the nucleus proprius cornus ventralis, pars centralis; PVdm, the nucleus proprius cornus ventralis, pars dorsomedialis; and PVvm, the nucleus proprius cornus ventralis, pars ventromedialis. The dorsolateral (PVld) and central (PVc) subnuclei contain smaller neurons than the more medial subnuclei. Consequently, the nucleus can be divided into a lateral *parvocellular* region (PVdl and PVc), dominated by medium and small cells, and a *magnocellular* medial region (PVdm and PVvm), dominated by large multipolar neurons (Figure 7). This organization is present in both the cervical and the lumbosacral enlargements. Some imposing aggregates of very large multipolar and deeply chromatophilic neurons are present in the cervical and lumbosacral enlargements in the magnocellular region of the PV. Their prominence has led the early anatomists to name this region "nucleus cornucommissuralis ventralis" (CCv; von Lenhossék, 1893). In thoracic spinal segments, approximately the dorsal half of the ventral horn is occupied by the PV. This region contains many medium and small neurons; the large and numerous multipolar neurons of the CCv apparently are lacking. The remaining neurons of the ventral horn, in all segments, are the large multipolar and deeply chromatophilic somatic alpha motor neurons (Figure 6; nucleus motorius medialis [Mm] and nucleus motorius lateralis [Ml]). The smaller gamma motor neurons innervating the intrafusal muscle fibers of the muscle spindles also are reportedly located in these nuclei (Nyberg-Hansen, 1965).

Degenerated dorsal root fibers that enter the ventral horn terminate predominantly on the medial and lateral groups of somatic motor neurons (nuclei motorius medialis et lateralis). Smaller amounts of dorsal root degeneration are scattered through the PV, but there does not appear to be any focal concentration of these fibers comparable to that observed in either the nuclei motorius lateralis et medialis or the PD and CB nuclear groups of the dorsal horn. As far as the modern data are concerned, there is no evidence to suggest a difference in the intensity of terminal degeneration in the medial or lateral somatic motor nuclei. Pronounced variations in peripheral innervation density of different muscle groups of the limbs may be reflected centrally by an increase in the innervation of homonymous motor neurons by IA afferent fibers.

Within each of the two major motoneuronal nuclear groups there are important subnuclear populations. It is beyond the scope of the present account to give a fuller explanation of the columnar arrangement of the subnuclear group of the nucleus motorius medialis and the nucleus motorius lateralis. The reader can consult the works on Romanes (1951), Sprague (1951), and Sterling and Kuypers (1967) for detailed considerations of this topic. For the present, it is sufficient to note that in a general way cranial segments innervate girdle musculature and proximal muscle groups in the limbs, progressively more caudal segments innervate more distal muscle groups of the limbs, and the neurons of the most caudal segments innervate the intrinsic muscles of the forefeet and hindfeet. The columnar subdivisions of the lateral motor nucleus are more clearly evident in mammals than in amphibians (Romanes, 1951; Cruce, 1974a). The columnar arrangement is far from fully understood, but there are suggestions that the somatic motor columns are arranged in functional groups, each corresponding with one-joint or two-joint muscle groups that produce flexion–extension, adduction–abduction, or rotation (Romanes, 1951). This view coincides with the analysis of muscle function as expressed by a biomechanical approach to movement and posture (Elftman, 1941; Gray, 1944, 1968).

Terminal dorsal root fibers converge on all subnuclear groups in the nucleus motorius medialis and the nucleus motorius lateralis of the cat and rhesus monkey. As far as I am aware, no mention is made in the neuroanatomical literature of a distinction in the number of dorsal root inputs into different subnuclear groups of somatic motor neurons.

In summary, dorsal root fibers in mammals converge on numerous cell groups of the dorsal horn and appear to have a more restricted distribution to ventral horn and intervening cell groups of gray matter. For example, dorsal root terminal degeneration is seen in the nucleus posteromarginalis of Waldeyer and the substantia gelatinosa and is massive in the nucleus proprius cornus dorsalis, the nucleus centrobasalis, and the column of Clarke. In the intermediate region of the gray matter, terminal degeneration is abundant in the nucleus cervicalis centralis and nucleus intermediomedialis. Direct dorsal root connections are present among the intranuclear dendritic segments of medial (axial) and lateral (limb) motor neurons. The patterns of axon degeneration suggest numerous direct connections on intranuclear dendrites of somatic limb and axial motor neurons.

Pinnipedia and Perissodactyla

Before the spinal anatomy of lower tetrapods is considered, it may be well to discuss additional preliminary mammalian data on dorsal root connections that have been obtained in the horse and harbor seal. Dorsal root lesions in the harbor seal *Phoca vitulina* (Petras, unpublished observations), and horse *Equus caballus* (Cummings and Petras, unpublished observations), produces degeneration remarkably similar to that in the cat and rhesus monkey. Degenerated axons are distributed in the zone of Lissauer and the dorsal funiculus. Fibers penetrating the dorsal horn are present in the nucleus posteromarginalis and substantia gelatinosa. Massive degeneration is seen in the nucleus proprius cornus dorsalis and in the nucleus centrobasalis. Abundant degeneration in Clarke's column (seal) and the nucleus intermediomedialis is also seen. In the horse the number of fibers in Clarke's column is not impressive, but this can be attributed to a limitation of the stain and short survival time. Direct dorsal root–motoneuronal connections and the likelihood of monosynaptic synapses are indicated by the presence of abundant fiber degeneration among neurons in the nucleus motorius medialis and the nucleus motorius lateralis. The striking similarity in the findings in mammals of such wide taxonomic and adaptive variation (*Felis catus, Macaca mulatta, Phoca vitulina,* and *Equus caballus*) suggests that dorsal root–mono-synaptic connections with somatic motor neurons are a typical mammalian condition. Perhaps it is an expression of the adaptation for fully erect quadrupedal standing that, once secured in primitive mammals, is retained among all higher mammalian taxa. It is not known whether secondary loss of this neural connection occurs in surviving species that have adapted to ecological niches not requiring upright posture.

AMPHIBIANS

The term "lower tetrapods" is used here to identify all amphibians and reptiles and to distinguish them from birds and mammals. Living amphibians are subdivided into three subclasses. The Urodela (= Caudata) includes the newts and salamanders. These are quadrupedal, long-bodied, tailed animals. The Anura includes the frogs and toads, which are short-bodied, tailless, and typically jumping amphibians possessing numerous skeletal specializations. The very specialized fossorial Apoda, the caecilians, are blind and limbless. Both urodeles and anurans were present in the Mesozoic era. Urodele fossils have been recovered from Cretaceous rocks, approximately 130 million years of age. Anuran remains occur in Triassic rocks, some 230 million years old. The class itself has evolved from Paleozoic piscine crossopterygian ancestors and can be traced to the late Devonian period, 400 million years ago.

Three major amphibian radiations have developed: the †Labyrinthodontia, the †Lepospondyli, and the Lissamphibia. The lepospondyls are an entirely extinct group, whereas the lissamphibians have given rise to the recent caudata (apodans and urodeles), and the salientians (anurans). Primitive labyrinthodont amphibians have separated into two major groups, the temnospondyls and the primitive anthracosaurian labyrinthodonts. From this latter radiation have developed several amphibian groups, including the embolomers and seymouriomorphs. From the Pennsylvanian seymouriomorph-like anthracosaurs, the reptiles have arisen. The long complex history of amphibians remains unsettled. What is clear, however, is that the living orders are highly specialized amphibians and far from anything remotely resembling reptilian ancestors. Perhaps the salamanders best typify some of the general characters of the class Amphibia, yet they too have diverged from early labyrinthodont ancestors. Although the limitations of studying surviving families must be recognized and acknowledged, it also must be appreciated that efforts to identify characters general to all or most species can help identify which characters are primitive from those which are specialized. By inference it may be supposed theoretically possible to establish the patterns of both primitive and specialized amphibian characters. Comparisons of the brains and spinal cords of extant amphibians with extant reptiles may lead to descriptive accounts on which speculations may be based regarding which morphological features of the central nervous system (coupled with the coevolution of other organ systems) have enabled seymouriomorph-like reptiles to invade terrestrial habitats.

Amphibian Spinal Cord

The gray matter of the spinal cord forms a large central core with a relatively small surrounding field of white matter (Figure 16). The two sides are joined by very large gray masses dorsal and ventral to the central canal. Dorsal and ventral horns are distinguishable but are not tall and slender columns. The zona intermedia is very large in all dimensions, its size surpassing the horns of the gray matter. The dorsal columns (funiculi) are proportionally larger than those of fishes but smaller than those of reptiles and much smaller than those of mammals. The lateral funiculi are larger and the ventral funiculi do not seem as large as in reptiles. A substantia gelatinosa is identified by some (Keenan, 1929; Ariëns-Kappers, Huber, & Crosby, 1936) along with a ventrally adjacent sensory cell group perhaps comparable to the nucleus proprius cornus dorsalis of mammals. Not all authors distinguish the same cell groups; some prefer instead to describe a general dorsal neuropil in the dorsal horn (Kennard, 1959) without further subdivision. A slightly different subdivision of the dorsal horn has been given by Joseph and Whitlock (1968a). These authors identify two areas: a small lateral "submarginal zone" and a larger medial "subfascicular zone".

The commissural region dorsal and ventral to the spinal canal is wide in the dorsoventral direction, as mentioned above, resulting in minimal separation of

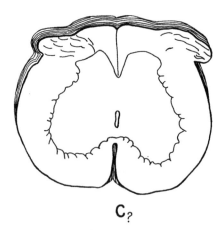

16

C?

FIG. 16 Outline sketch of a transverse section of the frog spinal cord, *Rana mugiens,* illustrating the general contour of the gray matter and its proportional volume to white matter. Modified from Ariëns-Kappers *et al.* (1936).

the dorsal horns. In reptiles, for example, the large dorsal columns result in a greater separation of the dorsal horns (see Figure 18).

The cells of the ventral horn can be divided into central cells and somatic motor neurons, a pattern typical of all limbed tetrapods. The intumescentia of cervical and lumbosacral spinal cord make their appearance in amphibians with the formation of the lateral motor nuclei that innervate limb muscles. The somatic motor neurons fall into two major populations: the medial motor neurons innervating axial (epaxial and hypaxial) muscle groups, and the lateral motor neurons innervating girdle and limb muscles. The nucleus motorius lateralis contains two or more subnuclear columns. Cruce (1974a) has recently described three major subnuclei in the lumbar enlargement of the frog, *Rana catesbiana.*

Somatic motor neurons in amphibians display a long and elaborate branching of their dendrites (Sala, 1892; Silver, 1942). These dendrites extend far beyond the limits of their parent nuclei (nuclei motorius medialis and lateralis) and contribute to the formation of a lateral dendritic or marginal plexus (Figure 17). Dorsally extending dendrites enter the zona intermedia and the dorsal horn. Silver (1942) has measured dendritic length in the frog (species not identified). Dorsal dendrites are as long as 500 μm, lateral dendrites are 300–400 μm long, and commissural dendrites appear longest of all, up to 1000 μm.

The terminal connections of incoming dorsal root fibers have been studied experimentally in *Rana pipiens, Rana catesbiana,* and *Bufo marinus* and the findings appear similar in all species (Chambers et al., 1960; Joseph & Whitlock, 1968a). A medial and lateral division of dorsal root fibers confirms the description by earlier authors using the Golgi method to histologically prepare the spinal cord for microscopic study. The lateral division is composed of small-caliber fibers ascending and descending in one of two fasciculi: one is adjacent to the head of the dorsal horn and located in the dorsal funiculus, and another is located laterally. This latter bundle, named the "fasciculus lateralis," is found in the dorsal quadrant of the lateral funiculus. Fibers of the lateral division may converge on the subfascicular region of Joseph and Whitlock (1968a). Fibers of the medial division are larger and give rise to terminals in the dorsal neuropil or subfascicular zone of Joseph and Whitlock (1968a). Additional fiber degeneration continues along a ventral course to terminate in a small plexus of fibers in the intermediate zone (Chambers *et al.,* 1960; Joseph & Whitlock, 1968a). Dorsal root fibers do not penetrate the ventral horn of the amphibian spinal cord. On this point the Golgi and fiber degeneration data are in agreement.

REPTILES

All living and extinct reptiles have been classified into six subclasses: the Anapsida, Lepidosauria, Archosauria, †Euryapsida, †Ichthyopterygia, and †Synapsida. Within these major groups there are 17 orders. Fourteen of these are

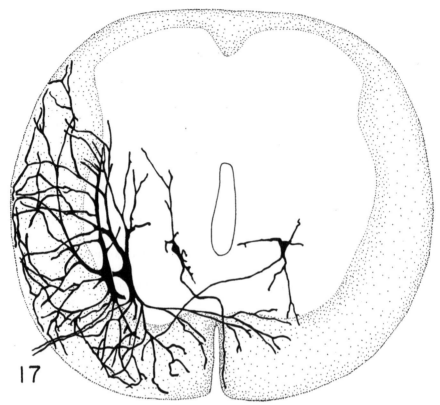

FIG. 17   Motor neurons in the spinal cord of *Bufo*. The two large cells on the left are motor neurons and the remaining cells are propriospinal neurons. Note the great length of the lateral dendritic processes of motor neurons and their contribution to the formation of the marginal dendritic plexus. The dorsal dendrites also extend for long distances along the border of the gray matter to reach the zona intermedia and dorsal horn. Modified from a figure by Sala which is published in Ariëns-Kappers *et al.* (1936).

extinct. The Archosauria contains the crocodilia of tropical Africa, Asia, and the New World. The lepidosaurian order Rynchocephalia has but one surviving and rare species, the tuatara (*Sphenodon*) of New Zealand. The snakes are a more recent group, having evolved from lacertilian ancestors. The *Sphenodon*, lizards, and snakes are the only surviving families of lepidosaurs. The turtles are an archaic group that predate the Mesozoic era. A distinctly separate lineage, the synapsid pelycosaurs, has produced the therapsids of the Carboniferous and Permian periods and Mesozoic era. The mammals have arisen from this abundant therapsid stock, probably from Permian families.

The array of adaptations and families of reptiles is staggering. The success of therapsids also has been remarkable and far outranks the early limited terrestrial successes of which the amphibians have been capable. Their bonds to the water have limited amphibian successes on land with the exception of the advanced anthracosaurs. Even more striking than the limited extant amphibians, the study of living reptiles represents but a small assemblage of a highly successful class. The living species are the products of the evolution of many millions of years, and their study has not provided types transitional to higher taxa. Hecht (1969) amply specifies the pitfalls of studying such a limited sample of this once great array of reptilian families. The functional morphology of the surviving groups should, nevertheless, provide a basis for describing the anatomy and functions of the reptilian brain. The comparisons that follow can be contrasted with other vertebrate classes and the resulting data should give some insights regarding the adaptive and evolutionary changes accompanying the origin of mammals.

The Reptilian Spinal Cord

The spinal gray matter of reptiles, as in other tetrapods, is divided into dorsal horns, intermediate zone, and ventral horns. The boundary between gray and white matter is distinct and numerous strands of gray matter penetrate deeply into the funiculi (Figure 18). Cervical and lumbar enlargements are evident in all reptiles except the limbless snakes. The spinal funiculi are relatively larger than those of amphibians, and the dorsal funiculi are continuous cranially with distinct dorsal column nuclei (Ariëns-Kappers *et al.*, 1936). The increased size of the dorsal funiculi causes the further separation of the dorsal horns of the gray matter.

The dorsal horns possess a distinct substantia gelatinosa (Keenan, 1929) and a subjacent dorsal neuropil. The intermediate zone is narrower dorsoventrally than in amphibians, and the spinal commissures are not as large dorsoventrally (Ramón y Cajal, 1952; Retzius, 1894, 1898a,b). The ventral horns are large and

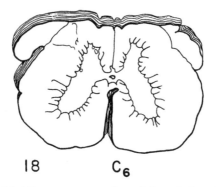

18   C_6

FIG. 18 Outline sketch of a transverse section of the spinal cord of a turtle. Note the relatively large area of white matter compared with the amphibian spinal cord, together with the more elongate character of the dorsal and ventral horns. Modified from Ariëns-Kappers *et al.* (1936).

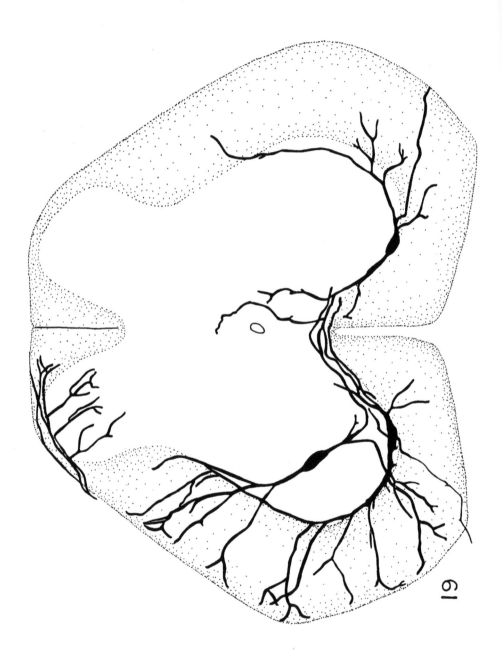

distinct and contain central cells and somatic motor neurons. Medial and lateral motor neuron populations are present in lacertalians and turtles (Ariëns-Kappers *et al.,* 1936; Nieuwenhuys, 1964). The medial motor group is absent in thoracic segments of turtles because of the development of a body shield, the plastron and carapace, and the accompanying loss of thoracic musculature. Snakes lack lateral motor neurons in keeping with the loss of all four limbs (Ariëns-Kappers *et al.,* 1936). Limb motor neurons possess long dorsal dendrites and funicular dendrites. The dorsal dendrites extend for long distances along the lateral margin of the gray matter and reach into the intermediate zone and the dorsal horn (Figure 19). A marginal plexus, situated in the lateral and ventrolateral portions of the lateral funiculus, is formed in part by motor neuron lateral dendrites (Figure 19). Incoming dorsal root fibers, revealed with the Golgi technique, form lateral and medial bundles (Banchi, 1903; de Lange 1917; Retzius, 1894, 1898a,b; van Gehuchten, 1897; Ramón y Cajal, 1891). The axons comprising both bundles divide and then give off collaterals to the dorsal horn. The lateral division is composed of smaller caliber fibers and these may be comparable to the tract of Lissauer in mammals. Fibers of the medial division are larger caliber axons and supply terminals to dorsal horn cells (Retzius, 1894; Ramón y Cajal, 1891) and to dorsal horn and ventral horn commissural cells. The so-called "reflexomotor" collaterals to somatic motoneuronal dendrites are present also. According to the older literature, dorsal root axons in such ophidians as *Natrix natrix* (= *Tropidonotus natrix,* Retzius, 1894; *Gongylus ocellatus,* Terni, 1920) enter the cord in a manner distinct from that of amphibians. A medial coarse-fibered population with collaterals to dorsal horn cells and ventral motor neurons is similar, but a lateral bundle named the "fasiculus lateralis" also is present in the dorsal part of the lateral funiculus (Retzius, 1894). Coarse fibers comprise the bundle and terminals are given off to dorsal horn cells, ventral horn commissural cells, and ventral horn motor neurons. According to the recent findings by Joseph and Whitlock (1968a), a bundle of lateral funicular fibers is present in amphibians as well.

A cytoarchitectonic study of the reptilian cord comparable to a number of detailed analyses of mammalian spinal cords has not been published. Cruce (1974b) is now studying the spinal cord of the lizard *Tupinambis nigropunctatus,* and he has subdivided the gray matter into layers following the system of Rexed (1954).

Joseph and Whitlock (1968b,c) describe the results of lumbar rhizotomy in the caiman, false iguana, and iguana. Lumbar dorsal root degeneration is found in

---

FIG. 19 Motor neurons in a lizard. Afferent dorsal root fibers are seen entering the dorsal funiculus and dorsal horn. Commissural, ventral, lateral, and dorsal dendritic categories can be identified. A marginal dendritic plexus formed by lateral dendrites is present in the lateral funiculus. Long dorsal dendrites course along the lateral margin of the zona intermedia and penetrate the basal aspect of the dorsal horn. Modified from a figure of the spinal cord of *Lacerta agilis* by Ramón y Cajal (1952).

two fields: the dorsal horn and the intermediate zone of *Caiman sclerops* (caiman), *Ctenosaura hemilopha* (false iguana), and *Iguana iguana* (iguana). In *Caiman* dorsal root fibers do not enter in the ventral horn. This is similar to the findings in the frog (*Rana pipiens*) and the toad (*Bufo marinus*). Iguanids (*Iguana iguana* and *Ctenosaura hemilopha*), however, possess very long dorsal root collaterals that cascade into the ventral horn. These axons course along the lateral margin of the intermediate zone and ventral horn. These primary afferent fibers are present along the dorsolateral border of limb motor neurons (nucleus motorius lateralis) close to the origin of dorsal dendrites from their cell bodies. A distribution of these fibers within the nucleus motorius lateralis does not appear to be supported by current evidence.

The dorsal dendrites of somatic motor neurons are arrayed in parallel bundles along the margin of the gray matter in all lower tetrapods studied so far. The extension of these dendrites through the intermediate zone and the ventral aspect of the dorsal horn places them within two areas where terminal dorsal root fibers are abundant. The Golgi data and the results of Joseph and Whitlock (1968a,b,c), taken together, although they do not directly demonstrate the synaptic boutons involved point strongly in the direction of monosynaptic connections of dorsal root fibers with the dorsal dendrites of limb motor neurons in lower tetrapods (Joseph & Whitlock, 1968c). A significant shift toward an increase in synaptic density on a large proportion of dorsal dendrites is seen in iguanids.

BIRDS

The origin of birds is poorly documented in the fossil record. *Archeoptyrix* was discovered in the Jurassic rocks of Germany. The skeleton was typically reptilian in character and except for the clear impression of feathers preserved in fine lithographic limestone, *Archeoptyrix* would have been classified as a reptilian species. Birds are believed to have been derived from advanced bipedal pseudo-suchian reptiles, although there are no fossil species among thecodont archosaurs with unequivocally demonstrable ancestral affinities to birds (Romer, 1966).

The Avian Spinal Cord

The appearance of the bird spinal cord is in general reminiscent of the reptilian cord. The avian cord differs by the possession and curious formation of the lumbar rhomboid fossa and the greater prominence of Hoffman-Koelliker nuclei, which become visible macroscopically as lobuli accessori. The outline of the gray matter with the funiculi is perhaps more distinct than in reptiles, and the dorsal and ventral horns are more prominent (Figure 20). The number of neurons appears high by comparison with amphibians. The dorsal funiculi are relatively

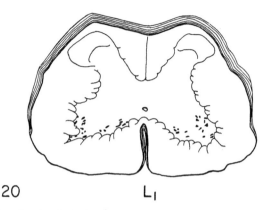

20   L₁

FIG. 20 Outline sketch of the L₁ segment from the spinal cord of the pigeon *Columba livia*. Motor neurons are represented as solid black triangles or spindles and were drawn from a Nissl stained section.

smaller in birds than in reptiles despite the presence of dorsal column nuclei in both classes. It has been suggested that the smaller dorsal funiculi of birds is correlated with feathering of the body (Ariëns-Kappers *et al.*, 1936). The presence of cutaneous receptors and nerve endings in feather follicles casts some doubt on this generalization.

Streeter (1904) and Huber (1936) have studied the cytoarchitectonics of the ostrich and pigeon spinal cords, respectively. They classify few cell groups compared with the subdivisions typical of mammalian studies. Matsushita's (1968) description of the chicken spinal cord more closely resembles the cytoarchitectonic divisions familiar to mammalian specialists. His description of spinal cord nuclei relies on a typically mammalian nomenclature. In view of the archosaurian heritage of birds and their greater phylogenetic affinities to surviving archosaurs, it may be best to base an analysis of their spinal architecture on an entirely new set of names unbiased by the carryover of terms from mammalian anatomy. Otherwise, the use of mammalian nomenclature engenders skepticism in regard to implied homologies. A comparison of the spinal cords of extant archosaurs and lepidosaurs with those of birds may better delineate the morphological adaptations of birds as distinct from mammals.

The avian (pigeon, ostrich, chicken) substantia gelatinosa is large and distinct (Streeter, 1904; Huber, 1936; Matsushita, 1968). A nucleus proprius cornus dorsalis in the chicken is described by Matsushita (1968), together with the column of Clarke (nucleus dorsalis) and a column of Terni for preganglionic sympathetic neurons. Of great importance to the present subject is that an avian equivalent of the nucleus centrobasalis (CB) apparently has not been described. The somatic motor neurons are similar to other tetrapods, i.e., an organization into a nucleus motorius medialis and a nucleus motorius lateralis. The subnuclear division of the lateral nucleus requires careful study using experimental tech-

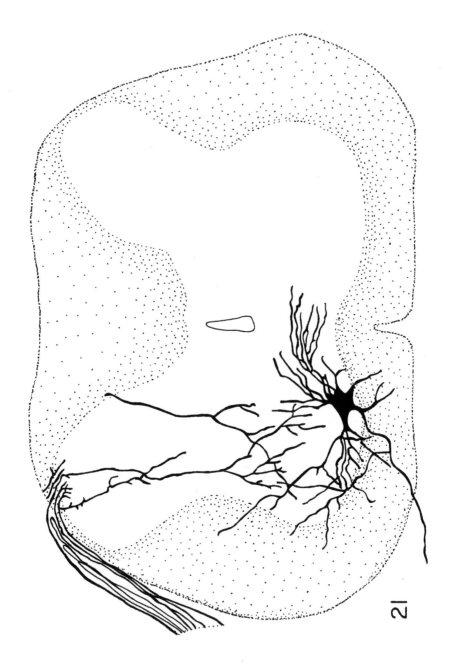

21

niques to define the morphological and functional relationships of the longitudinal subnuclear columns. The dendrites of somatic motor neurons appear less extensive than in amphibians and reptiles; these processes appear to be more greatly confined to the gray matter and to their parent nuclei (Figure 21).

Van Gehuchten (1893) and Ramón y Cajal (1952) have demonstrated with the Golgi technique primary afferent fiber connections to dorsal horn cells with "reflex collaterals" to the dendrites of somatic motor neurons (Figure 21).

Experimental evidence regarding the central connections of dorsal root fibers in the pigeon comes from the work of Leonard and Cohen (personal communication, a, b) together with the preliminary findings of Petras (unpublished observations, Figure 22). These authors confirm the separation of dorsal root fibers into a small lateral division coursing through the zone of Lissauer and a large-caliber medial division occupying the dorsal funiculus. Degenerated dorsal root fibers arborize in the dorsal horn in the nucleus pericornualis dorsalis, the substantia gelatinosa, and the adjoining nucleus proprius cornus dorsalis (Petras, unpublished observations). There is scattered fiber degeneration in the zona intermedia. A dense area of dorsal root fiber degeneration comparable to the area of the CB in mammals has not been verified (Petras, unpublished observations). Additional fibers reach the ventral horn and arborize among neurons of the nucleus motorius medialis and abundantly in the nucleus motorius lateralis. This degeneration is more abundant perhaps in the motor neuronal cell groups of cervical segments than the lumbosacral segments (Leonard & Cohen, personal communication, a, b). This suggests a greater afferent connection to wing motor neurons (Leonard & Cohen, personal communication, a). A further impression gained by Leonard and Cohen is that the afferent dorsal root fibers to limb motor neurons are not as abundant as described for mammals.

The above findings would indicate a synaptic invasion on the intranuclear dendrites of motor neurons. Synaptic connections on the dorsal (extranuclear) dendrites may also be present, and this character would be shared in common with all lower tetrapods and perhaps mammals. The presence of monosynaptic dorsal root–motoneuronal connections with *intranuclear dendrites* suggests that this connection evolved independently in birds and in mammals. Mammals were derived from the therapsid line of reptiles. Birds arose from advanced thecodont archosaurian reptiles. Their nearest living archosaurian relatives, the crocodiles, have not evolved the character (Joseph & Whitlock, 1968a,b), whereas terrestrial lizards appear to have established connections on a greater area of the dendritic network than crocodilians. These observations favor an independent evolution of the character in the lineage from which mammals have arisen and again in the archosaurian radiation leading toward birds. The greater synaptic density of

FIG. 21 Drawings of dorsal root fibers and a somatic motor neuron of the chick. In contrast to amphibians, the motoneuronal dendrites of birds are more confined to the gray matter. Modified from a figure by van Gehuchten (1893).

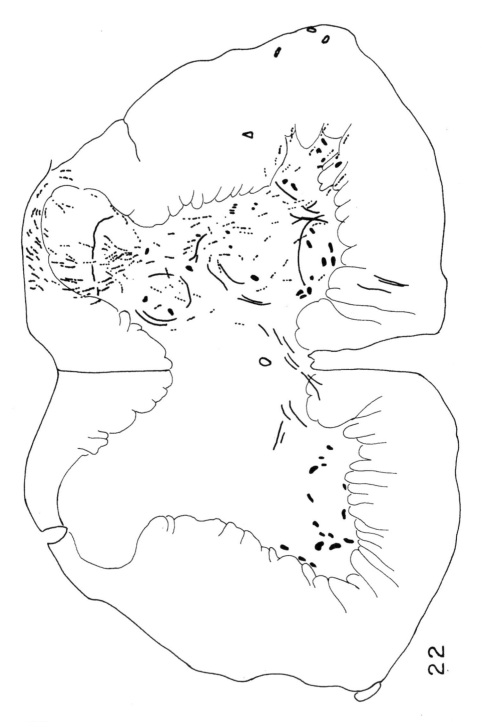

dorsal root connections in birds may be an adaptation to upright bipedal posture and ambulation and the selection for static and phasic muscle spindle information during flight.

An important structural difference can be identified by comparing the dendritic ramifications of ventral horn somatic motor neurons of lower tetrapods with those of birds and mammals. There is an elaborate dendritic array extending into the gray matter beyond the nuclear limits of the motor neuronal population, and another great display of dendritic branching in the white matter in the form of a marginal plexus in the lateral funiculus of lower tetrapods (compare Figures 17 and 19). The confinement of a greater proportion of the dendritic tree to the motor nucleus, and to the gray matter in birds and mammals (compare Figures 21, 22, 23, and 24), raises the question whether the occurrence of intranuclear dorsal root terminals is functionally significant. The available experimental evidence (Joseph & Whitlock, 1968a,b) suggests that the target of dorsal root fibers is dorsal dendrites and not the marginal plexus of archosaurian and lepidosaurian reptiles. I know of no experimental demonstration that the marginal dendritic plexus serves either as a major or exclusive synaptic surface for dorsal root axons. Perhaps the marginal plexus functions to receive multiple-afferent messages and these are derived from propriospinal and from long descending brain tracts to the spinal cord, whereas in birds and mammals the input to motoneuronal dendrites is more specific. If tegmentospinal-, tectospinal-, and telencephalospinal–motoneuronal connections can be demonstrated on the marginal plexus, then the monsynaptic dorsal root connections with intranuclear dendrites seen in birds and mammals may be considered a significant change in vertebrate morphology.

MORPHOLOGICAL ADAPTATION

Anatomically, living salamanders provide a means of studying an important theoretical transition in the sequence of evolutionary changes in the locomotor system toward a fully terrestrial life. Urodeles are capable of swimming and two modes of terrestrial locomotion (Evans, 1946). Slow locomotion is accomplished by raising the body off the ground, followed by the movements of alternate limbs forward and rearward, the action of the legs themselves serving to propel the animal forward. Rapid locomotion is accomplished by swift rhythmic undulations of swimming-like movements of the body with little assistance from the limbs. Urodeles have undergone an abduction of the proximal limb sigments (humerous and femur) so that they extend at right angles from the body and are

---

FIG. 22 Drawing of the course and termination of dorsal root axon degeneration (stippling) following lumbar dorsal rhizotomy in the pigeon. See the text for an explanation of the findings.

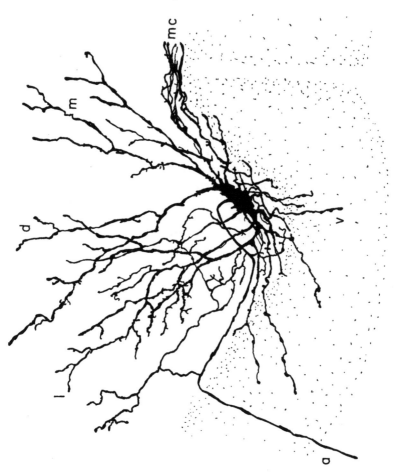

372

23

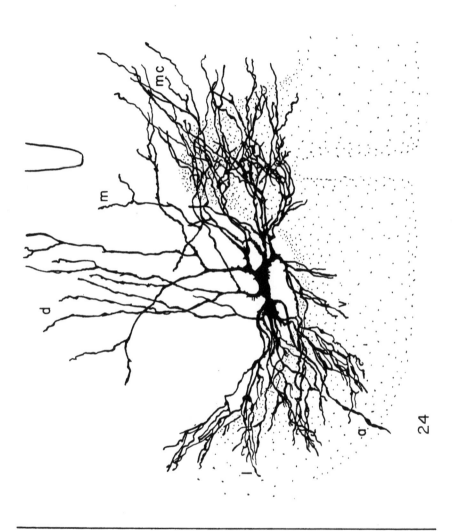

FIGS. 23 and 24 The dendritic ramification of ventral horn motor neurons in the newborn dog (Fig. 24) and fetal cat (Fig. 23). Note the great length of dendrites extending through gray matter beyond the limits of the ipsilateral parent motor column. Commissural dendritic branches cross to the contralateral side through the spinal commissures. The dendrites may be classified according to their position as ventral (v), dorsal (d), lateral (l), medial (m), extranuclear and intranuclear (longitudinal) dendrites. Modified and redrawn from illustrations by Ramón y Cajal (1952). The lower-case a identifies the axons of these neurons.

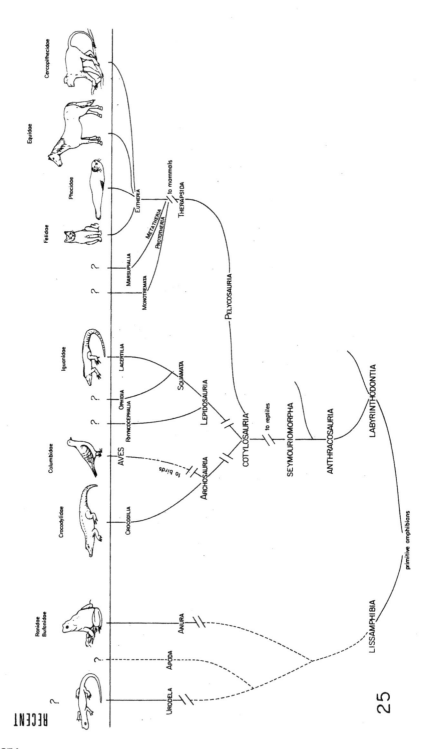

25

horizontal. The elbows point caudally and the knees rostrally, whereas the middle segments, the forearms and legs (radius and ulna, and tibia and fibula), are vertically oriented. The forefeet and hindfeet are rotated forward so that the digits are directed rostrally. Salamanders inhabit moist or wet lands and are found close to the water. While in the water, they place reliance on the trunkal musculature for propelling the body. *Bufo* and *Rana* are very specialized amphibians typified by the extreme foreshortening of the body, bone fusion in limbs, and axial skeleton. Ambulation is accomplished by walking or by the more specialized anuran means of locomotion, jumping. In some respects a saltatory gait is a relatively invariable or stereotyped movement; locomotion proceeds from point to point in ballistic fashion, its owner commonly bound to a fixed flight path. *Rana* also swims. This it does by rapid extension of its hindlegs and open webbed feet. When stationary *Bufo* and *Rana* assume a semiprone posture, the abdomen resting on the ground with fore- and hindlimbs flexed. Urodeles also utilize a prone resting posture. The demands placed on amphibians to assume and maintain the semiprone posture must differ significantly from the stringent requirements of keeping an upright mammal in balanced equilibrium.

More advanced and varied are the locomotor adaptation of mammals to a terrestrial life. Most mammals are capable of standing, a form of posture possessed by their advanced therapsid forebearers. In standing, the limbs are brought beneath the body, as opposed to the laterally abducted femur and humerus of limbed quadrupedal amphibians and reptiles. Although the limb bones bear much of the body's weight, these are jointed organs and collapse under this weight outside of a theoretical equilibrium state. Capsular muscle groups, together with ligamentous attachments, serve to keep the joint surfaces apposed and additional muscle groups acting across the adductor, abductor, flexor, and extensor surfaces not only stabilize joints preventing collapse of the limbs, but move one segment on the other. Muscle spindle organs serve to signal the degree of muscle tonicity and stretch and, through spinal reflex pathways, maintain a state of equilibrium to secure the standing position. Quadrupedal standing incurs great risks compared with the prone or semiprone resting posture seen in lower tetrapods. Mammals may be characterized by the wealth of monosynaptic connections made by the Group IA spindle afferents on motoneuronal dendrites. This is indicative of a greater influence of dorsal root fibers on the electrical activity of motor neurons and perhaps is closely tied to the

---

FIG. 25   Phylogenetic outline of some major vertebrate lineages from which recent amphibians, birds, reptiles and mammals have arisen. Very few major extant lower taxa are represented. The vertebrate families for which there are experimental data on dorsal root connections are indicated at the top of the diagram above the horizontal line symbolizing Recent geological time. Question marks indicate gaps in our knowledge of other vertebrate lineages.

evolutionary advances of the mammalian tetrapod posture, the facility to assume a fully standing position maintained over a long temporal span.

CONCLUDING COMMENTS

This brief examination of the spinal cord anatomy of tetrapods has led to a more detailed consideration of the comparative anatomy of monosynaptic connections with somatic motor neurons. Considering first the number of families and the adaptations exhibited by different amphibians, reptiles, birds, and mammals, and second, the incompleteness of the neurological literature, any functional generalizations regarding central connections of dorsal root ganglion cells seems hazardous. The available morphological data, nevertheless, tempt me to speculate on the functional implications of these connections.

All tetrapods possess kinesthetic afferents that form monosynaptic terminals on the dendritic surfaces of somatic motor neurons. Among tetrapods there is a wide divergence in synaptic density on dorsal and on intranuclear dendrites. Intranuclear synaptic density may be correlated with the establishment of a myotatic reflex and be further correlated with two factors: fully upright quadrupedal standing and complex ambulation. The morphological hypotheses and implications and their functional correlates may be summarized as follows.

1. Anuran amphibians possess monosynaptic dorsal root—motoneuronal connections limited to the distal segments of dorsal dendrites (Figure 26). Urodeles may be similarly constituted, although this is not actually known. Suprasegmental and propriospinal impulses, when coupled with muscle nerve afferents can cause firing of ventral horn motor neurons, whereas muscle afferent stimulation alone fails to fire (Holemans, Meij, & Meyer, 1966; Meij, Holemans, & Meyer, 1966) somatic motor neurons. The dorsal root connections described here suggest that the myotatic reflex has not been developed in anurans. This limited input may be correlated with quadrupedal semiprone standing and relatively invariate patterns of ambulation and swimming.

2. Archosaurian and lepidosaurian reptiles exhibit differences in the connectivity of dorsal root axons with appendicular somatic motor neurons. Crocodilians more closely resemble anuran amphibians in the absence of dorsal root fibers in the ventral horn. Afferent dorsal root connections appear limited to the distal ends of dorsal dendrites that project into the zona intermedia and dorsal horn (Figure 27). Terrestrial lizards possess dorsal root axons that synapse on a larger area of the dorsal dendritic arbor of ventral horn motor neurons (Figure 29). The increasing invasion and synaptic spread of dorsal root connections in the ventral horn of iguanids may reflect the better terrestrial adaptations and climbing abilities of these lizards compared with the principally amphibious adaptation of crocodilians.

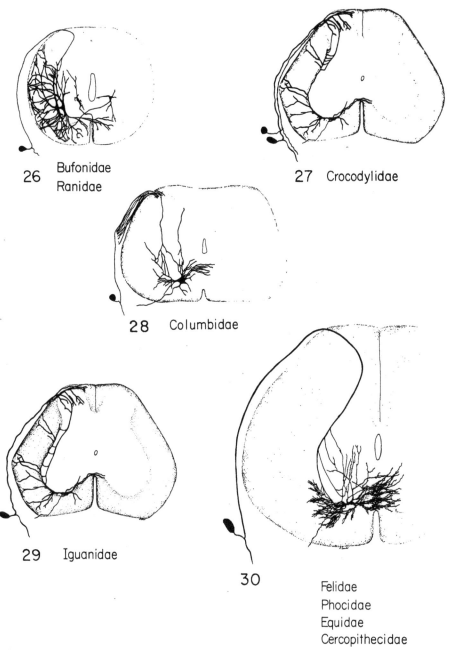

26 Bufonidae
    Ranidae

27 Crocodylidae

28 Columbidae

29 Iguanidae

30

    Felidae
    Phocidae
    Equidae
    Cercopithecidae

FIGS. 26–30 Schematic summary diagramming the supposed relationship of monosynaptic dorsal root motoneuronal connections in various families of lower tetrapods, mammals, and birds. See section on Concluding Comments.

3. Birds possess dorsal root afferent fibers (probably kinesthetic in function) diffusely distributed among the intranuclear dendrites of somatic motor neurons innervating muscles of the pectoral and pelvic limbs (Figure 28). This may represent an advance over their archosaurian ancestors and in part reflect the taxon's adaptation to flight in the pectoral limbs and to upright bipedal standing and ambulation in the pelvic limbs. Modern birds perhaps have advanced over their thecodont ancestors by an increase in the number of afferent connections to motoneuron dendrites. Theoretically, then, a myotatic reflex has evolved somewhere in the interval between ancestral pseudosuchian archosaurs and modern birds. As the class Aves is well advanced by the Cretaceous, this suggests the myotatic reflex may have been present for 130 million years.

4. The phyletic line leading from therapsid reptiles to primitive stem mammals has witnessed a major shift of kinesthetic inputs. Muscle afferent fibers probably occupy a greater area on both the intranuclear and extranuclear dendritic surfaces of somatic motor neurons (Figure 30). This may be correlated with the development in primitive insectivorous mammals (or in the interval between advanced therapsids and early mammals) of a fully extended quadrupedal posture combined with more complex patterns of gait. The important morphological–functional event is the establishment of the myotatic reflex in antigravity muscles: the stretching of antigravity muscles alone is sufficient to fire ventral horn motor neurons (appendicular and axial) and to cause contraction of homonymous striated muscles. Coupled with the evolution of other organ systems, the establishment of the myotatic reflex necessary for antigravity postures and the importance of phasic and static muscle receptors during ambulation better enabled mammals to occupy varied terrestrial habitats. The myotatic reflex evolving in the therapsid–mammalian lineage may have been present either in the Late Triassic or Jurassic (200–180 million years ago), if developed in advanced therapsid reptiles, or the Cretaceous (130 million years ago), if developed in primitive mammals.

5. Theoretically, the potential or actual acquisition of a myotatic reflex mechanism has occurred independently and perhaps several times in the course of vertebrate evolution: in the thecodont lineage toward birds and the therapsid lineage toward mammals. Whether terrestrial lizards similarly have developed muscle spindle IA afferents that can cause contraction of their homonymous muscles remains to be demonstrated.

REFERENCES

Ariëns-Kappers, C. U., Huber, G. C., & Crosby, E. *The comparative anatomy of the nervous system of vertebrates, including man.* Vol. I. New York: Macmillan, 1936.

Banchi, A. La minuta struttura della midolla spinale dei Chelonii (*Emys europaea*). *Archives Italiennes di Anatomie e di Embriologie*, 1903 2, 291 (cited by Ariëns-Kappers, Huber & Crosby, 1936).

Carpenter, M. B., Stein, B. M., & Shriver, J. E. Central projections of spinal dorsal roots in the monkey. II. Lower thoracic, lumbosacral and coccygeal dorsal roots. *American Journal of Anatomy*, 1968, **123**, 75–118.

Chambers, W. W., Sprague, J. M., & Liu, C. N. Anatomical organization of the frog and cat spinal cord, dorsal root and propriospinal pathways. *American Journal of the Medical Sciences*, 1960, **240**, 122–124.

Cruce, W. L. R. The anatomical organization of hindlimb motoneurons in the lumbar spinal cord of the frog. *Journal of Comparative Neurology*, 1974, **153**, 59–76. (a)

Cruce, W. L. R. Supraspinal projections to the spinal cord of the Tegu lizard *(Tupinambis nigropunctatus)*. *Anatomical Record*, 1974, **178**, 337. (b)

de Lange, S. J. Das Hinterhirn, das Nachhirn und das Rüchenmark der Reptilien. *Folia Neuro-biologica*, 1917, **10**, 385.

Elftman, H. The action of muscles in the body. *Biological Sympsia*, 1941, **3**, 191–209.

Evans, F. G. Anatomy and function of the foreleg in salamander locomotion. *Anatomical Record*, 1946, **95**, 257–281.

Gray, J. Studies in the mechanics of the tetrapod skeleton. *Journal of Experimental Biology*, 1944, **20**, 88–116.

Gray, J. *Animal locomotion.* London: Weidenfeld and Nicolson, 1968.

Hecht, M. K. The living lower tetrapods: Their interrelationships and phylogenetic position. *Annals of the New York Academy of Sciences*, 1969, **167**(1), 74–79.

Heimer, L., & Wall, P. D. The dorsal root distribution to the substantia gelatinosa of the rat with a note on the distribution in the cat. *Experimental Brain Research*, 1968, **6**, 89–99.

Holemans, K. C., Meij, H. S., & Meyer, B. J. The existence of a monosynaptic reflex arc in the spinal cord of the frog. *Experimental Neurology*, 1966, **14**, 175–186.

Hopson, J. A. The origin and adaptive radiation of mammal-like reptiles and nontherian mammals. *Annals of the New York Academy of Sciences*, 1969, **167**, 199–216.

Huber, J. F. Nerve roots and nuclear groups in the spinal cord of the pigeon. *Journal of Comparative Neurology*, 1936, **65**, 43–91.

Joseph, B. S., & Whitlock, D. G. Central projections of selected spinal dorsal roots in anuran amphibians. *Anatomical Record*, 1968, **160**, 279–288. (a)

Joseph, B. S., & Whitlock, D. G. Central projections of brachial and lumbar dorsal roots in reptiles. *Journal of Comparative Neurology*, 1968, **132**, 469–484. (b)

Joseph, B. S., & Whitlock, D. G. The morphology of spinal afferent–efferent relationships in vertebrates. *Brain, Behavior & Evolution*, 1968, **1**, 2–18. (c)

Keenan, E. The philogenetic development of the substantia gelatinosa Rolandi. Part II. Amphibians, reptiles and birds. *Koninklijke Nederlandse Akademie van Wetenschappen te Amsterdam*, 1929, **32**, 299.

Kennard, D. W. The anatomical organization of neurons in the lumbar region of the spinal cord of the frog *(Rana temporaria)*. *Journal of Comparative Neurology*, 1959, **111**, 447–468.

Leonard, R. B., & Cohen, D. H. Spinal projections of brachial dorsal root fibers in the pigeon *(Columba livia)* (personal communication). (a)

Leonard, R. B., & Cohen, D. H. Spinal projections of lumbar dorsal root fibers in the pigeon *(Columba livia)* (personal communication). (b)

Liu, C. N., & Chambers, W. W. Experimental study of anatomical organization of frog's spinal cord. *Anatomical Record*, 1957, **127**, 326.

Massazza, A. La citoarchitettonica del midollo spinale umano. *Archives d'anatomie, d'histologie et d'embryologie*, (Strasbourg), 1922, **1**, 323–410.

Massazza, A. La citoarchitettonica del midollo spinale umano. *Archives d'anatomie, d'histologie et d'embryologie*, (Strasbourg), 1923, **2**, 1–56.

Massazza, A. La citoarchitettonica del midollo spinale umano. *Archives d'anatomie, d'histologie et d'embryologie*, (Strasbourg), 1924, **3**, 115–189.

Matsushita, M. Zur Zytoarchitektonik des Hühnerrückenmarkes nach Silberimpragnation. *Acta Anatomica*, 1968, **70**, 238–259.

McKenna, M. C. The origin and early differentiation of therian mammals. *Annals of the New York Academy of Sciences*, 1969, **167**(1), 217–240.

Meij, H. S., Holemans, K. C., & Meyer, B. J. Monosynaptic transmission from afferents of one segment to motoneurons of other segments in the spinal cord. *Experimental Neurology*, 1966, **14**, 496–505.

Melzack, R., & Wall, P. D. Pain mechanisms: A new theory. *Science*, 1965, **150**, 971–979.

Nieuwenhuys, R. Comparative anatomy of the spinal cord. In J. C. Eccles & J. P. Schadé (Eds.), *Organization of the spinal cord: Progress in brain research*, Vol. 11. Amsterdam: Elsevier, 1964. Pp. 1–57.

Nyberg-Hansen, R. Anatomical demonstration of gamma motoneurons in the cat's spinal cord. *Experimental Neurology*, 1965, **13**, 71–81.

Olszewski, J. On the anatomical and functional organization of the spinal trigeminal nucleus. *Journal of Comparative Neurology*, 1950, **92**, 401–413.

Patterson, B. Early Cretaceous mammals and the evolution of mammalian molar teeth. *Fieldiana: Geology*, 1956, **13**, 1–105.

Petras, J. M. Afferent peripheral nerve fibers to the spinal cord and dorsal column nuclei in the cat. An analysis and comparison with the distribution of terminal efferent brain fibers to the spinal cord. *Anatomical Record*, 1965, **151**, 399–400.

Petras, J. M. Afferent fibers to the spinal cord. The terminal distribution of dorsal root and encephalospinal axons. In A. T. Jousse (Ed.), *Fourteenth Spinal Cord Injury Conference of the Veterans Administration, U. S. A., Medical Services Journal of Canada*, 1966, **22**, 668–694.

Petras, J. M. Cortical, tectal and tegmental fiber connections in the spinal cord of the cat. *Brain Research*, 1967, **6**, 275–324.

Petras, J. M. The substantia gelatinosa of Rolando. *Experientia*, 1968, **24**, 1045–1047.

Petras, J. M. Some efferent connections of the motor and somatosensory cortex of simian primates and felid, canid and procyonid carnivores. *Annals of the New York Academy of Sciences*, 1969, **167**(1), 469–505.

Petras, J. M., & Cummings, J. F. Autonomic neurons in the spinal cord of the rhesus monkey: A correlation of the findings of cytoarchitectonics and sympathectomy with fiber degeneration following dorsal rhizotomy. *Journal of Comparative Neurology*, 1972, **146**, 189–218.

Ralston, H. J. III. Dorsal root projections to the dorsal horn neurons in the cat spinal cord. *Journal of Comparative Neurology*, 1968, **132**, 303–330.

Ramón y Cajal, S. La medulla espinal de los reptiles. Barcelona, 1891 (cited by Ariëns-Kappers, Huber, & Crosby, 1936).

Ramón y Cajal, S. *Histologie du systeme nerveux de l'homme et des vertebrés*. Tome I. L. Azoulay (Trans.). Madrid: Consejo Superior de Investigaciones Cientificas, Instituto Ramón y Cajal, 1952.

Retzius, G. Die embryonale Entwicklung der Rückenmarkselemente bei den Ophidiern. *Biologische Untersuchungen*, 1894, **6**, 41.

Retzius, G. Weiteres über die embryonale Entwicklung der Rückenmarkselemente der Ophidiern. *Biologische Untersuchungen*, 1898, 8, 105. (a)

Retzius, G. Zur Kenntniss der Entwicklung der Elemente des Rückenmarks von *Anguis fragilis*. *Biologische Untersuchungen*, 1898, **8**, 109. (b)

Rexed, B. The cytoarchitectonic organization of the spinal cord in the cat. *Journal of Comparative Neurology*, 1952, **96**, 415–496.

Rexed, B. A cytoarchitectonic atlas of the spinal cord in the cat. *Journal of Comparative Neurology*, 1954, **100**, 297–379.

Rexed, B. Some aspects of the cytoarchitectonics and synaptology of the spinal cord. In J.

C. Eccles & J. P. Schadé, (Ed.), *Organization of the spinal cord. Progress in brain research,* Vol. 11. Amsterdam: Elsevier, 1964. Pp. 58–92.

Romanes, G. J. The motor cell columns of the lumbosacral spinal cord of the cat. *Journal of Comparative Neurology,* 1951, **94,** 313–364.

Romer, A. S. *Vertebrate paleontology.* Chicago: The University of Chicago Press, 1966.

Sala, L. Estructura de la medula espinal de los batracios. Barcelona: 1892 (cited by Ariëns-Kappers, Huber, & Crosby, 1936).

Selzer, M., & Spencer, W. A. Interactions between visceral and cutaneous afferents in the spinal cord: reciprocal primary afferent fiber depolarization. *Brain Research,* 1969, **14,** 349–366.

Shriver, J. E., Stein, B. M., & Carpenter, M. B. Central projections of spinal dorsal roots in the monkey. I. Cervical and upper thoracic dorsal roots. *American Journal of Anatomy,* 1968, **123,** 27–74.

Silver, M. L. Motoneurons of the spinal cord of the frog. *Journal of Comparative Neurology,* 1942, **77,** 1–39.

Simpson, G. G. The principles of classification and a classification of mammals. *Bulletin of the American Museum of Natural History,* 1945, **85,** 1–350.

Sprague, J. M. Motor and propriospinal cells in the thoracic and lumbar ventral horn of the rhesus monkey. *Journal of Comparative Neurology,* 1951, **95,** 103–123.

Sprague, J. M. The distribution of dorsal root fibers in motor cells in the lumbosacral spinal cord of the cat, and the site of excitatory and inhibitory terminals in monosynaptic pathways. *Proceedings of the Royal Society, London,* 1958, 149B, 534–556.

Sprague, J. M., & Ha, H. The terminal fields of dorsal root fibers in the lumbosacral spinal cord of the cat, and the dendritic organization of the motor nuclei. In J. C. Eccles & J. P. Schadé (Eds.), *Organization of the spinal cord: Progress in brain research,* Vol. 11. Amsterdam: Elsevier, 1964. Pp. 120–154.

Sterling, P., & Kuypers, H. G. J. M. Anatomical organization of the brachial spinal cord of the cat. I. The distribution of dorsal root fibers. *Brain Research,* 1967, 4, 1–15.

Streeter, G. L. The structure of the spinal cord of the ostrich. *American Journal of Anatomy,* 1904, 3, 1.

Terni, T. Sulla correlazione fra ampiezza del territorio di innervazioni e grandezza delle cellule gangliari: 2 Ricerche sui gangli spinali che innervano la coda rigenerata, nei Sauri (*Gongylus ocellatus*). *Archives Italiennes di Anatomie e di Embriologie,* 1920, 17, 507 (cited by Ariëns-Kappers, Huber, & Crosby, 1936).

Terni, T. Ricerche istologiche sui midollo spinale dei rettili, con particolare riguardo ai componenti spinali del fascicolo longitudinale mediale (Osservazioni in *Gongylus ocellatus* Wagl.) *Archives Italiennes di Anatomie e di Embriologie,* 1922, 18(Suppl.), 183 (cited by Ariëns-Kappers, Huber and Crosby, 1936).

Terni, T. Ricerche istologische sul midollo spinale dei rettili. *Archives Italiennes di Anatomie e di Embriologie,* 1922, 18(Suppl.) 183 (cited by Ariëns-Kappers, Huber, & Crosby, 1936).

Trevino, D. L., Coulter, J. D., & Willis, W. D. Location of cells of spinothalamic tract in lumbar enlargement of the monkey. *Journal of Neurophysiology,* 1973, **36,** 750–761.

van Gehuchten, A. Les éléments nerveaux moteurs des racines postérieurs. Anatomischer Anzeiger, 1893, 8, 215.

van Gehuchten, A. Contribution á l'étude de la moelle épinere chez les vertebrés *(Tropidonotus natrix).* La Cellule, 1897, **12:**115 (cited by Ariëns-Kappers, Huber & Crosby, 1936).

von Lenhossék, M. *Der feinere Bau des Nervensystems im Lichte neuster Forschungen.* Berlin: H. Kornfield, 1893.

Wall, P. D. The laminar organization of dorsal horn and effects of descending impulses. *Journal of Physiology (London),* 1967, ĺ88, 403–423.

16

Primate Radiations and
the Origin of Hominoids

Elwyn L. Simons

Yale University

In 1969 I published a study estimating times for the principal radiations and points of dichotomy among Order Primates. This contribution will provide an interim report on the newer information about splitting times in Primates. Such studies about dating primate fossils are partly in response to the so-called "molecular clock" dating method of Sarich and Wilson (1968), see also Sarich (1968a, b, 1970). The method extrapolates from the amount of immuno-chemical difference that can be determined to exist between the serum albumins of different living primate genera. The degrees of such difference are expressed in terms of a numerical evaluation (immunological distance or ID number).

A basic assumption is that the immunological distance that separates the various end members of the Order is a consequence of a slow and constant accumulation of differences in each separate lineage that has led to a living species. Such determinations require that in each branch of the primate family tree the rate of accumulation of differences has also been constant, that is, that body chemistry is under neutral selection. Only with such constancy would there be a direct relationship between the number of serum albumin differences separating two particular primate genera and species and the time since separation of their ancestral stock. Operating on this assumption and given one absolute date, such as the time of the basal radiation of this, or any order, it is assumed that all subsequent dates of divergence or radiation can be calculated.

The effectiveness of this method as a clock has been subjected to criticism, see Simons (1969), Uzzell and Pilbeam (1971), Lovejoy and Meindl (1972), Lovejoy, Burstein and Heiple (1972), Farris (1972), Van Valen (1974) and Ayala (1974), to mention only a few of the principle papers. All the branch point dates so far provided by the "molecular clock" dating method of Sarich are younger than the dates of divergence among primates based on dated fossils. Some of these dated radiations of fossil primates might be due to parallelism within the order being mistaken for actual relatedness, but since 10 different radiations are involved, it seems highly unlikely that all these primates are wrongly assigned or that the ID dates determined by Sarich (ref. cit.) are correct. However, it is not my intention to deal further here with the validity of ID dates, because the wide discrepancies between ID calculated times of divergence and those indicated by the primate fossil record as assessed by paleontologists and dated by geochemists as reasonably well known. It is somewhat ironic that the period of strongest promulgation of immunochemical dating and "molecular clocks" has closely coincided with the research that has accumulated most of the more accurate, geochemically determined, dates of the rock strata that contain fossil primates.

The first good summary of such potassium/argon dates, with respect to North America and to some extent Europe and Africa, is that of Evernden, Savage, Curtis and James (1964), but in the decade since then many more such studies have been published. Such information has made possible a much more precise placement of the time periods in which extinct primates have lived, a placement not possible previously (see Simons, 1963, 1965). Nevertheless, most relatively recent papers about early primates include little information on dates of divergence between the major taxonomic divisions of the order. Such reviewers of early primates include Van Valen and Sloan (1965), McKenna (1966, 1967), Szalay (1968), and Simons (1972).

The degree of unfamiliarity of geochemical dating results that obtained when Sarich and Wilson first published was embodied in the very date they took for the time of the initial radiation of the order: 60 million years. Combination of the dates given by Evernden et al. (1964) with the information supplied by Van Valen and Sloan (1965) on the Late Cretaceous occurrence of the earliest primate, *Purgatorius ceratops*, would have, in 1965, put the time of first documentation of Order Primates before the Paleocene and thus no later than in the period between 65 and 70 million years ago. Moreover, differentiation of Order Primates in the Late Cretaceous was not then a novel idea. This view had been held since the early 1950s at least by Jepsen (personal communication) and by other paleontologists such as myself (Simons, 1963, p. 75) or as indicated by Van Valen and Sloan (1965). By middle Paleocene times in North America there were at least seven or eight different genera of primates in existence, see Simons (1969). Some of these, such as *Plesiolestes*, *Elphidotarsius*, and *Picrodus*, are so different from each other and the others that their very assignment to Order Primates has been questioned. It seems impossible that all the species of these genera can have been derived from the species of a single genus existing only shortly before the actual time of which there is a fossil record of primate diversity (middle Paleocene). Middle Paleocene times according to Evernden et al. (1964, p. 167) must date to about 60–63 million years ago. Mammalian genera often persist for 5–10 million years, an order of magnitude of time duration that is also indicated for such early primates as *Plesiadapis*, *Protoadapis*,

Phenacolemur, Pelycodus, Notharctus, and the later Tertiary genera *Pliopithecus, Dryopithecus, Gigantopithecus, Ramapithecus* and perhaps *Australopithecus.* Therefore, in order to follow back the several widely divergent middle Paleocene primates to a common ancestral species, it is not excessive to postulate two or more annectant, earlier and unknown genera. From these, species could have branched out to produce the documented diversity of the middle Paleocene. A minimum projection backward in time to reach a common stock for all these primates might well be 10 to 20 million years. Therefore, the record of diversity of the North American middle Paleocene does not indicate a radiation then, but more likely one that occurred some time between 70 and 83 million years ago. If so, all Sarich ID dates would need to be increased. Thus Sarich's 36 ± 3 million years platyrrhine-catarrhine splitting time would be converted to 40–46 million, and so forth. This correction would bring his data somewhat more nearly in line with the geological evidence, but even so there would be wide discrepancies between the two dating systems.

The vagueness of the foregoing calculations emphasize what I believe is an important point. The early history of placental mammals is still too poorly known for anyone to be able to state when the initial radiation of the Order Primates actually occurred. Perhaps it was about 80, but the date might even have been nearer 90 million rather than 60. Not only would it have taken a considerable (but difficult to estimate) amount of time to reach the middle Paleocene primate level of primate diversification, but the fact remains that we cannot know the amount of taxonomic diversity which then obtained. These earliest well known primates come from only a tiny fraction of the Holarctic region as it was then constituted. It is entirely possible that many more genera and species of Order Primates had differentiated by that time but remain undiscovered. Today's entire remaining Paleocene deposits of Europe (that were deposited on land) cover only a few square miles. Moreover, no Paleocene primate-yielding beds are known anywhere in the middle latitudes.

To this insufficiency of sampling must also be added the fact that the earliest primate radiation includes several lineages of rodent-like animals that seem unrelated to later forms: the Plesiadapiformes. It has recently, been suggested, however, that some of these plesiadapiform primates are in or near the ancestry of Eocene and later primates, see Bown and Gingerich (1973), Gingerich (1975), and Bown and Rose (in press). Nevertheless, definite ancestors of the second major primate radiation during the early Eocene (Sparnacian-Wasatchian time) have not been located in the middle or late Paleocene, and therefore almost nothing about the time of initiation of this apparent second radiation is known. It is widely assumed that the great variety of earliest Eocene primates represent a second, later radiation but this has not really been proved from fossils. It is equally possible to interpret the considerable variety of Eocene primates as having derived from two or more stocks that have separated from other primate lineages at the same time as the middle Paleocene radiation discussed above. It is only necessary to assume that the apparent early Eocene radiation is derived from undiscovered Paleocene lineages. This possibility is not unlikely, because the structure of known Paleocene primates does not suggest them to me as ancestral candidates for the early Eocene ones. Moreover, the fossil faunas of the period, even in North America where they are best known, come from restricted areas: a relatively narrow corridor running north to south through Montana, Wyoming, Colorado and New Mexico. Early Eocene primate ancestors were likely elsewhere.

It is because of considerations such as the foregoing that I differ from Radinsky (this volume) in dating the so-called "second primate radiation", because the dates he gives, from 55 to 35 million years ago, are the dates when the full diversification of this radiation is taking place, not when it begins. The latter date is unknown. It is true that the earliest members of this radiation date to the period of 58–55 million years, so that it can be said that this period brackets the earliest evidence of the "second radiation", but the fossil evidence does not tell when diversification of "primates of modern aspect" (Simons, 1963, 1972) begins. Earliest Eocene genera include *Teilhardina, Anemorhysis, Loveina, Tetonius, Omomys, Pseudotetonius, Agerina, Pelycodus,* and *Protoadapis,* among others. According to current classification these genera are assignable to at least two and possibly three different families. By the end of the first third of the Eocene a considerable number of additional genera appear and it is not clear that they all can be directly derived from these earliest Eocene genera. This, of course, also implies the former existence of even more undiscovered early Eocene stocks. Moreover, in the early Eocene, known faunas come from such a restricted areal sampling that there can be no certainty that even a third of the genera existing at the time are known. It is also interesting that at the very time of the radiation of primates of modern aspect, 58–55 million years, the faunas of Europe and North America are the most nearly identical ever. Data summarized by McKenna (1972) show that up until the end of the Sparnacian-Wasatchian there must have been a very extensive land connection across the North Atlantic between Western Europe and North America. According to Russell (1975) 50% of the genera of the two areas were then common to both. The continuity of this European North American continent was apparently disrupted by North Atlantic rifting at about the end of the early Eocene, at about 49 million years ago, see McKenna (1972). E. D. Cope (1878) first noted the occurrence of vertebrate genera in common in the early Eocene of Europe and Western North America. The similarity between European and North American early Eocene faunas is further discussed by Simpson (1947). It is only with the recent work of Russell (1975) and others that the extent of commonality of faunas has become clearly understood. In my opinion this same period, covered by the Sparnacian-Wasatchian provincial ages, between 58 and about 49 million years ago, is the last interval when hominoids, cercopithecoids, and ceboids could have shared a common ancestor. As Russell and McKenna have indicated both the paleogeography and the warmer, tropical climates of those times would have allowed that ancestor to have had a New and Old World distribution, ranging from areas currently as far-flung as France and Wyoming. Such distributions are known to have been the case then for the primate genera *Pelycodus, Phenacolemur, Teilhardina* and *Plesiadapis*. Thus, it is simplest to posit that the catarrhines and platyrrhines were carried into the New and Old Worlds by continental rifting between Euronearctica in the middle Eocene rather than having had to migrate by water or land to achieve their hemispheric distribution.

It is now well demonstrated that both primates and rodents were deployed in the two southern continents of Africa and South America by the early Oligocene. At Salla, Bolivia in a fauna of Deseadan provincial age (early Oligocene) there are various hystricomorph rodents and a primate, *Branisella*. The meaning of this distribution, with respect to rodents, is outlined by Wood and Patterson (1970). At about the same time, say 30–38 million years ago, a diversified primate fauna, together with phiomyid rodents has been recovered from the (early Oligocene) Egyptian Fayum badlands (Simons 1963, 1965, 1967a and b). The conventional view is that the introduction of primates and rodents into the two southern continents, has occurred by island hopping or rafting southward from the two respective continents to the north: Europe and North America. Just when in the Eocene these groups are introduced into Africa is not certainly known because of the paucity of Eocene continental deposits there, and the consequent great scarcity of African Eocene mammals, but rodent incisors have recently been reported from late Eocene deposits in southern Libya by Savage (1971). The evidence of first occurrence of primates and rodents in South America is somewhat different from that in Africa because earlier and extensive Eocene mammalian faunas are known in South America. However, these do not contain rodents or primates identified to date. In both cases the introductions are similar, in that the primates and rodents are accompanied either by no or by only a few other mammals arriving simultaneously from elsewhere. For this reason, it has become a textbook example that these two orders are able to achieve unusual distributions by "rafting". Almost alone among mammals they contain arboreal species able to cling to masses of floating vegetation, and thus could "raft" across open ocean. Washed out to sea on tangled vegetative masses during floods they would be carried by stages across island arcs. In the Americas at least, these are presumed to have spanned the gulf between the northern and southern continents in Eocene times, for instance, see Simpson (1947) or Weyl (1974).

When the evidence of continuous primate distribution from Western Europe to North America in the early Eocene is combined with the above Oligocene distributions, certain limits are put on the time when the three groups of higher primates can have diverged. As I have outlined, Simons, 1969, the evidence is clear that hominoid- and cercopithecoid-like primates have already diverged in the faunas documented in the early Oligocene of North Africa. In South

America the early Oligocene (Deseadan age) primate *Banisella* from Bolivia shows that Primates were by then in South America. If *Branisella* is a primate unrelated to the Ceboidea, then the latter group would have to represent a second introduction into South America: an unlikely event. Nevertheless, there seems to be no good reason to put *Branisella* outside Ceboidea.

In sum, the evidence from the Fayum, Egypt, and La Salla, Bolivia, proves that primates had reached the southern continents by the earliest Oligocene about 33–38 million years ago. The date of the branchings that produced the three superfamilies of suborder Anthropoidea: Ceboidea, Cercopithecoidea, and Hominoidea, has to have happened much earlier. How much earlier it actually happened is not clear, since at least two branchings are necessary to produce the tripartite division of Anthropoidea. The date given by Sarich (1968), 36 ± 3 million years as the time of catarrhine-platyrrhine separation is in the early Oligocene. But by this time the Fayum monkeys and the Fayum apes are well differentiated from each other; not to speak of the differentiation of their combined ancestor still earlier from that of Platyrrhines. To be where they are in the early Oligocene, these three groups of primates must have derived from an originally Holarctic ancestor species, the descendants of which have produced, at different times, at least two subsequent branchings. Then the end members must already have migrated to points halfway around the world. From a whole range of anatomical and biochemical evidence it is known that hominoids and cercopithecoids are closer to each other than either is to the ceboids. Therefore, divergence of the basal species ancestors of these superfamilies must come long after that of the catarrhine-platyrrhine separation. Reasonable time projections for these two prior branchings should be consistent with the diversity of Fayum primates. There are 10+ species of six genera, belonging to at least two families. Considering all these factors one could not put the cercopithecoid-hominoid split much later than 40 million years ago. The Platyrrhine-catarrhine divergence must come considerably earlier, perhaps in the period around 50 to 55 million years ago.

In the early Oligocene, at 36 ± 3 million years a diverse group of primates, all Anthropoidea, occurs in the Fayum Africa. A split at this time requires, in an obligatory sense, two things: (1) That the newly split platyrrhine branch cross into South America after that date. (2) That primitive catarrhines remaining in Africa did not diverge into separate monkey and ape groups until much later in order that their closer biochemical similarity to each other than to platyrrhines be explainable. In order to believe that this is what happened the burden of proof lies on establishing a plausible route into South America and in showing that the Fayum primates are so generalized that none resembles specific catarrhine groups. In consequence of Sarich's branch point time being pene-contemporary with them, only one of them (at the very most) could be an acceptable ancestor to later forms. All the evidence is against these two conclusions. *Aegyptopithecus* and *Propliopithecus* bear extensive, specific resemblances to two different genera of the later (Miocene) apes of East Africa and seem to be good ancestors of each, respectively *Dryopithecus* and *Limnopithecus*. Species of all four genera, *Aegyptopithecus, Aeolopithecus, Oligopithecus* and *Propliopithecus* are unlike monkeys. In effect, however, anyone supporting a post-early Oligocene divergence for these groups must maintain that ceboids and cercopithecoids are derived from Oligocene forms that are in dental anatomy like Miocene-Recent apes. Such an interpretation takes no account of the fact that the teeth of modern Ceboidea most closely resemble in their anatomy those of North American Eocene primates not those of apes: compare, for instance, *Omomys: Saimiri* and *Alouatta: Notharctus.*

If it is desired to posit, as Sarich does, a late splitting of the New and Old World Anthropoidea, then as we have seen, one is immediately confronted with difficult zoogeographic considerations. Hominoids, ceboids, and cercopithecoids are more closely related to each other than any is to the prosimian primates. If Sarich's (1968) estimate is accepted for the time of platyrrhine-cattarrhine splitting, then it is obligatory to posit the rafting of one of the two groups across the South Atlantic. As discussed above and by McKenna (1972) after 49/50 million years ago direct overland crossings between Europe and North America were no longer possible. Rafting across the South Atlantic to account for the distribution of South American hystricomorph rodents and primates has been suggested by Hoffstetter (1972, 1974). The origin of the idea of a South Atlantic rafting for primates and rodents is the supposition of a narrower width of ocean at that time. The latest evidence from geophysical studies regarding the width of the South Atlantic at 35 to 40 m.y. does make it significantly narrower but still too wide for rafting. Funnel and Smith (1968), Smith and Hallam (1970) and Ladd, Dickson and Pitman (1973) have presented evidence on this point. It appears that at 40 million years BP the South Atlantic was at least 70% of its present width, a distance far too great for a plausible small mammal rafting. Fallen and tangled trees are implausible reservoirs of water. Covered by salt spray their leaves wilt and the raft would not move faster than ocean currents could carry it. Long before such a raft with its posited primate passengers had traveled 500 km they would have died of thirst. In spite of the fact that a paleontologist, Hoffstetter, has called on South Atlantic rafting to explain certain zoogeographical facts it remains a highly implausible hypothesis. Right up to the present time no scholar has succeeded in explaining away the numerous possible objections to it, or has proven any cases of primate long distance rafting.

The earliest catarrhines from the Fayum Oligocene are considerably advanced beyond the general prosimian condition (Simons, 1967a, b, 1972). Both the group represented by *Parapithecus* and *Apidium* and that represented by *Aegyptopithecus*, the oldest well-known hominoid have reached a definite higher primate grade. None can be explained away as not being Catarrhini, see Simons (1972). The Fayum species have no special dental resemblance to platyrrhines other than primitive characters such as retention of P_2. Species of all these primates show mandibular symphyseal fusion in juveniles, metopic closure at an early age, and *Apidium* and *Aegyptopithecus* have full postorbital closure. However, the dental differences between the parapithecids and the other Fayum genera (dryopithecine pongids or hylobatids) strongly suggest that their latest common ancestor in Africa had lived many millions of years before the time of their 30–38 million year old occurrence.

The best conjecture that can be made at present is that the splitting-time which produced Hominoidea and Cercopithecoidea was about 45 ± 5 million years B.P. and that it was in Africa. The split of Catarrhini and Platyrrhini was probably at about 10 million years earlier in the European-North American land mass. In sum, Hominoidea are best thought to have had their origin in Africa in middle Eocene times.

REFERENCES

Ayala, F. J. Biological evolution: natural selection or random walk? *American Scientist,* 1974, **62**, 692–701.

Bown, T. M., & Gingerich, P. D. The Paleocene primate *Plesiolestes* and the origin of Microsyopidae. *Folia Primatologia,* 1973, **19**, 1–8.

Bown, T. M., & Rose, K. D. New early Tertiary primates and a reappraisal of some Plesiadapiformes. *Folia Primatologia,* 1975, in press.

Cope, E. D. Sur les relations des niveaux de vertebres eteints dans l'Amerique du Nord et en Europe, *Comptes Rendus Congres International de Geologie* (Paris), 1878.

Evernden, J. F., Savage, D. E., Curtis, G. H., & James, G. T. Potassium-argon dates and the Cenozoic mammalian chronology of North America. *American Journal of Science,* 1964, **262**, 145–198.

Farris, J. S. Estimating phylogenetic trees from distance matrices. *American Naturalist,* 1972, **107**(951), 645–668.

Funnel, B. M., & Smith, A. G. Opening of the Atlantic Ocean. *Nature,* 1968, **219**, 1328–1331.

Gingerich, P. D. Anatomy of the temporal bone in the Oligocene anthropoid *Apidium* and the origin of Anthropoidea. *Folia Primatologia,* 1973, **19**, 329–337.

Gingerich, P. D. Systematic position of *Plesiadapis.* Nature, 1975, **253**, 111–113.

Hoffstetter, R. Relationships, origins, and history of the ceboid monkeys and caviomorph rodents: a modern reinterpretation. In T. Dobzhansky, M. K. Hecht, & W. C. Steere, (Eds.). *Evolutionary Biology,* New York: Appleton-Century Crofts 1972, **6**, 323–347.

Hoffstetter, R. Phylogeny and geographical deployment of the Primates. *Journal of Human Evolution,* 1974, **3**, 327–350.

Lavocat, R. La systematique des rongeurs hystricomorphes et la derive des continents. *Comptes Rendus Academie des Sciences (Paris), Series D,* 1969, **269**, 1496–1497.

Ladd, W. J., Dickson, G. O., & Pitman, W. C., III. The age of the South Atlantic. In A. M. Nairn & F. G. Stehli (Eds.), *The ocean basins and margins,* Vol. I. New York: Plenum Press, 1973, pp. 555–573.

Lovejoy, C. O., Burstein, A. H., & Heiple, K. G. Primate phylogeny and immunological distance. *Science,* 1972, **176**, 803–805.

Lovejoy, C. O., & Meindl, R. S. Eukaryote mutation and the protein clock. *Yearbook of Physical Anthropology,* 1972, **16**, 18–30.

McKenna, M. C. Paleontology and the origin of the primates. *Folia Primatologia,* 1966, **4**(1), 1–25.

McKenna, M. C. Classification, range and deployment of the prosimian primates. In: Problems Actuels de Paleontologie (Evolution des Vertebres). *Colloques Internationaux Centre National de la Recherche Scientifique (Paris)*, 1967, **163**, 607–610.

McKenna, M. C. Eocene final separation of the Eurasian and Greenland–North American land masses. *24th International Geological Congress*, 1972, **7**, 275–281.

Russell, D. E. Paleoecology of the Paleocene-Eocene transition in Europe. In F. S. Szalay (Ed.), *Contributions to primatology*. Vol. 5. Switzerland: Karger, 1975. Pp. 28–61.

Sarich, V. M. The origin of the hominids: An immunological approach. In S. L. Washburn & P. C. Jay (Eds.), *Perspectives on human evolution*. Vol. 6. New York: Holt, Rinehart & Winston, 1968. Pp. 94–131. (a)

Sarich, V. M. Quantitative immunochemistry and the evolution of Old World primates. In B. Chiarelli (Ed.), *Taxonomy* and *phylogeny of Old World primates with references to the origin of Man*. Torino: Rosenberg & Sellier, 1968. Pp. 139–140. (b)

Sarich, V. M. Primate systematics with special reference to Old World monkeys, In J. R. Napier & P. H. Napier (Eds.), *Old World Monkeys*. New York: Academic Press, 1970. Pp. 175–226.

Sarich, V. M., & Wilson, A. C. Immunological time scale for hominid evolution. *Science* 1968. **158**, 1200–1202.

Savage, R. J. G. Review of fossil mammals of Libya. In *Symposium on the geology of Libya*. Faculty of Sciences, University of Libya, Tripoli, 1971. Pp. 217–225.

Simons, E. L. A critical reappraisal of Tertiary primates. In J. Buettner-Janusch (Ed.), *Evolutionary and genetic biology of primates*. Vol. 20. New York: Academic Press, 1963. Pp. 65–129.

Simons, E. L. New fossil apes from Egypt and the initial differentiation of the Hominoidea. *Nature (London)*, 1965, **205**, 135–139.

Simons, E. L. New evidence on the anatomy of the earliest catarrhine primates. In D. Starck, R. Schneider & H. J. Kuhn (Eds.), *Neue Ergebnisse der Primatologie* (Progress *in Primatology*). Stuttgart: Fischer. 1967. Pp. 15–18. (a)

Simons, E. L. Review of the phyletic interrelationships of Oligocene and Miocene Anthropoidea. In *Problemes actuels de paleontologie (Evolution des Vertebres)*. *Colloques Internationaux des Centre National de la Recherche Scientifique (Paris)*, 1967. **163**, 597–602. (b)

Simons, E. L. The origin and radiation of the primates. *Annals of the New York Academy of Sciences*, 1969, **167**, 319–331.

Simons, E. L. *Primate evolution*. New York and London: Macmillan, 1972.

Simpson, G. G. Holarctic mammalian faunas and continental relationships during the Cenozoic. *Bulletin, Geologic Society of America*, 1947, **58**, 613–688.

Smith, A. G., & Hallam, A. The fit of the Southern Continents. *Nature*, 1970, **139**, 139–144.

Szalay, F. S. The beginning of primates. *Evolution*, 1968, **22**, 19–36.

Uzzell, T., & Pilbeam, D. Phyletic divergence dates of hominoid primates: A comparison of fossil and molecular data. *Evolution*, 1971, **25**, 615–635.

Van Valen, L., & Sloan, R. E. The earliest primates. *Science*, 1965, **150**, 743–745.

Van Valen, L. Molecular evolution as predicted by natural selection. *Journal of Molecular Evolution*, 1974, **3**, 89–101.

Weyl, R. Die palaogeographische Entwicklung Mittelamerikas. *Zentralblatt für Geologie und Paläontologie*, 1974, Teil I, 5/6, 432–466.

Wood, A. E. An Eocene hystricognathous rodent from Texas. Its significance in interpretations of continental drift. *Science*, 1972, **175**, 1250–1251.

Wood, A. E., & Patterson, B. Relationships among hystricognathous and hystricomorphous rodents. *Mammalia*, 1970, **34**, 628–639.

17
Brain Evolution
in the Order Primates

C. B. G. Campbell

Indiana University

In higher mammals, such as carnivores, primates, and man, the brain has developed sufficiently to make effective use of sight and hearing. The areas of neocortex allocated to these senses have increased in size and influence—as has even the area for general body sensation—till they dominate the hemispheres, while the olfactory areas have regressed. At the same time, the superior colliculi have become vestigial, dealing only with some visual reflexes. Thus, destruction of the visual cortex causes total blindness; but destruction of the superior colliculi causes only disorder in pupillary and focal adjustment. As for the inferior colliculi, they have never attained any great prominence. Auditory discrimination is not much needed till intelligence has reached a level where sounds take on a symbolic meaning, as in animal cries, speech and music; and at such a level, the cerebral cortex has already incorporated the perception of hearing. The inferior colliculi are therefore insignificant in all vertebrates and act only in some reflex activity and to relay messages to the thalamus [p. 105].

This quotation, from a textbook for medical students (Elliott 1963), incorporates an unnecessary amount of misinformation on the organization and evolution of the brain; some of it is blatant and some more subtle. Some of this misinformation is factual and is a product of the time in which it was written. More importantly, some of it is conceptual and is a result of the milieu in which much of neurobiological research has been performed. Most research on the vertebrate nervous system is done in the schools and research institutes of medicine, where it has been isolated from the theory and methods of zoology. Its human orientation has tended to make it narrow in outlook and it has suffered from a certain amount of theological influence as well.

As a consequence, it has been customary to think of the animal kingdom in terms of a *scala naturae* with man as a supreme achievement. His nearest relatives are somewhat imperfect versions of him. Because the brain and its functions figure so prominently in human achievements it plays a large role in these notions. There has been a consistent tendency to attempt discovering some characteristics of primate brains that make them superior to those of other mammals and others that clearly separate man from nonhuman primates. In spite of a continuing effort to find such characteristics, the search has not been entirely successful so far. A number of what have been considered to be progressive trends have been described that supposedly set primate brains apart from those of other mammals. Perhaps the most lucid presentation of these is to be found in the writings of Le Gros Clark (1971):

1. "Undoubtedly the most distinctive trait of the Primates, wherein this order contrasts with *all other mammalian orders* in its evolutionary history, is the tendency towards the development of a brain which is *large in proportion to the total body weight,* and which is particularly characterized by a relatively extensive and often richly convoluted cerebral cortex [Le Gros Clark, 1971; italics mine]."

2. The visual system is highly developed in relation to the need for a high degree of visual acuity in arboreal life.

3. There is an enhancement of tactile sensibility and, therefore, of its central nervous system apparatus.

4. The olfactory regions of the brain undergo a retrogression, whereas the nonolfactory regions, particularly the visual cortex, are expanded.

5. There is a progressive structural differentiation of the cerebral cortical areas in the primates, particularly in the anthropoid primates.

6. More of the sulci found in the cerebral cortex of primates are limiting sulci or axial sulci, rather than merely mechanical sulci as found in nonprimate mammals.

7. Conspicuous development of the occipital and temporal lobes of the brain is found in primates.

8. A progressive expansion of the frontal lobes of the brain occurs in the Anthropoidea.

9. The cerebellum, particularly its middle lobe, undergoes expansion.

In this chapter, I undertake to examine most of these characteristics, although not in the precise sequence in which they have been presented.

THE OLFACTORY SYSTEM

The degree of elaboration of this system is customarily judged by examining the size of the olfactory bulbs in relation to the size of the brain as a whole, and estimating the relative extent of the pyriform lobe by how far down on the lateral aspect of the cerebrum the rhinal sulcus is found. This sulcus marks the boundary between neocortex and the paleocortex of the pyriform lobe. The paleocortex receives secondary olfactory fibers from the olfactory bulb via the lateral root of the olfactory tract.

It does appear to be true that living primates have olfactory bulbs that are small in relation to the brain as a whole when contrasted with some other living mammals, e.g., *Rattus, Canis, Tenrec,* and *Erinaceus.* Radinsky (1970) notes that the endocast of *Tetonius,* an Early Eocene prosimian, and the oldest primate endocast, reveals olfactory bulbs that are reduced in comparison with those of other early mammals but are larger than those of later primates. Most of the Early Tertiary and Recent prosimian primate endocasts shown by Radinsky have reduced olfactory bulbs. Presumably a primate trend toward having olfactory bulbs that are small in relation to total brain size is at least 55 million years old (the age of *Tetonius*).

It also appears to be true that in extant primates and the fossil primates the endocasts of which are available the rhinal sulcus is low on the lateral aspect of the cerebral hemisphere. This indicates that a larger proportion of the hemisphere is occupied by neocortex than by the paleocortex associated with the olfactory system. It should be noted that using these characteristics as indicators, reduction in the olfactory system has occurred independently in other mammalian orders as well. Consequently this is not an exclusively primate characteristic. Some, but not all, chiropterans have reduced olfactory bulbs, as do most aquatic mammals, the sireneans and cetaceans. Among rodents, the arboreal squirrels have somewhat reduced olfactory bulbs and a moderately low rhinal sulcus as well (Campbell, 1966). Downwardly displaced rhinal sulci have been found in mammals of a number of genera in the Marsupialia and Rodentia as well as in the Lagomorpha (Campbell, 1966). This feature is particularly striking in rabbit brains where the olfactory bulbs are, nevertheless, rather large. I wonder whether the downward displacement of the rhinal sulcus is really indicative of a reduction in the olfactory system at all. It may simply be a result of an expansion of the visual areas of the occipital and temporal lobes.

THE VISUAL SYSTEM

A highly elaborated visual system is certainly a prominent feature in primate brains as demonstrated in living animals and fossil endocasts. It is almost certainly a characteristic associated with adaptation to an arboreal life in this instance, as similar elaborations are found in other arboreal mammals. Elaborate visual systems are found in mammalian groups with other particular adaptations as well. For example, the lateral geniculate nucleus of primates is composed of cytoarchitecturally differentiated aggregations of cells forming layers or laminae separated by cell-free gaps. This nucleus receives its input from the retina of the eye in a highly ordered fashion and the axons of its cells project to the primary visual area, Area 17 of Brodmann.

In many other mammals, especially those generally considered to be "primitive," this nucleus is not differentiated into layers and its intrinsic organization is not revealed without experimental study. The contrast between the appearance of the nucleus in primates and its seeming lack of organization in these other animals is most impressive and has engendered considerable speculation about the importance of the characteristic of lamination. It has also been known to be a characteristic of some nonprimate mammals, e.g., the cat, but in these animals the number of laminae is less than in most primates. Consequently, it has still been possible to consider them as representing a more primitive state.

Although ten placental orders have not been examined as yet, it is now known that highly organized lateral geniculate nuclei, which are laminated in some degree, are found in placental mammals of the orders Insectivora, Carnivora, Artiodactyla, Perissodactyla, Primates, Rodentia, and Chiroptera (Campbell, 1972). A number of marsupials also possess this characteristic (Hayhow, 1967; Johnson & Marsh, 1969), whereas the monotremes do not (Campbell & Hayhow, 1971, 1972). Animals within these orders possessing highly organized nuclei are specialized for flight, gliding, rapid movement in the trees, or rapid movement on the ground. In most cases animals with seemingly unspecialized nuclei exist within the same order and are either more generalized or possess some other specialization, e.g., echolocation in the microchiropterans.

Considerable emphasis is usually placed on the greater extent and degree of differentiation of the striate or primary visual cortex (Area 17) in primates. In addition to this cortical area, which receives an input from the homolateral lateral geniculate nucleus, there is also, in the primates examined so far, a circumstriate or prestriate belt of cortex comprising Brodmann's Areas 18 and 19 (except in *Saimiri* which appears to lack an Area 19). Separate representations of the contralateral visual field can be shown to be present in each of Areas 17, 18, and 19. Another cortical area (Area MT) located in the middle temporal gyrus or area of the temporal lobe has been found photically responsive in *Callithrix jacchus* (Woolsey, Akert, Benjamin, Leibowitz, & Welker, 1955) and *Saimiri sciureus* (Doty, Kimura, & Morgenson, 1964). Allman and Kaas (1971) have demonstrated that this area is retinotopically organized and is a representation of the contralateral visual hemifield in *Aotus trivirgatus,* the owl monkey. A number of workers have shown that this area receives an input from the striate cortex (Area 17) in several primates, including *Macaca, Saimiri, Callithrix,* and *Galago* (A prosimian). Spatz and Tigges (1972a) have studied the efferent projections of this area in *Callithrix* and have found them to be extensive. On the ipsilateral hemisphere, projections have been found to the frontal eye field and eight regions of the temporal, parietal, and occipital lobes. Fibers crossing via the corpus callosum projected to the contralateral MT and two other areas immediately adjacent to the MT and corresponding with two of the ipsilateral terminal sites.

Spatz and Tigges (1972b) describe a difference in the organization of the striate–prestriate association between the cercopithecoid *Macaca mulatta* and the ceboid *Saimiri sciureus.* In *Macaca* Area 17 appears to project topographically to both an Area 18 and an Area 19. In *Saimiri* no Area 19 has been identifiable and a topographic projection to an area immediately adjacent to Area 17 has been the only such projection found. This area is labeled "Area 18" by Spatz and Tigges because of its position. A projection to the posterior bank of the superior temporal sulcus has also been noted. It is not yet clear whether or not this is likely to be a consistent finding in the New World monkeys as a group; however, because the New World and Old World monkeys are thought to have a separate ancestry it is not likely to be surprising if some differences in the organization of their central nervous pathways are found.

To add a further complication to the once relatively simple picture of cortical visual areas, Allman, Kaas, Lane, and Miezin (1973) have found what appears to be a crescent-shaped visual area surrounding the middle temporal visual area in the owl monkey, *Aotus trivirgatus*. It contains yet another representation of the contralateral half of the visual field and relates to the MT much as Area 18 does to Area 17. Allman *et al.* (1973) note that Spatz and Tigges (1972a) have made small lesions in the center of the MT in the marmoset *Saimiri* and have found two foci of degeneration adjacent to the MT. These two foci are apparently located within an area corresponding to the crescent surrounding the MT. This suggests to them that the MT projects onto the crescent just as visual area I (VI) (Area 17) projects onto VII (Area 18). This also suggests that the crescent-shaped area described by Allman and his co-workers, which is cytoarchitecturally distinct from the surrounding cortex, is possibly also present in at least one other New World monkey, and further work may establish it in others.

Such a complex series of visual hemifield representations has not been found in mammals other than primates. The hedgehog, *Erinaceus europaeus*, is found by Kaas, Hall, and Diamond (1970) to have only two neocortical visual areas, VI (Area 17) and VII (Area 18). All of the mammals examined so far, including those discussed above as well as the rabbit, rat, grey squirrel, tree shrew, opossum, and domestic cat, have at least these same two cortical visual fields. In addition, some have been found to have another visual field in the temporal lobe, for example, the domestic cat (see Burrows & Hayhow, 1971); the tree shrew, *Tupaia glis* (Killackey, Snyder, & Diamond, 1971); and the grey squirrel, *Sciurus carolinensis* (Hall, Kaas, Killackey & Diamond, 1971). The grey squirrel resembles the domestic cat and primates in having three occipital lobe visual areas, Areas 17, 18, and 19 (Hall *et al.*, 1971). An exception to all of this may be the mole, *Scalopus aquaticus*, which seems to lack a visual cortex (Allison & Van Twyver, 1970).

With the information presently available it does seem that there has been a trend in the primate lineage to elaborate on a basic pattern and increase in extent that portion of the neocortex associated with vision and to parcellate it into a number of representations of the contralateral hemifield. It is not surprising that the occipital lobe in primates seems more prominent than in most other mammals. It is of course possible that more careful examination of nonprimate mammals can reveal homologs of these areas found in primates. There is a great deal of work ahead in determining the range of patterns of visual system organization present in the visual systems of primates and other mammals as well as in understanding the functional significance of these patterns.

THE TEMPORAL LOBE

Radinsky (1970), on the basis of the endocranial casts of fossil prosimians available to him, states that as far back as the Early Eocene (55 million years ago) primates have relatively large occipital and temporal lobes compared to those of contemporary mammals. Most living mammals also, with the possible exception of proboscideans (Radinsky, this volume) and a few others, do not have really well-developed temporal lobes. As has been seen, the temporal lobe in primates is the site of extraoccipital visual cortex. It is also the site of auditory cortex and has connections with all the other lobes of the hemisphere as well as an input into deeper lying limbic structures (Pandya & Kuypers, 1969).

Le Gros Clark (1971 and previous publications) uses the slight downward projection of the temporal neocortex in *Tupaia* as one of his arguments that the tupaiids more closely resemble primates than insectivores and should be classified with the former group. Le Gros Clark consistently makes comparisons of tree shrews only with terrestrial insectivores, on the one hand, and arboreal primates, on the other. This practice has been frequently criticized (Campbell, 1966; McKenna, 1966; Martin, 1968; Van Valen, 1965). When a wider range of mammals is examined, a similar downward displacement of the temporal neocortex can be found in mammals of a number of other orders, including marsupials, rodents, and lagomorphs (Campbell, 1966).

SULCI AND DIFFERENTIATION OF CORTICAL AREAS

Le Gros Clark (1971) suggests that the pattern of convolutions and intervening sulci seen on the surface of the cerebral hemisphere are a result of two main factors. The first, a general mechanical factor, is a result of the stresses imposed on the developing and expanding cerebral hemisphere during its development. The second factor is the growth and expansion of individual cortical areas in relation to each other. He further suggests that the two factors may interact in determining the precise location of an individual sulcus.

Le Gros Clark states that nonprimate mammals have cortical convolution patterns that result primarily from the general mechanical factor. These are usually oriented longitudinally and he illustrates the brain of a carnivore, the domestic cat, as an example of this type of brain (Le Gros Clark, 1971, p. 229). Although sulci having the same mechanical basis contribute to the patterns found in the cortex of anthropoid primates, he believes that more of the sulci are developed as foldings of the cortex along the boundary lines between adjacent cortical areas (limiting sulci) or as foldings along the middle of a specific area (axial sulci).

Prosimian primates, living and fossil, have relatively few (or no) cerebral cortical sulci when compared with anthropoids and many of them have a longitudinal orientation. Really strikingly gyrencephalic brains among primates are only found in the Anthropoidea, especially among hominoids. According to the so-called "law" of Baillarger-Dareste (Ariëns-Kappers, Huber, & Crosby, 1936), gyrification is a function of brain size. Expansion of particular cortical areas is an additional factor. Primates as a group are large animals when compared to the bulk of vertebrates. The hominoids are particularly large as a group. Gyrencephaly is found in a number of mammalian orders, including the Perissodactyla, Artiodactyla, Proboscidea, Carnivora, Cetacea, Monotremata, and, of course, the Primates. Most of these orders contain large animals. It has been a continuing source of puzzlement to those who expect "primitive" animals to have "primitive" features that the echidnas (*Tachyglossus* and *Zaglossus*) have markedly gyrencephalic brains whereas the platypus (*Ornithorhynchus*) is lissencephalic.

It has been shown by Woolsey and his co-workers (Woolsey, 1959), Welker and Campos (1963), and Welker and Seidenstein (1959) that mammals other than primates, including carnivores, have limiting and axial sulci as well. Radinsky (1968) used some of this data to infer the boundaries of certain cortical areas in fossil otters. Lende (1964) has shown that many of the sulci found in the echidna, *Tachyglossus,* are limiting sulci for the visual, somatic sensory, motor, and auditory cortexes. Surely a similar situation exists in some large brains with gyrencephaly found in nonprimate orders.

RELATIVE BRAIN SIZE

I will not discuss relative brain size at any great length. Other contributors to this volume are far better equipped to pursue it than I am. I think that Le Gros Clark's (1971) statement that the development of a large brain in relation to body weight is the most distinctive trait of the order Primates and contrasts with all other mammalian orders is perhaps a bit strong. He further qualifies and tempers the statement by adding that primate brains are characterized by an extensive and highly convoluted cerebral cortex. A number of mammals have brain weight–body weight ratios within the primate range (Cobb, 1965). Most of them are small animals and are not gyrencephalic. The porpoise closely approximates man in this regard and is gyrencephalic. Le Gros Clark admits that a similar progressive expansion of the brain has occurred in many other groups of mammals in the Tertiary (there has been a general trend toward increasing body size in the Mammalia), but the expansion begins earlier in the primates and advances much farther.

Cobb (1965) has discussed many of the pitfalls in the use of the brain weight–body weight ratio and, indeed, there has been a search for a better measure of brain "progress." Radinsky (1967) has proposed a new measure utilizing the ratio between cranial volume and foramen magnum area. Wanner (1971) has criticized the measure and suggests that it is only useful at the ordinal level. Holloway (1966, 1968) has discussed at length the overemphasis on brain size.

Radinsky (1970) has noted that endocranial casts of fossil primates indicate that expansion of the frontal lobe lagged behind that of the rest of the cortex, for as recently as the Early Oligocene (35 million years ago) the frontal lobes were relatively smaller than those in almost all modern prosimians. Presumably this is because the expansion of the occipital and temporal lobes is associated with the expansion of cortical visual areas. This accompanied the invasion of the arboreal habitat, which apparently occurred early in primate evolution. There has been somewhat less emphasis on the uniqueness of the primate frontal lobe in recent years. More interest has been centering on the parietal and, particularly, the temporal lobes.

A large and elaborate cerebellum seems a useful accompaniment to a life in the trees, and there has been a trend toward an enlargement of that organ, particularly that portion related to the expanding cerebral cortex. The general trend toward increasing body size and the upright posture in *Homo* enhances this effect. Again, arboreal mammals generally show trends in this direction.

Essentially all of the trends noted in the brain of primates can be found occurring in brains of animals of other mammalian groups. It is perhaps the particular combination of trends present, as well as the degree to which they have been pursued, that sets the primate brain apart. However, it is most often a combination of characteristics that makes any animal group distinguishable from its fellows.

THE UNIQUENESS OF THE HUMAN BRAIN

"All brains of primates seem to be models of each other. In fact, no new structure *per se* is found in man's brain that is not found in the brain of other primates." (Noback & Moskowitz, 1963.) In spite of statements of this order, the urge to find some concrete difference between the brains of *Homo* and nonhuman primates is still strong. A recent paper (Geschwind, 1965) that purports to present a unique feature of human brains seems to have made some impact on the thinking of anthropologists and others. Because the finding relates to speech and language, always considered uniquely human, it has seemed to be a really likely candidate.

The following quotations from Geschwind (1965), I believe, sum up his argument:

The preceding parts of this paper have cited the evidence that in lower mammals, the primary projection areas of the cortex subserve certain functions which tend subsequently to be separated in the primates. In keeping with this relatively minor degree of separation of functions, only a few regions of differing cytoarchitectonic structure are distinguishable. As we ascend the phylogenetic scale, the associative activities become separated to a great extent from the receptive. Large association areas more clearly separable from primary projection areas appear, and cytoarchitectonic differentiations increase. In accordance with the principle of Flechsig (which is applicable to man and the other primates but not to subprimate forms), the primary projection areas now send their connexions primarily to the immediately adjacent association cortex (parakonio-cortex); the long connexions (either within a hemisphere or between hemispheres) between different cortical regions take place predominantly between parts of the association cortex. To a great extent the most important connexions of the association cortex are with the neocortex of the temporal lobe (and perhaps also of the insula) which in turn feeds into limbic structures. In keeping with this, connexions involving linkages between any one sensory modality and the limbic system tend to be powerful (these connexions subserve emotional and autonomic responses to sensory stimuli, associations between one sensory modality and gustatory or olfactory stimuli, etc.) while other non-limbic sensory–sensory connexions tend to be weak. I have, in the first part of this paper, discussed in detail the effects of lesions separating the primary sensory modalities from the limbic structures in the primate.

The situation in man is not simply a slightly more complex version of the situation present in the higher primates but depends on the introduction of a new anatomical structure, the human inferior parietal lobule, which includes the angular and supramarginal gyri, to a rough approximation areas 39 and 40 of Brodmann. In keeping with the views of many anatomists Crosby et al. (1962) comment that these areas have not been recognized in the macaque. Critchley (1953), in his review of the anatomy of this region, says that even in the higher apes these areas are present only in rudimentary form. In keeping with the late evolutionary development of this region are certain other findings. The gyral structure of this area tends to be highly variable. In addition this area is one of the late myelinating regions or "terminal zones" as Flechsig termed them. In fact, this region was, in Flechsig's map, one of the last three to myelinate. DeCrinis (cited by Bonin and Bailey, 1961) showed that part of this region is one of the last cortical areas in which dendrites appear. Yakovlev (personal communication) has pointed out that this region matures cytoarchitectonically very late, often in late childhood. In addition, he has pointed out that preliminary studies suggest that this region receives very few thalamic afferents. In this respect it is similar to part of the frontal association area which is also largely athalamic; this part of the frontal lobe is also phylogentically new, myelinates late and forms dendrites late. The afferent connexions of this new parietal association area may therefore be predominantly from other cortical regions. As an association area, this region is also different from the older association areas in not being essentially concentric with one of the primary projection centres. . . .

We thus have this extensive, evolutionarily advanced, parietal association area developing not in apposition to the primary projection areas for vision, somesthetic sensibility, and hearing but rather at the point of junction of these areas as Critchley (1953) has indicated. This region possibly being one of few thalamic connexions may well receive most of its afferents from the adjacent association areas; it is thus an association area of association areas. In more classical terms, it would be called a secondary association area. The probable significance of this anatomical location is heightened by reference to our earlier discussion of subhuman forms. In these it appears as if association areas feed into temporal neocortex relaying in turn to limbic and rhinencephalic structures. As I pointed out in the earlier discussion, cross-connexions between primary nonlimbic sensory modalities are weak in subhuman forms. In man, with the introduction of the angular gyrus region, intermodal associations become powerful. In a sense the parietal association area frees man to some extent from the limbic system. This independence is only relative since ultimately learning still depends even in man, on intact connexions with limbic structures. The well-known permanent severe disturbance of new learning resulting from bilateral lesions of the hippocampal region attests to this fact (Scoville & Milner, 1957).

Geschwind's paper was written at a time when almost all information on corticocortical connections was based on older, less reliable techniques. Indeed, a lot of it was based on normal descriptive anatomy from the turn of the century. A whole series of papers has appeared since then that I believe seriously damage his argument. These papers are based on modern experimental anatomical techniques utilizing the Nauta-Gygax (Nauta & Gygax, 1954) and Fink-Heimer (Fink & Heimer, 1967) methods.

First of all, Flechsig's rule, which states that the primary sensory areas of the neocortex neither send nor receive long corticocortical connections and therefore are devoid of callosal connections, appears to be erroneous. Pandya and Vignolo (1969) have shown that the primary somatosensory cortex of the postcentral gyrus in the rhesus monkey connects with its counterpart in the opposite hemisphere via the corpus callosum. Berlucchi (1972) points out that there are now several studies demonstrating callosal connections between visual cortexes, including the striate area (Area 17) in primates and other mammals. Pandya, Hallett, and Mukherjee (1969) have found similar callosal connections between the primary auditory cortex of the two hemispheres of the rhesus monkey.

Pandya and Kuypers (1969), in an extensive study of corticocortical connections in the rhesus monkey have found precisely the situation described by Geschwind as an exclusively human characteristic. They state:

> According to our findings projections from several sensory areas converge upon the premotor area. Some such convergence also seems to occur in the inferior parietal lobule. Here somatosensory, auditory and visual projections abut, the visual projections being dominant. The two areas in question might function in integrating information of different sensory modalities. In this respect it may be relevant to observe that these areas are interconnected, that both project to the cingulate gyrus and that their ablations seem to result in a somewhat similar unilateral neglect [p. 31].

Lancaster (1968) relies heavily on Geschwind's postulate of the role of the "unique" inferior parietal lobule. She, as have some other anthropologists, has placed the essential substrate for speech solely in the brain. She points out that she deliberately omits mention of the anatomical mechanisms used in making sounds because of her conviction that no evolution in the mouth, tongue, or larynx has been necessary to initiate the origin of human language. After all, the nonhuman primates make a wide variety of sounds. I find it difficult to believe that nonhuman primates are capable of making the wide range of sounds, and especially the more subtle ones of which *Homo* is capable. This capacity in humans has been attributed to altered laryngovelar relations correlated with upright posture (Du Brul, 1958).

The processes of natural selection operate on populations of whole animals not just on brains. The finest brain in the world must be severly handicapped without effector systems with which to operate on the external environment. The nonhuman primates have not developed humanoid speech because the selection pressures to which they have been subjected have not been the same, and because of the vagaries of chance. Under precisely similar pressures it is not unlikely that other groups of primates can have evolved a similar kind of speech. Parallelism is a well-known phenomenon in the order Primates. It is certainly not surprising that the living nonhuman primates, lacking upright posture and adapted for their own particular modes of life as they are, are not capable of humanoid speech.

I began this chapter by quoting a passage from a scientific book that purported to tell something about nature. I will end it with another quotation. This one, from a nonscientific source, I believe, tells more scientific truth about nature than does the other.

> We need another and a wiser and perhaps a more mystical concept of animals. Remote from universal nature, and living by complicated artifice, man in civilization surveys the creatures through the glass of his knowledge and sees thereby a feather magnified and the whole image in distortion. We patronize them for their incompleteness, for their tragic fate of having taken form so far below ourselves. And therein we err, and greatly err. For the animal shall not be measured by man. In a world older and more complete than ours they move finished and complete, gifted with extensions of the senses we have lost or never attained, living by voices we shall never hear. They are not brethren; they are not underlings; they are other nations, caught with ourselves in the net of life and time, fellow prisoners of the splendour and travail of the earth [Henry Beston, *The Outermost House*, 1949, p. 25].

REFERENCES

Allison, T., & Van Twyver, H. Sensory representation in the neocortex of the mole, *Scalopus aquaticus. Experimental Neurology,* 1970, **27**, 554–563.

Allman, J. M., & Kaas, J. H. A representation of the visual field in the caudal third of the middle temporal gyrus of the owl monkey *(Aotus trivirgatus). Brain Research,* 1971, **31**, 85–105.

Allman, J. M., Kaas, J. H., Lane, R. H., & Miezin, F. M. A crescent-shaped cortical visual area surrounding the middle temporal area (MT) in the owl monkey *(Aotus trivirgatus). Anatomical Record,* 1973, **175**, 263–264.

Ariëns-Kappers, C. U., Huber, G. C., & Crosby, E. C. *The comparative anatomy of the nervous system of vertebrates, including man.* New York; Macmillan, 1936.

Berlucchi, G. Anatomical and physiological aspects of visual functions of corpus callosum. *Brain Research,* 1972, **37**, 371–392.

Beston, H. *The outermost house: A year of life on the great beach of Cape Cod.* New York: Holt, Rinehart & Winston, 1949.

Burrows, G. R., & Hayhow, W. R. The organization of the thalamocortical visual pathways in the cat. *Brain, Behavior, & Evolution,* 1971, **4,** 220–272.

Campbell, C. B. G. The relationships of the tree shrews: the evidence of the nervous system. *Evolution,* 1966, **20,** 276–281.

Campbell, C. B. G. Evolutionary patterns in mammalian diencephalic visual nuclei and their fiber connections. *Brain, Behavior, & Evolution,* 1972, **6,** 218–236.

Campbell, C. B. G., & Hayhow, W. R. Primary optic pathways in the echidna, *Tachyglossus aculeatus:* an experimental degeneration study. *Journal of Comparative Neurology,* 1971, **143,** 119–136.

Campbell, C. B. G., & Hayhow, W. R. Primary optic pathways in the duckbill platypus *Ornithorhynchus anatinus*: an experimental degeneration study. *Journal of Comparative Neurology,* 1972, **145,** 195–208.

Cobb, S. Brain size. *Archives of Neurology,* 1965, **12,** 555–561.

Critchley, M. *The parietal lobes.* London: Edward Arnold, 1953.

Crosby, E. C., Humphrey, T., & Lauer, E. W. *Correlative anatomy of the nervous system.* New York: MacMillan, 1962.

Doty, R. W., Kimura, D. S., & Morgenson, G. J. Photically and electrically elicited responses in the central visual system of the squirrel monkey. *Experimental Neurology,* 1964, **10,** 19–51.

DuBrul, E. L. *Evolution of the speech apparatus.* Springfield, Illinois: Charles C Thomas, 1958.

Elliott, H. C. *Textbook of neuroanatomy.* Philadelphia: Lippincott, 1963.

Fink, R. P., & Heimer, L. Two methods for selective silver impregnation of degenerating axons and their synaptic endings in the central nervous system. *Brain Research,* 1967, **4,** 369–374.

Geschwind, N. Disconnection syndromes in animals and man. *Brain,* 1965, 88, 237–294, 585–644.

Hall, W. C., Kaas, J. H., Killackey, H., & Diamond, I. T. Cortical visual areas in the grey squirrel *(Sciurus carolinensis)*: a correlation between cortical evoked potential maps and architectonic subdivisions. *Journal of Neurophysiology,* 1971, **34,** 437–452.

Hayhow, W. R. The lateral geniculate nucleus of the marsupial phalanger, *Trichosurus vulpecula.* An experimental study of cytoarchitecture in relation to the intranuclear optic nerve projection fields. *Journal of Comparative Neurology,* 1967, **131,** 571–604.

Holloway, R. L., Jr. Cranial capacity, neural reorganization, and hominid evolution: a search for more suitable parameters. *American Anthropologist,* 1966, **68,** 103–121.

Holloway, R. L., Jr. The evolution of the primate brain: some aspects of quantitative relations. *Brain Research,* 1968, 7, 121–172.

Johnson, J. I. Jr., & Marsh, M. P. Laminated lateral geniculate in the nocturnal marsupial *Petaurus breviceps* (sugar glider). *Brain Research,* 1969, **15,** 250–254.

Kaas, J., Hall, W. C., & Diamond, I. T. Cortical visual areas I and II in the hedgehog: relations between evoked potential maps and architectonic subdivisions. *Journal of Neurophysiology,* 1970, **33,** 595–615.

Killackey, H., Snyder, M., & Diamond, I. T. Function of striate and temporal cortex in the tree shrew. *Journal of Comparative Physiology & Psychology,* 1971, 74, 1–29.

Lancaster, J. B. Primate communication systems and the emergence of human language. In P. C. Jay (Ed.), *Primates: Studies in Adaptation and Variability.* New York: Holt, Rhinehart & Winston, 1968, pp. 439–457.

Le Gros Clark, W. E. *The Antecedents of Man.* Chicago: Quadrangle, 1971.

Lende, R. A. Representation in the cerebral cortex of a primitive mammal, sensorimotor, visual, and auditory fields in the echidna *(Tachyglossus aculeatus).* *Journal of Neurophysiology,* 1964, **27,** 37–48.

Martin, R. D. Towards a new definition of primates. *Man,* 1968, **3,** 377–401.

McKenna, M. C. Paleontology and the origin of the primates. *Folia Primatologica,* 1966, **4,** 1–25.

Nauta, W. J. H., & Gygax, P. A. Silver impregnation of degenerating axons in the central nervous system: a modified technic. *Stain Technology,* 1954, **29,** 91–93.

Noback, C. R., & Moskowitz, N. The primate nervous system: functional and structural aspects in phylogeny. In J. Buettner-Janusch (Ed.), *Evolutionary and genetic biology of primates.* Vol. I. New York: Academic Press, 1963. Pp. 131–177.

Pandya, D. N., Hallett, M., & Mukherjee, S. K. Intra- and interhemispheric connections of the neocortical auditory system in the rhesus monkey. *Brain Research,* 1969, **14,** 49–66.

Pandya, D. N., & Kuypers, H. G. J. M. Cortico-cortical connections in the rhesus monkey. *Brain Research,* 1969, **13,** 13–36.

Pandya, D. N., & Vignolo, L. A. Interhemispheric projections of the parietal lobe in the rhesus monkey. *Brain Research,* 1969, **15,** 49–65.

Radinsky, L. B. Relative brain size: a new measure. *Science,* 1967, **155,** 836–838.

Radinsky, L. B. Evolution of somatic sensory specialization in otter brains. *Journal of Comparative Neurology,* 1968, **134,** 495–505.

Radinsky, L. B. The fossil evidence of prosimian brain evolution. In C. R. Noback & W. Montagna (Eds.), *The primate brain: Advances in primatology* Vol. 1. New York: Appleton-Century-Crofts, 1970, Pp. 209–224.

Scoville, W. B., & Milner, B. Loss of recent memory after bilateral hippocampal lesions. *Journal of Neurology, Neurosurgery, & Psychiatry,* 1957, **20,** 11–21.

Spatz, W. B., & Tigges, J. Experimental-anatomical studies on the "middle temporal visual area (MT)" in primates. I. Efferent cortico-cortical connections in the marmoset *Callithrix jacchus. Journal of Comparative Neurology,* 1972, **146,** 451–463.   (a)

Spatz, W. B., & Tigges, J. Species difference between Old World and New World monkeys in the organization of the striate-prestriate association, *Brain Research,* 1972, **43,** 591–594.   (b)

Van Valen, L. Tree shrews, primates and fossils. *Evolution,* 1965, **19,** 137–151.

von Bonin, G., & Bailey, P. *The neocortex of Macaca mulatta.* Urbana: University of Illinois Press, 1947.

Wanner, J. A. Relative brain size: a critique of a new measure. *American Journal of Physical Anthropology,* 1971, **35,** 255–258.

Welker, W. I., & Campos, G. B. Physiological significance of sulci in somatic sensory cerebral cortex in mammals of the family Procyonidae. *Journal of Comparative Neurology,* 1963, **120,** 19–36.

Welker, W. I., & Seidenstein, S. Somatic sensory representations in the cerebral cortex of the racoon. *Journal of Comparative Neurology,* 1959, **111,** 469–499.

Woolsey, C. N. Some observations of brain fissuration in relation to cortical localization of function. In D. B. Tower and J. P. Schadé (Eds.) *Structure and function of the cerebral cortex.* Amsterdam: Elsevier, 1959.

Woolsey, C. N., Akert, K., Benjamin, R. M., Leibowitz, H., & Welker, M. J. Visual cortex of the marmoset. *Federation Proceedings,* 1955, **14**(1), Part. 1, 166.

18
Tool Use in Mammals

J. M. Warren

The Pennsylvania State University

INTRODUCTION

This chapter surveys the data on the use of tools by mammals in the wild, in captivity, and in the laboratory. It is primarily concerned with assessing the importance of tool use in the adaptation of wild mammals and with evaluating the major theories of tool-using behavior in nonhuman primates.

The argument consists of three main points:

1. Feral monkeys and apes spontaneously use tools, frequently and in a wide variety of situations. Tool use in nonprimate mammals is rare, in every sense of the word, being observed generally only in a few individuals of scattered species, and usually only in the form of a single response.

2. Primates must learn to use tools, and cultural transmission of tool-using skills is important, particularly in chimpanzees.

3. Tool using in primates depends more importantly on such cognitive processes as insight and imitation than is suggested by several influential treatments of the subject.

OCCURRENCE OF TOOL USE IN MAMMALS

Nonprimates

A tool is any object extraneous to the bodily equipment of an animal that serves as a functional extension of the organism, permitting it to enlarge the range of its movements or to increase their efficiency in manipulating the environment (Hall, 1963).

Only one nonprimate mammal is known regularly to use a tool. The sea otter feeds by floating on its back and hammering shellfish against a stone resting on its chest. It is not known whether this behavior pattern is dependent on learning, nor what sort of learning may be involved (Hall, 1963; van Lawick-Goodall, 1970).

There are a few reports that individual horses, elephants, and goats use sticks as scratching tools (van Lawick-Goodall, 1970). These rare observations of tool use in nonprimate mammals are important because they show that the infrequent use of tools by mammals other than primates cannot be attributed solely to their limited motor capacities. The wide variety of manipulatory responses nonprimates can be taught in circuses and laboratories also argues against the view that their low level of spontaneous tool use is entirely because of restricted motor capacities.

Primates

Some of the sorts of tool-using behavior observed in the spontaneous behavior of apes and monkeys are listed in Tables 1 and 2, where the letters W and C indicate whether a given type of tool use has been seen in a wild or a captive specimen. When no letter is given, the original source may be found in van Lawick-Goodall's (1970) comprehensive review; footnotes refer to papers published since van Lawick-Goodall's (1970) review was prepared.

Some of the classes listed in Tables 1 and 2, such as Types 4 (use of branches as whips) and 9 (use of branches as ladders), refer to a single sort of action. Others encompass a variety of diverse responses. Type 6 (manipulation of objects to obtain or prepare food) includes the greatest number of different actions:

1. Cebus monkeys peel bark from twigs, which they then use to pry insects from under bark (this behavior also qualifies as object modification or tool making)

2. Rhesus macaques use leaves to rub dirt from food

3. Chimpanzees have been observed spontaneously to use various objects in connection with feeding or drinking in the following ways:

a. Rocks to open nuts and fruits

b. Sticks to dig up edible roots

c. Branches as levers to open a box containing bananas

d. Sticks, twigs, and grasses to obtain honey, ants or termites

e. A stick to knock a banana from the hand of a human that the chimp fears

f. A wad of leaves as a sponge to collect water that cannot be reached by mouth

g. A wad of leaves to gather the last scraps of brain tissue from within the calvarium of a prey animal (Teleki, 1973)

TABLE 1
**Types of Spontaneous Tool Use Observed in Apes and Monkeys
in Agonistic Contexts**

	Apes				Monkeys			
	Chimp	Gorilla	Orang.	Gibbon	Cebus	Baboon	Macaque	Other genera
1. Manipulation of objects to enhance aggressive displays	W	W		W				W–Colobus, langur patas, vervet
2. Dropping objects from trees to intimidate predators	W	W	W	W	W	W	W	W–howler, spider
3. Deliberately aimed throwing of objects at other animals	WC	C	WC		C	C		
4. Use of saplings and branches as whips to strike other animals	W							
5. Use of sticks or branches as clubs to beat other animals	WC		C		C			

TABLE 2
Types of Spontaneous Tool Use in Apes and Monkeys
in Nonagonistic Contexts

	Apes				Monkeys			
	Chimp	Gorilla	Orang.	Gibbon	Cebus	Baboon	Macaque	Other genera
6. Manipulation of objects to obtain or prepare food	W[a]C	W	WC		WC		W	
7. Use of sticks and twigs to investigate unfamiliar objects	WC							
8. Manipulation of objects in body care and grooming	WC[b]					W[c]		
9. Use of branches as ladders	C[d]					C		
10. Object modification or tool making	WC				W	W		

[a]Jones and Pi (1969), Struhsaker and Hunkeler (1971), Teleki (1973).
[b]McGrew and Tutin (1972), McGrew and Tutin (1973).
[c]van Lawick-Goodall, van Lawick, and Packer (1973).
[d]Menzel (1972, 1973).

h. Sticks as olfactory aids, to determine whether a termite nest is suitable for exploration or whether rotten branches contain edible grubs

Type 8 (use of objects in body care and grooming) tool-using responses include the following activities:

a. Baboons have been seen to use a stone and a maize kernel in self-grooming to remove, respectively, sticky congealed fruit juice and blood from the face

b. Wild chimpanzees use leaves as toilet paper, to wipe feces, mud, sticky fruit juice, urine, and blood from their bodies

c. A captive female chimp has been observed to use twigs in social grooming of a companion's teeth, and a wild female has been observed to use a twig in self-directed dental grooming (McGrew & Tutin, 1972, 1973)

The subject of tool making by primates deserves additional comment. Most of the tools made by nonhuman primates do not conform very closely to any particular standard. There are, however, two exceptions to this rule. The dimensions of the straw or twig used by chimpanzees to collect termites from their nest are quite critical, and young chimpanzees only slowly acquire facility in fabricating the right sort of tools (van Lawick-Goodall, 1968). Teleki (1974) tried to become skilled in the chimpanzees' technique for termiting during a recent stay at the Gombe Stream Reserve. In spite of weeks of diligent practice he failed to learn to select suitable tools for termiting nearly as well as an experienced adult chimpanzee. His lack of success rather dramatically confirms the proposition that selection of a tool with the proper characteristics requires responses to far more subtle stimulus properties than can be guessed from watching the skilled performance of sophisticated chimpanzees.

The second exceptional case concerns the chimpanzees' technique for collecting driver ants. McGrew (1974) describes the method as follows. The chimpanzee dips a stick into an ant nest, and waits until about 300 ants have crawled on the stick. Then

the hand holding the tool shifts it vertically such that its distal end is just below the mouth. Simultaneously the opposite hand (which may have been idle or supporting in a tripedal stance) grasps in a loose power grip the proximal tool end just above the other hand. Immediately the upper hand slides up the length of the tool in a rapid unbroken motion (the "pull-through"). This momentarily catches the ants in a mass which accumulates on the sides of the flexed thumb and forefinger and lateral surface of the moving hand. The mass is about the size of a hen's egg and contains an estimated 300 ants. As the hand moves up the tool, the mouth opens wide with lips partly retracted. As the hand leaves the wand it goes directly to the mouth, transferring the ant mass into it. Presumably the mass of ants is so jumbled that few can bite the predator before being consumed [McGrew, 1974, pp. 504–505].

This particular method of preying on insects imposes specific contraints on the sort of tool used by chimpanzees.

Unlike other wild chimpanzee tools which vary greatly in dimension and material, ant dipping tools are very stereotyped. More than any other known naturally used non-human tool, they are shaped to form, presumably based on some kind of cognitive model. A length of about 60–70 cm seems optimal. A tool that is too short would allow too many ants to climb onto the chimpanzee before they could be eaten. A tool that is too long would be awkward to manoeuvre. A tool of too narrow diameter would be floppy and fragile. A tool that is too thick would probably allow too many ants onto the tool at once and make it difficult to form the bite-sized mass during the pull-through. A smooth tool surface is necessary for an unimpeded pull-through. Naturally uniform bark is not a problem, but rough and uneven bark must accordingly be stripped away. Only fairly straight sticks are used: crooks, forks and side branches would slow down the pull-through motion and allow the ants to disperse. The limiting factors outlined above are only inferential, but all relate to dealing efficiently with a pain-inflicting prey which comes in small, mobile units [McGrew, 1974, p. 507].

McGrew also presents evidence that driver ants taken in this way constitute a significant source of food for chimpanzees.

Some students of human evolution have been fond of postulating uniquely human characteristics, found in man but not in other animals. For years it has been argued that only man spontaneously uses or makes tools. Once these claims were discredited, new ones were advanced. For example, it is now asserted that only man makes tools to a standard pattern and that only man derives a significant portion of his food from the use of tools. The observations by McGrew, Teleki, and van Lawick-Goodall now cast doubt on these recent assertions of man's uniqueness in respect to the use of tools. At present there seems to be only one defensible statement concerning uniquely human tool-using behavior: only man uses tools to make tools.

More information and illustrations of tool-using behavior by chimpanzees and other vertebrates are provided in van Lawick-Goodall's (1968, 1970, 1973) reviews. The material presented in this section suffices, however, to indicate that primates, and particularly the great apes, far surpass other mammals in respect to the number, variety, and quality of tool-using performances they emit spontaneously in the absence of human intervention.

ROLE OF LEARNING IN THE DEVELOPMENT OF TOOL USE BY PRIMATES

The evidence concerning the role of experience and learning in the development of spontaneous tool-using behavior in wild primates has been obtained in work with chimpanzees. Almost necessarily the evidence is indirect and inferential; it consists of the following kinds:

a. *Comparisons of trapped wild and laboratory-reared chimpanzees.* Chimps that had been reared under severe conditions of perceptual and social deprivation during the first 2 years of life were greatly inferior to controls that had been captured in infancy in learning to use a stick to rake in food, when both groups were aged 6–8 years (Menzel, Davenport, & Rogers, 1970). Captured wild chimpanzees were also markedly superior to lab-reared animals in respect to the frequency and quality of their spontaneous nest-building activity (Bernstein, 1962).

b. *Spontaneous use of human artefacts.* It is reported by van Lawick-Goodall that

One male chimpanzee at the Gombe National Park appeared to make deliberate use of abnormal objects to better his charging displays: this, in turn, probably led to his becoming the dominant male of the group. In 1964 he held a very low social status. In December that year he began to use empty 4-gallon paraffin cans during his charging displays. Initially he used one can only, hitting it ahead of him with alternate hands or occasionally kicking it as he ran. After a while he was able to keep three cans on the move at once without noticeably diminishing his speed. The effect of such performances on his conspecifics was dramatic; the noise of the cans was tremendous and, as he approached, the other chimpanzees hurried out of his way, including those who held a much higher status. Often he repeated the display three or four times, running straight toward one or more of the other chimpanzees present. When he finally stopped, the others usually approached and directed submissive gestures toward him. After 4 months we removed all cans, but by then he had acquired the number one position—which he still holds five years later. That his use of these cans was deliberate is suggested by the fact that, once the pattern was established, he would often walk calmly to the tent and select his cans. [van Lawick-Goodall, 1970, p. 210]

It is hardly necessary to point out that it is difficult to believe that banging 4-gallon paraffin tins can be an innate motor pattern in chimpanzees.

c. *Observed practice effects.* At the Gombe Stream site, van Lawick-Goodall (1970) established a feeding area where chimpanzees competed with other chimps and baboons for a limited supply of bananas, a highly preferred food. Under these socially stressful conditions, a few animals began to throw missiles at their competitors. In subsequent years there was a sharp increase in both the number of chimps that threw objects with deliberate aim and the total number of missiles dispatched.

d. *Individual differences within the same population.* By no means do all of adult males at the Gombe Stream throw sticks, rocks, or other objects at other animals during aggressive interactions. This suggests that the persistent occurrence or nonoccurrence of aimed throwing behavior may be determined by an individual's particular history of reinforcement or nonreinforcement for this behavior.

e. *Differences in tool using between populations.* Both chimpanzees that live in relatively open savanna regions and those that live in dense rain forests throw sticks at dummy leopards, but the frequency and accuracy of throwing is greater in the chimpanzees from the more open habitat. This datum may reflect the better opportunities for practice and consequent improvement of throwing behavior on the savanna compared to the thick forest.

There also appear to be different and presumably learned cultural traditions in tool using by chimpanzees. Neither the use of rocks to crack open nuts nor the use of twigs or grass in fishing for termites is universal in the species. Chimpanzees in Liberia and the Ivory Coast open nuts and hard fruits by pounding them with stones but do not probe for termites or other insects. Chimpanzees from Cameroon and Rio Muni and from Tanzania do not pound nuts and fruits, but both groups probe for insects and honey. The West and East African populations of insect collectors may be further differentiated. The Tanzania animals take both termites and ants; the Rio Muni and Cameroon populations take only termites. The populations also differ in their preferences for the materials from which to make termiting tools. In addition, the use of sponges to take water or brains has only been observed in chimpanzees in Tanzania (Teleki, 1974). Finally, chimpanzees in Uganda use leafy branches as fly whisks, but no chimpanzee at the Gombe Stream Reserve has done so during 12 years of intensive study (van Lawick-Goodall, 1973).

The reasonable inference that learning is probably critical for the development of spontaneous tool use in feral chimpanzees is supported by experimental work with laboratory-reared chimpanzees. Birch (1945) showed that young chimpanzees born in the laboratory fail to solve very simple problems of securing food with a stick until they have considerable experience using sticks in other situations. Additional evidence on the role of learning in tool using by chimpanzees is presented in the next section.

THEORIES CONCERNING THE PROCESSES UNDERLYING TOOL USE

Köhler

The classical account of tool-using behavior in captive chimpanzees is Köhler's (1925) monograph, *The Mentality of Apes*. Animals were confronted with situations in which a banana was presented beyond reach and they were obliged to bridge the gap by constructing a tower of boxes or by joining together two short sticks to make one long enough to rake in the banana. The chimpanzees typically behave in a rather confused fashion for a variable period of time and then suddenly initiate a smooth and continuous series of actions that led to their securing the incentive. Characteristically, retention of the solution to a problem was very good.

From these and similar behavioral observations, Köhler inferred that the crucial event underlying insightful behavior in chimpanzees is a sudden reorganization of the perceptual field that allows the subject to see the proper relationships among the relevant elements in the problem situation. Given this restructuring of the perceptual field, the appropriate behavior is assumed to follow in a continuous well-organized sequence almost automatically.

Köhler's observations have been repeated a number of times (Birch, 1945; Yerkes, 1943). No one has seriously questioned the validity of his results nor the accuracy of his descriptions, but his view that the behavior observed in such situations necessarily implies insight has been questioned most severely by Schiller and others.

Schiller and Fellow Travelers

Ethologists maintain that an essential step in studying the behavior of a species is to make a complete inventory of the unlearned motor patterns comprising its behavioral repertory. Schiller (1952, 1957) adopted this approach in studying the instrumentation behavior of chimpanzees. He provided chimpanzees with some of the same materials Köhler had used in his work, but in the absence of bananas or any other external incentive. He found that the construction of a tower, climbing to the top, and jumping up in the air is an intrinsically motivated, spontaneous motor pattern in chimpanzees:

> I tested 12 chimpanzees of 6 to 10 years of age in two 15 minute periods with two of the smallest standardized boxes used by Yerkes and associates. All of them dragged the boxes along the floor, sat and stood on them, rolled them over, carried them carefully balanced to some preferred corner, and used them as pillows. Six of the animals actually stacked them and climbed on the tower, jumping upward from the top repeatedly, with arms lifted above the head and stretched toward the ceiling. For the human observer it was hard to believe that there was no food above them to be reached. Needless to say none of these animals had ever been tested in box-stacking problem situations. [Schiller, 1952, p. 186]

Again,

> Köhler gave a classical description of a male chimpanzee, Sultan, aged about 6 years, who solved the drawing in problem by putting together two bamboo sticks that matched on either end. Sultan's performance was first produced in play but was utilized "immediately" and was repeated the same and the next day several times even with three sticks. Moreover, he pulled out a stopper that prevented joining the sticks before he attempted an insertion and bit off the too-broad end of a board to make it match the hole.
>
> None of these activities need be regarded as having any reference to the problem situation. The manipulation of sticks by older animals that have had no prior opportunity of handling jointed sticks when there is no food to be reached shows all of these varieties of activity. Licking, chewing, stroking and splitting the stick, banging, poking, hammering with it, and thrusting the end into any available openings are responses that occur frequently and constitute the basis of complex motor patterns of utilizing sticks as tools [Schiller, 1952, p. 184]

After the chimpanzees had considerable experience stacking boxes and joining sticks, Schiller tested them for the ability to obtain food by means of these acts.

Introduction of food delayed and interfered with box piling and the joining of sticks compared with performance in play. In other words, there was no evidence of insight about the relation between the tools and food.

Schiller maintained that the solution of instrumentation problems by chimpanzees essentially entails the fixation of sequences of species-typical motor patterns by reinforcement, with food or some other desirable consequence:

> That a chimpanzee breaks off a branch if excited, has nothing to do with his desire to get at the food. Once he has the stick in his hand, he will use it sooner or later. Such a sequence can easily be reinforced in a couple of trials and then it appears to be a coherent, continuous pattern. All problem-solving patterns look like this, but they are really composites of originally independent reactions ... [Schiller, 1957, pp. 275–276]

Schiller's idea that tool-using performances in chimpanzees are based on contingent reinforcement of species-typical fixed motor patterns is parsimonious and rightly stresses the importance of fixed motor patterns in many instances of complex problem-solving behavior. Schiller's views have been extended to provide a general account of the derivation of tool using by monkeys and apes.

The ten types of tool-using behavior listed in Tables 1 and 2 are divided into groups seen in agonistic and nonagonistic situations. This division reflects the belief (Hall, 1963; van Lawick-Goodall, 1970) that aggressive and nonaggressive tool-using performances derive from two different classes of species-typical motor patterns.

The several uses of tools as weapons by primates are regarded by van Lawick-Goodall as learned derivatives of their inherent threat-gesture repetoire, their innate tendency to manipulate objects, and their capacity to modify their behavior as a consequence of differential reinforcement contingencies:

> Many species of monkeys and all the apes may hit at or shake branches during aggressive displays directed toward enemies. As Hall (1963) has pointed out, it seems logical to suppose that if a branch shaken down by a monkey actually hits the intruder below, this will be more rewarding to the monkey concerned than if this did not happen. In the same way, a large branch which hits the enemy will probably produce a more rewarding response than would a small twig. It seems possible, therefore, that in species known for their learning ability, repeated experiences rewarded in this way might cause the slight modification of the threat gesture repertory which is necessary in order for them to break off objects and drop them purposefully . . . .
>
> That throwing behavior has been recorded in baboons is not surprising. As Hall (1963) pointed out, not only do baboons hold and shake branches in threat directed at an enemy, but they also manipulate stones as they turn them over in feeding and, when suddenly startled by a small noxious insect, these primates may make a swift, underarm hitting away movement. Thus the baboon has available the motor patterns necessary to direct objects as well as gestures toward predators or intruders.
>
> The apes are anatomically better adapted to make throwing movements than are the monkeys: the ape shoulder girdle, like that of man, enables him to throw with power (Washburn & Jay, 1967), and he is also anatomically adapted for standing upright, a good posture for forceful throwing. In addition, the chimpanzee, when threatening a conspecific, baboon, or human, may make arm movements very similar, or exactly

similar, to those seen during aimed throwing. On some occasions at the Gombe Stream, a chimpanzee let fly (almost by accident it seemed) some object that it happened to be holding—such as a banana—during an aggressive encounter with a baboon. Because the threat gesture was directed toward the baboon, the object also traveled in that direction.

Some chimpanzees at the Gombe Stream threw far more frequently than others: there are a few adult males who have never been observed to throw in aggressive encounters. Possibly, therefore, an important factor in the development of aimed throwing lies in the experience of the individual: the success of an "accidental" throw might well reinforce the behavior so that it is repeated in a similar aggressive context . . . .

A similar hypothesis may be put forward regarding the use of sticks as weapons in chimpanzees. It is but a short step from the violent shaking of a branch to the deliberate swaying of a branch which causes the ends to touch or whip the object which has aroused the individual's hostility. Similarly, the random waving of branches during charging displays frequently causes other chimpanzees or baboons to rush out of the way—sometimes they may actually be hit. The screaming, running away, or cringing responses induced by such random branch-waving might well be sufficient "reward" to reinforce the branch-waving pattern. Moreover, some movements shown by a chimpanzee during normal threat or attack are very similar to the hitting movements when a stick is wielded [van Lawick-Goodall, 1970, pp. 221–223]

The provenance of nonagonistic tool use in primates is thought to be quite different. The use of objects in connection with feeding and other nonaggressive behaviors probably evolved from the combination of species-typical patterns of investigative and manipulatory behaviors (Schiller, 1952, 1957; van Lawick-Goodall, 1968) into sequences that are reinforced by such rewards as food and relief from bodily discomfort:

The hands of living primates are well adapted to grasp and handle a variety of objects (Napier, 1960). Furthermore, monkeys and apes, both in the wild and in captivity, do manipulate things constantly, not only during feeding, grooming, and so on, but also, particularly in the case of youngsters, during play and during exploratory behavior, when new or unusual features of the environment are carefully examined (e.g., Köhler, 1925; Yerkes & Yerkes, 1929; Menzel, 1964; Butler, 1965; van Lawick-Goodall, 1968). Schiller (1957) has suggested that many of the manipulative patterns observed in primates may be "innate" and that from such patterns adaptive behavior such as tool-using may be derived.

In captivity *Cebus* monkeys, as we have seen, may hammer open hard food objects against the ground or a rock. In addition, these monkeys frequently pick up stones and other objects and bang them against the ground, apparently as a form of play activity (Hill, 1960). Thus *Cebus* monkeys show a tendency to use objects as hammers from which purposeful tool-using may well be derived.

Young chimpanzees in captivity, if allowed to play with sticks, normally use these to poke or tap at a variety of objects before touching them with either hands or lips (Menzel, 1964, Butler, 1965). In addition, chimpanzees, as well as other primates, may actually manipulate levers, buttons, and so on in test situations for no other reward than the performance of the activity (Schiller, 1957; Harlow, 1950). It is not difficult, then to imagine that a young chimpanzee in the wild, as he played by a termite nest, might first of all scratch inquiringly at a spot of damp earth sealing a termite passage, and secondly poke a grass stem or twig into the hole which he has thus revealed. Provided the youngster was familiar with termites (and all of the primates at the Gombe Stream feed on the winged forms of these insects) he would undoubtedly eat the insects which he found clinging to his "tool." Such a reward would undoubtedly induce him to push the twig once again into the hole. [van Lawick-Goodall, 1970, pp. 240–241]

Observations on both wild (van Lawick-Goodall, 1970) and captive chimpanzees (Schiller, 1952, 1957) therefore suggest that much of tool-using behavior in nonhuman primates results from the selection, ordering, and fixation of appropriate motor patterns, already present in the repertory of normally reared animals, through the effects of contingent reinforcement. As can be seen in the next section, however, some tool-using behaviors seem to defy explanation in terms of instrumental conditioning.

Contemporary Revisionists

In spite of the range of phenomena encompassed by the view expounded in the preceding section, it appears progressively more improbable that all of the tool-using behavior of chimpanzees can be explained solely in terms of instrumental conditioning principles. Three sorts of evidence for this view are considered here:

1. Chimpanzees solve some implementation problems by responses that are not species-typical motor patterns.
2. The possession of the correct motor pattern does not always insure adequate performance on tool-using problems.
3. There is evidence that tool using is often learned by imitation rather than by trial and error.

Novel responses. Three examples are provided to illustrate the point that chimpanzees do not rely exclusively on preformed species-typical motor patterns in solving implementation problems. After Köhler's chimpanzees had developed some skill in using boxes to gain otherwise inaccessible food, he tested them with a box full of rocks, which weighted the box so that it was very difficult to take to the food. Removal of the rocks from the box seems a most improbable innate motor pattern and almost certainly was not one practiced in the forest prior to capture; yet it was emitted when required for problem solution.

The second example is van Lawick-Goodall's (1970) observation that an adult male chimpanzee, too frightened to take a banana from the observer's hand, secured the banana by knocking it from the observer's hand with a stick. "This observation is of interest since it was the only time when it was possible to observe what was probably an original solution to a completely new problem involving tool use in the wild" (van Lawick-Goodall, 1970, p. 232). Menzel (1972, 1973) made extensive observations on a group of captive chimpanzees that learned to use branches as ladders for gaining access to forbidden areas and escaping from their compound. Only twice before had the use of branches in this fashion been reported. In both cases (Köhler, 1925; Bolwig, 1961), the ladders were used in order to reach food, not in spontaneous play. This response therefore appears to be a new one that emerges in problem-solving situations and not one of Schiller's predictable motor patterns.

This list of apposite examples is short but suffices to indicate that some tool-using performances in chimpanzees involve the development of new adaptive response sequences rather than the selection of a preformed species-typical motor pattern.

Insufficiency of motor patterns hypothesis. Chimpanzees reared under impoverished conditions in the laboratory are grossly inferior to controls, captured in the wild during infancy, in developing skill in taking food with a stick (Menzel, *et al.,* 1970). The chimpanzees who spent their first 2 or 3 years in restricted environments had spent the time since release from restriction in conditions like those in which the wild-born controls were kept. The restricted chimpanzees played with sticks and other novel objects. They were physically and emotionally capable of the arm, hand, finger, and other pure motor patterns required for solution of the stick problem. Menzel, Davenport, and Rogers concluded that the essential deficit in the restricted chimpanzees was a lack of the ability to apply their skills and motor capacities quickly to new situations, the absence of adaptability.

The work of Menzel and his colleagues, which compares directly the performance of lab-reared chimpanzees and animals captured in the field as infants, has several interesting implications. Schiller (1952, 1957), for example, may have failed to obtain evidence of insight in his subjects because they were laboratory-reared animals, while Köhler observed trapped wild specimens. More relevant in the present context, however, is the demonstration that the presence of the relevant motor patterns in a chimpanzee does not automatically result in appropriate application of the patterns to solution of instrumentation problems. Schiller's (1957) dictum that once a chimpanzee has a stick in his hand he will use it sooner or later is not universally valid.

Menzel's (1972, 1973) observations on a captive group of chimpanzees that spontaneously invented the ladder presents a strong challenge to the explanation of tool use in chimpanzees in terms of operant conditioning. The group, observed daily for 5 years, showed several stages in the ways they used sticks and poles as climbing instruments. At first, the animals simply set the pole upright, quickly climbed to the top, and jumped before the pole toppled over. Years later the chimpanzees learned to brace branches against a wall, in order to enter the windows of the experimenters' observation house; still a year later, the chimpanzees adapted this skill to get over the walls of their compound and to escape from captivity.

A single chimpanzee initiated a more complicated routine for using poles. He used a pole as a ladder to climb onto a narrow elevated runway, pulled up the pole, set its base on the runway and the opposite end on a tree trunk, and then climbed the pole into the treetop, avoiding electric shock wires intended to keep the chimpanzees out of the trees. Eventually all of the chimpanzees in the group learned to use poles as ladders to the tree tops. Careful observations of the unsucessful early efforts of the other animals to duplicate the innovator's performance showed that the chimpanzees had no difficulty in reproducing the successful movement patterns. The major problem was to adjust the motor patterns to the stringent physical requirements of the situation: placing the butt of the pole on, not next to, the runway.

Menzel's work is, in several important respects, contradictory of the interpretation of tool use advanced by Schiller. The early unsuccessful efforts of the chimpanzees who tried to use poles to get up the trees lacked none of the necessary motor responses; their failures resulted from neglect of the pole base–runway connection. The fact that difficulties in learning are perceptual–motor rather than purely motor is more compatible with the position of Köhler than that of Schiller.

The observation that a year elapsed from the time when the chimpanzees used ladders to enter the observation house until they used ladders to escape over the walls of their field cage provides yet another illustration of the importance of nonmotor factors in tool using by chimpanzees. The animals clearly had the necessary motor pattern but did not use it until they eventually saw it could be used to scale walls as well as to get into the observation house.

Imitation.   Thorpe (1956, p. 411) has defined intelligent or true imitation as an animal's "copying a novel or otherwise improbable act—some act for which there is clearly no instinctive tendency." Imitation in this sense has clearly been shown by the chimpanzees in Menzel's study of the use of ladders. "Once a single chimpanzee performed a given new behavior, the others watched closely and attempted the same thing almost immediately; and eventually all members of the group shared the behavior" (Menzel, 1972, p. 456). The most dramatic illustration of social learning in this group is the fact that all eight chimpanzees use ladders to escape from their field cage within a day after the first escape.

Chimpanzees reared in human homes are apt mimics and develop considerable proficiency in imitating the behavior of their human caretakers (Gardner & Gardner, 1969; Hayes & Hayes, 1952). The Hayeses report several cases of immediate imitation of tool-using performances, such as opening paint cans with a screw driver, hammering a stake into the ground, and using a stick to push a food lure from a wire tunnel.

It is difficult to categorize accurately the types of imitation seen in feral primate groups. However, the data are sufficient to show that the social modification of individual behavior plays an important role in the development and diffusion of tool-using and other behaviors. Immature chimpanzees closely observe and duplicate, at least in part, a number of patterns of tool use by adults: fishing for termites and other insects, drinking with a leaf sponge, wiping the body with leaves and nest building (van Lawick-Goodall, 1970, 1973).

The most comprehensive studies on imitative learning and the development of cultural traditions in nonhuman animals have been carried out with free-ranging groups of Japanese macaques. Individuals in these troops have developed such new habits as washing sweet potatoes in water to remove sand and in sea water to season them with salt, separating wheat and sand by a placer mining technique, and swimming in the sea (Itani & Nishimura, 1973). Later many other members of the troop acquired the novel behaviors. The diffusion of these novel habits clearly demonstrates the capacity of primates to acquire new and improbable responses by observing their companions.

The current literature on imitation is a conceptual and empirical muddle (Davis, 1973) that I have not tried to sort out here. For my present purposes, it is enough to point out that whatever processes may be involved in different cases, imitation plays an important role in the determination of many behaviors, including tool use in primates, and that neither operant nor conventional reinforcement theories of learning provide a satisfactory model of imitative learning (Davis, 1973). These conclusions in turn imply another limitation on the generality of Schiller's interpretation that tool use in primates reflects no more than the fixation of innate motor patterns by selective reinforcement.

I have deliberately emphasized the evidence for a cognitive rather than an operant learning theory of tool use by primates because the operant theory is currently more fashionable. Actually, it must be admitted, neither view provides a very good or complete model of tool use by primates. Köhler (1925) was chiefly interested in how chimpanzees use existing perceptual and motor skills to solve instrumentation problems, and he was rather unconcerned with the genesis of these skills. His critics have shown that competence in using tools critically depends on prior learning of specific skills, however. The demonstration that appropriate experience and learning is necessary for effective tool use was an important contribution. However, Schiller and his successors have erred in suggesting that possession of the required learned skills is sufficient for adequate tool use. Chimpanzees may have the requisite skills and still fail to solve tool problems; appropriate skills are therefore necessary but insufficient for competent instrumental performance (Menzel, 1972, 1973; Menzel et al., 1970).

One major reason there is no satisfactory comprehensive theory of tool use by primates surely must be that the current theories are based on laboratory tests made to answer questions about learning theories. Köhler's (1925) studies were undertaken to refute Thorndike's position that learning consists of the stamping in of stimulus—response bonds by selective reinforcement. Schiller's (1952, 1957) work was carried out primarily to show the importance of fixed motor patterns in primate behavior.

I can offer no resolution of this theoretical difficulty, but I can suggest that one major hope for significant advances in this field is to increase the amount of research on wild and on unrestrained captive primates, and to pay more attention to problems posed by chimpanzees than by learning theorists.

Observations on termiting by chimpanzees afford some hints of the gains to be realized from a more balanced approach that relies on naturalistic as well as experimental investigations. Infant chimpanzees make their first attempts to fish for termites in imitation of adults but gain adult levels of proficiency only after years of trial-and-error practice (van Lawick-Goodall, 1968, 1970). Note that both imitation and trial-and-error learning contribute to the development of this skill and that imitation seems to come first. This suggests that operant and more cognitive learning processes are complementary in the real world, as most products of natural selection are, and that operant learning and imitation may be opposites or hierarchically ordered only in the speculations of learning theorists.

Teleki (1974) spent several weeks during two termiting seasons in the Gombe Stream Reserve trying to learn the skills used by adult chimpanzees in fishing for termites. He failed completely to learn how chimps located the concealed entrances to tunnels on the outside of termite mounds. On the assumption the chimpanzees were not using sensory cues that were not available to him, Teleki was forced to speculate that chimps memorize the location of one hundred or more tunnels in familiar termite mounds. This hypothesis in turn led to a series of interesting experimental predictions that almost certainly would not have occurred to workers in the laboratory.

SUMMARY AND CONCLUSIONS

Primates greatly exceed all other mammals in respect to the number and variety of spontaneous tool-using behaviors they exhibit. Primates appear to be the only nonhuman mammals that make tools.

Several sorts of evidence strongly suggest that tool use in feral primates is the result of learning rather than of innately determined factors.

The currently prevalent view that tool use in primates represents the relatively automatic selection and fixation of preformed motor patterns by differential reinforcement is inadequate. This model requires drastic revision to accommodate the facts that chimpanzees (1) use novel motor patterns in solving some tool problems, (2) sometimes have perceptual rather than motor difficulties in solving instrumentation problems, and (3) occasionally learn to use tools by imitation rather than trial and error.

It must be conceded, however, that the insight theory of tool use is also incomplete, for it fails to deal satisfactorily with the role of past experience in the development of problem-solving skills. The theoretical approaches deriving from Köhler and from Schiller are both limited in scope, because they are based on observations of laboratory chimpanzees, undertaken to investigate problems of learning or behavior theory. Consequently, because neither theory has been formulated in the light of sound information about the range of tool-using skills seen in feral primates, neither provides a comprehensive account of such behaviors.

The need for a thorough revision of current theories concerning the use of tools by chimpanzees is emphasized by recent research on other aspects of behavior in chimpanzees. Chimpanzees not only make and use tools but they also show several additional kinds of behavior once regarded as uniquely human: cooperative hunting and food sharing (Teleki, 1973), language-like behavior, and intermodal transfer (van Lawick-Goodall, 1973). These observations strongly support the idea that attempts to explain the tool-using behavior of chimpanzees solely in terms of instrumental conditioning are almost certain to fail.

Research on tool use has contributed substantially to the view that the behavior of chimpanzees is very much more like that of humans than anyone imagined a decade ago.

ACKNOWLEDGMENTS

Preparation of this paper was supported by Grant MH-04726, from the National Institute of Mental Health, U.S. Public Health Service.

REFERENCES

Bernstein, I. S. Response to nesting materials of wild born and captive born chimpanzees. *Animal Behaviour,* 1962, **10,** 1–6.

Birch, H. G. The relation of previous experience to insightful problem solving. *Journal of Comparative Psychology,* 1945, **38,** 367–383.

Bolwig, N. An intelligent tool-using baboon. *South African Journal of Science,* 1961, **57,** 147–152.

Butler, R. A. Investigative behavior. In A. M. Schrier, H. F. Harlow, & F. Stollnitz (Eds.), *Behavior of nonhuman primates,* Vol. 2. New York: Academic Press, 1965, Pp. 463–493.

Davis, J. M. Imitation: a review and critique. In P. P. G. Bateson & P. H. Klopfer (Eds.), *Perspectives in ethology.* New York: Plenum Press, 1973, Pp. 43–72.

Gardner, R. A., & Gardner, B. T. Teaching sign language to a chimpanzee. *Science,* 1969, **165,** 664–672.

Hall, K. R. L. Tool-using performances as indicators of behavioral adaptability. *Current Anthropology,* 1963, **4,** 479–487.

Harlow, H. F. Learning and satiation of response in intrinsically motivated complex puzzle performance by monkeys. *Journal of Comparative & Physiological Psychology,* 1950, **43,** 289–294.

Hayes, K. J., & Hayes, C. Imitation in a home raised chimpanzee. *Journal of Comparative & Physiological Psychology,* 1952, **45,** 450–459.

Hill, W. C. O. *Primates,* Vol. 4. Edinburgh: University of Edinburgh Press, 1960.

Itani, J., & Nishimura, A. The study of infrahuman culture in Japan. *Symposia of the Fourth International Congress of Primatology.* Basel: Karger, 1973, Vol. 1. pp. 26–50.

Jones, C., & Pi, J. S. Sticks used by chimpanzees in Rio Muni, West Africa. *Nature (London),* 1969, **223,** 100–101.

Köhler, W. *The mentality of apes.* New York: Harcourt, Brace, 1925.

McGrew, W. C. Tool use by wild chimpanzees in feeding upon driver ants. *Journal of Human Evolution,* 1974, **3,** 501–508.

McGrew, W. C., & Tutin, C. E. G. Chimpanzee dentistry. *Journal of the American Dental Association,* 1972, **85**, 1198–1204.

McGrew, W. C., & Tutin, C. E. G. Chimpanzee tool use in dental grooming. *Nature* (London), 1973, **241**, 477–478.

Menzel, E. W. Patterns of responsiveness in chimpanzees reared through infancy under conditions of environmental restrictions. *Psychologische Forschung,* 1964, **27**, 337–365.

Menzel, E. W. Spontaneous invention of ladders in a group of young chimpanzees. *Folia Primatologia,* 1972, **17**, 87–106.

Menzel, E. W. Further observations on the use of ladders in a group of young chimpanzees. *Folia Primatologia,* 1973, **19**, 450–457.

Menzel, E. W., Davenport, R. K., & Rogers, C. M. The development of tool using in wild-born and restriction-reared chimpanzees. *Folia Primatologia,* 1970, **12**, 273–283.

Napier, J. R. Studies of the hands of living primates. *Proceedings of the Zoological Society of London,* 1960, **134**, 647–657.

Schiller, P. H. Innate constituents of complex responses in primates. *Psychological Review,* 1952, **59**, 177–191.

Schiller, P. H. Innate motor action as a basis of learning. In C. H. Schiller (Ed.), *Instinctive behavior.* New York: International Universities Press, 1957, pp. 264–287.

Struhsaker, T. T., & Hunkeler, P. Evidence of tool-using by chimpanzees in the Ivory Coast. *Folia Primatologia,* 1971, **15**, 212–219.

Teleki, G. The omnivorous chimpanzee. *Scientific American,* 1973, **228**(1), 32–42.

Teleki, G. Chimpanzee subsistence technology: Materials. *Journal of Human Evolution,* 1974, **3**, 575–594.

Thorpe, W. H. *Learning and instinct in animals.* London: Methuen, 1956.

van Lawick-Goodall, J. The behaviour of free-living chimpanzees in the Gombe Stream Reserve. *Animal Behaviour Monographs,* 1968, **1**, 161–311.

van Lawick-Goodall, J. Tool-using in primates and other vertebrates. *Advances in the Study of Behavior,* 1970, **3**, 195–249.

van Lawick-Goodall, J. Cultural elements in a chimpanzee community. *Symposia of the Fourth International Congress of Primatology.* Vol. 1. Basel: Karger, 1973. Pp. 26–50.

van Lawick-Goodall, J., van Lawick, H., & Packer, C. Tool use in free living baboons in the Gombe National Park, Tanzania. *Nature* (London), 1973, **241**, 212–213.

Washburn, S. L., & Jay, P. More on tool-use among primates. *Current Anthropology,* 1967, **8**, 253–254.

Yerkes, R. M. *Chimpanzees.* New Haven: Yale University Press, 1943.

Yerkes, R. M., & Yerkes, A. M. *The great apes.* New Haven: Yale University Press, 1929.

19

Primate Social Behavior: Pattern and Process

William A. Mason

University of California, Davis, California

Monkeys and apes, as is well known, are devoutly gregarious animals. In contrast to many creatures that join together only at certain seasons of the year to perform the essential functions of procreation and to participate, perhaps, in the rearing of young, most primates spend their lives from start to finish as members of an organized group—as part of a complex social system that moves through space, that endures, that manages somehow to hang together in spite of internal wrangling, outside threats to its integrity, and a steady turnover in personnel.

The point to be emphasized is that the pattern abides, even though the individuals do not. Moreover, it has been known for many years that the patterns are different for different species (Carpenter, 1942). In actuarial terms, at least, a characteristic grouping pattern can be described for each species of monkey or ape. For the patas monkey, the gelada, and the hamadryas baboon, for example, the basic social unit is the "harem" or unimale group. The South American howler monkey, the rhesus, and the savannah baboon are a few of the many species that are usually organized in multimale groups containing several adult males and females, as well as immature animals in all stages of development. A handful of species—the gibbon is the most familiar example—is characteristically arranged in monogamous families consisting of an adult male and female and one or two young.

Going beyond the obvious contrasts in group size and composition more subtle evidence is encountered of wide differences in the organization of social life, even among species that favor similar grouping patterns. The hamadryas baboon (*Papio hamadryas*) and the patas monkey (*Erythrocebus patas*), are two species that organize themselves in unimale groups. The hamadryas male is the undisputed ruler of the roost and he keeps his female followers in line by biting their necks if they lag behind or show a wayward fancy for some prowling bachelor on the lookout for a mate (Kummer, 1968a). In contrast, the patas male seems to exercise little if any physical control over the females in his group and, in fact, is often the target of their aggression, in spite of his obvious superiority in size and strength (Hall, 1965). Relations between groups are another point of difference between these species. Hamadryas unimale units congregate at night on cliffs and sleeping rocks where groups settle down peaceably, separated from each other by no more than a few meters. Relations between patas unimale groups appear at all times strained. Groups are so widely dispersed that they seldom meet. During the dry season, when several groups may be drawn to a common waterhole, encounters between groups are plainly agonistic (Struhsaker & Gartlan, 1970).

The list can be greatly extended, of course, but it serves no useful purpose to do so now because my aim at this point is simply to provide a reminder that species differences in social grouping patterns are real and pervasive. This is hardly news but it is an essential starting point for the perspective that is to be developed in this chapter. My goals are to arrive at an overview of primate societies as social systems, to discover a common ground for dealing with the patent diversity among species in patterns of social organization, and to find a place within this picture for the individual primate as a competent and contributing member of the body social.

CLASSIFICATION OF PRIMATE SOCIAL SYSTEMS

A first step toward coming to grips with any complex natural phenomena is to order them according to some rational scheme. The obvious interspecies diversity in group size and composition has led to several recent attempts to construct taxonomies of primate social systems. Crook (1970), for example, distinguishes five main types, largely on the basis of differences in group size, age and sex composition, and relations between groups. At one extreme he places the small unimale group with strong territorial proclivities, represented by the titi monkey (*Callicebus moloch;* Mason, 1968), the colobus (*Colobus quereza;* Marler, 1969), the lutong (*Presbytis cristatus;* Bernstein, 1968) and the gibbon (*Hylobates;* Carpenter, 1940). At the opposite end he places relatively open groups, showing little stability in composition or internal structure, represented by the chimpanzee (*Pan troglodytes;* Goodall, 1965). Kummer (1971) has suggested a three-part classification scheme: the multimale group, the unimale group, and the monogamous family unit.

ANALYSIS OF SOCIAL SYSTEMS

Two kinds of questions have been asked about differences in social structure. One is a functional question, asking what useful purpose a particular grouping pattern serves. The question implies a concern with evolution and adaptation.

The guiding assumption is that the characteristic grouping pattern of a species is the product of natural selection, that it contributes to survival in determinable ways. Interest is focused on the "design features" of the social system in relation to the exigencies and constraints of the natural habitat, on social structure as a means of coping with the environment.

The second question is concerned with causation. Causal questions are concerned with how a particular grouping pattern is actually achieved. Through what social processes do members of a primate group create and maintain the social system in which they live? Whence comes the apparent predilection of this species for monogamy and that one for a more extended social system? What factors determine the size, the composition, and the cohesiveness of primate societies?

ADAPTATION AND FUNCTION

I shall first consider the problem of the evolutionary origins and adaptive characteristics of primate grouping patterns. This issue is the focus of lively concern among many field workers, some of whom have suggested correlations between ecological factors and specific features of social systems. Most biologists agree that the major ecological determinants in the evolution of social structure are the distribution and availability of food and other essential resources and predation pressure. These constitute limiting factors within which various organizational strategies may evolve, the "optimal" strategy being one that allows any particular adult group member to maximize its genetic contribution to succeeding generations. For example, in an area of abundant resources, no predation, and ready access to females, the male's optimum strategy is to impregnate as many females as possible, up to the limit of his sexual capabilities. In a less benign environment he can improve his chances of evolutionary success by establishing a more abiding relationship with several females that carry his children, extending to mothers and young some measure of protection, and perhaps insuring for them an adequate supply of resources. In an environment in which large predators are a persistent threat, some advantage may be gained by teaming up with other adult males, in spite of the potential competition for females that such an arrangement may introduce. Comparable considerations apply in the natural selection of female strategies, although they differ in detail, of course, owing to the greater female investment in the young. A female not only gives more to each of her children than does a male; her investment per child is also increased because she is usually limited, as he is not, to producing just one offspring in a breeding season (Goss-Custard, Dunbar, & Aldrich-Blake, 1972).

Among the first efforts to sort out the relations between social structure and ecology was DeVore's suggestion that large organized groups, low population density, intergroup spacing by visual rather than vocal means, prominent dominance relations, and patent physical and behavioral dimorphism were more characteristic of terrestrial than of arboreal primates and could be regarded as adaptations to predation and other selection pressures peculiar to life on the open ground (DeVore, 1963). In a more detailed analysis, Crook and Gartlan (1966) assigned species to five grades, each grade representing a relation between characteristics of social structure and habitat features. For example, the primates they assigned to Grade I were solitary, nocturnal insectivores; species in Grade II were said to live in pairs, to subsist on a diet of fruit or leaves, and to maintain intergroup dispersion using marking or territorial displays; Grade V included the patas monkey, the gelada (*Theropithecus gelada*), and the hamadryas—all grassland or dry savannah monkeys organized in unimale groups.

Crook and Gartlan regard their Grade V, the unimale group, as a specialized offshoot of the multimale pattern, evolved in response to the problems posed by the widely dispersed resources characteristic of a semidesert habitat—a view that others share (Crook, 1970; Gartlan & Brain, 1968; Kummer, 1967, 1968b, 1971).

The assumption that the unimale grouping pattern is a highly specialized form of social organization has been questioned by Eisenberg and associates. They suggest instead that the unimale group is a general pattern from which the multimale system has evolved (Eisenberg, Muckenhirn, & Rudran, 1972). The ancestral primates are characterized as primitive nocturnal insectivores living in a solitary state in which adult males and females have separate centers of activity and encounter each other primarily during the breeding period (a pattern displayed by some living prosimians). Evolution toward more permanent social arrangements has proceeded along two pathways, one leading to the monogamous family unit, shown by some prosimians, several species of South American monkeys, and the gibbon, and the other leading to a cohesive group of females and their progeny, with a single adult male in periodic contact. Either of these patterns can presumably have given rise to the cohesive unimale group, and this in turn to the multimale group (Figure 1). The major psychological requirement for this development, according to Eisenberg *et al.* (1972), is increased tolerance among males, and they regard the age-graded male troop (characterized by a single adult male at the top of a linear male dominance order, based on age) as an important intermediate stage between the strict unimale pattern and the multimale troop. The failure to appreciate the prevalence of this intermediate pattern, they suggest, is responsible for the popular view that the multimale system is general and widespread among primates, whereas it is more properly regarded as a specialized offshoot of the age-graded unimale troop and is limited to just a few taxa. Among most forest-dwelling species "... there is a strong tendency toward a polygynous unimale reproductive unit" (p. 867).

With regard to the factors shaping the evolution of primate social structures, Eisenberg *et al.* share with other theorists the belief that resource distribution,

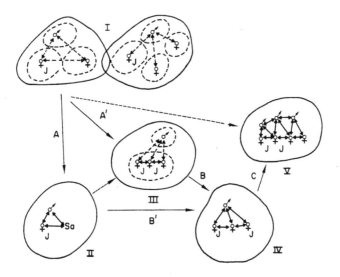

feeding strategies, and predation have been major selection pressures. Their classification relating social organization to feeding ecology is presented in Table 1.

INTRASPECIFIC VARIATIONS IN SOCIAL STRUCTURE

Apart from suggesting some of the evolutionary sources and adaptive benefits of primate social systems, such typological schemes have great heuristic value, for they indicate in categorical terms the phenomena—the social end points—for which explanations are sought. At the same time, however, it is evident that grouping patterns are the result of many diverse influences and some variation within each species must be expected. Each generation of animals builds its own society. Because of the complexity of causal factors involved, any classification of primate social structures should be viewed as a somewhat idealized description of modal tendencies, a statistical approximation, rather than as a hard and fast taxonomy expressing fixed species-typical traits. In fact, recent findings leave no doubt that substantial within-species variations can be found in group size, group composition, and many other prominent aspects of social organization.

Consider first variations in group composition. For several species, the tendency to form groups on the unimale or multimale pattern seems to depend on ecological conditions. In the Dharwar area of southern India, for example, a region characterized by dry deciduous forests, summer drought, and high popu-

TABLE 1

Range of Social Organization and Feeding Ecology for Selected Primate Species

Solitary species	Parental family	Minimal adult ♂ tolerance[a] (uni-male troop)[b]	Intermediate ♂ tolerance[c] (age-graded-male troop)[b]	Highest ♂ tolerance[d] (multi-male troop)[b]
Insectivore–frugivore	Frugivore–insectivore	Arboreal folivore	Arboreal folivore	Arboreal frugivore
		Cebidae	Colobinae	Indriidae
		Alouatta palliata	*Presbytis cristatus*	*propithecus verreauxi*
Lemuridae	Callithricidae (Hapalidae)		*Presbytis entellus*	Lemuridae
Microcebus murinus	*Saguinus oedipus*	Colobinae		*Lemur fulvus*
Cheirogaleus major	*Cebuella pygmaeus*	*Colobus guereza*	Cebidae	
	Callithrix jacchus	*Presbytis senex*	*Alouatta palliata*	Semiterrestrial
Daubentoniidae		*Presbytis johni*		frugivore-omnivore
Daubentonia madagascarensis		*Presbytis entellus*	Arboreal frugivore	Cercopithecidae
	Cebidae			*cercopithecus aethiops*
	Callicebus moloch	Aboreal frugivore	Cebidae	*Macaca fuscata*
Lorisidae	*Aotus trivirgatus*		*Ateles geoffroyl*	*Macaca mulatta*
Loris tardigradus		Cebidae	*Saimiri sciureus*	*Macaca radiata*
Perodicticus potto		*Cebus capucinus*		
			Ceropithecidae	

Folivore	Folivore-frugivore	Semiterrestrial frugivore	Semiterrestrial frugivore-omnivore	Terrestrial folivore-frugivore
Lemuridae *Lepilemur mustelinus*	Indriidae *Indri indri* Hylobatidae *Hylobates lar* *Symphalangus syndactylus*	Cercopithecidae *Cercopithecus mitis* *Cercopithecus campbelli* *Cercocebus albigena* Cercopithecidae *Erythrocebus patas* *Theropithecus gelada* *Mandrillus leucocephalus* *Papio hamadryas*	*Miopithecus talapoin* Cercopithecidae *Cercopithecus aethiops* *Cercocebus torquatus* *Macaca sinica* *Papio cynocephalus* *Papio ursinus* *Papio anubis* *Macaca sinica*	Pongidae *Gorilla gorilla* Pongidae *Pan satyrus*

[a] Troop with one adult male and strong intolerance to maturing males.
[b] "Troop" refers to the basic social grouping of adult females and their dependent or semidependent offspring.
[c] Troop typically showing age-graded-male series.
[d] Troop with several mature, adult males and age-graded series of males.
After Eisenberg et al. (1972).

lation density, the common langur (*Presbytis entellus*) shows a strong tendency toward unimale groups. This pattern has been found in 73% of 63 troops sampled by Japanese field workers (Sugiyama, 1967; Yoshiba, 1968). In contrast, in the Orcha area of northern India, a region characterized by moist deciduous forests, moderate summer climate, and relatively low population density, the most common pattern among the same langur species is the multimale group (Jay, 1965; Yoshiba, 1968).

Vogel (1971), also working with Indian langurs, has compared composition of groups living in the foothills of the Himalayas at an elevation of 900–1500 m with troops living at a lower elevation (400 m) and has found that multimale groups are more numerous at the lower altitude. Similar data have been reported for the black spider monkey of South America (*Ateles paniscus*) by Durham (1971), who has found a mean of about four adult males per group at an elevation of 275 m, as compared with purely unimale groups at elevations of 889 m and above.

The howler monkey (*Alouatta palliata*) of Barro Colorado Island in Panama illustrates a different aspect of environmental effects on group composition. When these monkeys were first studied by Carpenter in the 1930's he found a total population of some 400–500 animals. In his census of 1933 he counted 28 groups, 18 of which contained three or more adult males. Only three groups had just one male. The next systematic census, by Collias and Southwick (1952), occurred several years after a drastic reduction in the howler population, presumably the result of an epidemic disease. In contrast to Carpenter's findings, none of the 30 groups censused by Collias and Southwick had more than two males, most had just one, and all had at least one (Carpenter, 1962). Unless it is assumed that providence conspired to spare at least one male in each group, the most reasonable interpretation is that the even distribution of males was brought about by the animal's own efforts. Eight years later, when the population was again censused, the grouping pattern resembled the preepidemic results: only six of the 44 groups included in the survey had just one male and most had three or more (Carpenter, 1962; see Chivers, 1969, for other changes with continued growth of the population).

Intraspecies diversity is equally evident in other aspects of social organization. Forest-living baboons (*Papio anubis*) differ from savannah-living baboons of the same species in patterns of group deployment, the stability of group membership, and intergroup relations (compare Hall & DeVore, 1965, with Rowell, 1966a). Vervet monkeys (*Cercopithecus aethiops*) occupying a habitat rich in food resources tend to have small, well-defined territories that are strongly defended against conspecifics, as compared with groups of the same species living in poor habitat (Gartlan & Brain, 1968). Northern and southern populations of the common Indian langur differ in the extent of home ranges, in the average size of groups, and in the nature of social relations within and between groups. Group stability is sharply different in the two forms. In the northern groups, little evidence of social change has been noted during 18 months of field observations. In contrast, violent social upheaval is relatively common among the southern groups observed by Sugiyama, Yoshiba, and their co-workers. The precipitating event is usually an attack on a bisexual troop carried forward by an all-male band. If these invading males succeed in driving off the "leader" they sometimes fight among themselves until only one male remains. On three occasions following a successful invasion, the new leader has turned on the infants in the group, attacked them while they are still clinging to their mothers, and inflicted serious injuries, while leaving the mothers unharmed (Sugiyama, 1967; Yoshiba, 1968). This behavior is the more remarkable for being so contrary to the protective role that is usually ascribed to adult male primates and which, indeed, is also displayed by the males of these same southern langurs when they are the established leaders of stable groups (Sugiyama, 1965).

CAUSAL ANALYSIS OF SOCIAL SYSTEMS: THE PROBLEM AND A PERSPECTIVE

The cautionary message in such findings is clear. At the very least, they erase any lingering tendency to regard social organization as a fixed, unitary, species-typical trait. However, I doubt that this has ever been a really serious temptation. From the beginning of primate field work, social systems have been anticipated to show variability (Carpenter, 1942), and with very good reason. After all, if it is assumed that primate groups are not only adaptive but adaptable systems, they should have the resources for coping with a changing environment because change is an inevitable part of the natural scene. The physical circumstances in which the group exists—climate and the availability of food, cover, and other essential commodities—varies with place and season. The cycle of births and deaths is also an essential part of this picture. The number of animals born into the group and the number that survive to maturity cause changes in group size and composition, as does the attrition in membership owing to predation, old age, and disease. In addition, grouping patterns may be influenced by the presence of neighboring groups of the same species indirectly, through competition for limiting resources, and directly, by harassment, invasion, or recruitment.

The issue, then, is not whether grouping patterns are species typical, for the terms in which this question will be answered are already clear. The issues are instead how a group manages to preserve some semblance of social order, to maintain a degree of cohesiveness and continuity, in spite of the vagaries of time and local circumstance. What are the sources of stability and change? How can the existence of modal grouping patterns be accounted for and at the same time the fact dealt with that deviations from these patterns are bound to occur?

The focal unit implied in these questions is the group, and what is called for is a causal analysis of social systems. Such an enterprise must consider three different causal levels. The first is the individual, the primary element in all primate societies, who brings to any situation a particular repertoire of social behaviors. This repertoire, which is influenced by a host of factors including previous experience, determines the individual's potential contribution to the social system. The second level is the social setting. This includes the membership of the group, expressed in terms of numbers of animals in each age–sex class, and, of course, their individual social repertoires. Social behavior—what the members of the group do to, for, and with each other—reflects in large part the reciprocal interactions among the repertoires of the participants. At any given moment in time such interactions are limited by the size and composition of the group, but in the long run they also help to determine its size and composition. They are therefore the social regulators of the system. The third causal level is the environment of the social system. This includes all socially relevant features of the habitat, whether they impinge directly on organismic processes and therefore on the social repertoire of individuals (as might nutrition, climatic factors, or disease) or limit the possibilities for interanimal social adjustment (as may the distribution of food, shelters, or the amount of usable space). The environment also includes predators and competitors, particularly competitors of the same species, which can pose multiple threats to the integrity of the social system.

The value of this tripartite division of the proximal influences on the social system is strategic rather than explanatory. It provides a perspective and a point of departure. By identifying three classes of causal factors and emphasizing the interplay between them, it establishes the major analytic tasks. These, of course, are to specify the relevant parameters for each class and to describe how the interactions between them result in the social processes that shape and maintain the system.

To complete the perspective, it is essential to recognize two additional dimensions: time and the level of the social unit. Some influential events occur almost in the blinking of an eye; others occupy days, months, or even an entire lifetime of an individual or a group. Likewise, some important interactions occur between just two individuals—mating, parental care, a contest over some coveted item—but most events, at least in natural circumstances, involve several individuals—kinship clusters, cohorts, coalitions, or even an entire social group.

Each of these dimensions has both a descriptive (objective) and an analytic (psychological) aspect. The descriptive aspect is the easiest to appreciate, for it is the one observers of the system use to locate and order significant social processes. The analytic aspect refers to the way time and the level of the social unit are represented within the animals themselves. That both the observers and the observed are making use of the same dimensions is not surprising, of course, but the fact that the two parties use these dimensions in quite different ways (which is often overlooked) is an unending source of uncertainty and confusion.

On the one hand, it is known that the nonhuman primates are adept at integrating information in complex ways across space and over time. They are equipped with excellent memories and are capable of forming concepts, responding to abstract relations, identifying configurations, and developing hypotheses, strategies, and sets (for reviews see chapters in Jarrard, 1971; Schrier, Harlow, & Stollnitz, 1965). There is good reason to suppose that these cognitive abilities play an important part in the regulation of social life, probably to a greater extent than for any other gregarious animal.

On the other hand, there is little systematic knowledge as to how this occurs in any species. Only sporadic attempts have been made to show how intellectual functions, demonstrated repeatedly with individual subjects in the controlled environment of the learning laboratory, actually operate in a social setting, probably the very situation that provides the most pressing needs and the most abundant opportunities for the development and exercise of high level cognitive skills.

Guesswork is not good enough and is particularly hazardous when extrapolations are made from the relatively well-studied baboons, macaques, and chimpanzees to other species, for it seems likely that wide differences obtain in the quality as well as the rate of learning among various primate taxa (e.g., Warren, 1965a, b; Rumbaugh, 1970, 1971). The fact is, however, that until a firmer bridge is laid between laboratory investigations of cognition in the individual primate and the complex realities of social life, educated guesses are about the best that can be done. At any rate, because I wish in the following sections to present the individual as a knowledgable as well as active participant in the social group, I must try to choose a path that lies somewhere between endowing the monkey or ape with near-human cognitive abilities, on the one hand, and on the other, treating them as mindless robots, devoid of personal history, responding with machinelike predictability to the immediate push and pull of external events.

SOCIAL SYSTEMS AND THE INDIVIDUAL: THE SOCIAL REPERTOIRE

Individuals are both the products and the producers of societies. Here the individual is viewed as a potential contributor to the social system, as one element in a network of relations; the factors outside the immediate situation that determine the character of his social responses can then be examined.

Each newborn monkey or ape is an on-going, organized enterprise, ready to function within a particular social niche. However, it is a far cry from the adult that it becomes. As its career is followed from birth to maturity, an orderly progression through various developmental stages is witnessed: the first clumsy efforts to find the nipple and to nurse, the growing dependence on the parent as a source of reassurance and support, the gradual waning of this bond, the early tentative contacts with siblings and with age mates, the emergence of abiding relationships and of status, and finally the achievement of an established position as a mature member of the community.

Therefore, in normal circumstances the individual acquires a "species-typical" repertoire and its structure and determinants can be explored in various ways. Behavioral differences between the sexes; the relevance of age, of brain functions, and of endocrine products; and other aspects of current physiological state have been established for many species.

Among macaques, for example, male and female show reliable differences in social responsiveness early in the first year of life. Males are more likely to initiate rough and tumble play, they are less likely to respond passively to social contact or withdraw from it, and they spend less time in contact with their mothers, are cradled and retrieved less often, and are more often punished by them than are females (Harlow, 1965; Jensen, Bobbitt, & Gordon, 1968; Mitchell, 1968). Such contrasts are partly dependent on the early organizing effects of humoral factors, for genetic females can be moved toward male patterns of social responsiveness if they are treated prenatally with androgens (Goy, 1970).

As may be expected, sex differences are more prominent in mature animals, but their particular expression depends on a number of factors, including current endocrine state. Females of several species of old world primates show changes in social responsiveness in relation to the reproductive cycle, the most prominent of which is an increase in assertiveness during the peak of sexual receptivity (e.g., Ball & Hartman, 1935; Birch & Clark, 1946; Rowell, 1967a, 1971). Social responsiveness of males is likewise influenced by hormonal factors. Aggressive behavior is significantly correlated with testosterone levels in male rhesus (*Macaca mulatta*); castrates of this species show a gradual loss of dominance, tend to associate with each other, and in some cases display increased interest in caring for immature monkeys (Rose, Holaday, & Bernstein, 1971; Wilson & Vessey, 1968).

Species differences are prominent and pervasive. The tendency to seek or tolerate social contact, the attraction between and within the sexes, and the specific patterns that appear in the regulation of social conduct may contrast sharply even between quite closely related species. For example, bonnet and pigtail macaques (*M. radiata, M. nemestrina*) differ in the amount of time they spend in social contact, in the quality of maternal care, and in the details of male sexual performance (Nadler & Rosenblum, 1971; Rosenblum & Kaufman, 1967; Rosenblum, Kaufman, & Stynes, 1964); among baboons, the neckbite is used as a specific herding technique by hamadryas males but never by males of closely related species; the male squirrel monkey (*Saimiri sciureus*) shows its erect penis—most often to other males—as part of an elaborate social display that is seen in no other South American monkey.

Experience is, of course, important. The macaque or chimpanzee raised apart from its natural mother and denied contact with age mates displays a host of deficiencies and aberrations, some of which persist into adult life. At maturity, for example, the isolation-reared macaque shows abnormalities in mating behavior and the quality of maternal care. It does not communicate effectively with its fellows, is easily provoked, and when aggressively aroused may bite itself so severely as to require medical attention. As is to be expected, symptoms vary with species and with the duration, amount, and kind of early social experience that is provided (for reviews see Berkson, 1967; Mason, Davenport & Menzel, 1968; Mitchell, 1970; Sackett, 1968). Beyond establishing the nature and range of effects that are produced by social deprivation, the rearing study holds promise of a more interesting and fundamental contribution. When the circumstances in which an individual develops are varied, a clearer view is gained of the early social repertoire and of how this is modified, sometimes in quite specific ways, by the interactions of the growing organism with its environment.

The important lesson to be drawn from such research is that the basic unit in the primate social system is indeed an *individual,* with all that this implies about the uniqueness, unity, and complexity of behavioral organization. At any stage of its career the primate's social repertoire is the product of multiple factors. Its potential contribution to the group is the result of complex interactions linked with age, sex, and physiological status—the particular effects of which vary with species and with experience. The complementary processes of assimilation and accommodation that Piaget identifies in every form of organismic growth are plainly at work. The individual is an open, self-regulating cybernetic system in continuous intercourse with its surroundings—sensitive to certain stimulus configurations; equipped with definite response biases, dispositions, and preferences; attuned to certain kinds of feedback—acting selectively on its environment and undergoing continuous changes in organization as a consequence of its acts.

THE SOCIAL SETTING:
GROUPING TENDENCIES AND REGULATORY PROCESSES

In view of the primate's rich potential for individual variation, how is it that species-typical social systems do in fact emerge? Carpenter was the first to draw attention to this question, and he suggested that the determining factors would be found in "interactive behavioral systems which regulate and maintain norms of ordered adjustment for individuals within structured groups . . . " (Carpenter, 1942, 1962). My present concern is the nature of these norms and the regulatory processes that produce and maintain them.

In principle, the behavior of any member of a social group can be influenced equally by the behavior of every other member. Actually, of course, this never occurs in primate societies. Every primate group shows a highly structured pattern of relationships, expressed in the spatial arrangement of individuals, and in who does what to whom, how often, and under what circumstances. From such information a particular set of individuals can be identified that move together and interact mostly with each other. This sociometric set, the "primary affiliative unit,"[1] is here regarded as a first-order structural outcome of socioregulatory mechanisms. A social group may include one or many primary affiliative units, usually worked out along lines described by the age, sex (male–male, male–female, female–female), and kinship of the participants. In view of the signal importance of mating and parental caretaking activities, it is not surprising that these functions are often linked with the primary affiliative units; the details, however, vary widely with species.

In macaque societies, kinship within the maternal line forms one basis for an important affiliative unit. This is often a sizeable subgroup, made up of animals of all ages and both sexes, embracing several generations and degrees of kinship (grandmothers, mothers, daughters, sons, grandchildren, siblings, "aunts," "uncles," "cousins"). Cohesion is strong. Members feed together amicably, groom together, play together, and support each other in quarrels (e.g., Yamada, 1963). Abiding kinship ties are also a significant factor in the social organization of chimpanzees and probably of savanna baboons. If they are present in langurs, patas monkeys, gelada and hamadryas baboons they are much attenuated (or obscured by incompatible tendencies), and they show minimal development in gibbons and titi monkeys, in which all ties between parent and young are apparently severed when the offspring reach maturity.

Relations between and within the sexes provide another important dimension for the formation of primary affiliative units. Variations in the strength of preferences and aversions offer rich possibilities for generating different structural outcomes. For example, the presence of strong intrasexual attraction in both sexes together with strong attraction between the sexes would be conducive to the formation of stable multimale groups. Within this pattern, a weakening of intersexual attraction (or its activation only during the mating season) can create a condition similar to that described for many ungulates (but so far, for no primate) in which male and female subgroups are only loosely joined, except during the breeding period. Strong intrasexual attraction in females coupled with weak attraction (or antagonism) among males may dispose toward the formation of unimale groups. In species in which high intersexual attraction is combined with low attraction within sexes, a tendency toward monogamous units can be expected. Other variations on this basic theme are easy to imagine. The impor-

---

[1] As far as I know this term has not been used previously in discussions of primate social behavior. To be sure, the concept is not entirely free of ambiguity but it seems workable, applicable to a wide range of species, and can be refined (or discarded) as warranted by additional information.

tant point is that even rather subtle differences in the pattern of attractions can have large effects on social organization.

Systematic data on grouping tendencies are few and far between, but field accounts clearly indicate that interspecies variations in attraction between and within the sexes are real and important factors in the development of primary affiliative units. A specific illustration is provided by a comparative investigation of the squirrel monkey (*Saimiri*) and the titi (*Callicebus*), two species of South American monkeys that show contrasting forms of social organization in the wild. The modal pattern of social organization for *Saimiri* is the multimale group, ranging in size from fewer than 20 animals up to several hundred (Baldwin & Baldwin, 1971; Thorington, 1968). Home ranges may be large and there is no evidence of territorial defense. In contrast, *Callicebus* is found in small monogamous "families," consisting of an adult male, an adult female, and one or two young, and shows strong territorial tendencies (Mason, 1966, 1968).

Field reports indicate that within a squirrel monkey social group, males and females often go about in separate parties. These unisexual grouping tendencies have been confirmed in experimental studies (Mason, 1971). Females housed with a single male companion have overwhelmingly preferred an unfamiliar female over the familiar male, and when several established male–female pairs have been released simultaneously into a large outdoor enclosure, the females quickly form a cohesive subgroup and interact mostly with each other. Males housed only with females have not shown a consistent preference for either sex in social-choice tests, although individual reactions to other males range from unequivocal avoidance to strong attraction. In the outdoor enclosure, intermale relations are initially unstable and some severe fighting occurs, but within a few days this all but disappears and the males spend most of their time together, making contact with each other more than twice as frequently as they do females (Figure 2).

Parallel experiments with the monogamous titi monkeys have yielded quite a different pattern of results. Captive titis housed in pairs show evidence of a strong and exclusive bond with a cage mate of the opposite sex (whereas squirrel monkeys housed under identical conditions do not) and they prefer this familiar companion over a stranger of either sex. Unfamiliar monkeys tend to be avoided by most animals, regardless of the sex of the subject or of the incentive monkey (the females are more cautious than the males). Moreover, when established male–female pairs of titi monkeys are introduced into a large enclosure, cage mates are far more likely to approach, follow, and make contact with each other than are unacquainted males and females (Figure 3).

The tasks of identifying primary affiliative units and of measuring the strength and direction of interpersonal attraction are fairly straightforward. At least it seems so in comparison with the problem of establishing the causal basis of these grouping tendencies, which presents some formidable and largely unresolved difficulties. What is required is an analysis of the historical antecedents, social

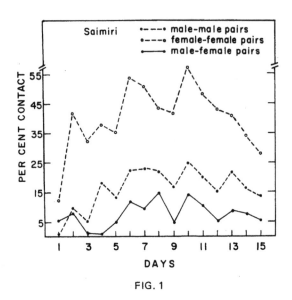

FIG. 1

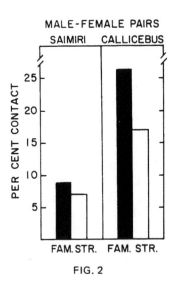

FIG. 2

motivations, and specific patterns of interaction that give rise to a particular structural outcome.

Current understanding of the socioregulatory processes operating between mother and infant is furthest along but even so is limited to the first stages in the development of the isolated mother–infant dyad and does not take into account the formation of the extended kinship group. Most likely an important factor in the development of this larger system is the ability of the participants to sustain and elaborate on the early mother–child relationship, making it the basis for enduring social ties. This ability no doubt receives support from the general tendency, present in some degree in adolescent and adult animals of both sexes (particularly females), to respond "maternally" toward infants (e.g., Hinde & Spencer-Booth, 1968; Itani, 1959; Lahiri & Southwick, 1966; MacRoberts, 1970).

Information on the sources of attraction and order in other types of affiliative units is scattered and more suggestive than systematic. For example, Harlow (1965) hypothesizes a number of different affectional systems, each characterizing the salient features of an individual's social repertoire during various phases of its life cycle and presumably disposing it to seek out and associate with others with kindred tastes and inclinations. Mother and infants are bound together by maternal and infant affectional systems, and males and females by a heterosexual affectional system. The positive responsiveness of adult males to immature animals is an expression of a paternal affectional system, and the possibility of male–male and female–female affectional systems is not overlooked. Harlow (1965) does nothing to allay the suspicion of arbitrariness when he writes with admirable candor that "The number of affectional systems in the primates can be expanded to an almost unlimited degree according to the whim of any man who classified . . ." (Harlow & Harlow, 1965, p. 288). As has been seen, however, the identification of primary affiliative units is an empirical matter and species differ broadly in the number and kind of units that they form. Why should this be? How can the fact that males form coherent subgroups in some species and not in others be explained? If intersexual associations are a salient feature of social life, how are they brought about and why do they persist in spite of pronounced seasonal changes in sexual behavior?

It seems likely that regulatory processes differ, even in species that show similar grouping patterns, as Kummer suggests in his comparative analysis of the dynamics of group formation and maintenance of one-male units in patas monkeys and gelada and hamadryas baboons (Kummer, 1971). He describes the hamadryas male as the primary agent in establishing the group and preserving its integrity. The male's relationship to the females in his group is genuinely proprietary and is enforced by a set of specialized herding techniques. His task is eased somewhat by the fact that other males inhibit approaches to a female that "belongs" to another. Kummer's experiments suggest that the female plays no special part in preserving group integrity. Hamadryas females transplanted into a multimale group of savanna baboons do not form an abiding and exclusive relationship with a single male; in contrast, when a female savanna baboon is introduced into a troop of hamadryas baboons she is taken over by a single male and responds to his herding techniques much as do hamadryas females. In the management of gelada unimale groups, the male is assisted by his females, each of whom actively prevents female members subordinate to her from interacting with outsiders and outsiders from entering the group. The integrity of the patas unimale group is apparently maintained by a tendency among all animals to react aggressively toward strangers of the same age and sex, which adult males display with particular vigor.

Changes in social context often provide information on regulatory processes. For example, comparison of segregated male–female pairs of normally troop-living squirrel monkeys with typically monogamous titis also living under paired conditions suggests many contrasts in socioregulatory behavior. Compared to *Saimiri,* much more of the titi monkey's activities are organized around its cage mate. Male and female titi monkeys spend much time grooming each other (about seven times more than squirrel monkeys); they sit quietly side by side for long periods, often with their tails intertwined (a pattern not shown by *Saimiri*) and they are distressed when they are apart. Their behavior is closely coordi-nated: they call together, move together (even in a fairly small cage where they have no place to go), and in contrast to the squirrel monkeys, in which infant care is exclusively a female responsibility, both sexes participate actively in the care of the young. Tests of social facilitation indicate that established pairs of male and female titi monkeys are much more likely to feed at the same time and in the same place than are pairs of squirrel monkeys, even though neither species is notably competitive over food.

Another way of using social context to get at regulatory processes is to remove key animals from an established group. This technique has been used frequently with various species of old world monkeys, mostly macaques, and attention has focused principally on mother–infant dyads and adult males. The mother–infant pair is a subgroup of particular interest, of course, and the question most often asked is how their relationship is affected by social context. The data present a consistent picture. Early in the relationship, group-living mothers tend to be more restrictive and protective with their infants and spend more time in contact with them than do mothers that are living alone with their babies. As the infant grows older, mothers in isolated dyads tend to be more rejecting and punitive, presumably a response to the high level of play overtures and other unwanted attentions from the infant, directed toward the mother because more appropri-ate companions are lacking (Castell & Wilson, 1971; Kaplan, 1972; Hinde & Spencer-Booth, 1967; Rowell, 1968; Wolfheim, Jensen, & Bobbitt, 1970).

Removal of adult males is also a common technique for exploring context effects. In most primate societies, adult males contribute to social regulation chiefly as suppressors and enforcers. To appreciate these functions it is necessary to consider the whole broad question of status or social power and how its expression and uses vary between species and within groups.

Power is potentially a factor in every social relationship. In its most primitive form it is expressed whenever an individual of superior strength or skill appropriates a commodity or prevents access to an object or state valued by another. This occurs even in relationships in which the superior animal is clearly a benefactor on most occasions. For example, the mother displays social power whenever she asserts herself as an individual whose interests are at variance with the immediate desires of her child. She grooms the infant when it does not seek grooming, embraces it when it does not wish to be held, restrains it when it seeks to approach a novel object, denies it access to the breast, takes food from it, and punishes it with bites and slaps. No matter whether these assertions of power are prompted by maternal solicitude or by a baser sort of self-interest, the immediate effects are the same; only minimal abilities for social learning are required for such experiences to give rise to stable differences in dominance and subordination.

The point is often made, however, that the term "dominance," with its implication of simple priority of access to the good things of life, does not begin to do justice to the complexity of social functions associated with high status in primate societies—particularly macaque and baboon societies, in which status seems to play an elaborate and specialized role in the regulation of social intercourse. In fact, in these species it is difficult to find a social relationship in which it does not take some part.

In rhesus and Japanese macaques (*M. fuscata*), status differences are evident among the members of a kinship subgroup as well as between the kinship group as a whole and other such units. Both aspects of status strongly depend on the mother (Marsden, 1968). Females are particularly affected by kinship ties. The social rank of all immature offspring, regardless of their sex, depends on the mother's rank, but daughters usually retain this dependent status into full maturity. Each daughter comes to rank within the kinship group just below her mother and just above her older sisters (Sade, 1969). Sons may also remain subordinate to their mothers even into maturity—a condition that seems to play a major part in creating the "incest taboo"—but kinship ties play a less prominent part in determining their adult status than is the case for their female siblings. To be sure, an adolescent male's rank within an entire group may be influenced by his mother's status, and she may facilitate his rise to a high position within the central hierarchy of adult males (Koford, 1963). Moreover, even at maturity, he may receive aid from his mother when he is attacked.

Nevertheless, a male's adult status is only loosely tied to his mother's rank, if only because males tend to be socially mobile. Many subadult and mature males live outside their natal groups as solitaries or become part of new affiliative units as members of a dominant-male set, an all-male band, or another bisexual troop

(Chance, 1956; Kawamura, 1965; Kawai, 1965a, b; Nishida, 1966). Among the rhesus on Cayo Säntiago, Koford (1966) has found that about a third of males 3 years of age and older change groups annually. Sade's 10-year observations of a single troop from the same population have shown that, with one exception, no male remains with the group for more than 4 years. He has concluded that adult females form the stable core of the social group (Sade, 1972).

In spite of the rather transient nature of the males' membership in a particular group, however, the indication is clear that they play a special part in the regulation of social life. Certain traits shown by high-status males (although by no means exclusively by them) seem to contribute significantly to the integrity of the group. Dominant males keep intragroup aggression under control; they are protective of other group members, particularly the young; and they are often at the forefront when danger threatens (e.g., Lindburg, 1971; Tokuda & Jensen, 1968). Bernstein (1964a, 1966) has demonstrated with captive groups that the most active and vigorous responses to a human intruder are by the dominant adult male. Furthermore, his reaction depends on the physical presence of the group (he attacks less vigorously and retreats more often when alone) and on whom is being harassed: a human is attacked more frequently and vigorously when he holds a group member than when his victim is a stranger of the same age and sex.

These specialized organizational functions of adult males in baboon and macaque societies are plainly tied to social power but the details of this connection are by no means clear. It is apparent, however, that two separate questions are involved—one having to do with the social repertoire of the adult male and how it is that under certain circumstances he displays the characteristic policing and protective behaviors described, and the other concerning why other members of the group respond to him as they do. As shall be seen many of their reactions are difficult to explain on the grounds of simple avoidance of an intimidator.

Consider first the repertoire of the adult male. Most likely, the majority of males are prepared to display the behavioral traits of the alpha or control animal but are inhibited from doing so by the presence of other males who are dominant over them. I do not suggest that this suppression is actively imposed (although this may indeed be a factor), but that it is a general consequence of being in a subordinate position.

What evidence can be mustered in support of such a view? First, behavioral data, such as the experiment showing that removal of the alpha male from a captive macaque group leads to increased social activity among the two remaining males (including alphalike behavior in one of them) and that the return of alpha brings about a prompt resumption of the status quo (Bernstein, 1964b). Similarly, the introduction of an adult male into an established group of juvenile rhesus produces a general inhibition of activity (Bernstein & Draper, 1964). Second, results of brain research demonstrate that activation of brain sites causing directed chasing and attack when the stimulated monkey occupies a superior rank do not provoke such behavior if the same animal is in a subordinate position (Delgado & Mir, 1969). Finally, endocrinological evidence indicates that ACTH response levels (a measure of stress) are related to social position and rise and fall in subordinate monkeys with the presence or absence of the dominant male. Other evidence shows that levels of testosterone are high in dominant or aggressive animals and decrease markedly in males experiencing a sudden and decisive defeat (Rose, Gordon, & Bernstein, 1972; Rose et al., 1971; Sassenrath, 1970). To be sure, all such effects have been demonstrated in caged groups and they are probably more extreme than those occurring in natural societies, in which there is more freedom to maneuver, but there is no reason to question their generality.

Consider now the subordinate animal. Although the fact is not always appreciated, status, like beauty, is in the eye of the beholder. Although the position of macaque or baboon in a social group may ultimately depend on its strength or fighting skills or on its having numerous or powerful allies, fighting power alone cannot account for the subtleties of status in these complex societies. Dominance is as much conferred as it is won. For example, when Bernstein and I released a mixed group of rhesus monkeys, all strangers to each other, into an outdoor enclosure, there were no contests between the largest male and the ten other animals in the group, in spite of the fact that the overall level of agonistic interaction was high (Bernstein & Mason, 1963).

As Rowell (1966b) has put it, the hierarchy of social rank is maintained largely by the behavior of subordinate animals. Taking this suggestion a step further, Chance and Jolly (1970) have proposed that a distinguishing feature of baboon, macaque, and other primate societies organized around a well-defined central hierarchy of adult males is the focus of the subordinate's attention on animals that rank above it, particularly on the supremely dominant male. This animal is the object of strong ambivalent tendencies within all other members of the group.

The evidence that the dominant male is the object of mixed reactions in macaque and baboon societies is considerable. One of the most reliable indicators of a high-ranking monkey is its ability to make space, to displace other animals. Yet among free-ranging rhesus dominant males have a greater number of associates in proximity than do other males, and they receive a disproportionate share of grooming from females and immatures (Conaway & Koford, 1965; Fisler, 1967; Kaufmann, 1967). Elsewhere, (1964) I have suggested that the ambivalent reaction toward a highly dominant animal may be based on an arousal mechanism similar to that involved in the response toward moderately novel inanimate objects. However, in contrast to the inanimate object, which loses its potency as the novelty "wears off," the dominant baboon or macaque is able to maintain a state of ambivalence in his associates because of his size, and particularly, his comportment, which includes such attention-provoking activities as branch shaking, loud vocalizations, rushing, chasing, threats—the effective-ness of which are perhaps reinforced by an occasional bite or slap (Mason, 1964). In chimpanzees the display of an adult male, which is notably elaborate and (to me, at least) awe inspiring, not only provokes attention but often causes other animals to approach and show some gesture of "friendship" or "concilia-tion" (Reynolds & Luscombe, 1969). It should be recalled too, that Mike, one of van Lawick-Goodall's chimps at the Gombe Stream Reserve, apparently improved his status with the help of an empty kerosene can that he banged around during his displays (van Lawick-Goodall, 1967).

ENVIRONMENTAL EFFECTS

Emphasis shifts in this section to the question of environmental influences on social processes. I have characterized the environment as including all socially relevant features of the habitat. Obviously this is too gross for present purposes. One useful refinement is to distinguish the effects of alien groups of the same species from other kinds of environmental influences.

A social group is a more or less stable assembly of animals that move together and interact mostly with each other. It exists not in geographical isolation but as part of a larger population of potentially interbreeding individuals, most of which are also living in groups. Segregation among these units is maintained by three basic mechanisms: attachment to place, intragroup affinities, and the active responses of group members to outsiders. This last mechanism is of main concern here.

Most primate societies are semiclosed systems. Members are recognized and have their established places in the web of social relations; aliens are responded to as such and are characteristically met with hostility. The form and intensity of this reaction varies with species and local conditions. Baboons, for example, display a high degree of intergroup tolerance. There are many accounts of groups associating closely with no indication of antagonism. Under some conditions, however—for instance, the lodging of several groups within the same grove of trees—there may be considerable agonistic behavior (Altmann & Altmann, 1970). Some species are at the opposite extreme; groups not only consistently respond to each other with aggression, but seem to go out of their way to create the occasion for an agonistic encounter. Neighboring groups of titi monkeys or gibbons regularly converge on certain areas where their home ranges overlap and engage in extended bouts of vigorous calling, chasing, and aggressive display (Ellefson, 1968; Mason, 1968). The gray langur of Ceylon, as do the gibbon and titi monkey, also seeks out its neighbors, but these encounters are not limited to any particular areas. In fact, a group may pass through the home range of one neighbor in order to make aggressive contact with another (Ripley, 1967).

In many species, face-to-face meetings between groups are apparently avoided whenever possible. Groups of howler monkeys are seldom in proximity and the usual assumption is that separation is maintained by the roaring vocalizations for which these monkeys are noted. Similar functions have been suggested for other species that produce vocalizations loud enough to carry over some distance and indicate a group's location (Marler, 1968). Groups of rhesus monkeys also tend to avoid contact but seemingly lack the means to accomplish this at a distance. Intergroup relations are predominantly agonistic and groups that meet regularly establish positions of dominance and subordination. An early experiment by Carpenter suggested that intergroup dominance is mainly determined by the number 1 male (Carpenter, 1942). This now seems unlikely. More important factors are probably the size of the group and the ability of its members to form effective fighting coalitions. Low-ranking males take a most active part in initiating intergroup aggression (Marsden, 1969; Morrison & Menzel, 1972; Vessey, 1968, 1971).

These few examples illustrate a general tendency among primates to distinguish sharply between group members and outsiders. Even if the problem of competition for limiting resources is put aside, the mere presence and number of extragroup conspecifics can be expected to have an effect on the quality of social life. The greater the number of groups to be avoided, the more restricted is the available habitat and the more time is required for intergroup adjustments. The more frequent and prolonged the encounters between groups, the less time is available for feeding and routine social activities. Beyond these commonsensical expectations, is there any concrete evidence that social processes and social organization are permanently altered by the density of groups? Very little actually, and most of it is rather circumstantial. One of the immediate effects of an intergroup encounter is to bring the members of a group closer together (in the literal sense). This can be seen in the large multimale groups of savanna baboons and rhesus, in the small unimale units of the hamadryas baboon, and in the family-type groups of titi monkeys (Altmann & Altmann, 1970; Kummer, 1968a; Mason, 1966; Vessey, 1968).

Presumably as a psychological correlate inferred from the observation that members of a group draw together physically in response to external challenge, some authors have suggested that the cumulative effect of intergroup encounters is to enhance social cohesion. This intriguing idea goes back at least to the writings of the sociologist Georg Simmel at the turn of the century, but it finds scant support in the nonhuman primates. Possibly, a little social stress strengthens solidarity, but there are indications that prolonged exposure to the pressure of alien groups increases internal wrangling and jeopardizes the stability of a social system. For example, Southwick (1969) has found the incidence of intragroup aggression among macaques living in a densely populated temple area to be about four times higher than the level among monkeys living in a living in a less crowded forest habitat. In another study, a group of rhesus occupying a midposition in the intergroup dominance hierarchy on Cayo Santiago was reduced to less than half its normal size by trapping. The diminished group dropped to the bottom of the hierarchy; it lost its customary boldness, became furtive, and showed a sharp reduction in play and grooming. These changes were accompanied by a marked increase in intragroup strife: "Little provocation was necessary to elicit savage attacks which often involved much of the group before ending." The captured portion of the group, released on an island previously uninhabited by monkeys, also showed pronounced changes in social behavior, but a sharp increase in intragroup aggression was not among them (Morrison & Menzel, 1972). There is a suggestion that aggressiveness may be persistently higher among individual rhesus living under generally stressful conditions. Monkeys captured in the urban bazaars of India fought violently when placed together for the first time—all received serious injuries and one was killed—whereas forest—caught monkeys under the same conditions were much less aggressive. Urban monkeys were also more aggressive toward rural monkeys when they were paired in competitive tests and were generally more successful in securing food (Singh, 1966, 1968).

These examples, which unfortunately are drawn entirely from the highly aggressive rhesus macaque, suggest that encounters with alien groups are a particularly potent source of stress on a social system. Most probably, however, the effects of population density differ mainly in degree from those produced by other adverse environmental conditions. It may be speculated that modest departures from optimum conditions, whether induced by social or physical agents, lead to an exaggeration or strengthening of established regulatory mechanisms. More extreme deviations, however, can be expected to cause a breakdown in these internal social controls. Although the information is meager, it is in accord with this thesis.

Evidence has been considered from several species indicating that the size of free-ranging groups and, in some cases, their composition are affected by environmental factors. The trend is for group size to become smaller in less favorable habitat, and for composition to shift from a multimale toward a unimale pattern. It seems likely that such changes are partly the result of active social processes instead of the simple attrition of personnel or differential mortality. In captive groups of macaques and baboons, a temporary reduction in space or the restriction of food sources to a single location tend to increase fighting, but these conditions also sharpen status differences. Furthermore, the largest effects are observed between, rather than within, primary affiliative units (Alexander & Roth, 1971; Rowell, 1967b; Southwick, 1967, 1969).

Under more prolonged or extreme adverse conditions signs of social disorganization appear. For example, under the stress of trapping and confinement of free-ranging rhesus to small holding cages, even close relatives attacked each other and mothers abandoned or actively rejected their infants. In the first few days following the release of these same monkeys on an uninhabitated island, the dominant males played little part in policing and coordinating the group, the monkeys broke up into small subgroups or foraged individually, juveniles seldom joined together in play groups, and there was a noticeable decrease in maternal solicitude and restraint: "The mothers were aloof, each seeming to merely tolerate its infant when with it and to ignore it when they were separated" (Morrison & Menzel, 1972). Baboons living under near-starvation conditions usually foraged singly or in very small parties, and social interaction of any sort was rarely observed: "The cohesive tendencies of mating, of mother–infant relations and the 'attractiveness' of them to others in a group, and of other social gestures of greeting, vocalization and so on, together with aggressive domination (also cohesive) of the adult males were never evident . . ." (Hall, 1963).

Naturally the "optimum" environment varies with species, as do the limits within which the social system can adapt and the nature of its response to environmental perturbations. Small numbers of squirrel monkeys released within a 100 by 400 ft enclosure developed an organization that was a good approximation of the pattern described for natural groups. In contrast, equivalent numbers of titi monkeys in the same situation did not form their characteristic territorial arrangements, in spite of the fact that previously established pair affinities persisted. However, neither did they engage in severe fighting, presumably because sufficient space was available to permit effective flight. When the amount of space was much reduced, however, fighting became a serious problem. Titi monkeys seem to lack an effective alternative to flight as a means of terminating aggression. Fights seldom if ever result in the ranking of animals as dominant and subordinate, which is so conspicuous a feature of the social relations of macaques and many other old world primates. Consequently, the reduction in space has more severe effects on social behavior in this species than in those that have evolved more refined methods for controlling aggression.

CONCLUSIONS

The naive hope of discovering a simple and satisfying evolutionary progression in the structure of primate social systems had to be abandoned almost as soon as adequate naturalistic data began to appear. Zuckerman's early study of the hamadryas baboon, Nissen's preliminary observations on the chimpanzee, and Carpenter's findings on howler monkeys, gibbons, spider monkeys, and rhesus macaques were clear indications of a diversity of social arrangements that would not be easily fitted within accepted phylogenetic schemes. Subsequent developments did not alter this outlook in any essential way but instead confirmed it. "Monogamous units," "harems," and "multimale groups" were found in most of the major divisions within the primate order.

The recent emphasis on the role of ecological pressures and constraints in the evolution of primate social systems has helped to account for this diversity. No doubt it will continue to provide a stimulus to thought on the origins and adaptive functions of primate societies. In evaluating the potential contribution of this orientation to our understanding of the evolution of primate social groups, however, it is well to keep two points in mind: first, similar structural outcomes can result from quite different social processes. Multiple instances of convergence are found within the order Primates; moreover, it is difficult to support the claim that there is anything truly distinctive in the organizational patterns that the primates have evolved. For every known variety of primate social structure, a similar pattern can be found in other vertebrate species. Second, natural selection operates on individuals. Primate social systems have evolved because of the selective advantages they confer on their individual constituents. It is also true, of course, that the society itself becomes a factor in selection and tends to favor those animals that survive and prosper within the system and therefore help to perpetuate it.

The point, then, is that the unusual and distinctive features in the evolution of primate social life are less likely to be found in the structure of the social system than in the social processes and social dynamics that produce it; and these depend ultimately on the psychological characteristics of the individual. In recent decades much has been learned about the natural history of primate societies, about primate "intelligence," and about the psychosocial development of the individual. The interdependence of these three traditions is already evident. A synthesis is clearly overdue.

REFERENCES

Alexander, B. K., & Roth, E. M. The effects of acute crowding on aggressive behavior of Japanese monkeys. *Behaviour,* 1971, 39, 6–90.

Altmann, S. A., & Altmann, J. *Baboon ecology.* Chicago: University of Chicago Press, 1970.

Baldwin, J. D., & Baldwin, J. I. Squirrel monkeys (*Saimiri*) in natural habitats in Panama, Colombia, Brazil, and Peru. *Primates,* 1971, 12, 45–61.

Ball, J., & Hartman, C. G. Sexual excitability as related to the menstrual cycle in the monkey. *American Journal of Obstertrics & Gynecology,* 1935, 29, 117–119.

Berkson, G. Abnormal stereotyped motor acts. In J. Zubin & H. F. Hunt (Eds.), *Comparative psychopathology—Animal and human.* New York: Grune & Stratton, 1967, pp. 76–94.

Bernstein, I. S. Role of the dominant male rhesus monkey in response to external challenges to the group. *Journal of Comparative Physiological Psychology,* 1964, 57, 404–406. (a)

Bernstein, I. S. Group social patterns as influenced by removal and later reintroduction of the dominant male rhesus. *Psychological Reports,* 1964, 14, 3–10. (b)

Bernstein, I. S. An investigation of the organization of pigtail monkey groups through the use of challenges. *Primates,* 1966, 7, 471–480.

Bernstein, I. S. The Lutong of Kuala Selangor. *Behaviour,* 1968, 32, 1–16.

Bernstein, I. S., & Draper, W. A. The behaviour of juvenile rhesus monkeys in groups. *Animal Behaviour,* 1964, **12,** 84–91.

Bernstein, I. S., & Mason, W. A. Group formation by rhesus monkeys. *Animal Behaviour,* 1963, **11,** 28–31.

Birch, H. G., & Clark, G. Hormonal modification of social behavior. II. The effects of sex-hormone administration on the social dominance status of the female-castrate chimpanzee. *Psychosomatic Medicine,* 1946, **8,** 320–331.

Carpenter, C. R. A field study in Siam of the behavior and social relations of the gibbon *(Hylobates lar). Comparative Psychology Monographs,* 1940, **16,** 1–212.

Carpenter, C. R. Societies of monkeys and apes. *Biological Symposia,* 1942, **8,** 177–204.

Carpenter, C. R. Field studies of a primate population. In E. L. Bliss (Ed.), *Roots of behavior.* New York: Harper (Hoeber), 1962, Pp. 286–294.

Castell, R., & Wilson, C. Influence of spatial environment on development of mother-infant interaction in pigtail monkeys. *Behaviour,* 1971, **39,** 202–211.

Chance, M. R. A. Social structure of a colony of *Mecaca mulatta. British Journal of Animal Behavior,* 1956, **4,** 1–13.

Chance, M. R. A., & Jolly, C. J. *Social groups of monkeys, apes and men.* New York: Dutton, 1970.

Chivers, D. J. On the daily behaviour and spacing of howling monkey groups. *Folia Primatologica,* 1969, **10,** 48–102.

Collias, N., & Southwick, C. A field study of population density and social organization in howling monkeys. *Proceedings of the American Philosophical Society,* 1952, **96,** 143–156.

Conaway, C. H., & Koford, C. B. Estrous cycles and mating behavior in a free-ranging band of rhesus monkeys. *Journal of Mammalogy,* 1965, **45,** 577–588.

Crook, J. H. The socio-ecology of primates. In J. H. Crook (Ed.), *Social behaviour in birds and mammals.* New York: Academic Press, 1970, pp. 103–106.

Crook, J. H., & Gartlan, J. S. Evolution of primate societies. *Nature (London),* 1966, **210,** 1200–1203.

Delgado, J. M. R., & Mir, D. Fragmental organization of emotional behavior in the monkey brain. *Annals of the New York Academy of Sciences,* 1969, **159,** 731–751.

DeVore, I. Comparative ecology and behavior of monkeys and apes. In S. L. Washburn (Ed.), *Classification and human evolution.* Chicago: Aldine, 1963, Pp. 301–319.

Durham, N. M. Effects of altitude differences on group organization of wild black spider monkeys *(Ateles paniscus). Proceedings of the 3rd International Congress of Primatology, Zurich,* 1971, **3,** 32–40.

Eisenberg, J. F., Muckenhirn, N. A., & Rudran, R. The relation between ecology and social structure in primates. *Science,* 1972, **176,** 863–874.

Ellefson, J. O. Territorial behavior in the common white-handed gibbon *Hylobates lar* (Linn.) In P. C. Jay (Ed.), *Primates: Studies in adaptation and variability.* New York: Holt, 1968, Pp. 180–199.

Fisler, G. F. Nonbreeding activities of three adult males in a band of free-ranging rhesus monkeys. *Journal of Mammalogy,* 1967, **48,** 70–78.

Gartlan, J. B., & Brain, C. K. Ecology and social variability in *Cercopithecus aethiops* and *C. mitis.* In P. C. Jay (Ed.), *Primates: Studies in adaptation and variability.* New York: Holt, 1968, Pp. 253–292.

Goodall, J. Chimpanzees of the Gombe Stream Reserve. In I. DeVore (Ed.), *Primate behavior.* New York: Holt, 1965, Pp. 425–473.

Goss-Custard, J. D., Dunbar, R. I. M., & Aldrich-Blake, F. P. G. Survival mating and rearing strategies in the evolution of primate social structure. *Folia Primatologica,* 1972, **17,** 1–19.

Goy, R. W. Experimental control of psychosexuality. *Philosophical Transactions of the Royal Society of London*, 1970, **259**, 149–162.

Hall, K. R. L. Variations in the ecology of the chacma baboon, *Papio ursinus*. *Symposia of the Zoological Society of London*, 1963, **10**, 1–28.

Hall, K. R. L. Behaviour and ecology of the wild Patas monkey, *Erythrocebus patas*, in Uganda. *Journal of Zoology*, 1965, **148**, 15–87.

Hall, K. R. L., & DeVore, I. Baboon social behavior. In I. DeVore (Ed.), *Primate behavior*. New York: Holt, 1965, Pp. 53–110.

Harlow, H. F. Sexual behavior in the rhesus monkey. In F. A. Beach (Ed.), *Sex and behavior*. New York: Wiley, 1965, Pp. 234–265.

Harlow, H. F., & Harlow, M. K. The affectional systems. In A. M. Schrier, H. Harlow, & F. Stollnitz (Eds.), *Behavior of nonhuman primates*, Vol. 2. New York: Academic Press, 1965, Pp. 287–334.

Hinde, R. A., & Spencer-Booth, Y. The effect of social companions on mother-infant relations in rhesus monkeys. In D. Morris (Ed.), *Primate Ethology*. Chicago: Aldine, 1967, Pp. 267–286.

Hinde, R. A. & Spencer-Booth, Y. The study of mother–infant interaction in captive group-living rhesus monkeys. *Proceedings of the Royal Society*, 1968, **169**, 177–201.

Itani, J. Paternal care in the wild Japanese monkey, *Macaca fuscata fuscata*. *Primates*, 1959, **2**, 61–93.

Jarrard, L. E. (Ed.) *Cognitive processes of nonhuman primates*. New York: Academic Press, 1971.

Jay, P. The common langur of North India. In I. DeVore (Ed.), *Primate behavior*. New York: Holt, 1965, Pp. 197–249.

Jensen, G. D., Bobbitt, R. A., & Gordon, B. N. Sex differences in the development of independence of infant monkeys. *Behaviour*, 1968, **30**, 1–13.

Kaplan, J. Differences in the mother-infant relations of squirrel monkeys housed in social and restricted environments. *Developmental Psychobiology*, 1972, **5**, 43–52.

Kaufmann, J. H. Social relations of adult males in a free-ranging band of rhesus monkeys. In S. Altmann (Ed.), *Social Communication Among Primates*. Chicago: University of Chicago Press, 1967, Pp. 73–98.

Kawai, M. On the system of social ranks in a natural troop of Japanese monkeys: I. Basic rank and dependent rank. In S. A. Altmann (Ed.), *Japanese monkeys: A collection of translations*. Published by the editor, 1965, Pp. 66–86. (a)

Kawai, M. On the system of social ranks in a natural troop of Japanese monkeys: II. Ranking order as observed among the monkeys on and near the test box. In S. A. Altmann (Ed.), *Japanese monkeys: A collection of translations*. Published by the editor, 1965, Pp. 87–104. (b)

Kawamura, S. Matriarchal social rank in the Minoo-B troop: A study of the rank system of Japanese monkeys. In S. A. Altmann (Ed.), *Japanese monkeys: A collection of translations*. Published by the editor, 1965, Pp. 105–112.

Koford, C. B. Rank of mothers and sons in bands of rhesus monkeys. *Science*, 1963, **141**, 356–357.

Koford, C. B. Population changes in rhesus monkeys: Cayo Santiago. *Tulane Studies in Zoology*, 1966, **13**, 1–7.

Kummer, H. Dimensions of a comparative biology of primate groups. *American Journal of Physical Anthropology*, 1967, **27**, 357–366.

Kummer, H. *Social organization of Hamadryas baboons: A field study*. Chicago: University of Chicago Press, 1968. (a)

Kummer, H. Two variations in the social organization of baboons. In P. C. Jay (Ed.), *Primates: Studies in adaptation and variability*. New York: Holt, 1968, Pp. 293–312. (b)

Kummer, H. *Primate societies.* Chicago: Aldine, 1971.

Lahiri, R. K., & Southwick, C. H. Parental care in *Macaca sylvana. Folia Primatologica,* 1966, **4,** 257–264.

Lindburg, D. G. The rhesus monkey in North India: An ecological and behavioral study. In L. A. Rosenblum (Ed.), *Primate behavior: Developments in field and laboratory research,* Vol. 2. New York: Academic Press, 1971, Pp. 1–105.

MacRoberts, M. H. The social organization of barbary apes *(Macaca sylvana)* on Gibraltar. *American Journal of Physical Anthropology,* 1970, **33,** 38–100.

Marler, P. Aggregation and dispersal: Two functions in primate communication. In P. C. Jay (Ed.), *Primates: Studies in adaptation and variability.* New York: Holt, 1968, Pp. 420–438.

Marler, P. *Colobus guereza:* Territoriality and group composition. *Science,* 1969, **163,** 93–95.

Marsden, H. M. Agonistic behaviour of young rhesus monkeys after changes induced in social rank of their mothers. *Animal Behaviour,* 1968, **16,** 38–44.

Marsden, H. M. Dominance order reversal of two groups of rhesus monkeys in tunnel-connected enclosures. *Proceedings of the 2nd International Congress of Primatology, Atlanta, Ga.,* 1969, **1,** 52–58.

Mason, W. A. Sociability and social organization in monkeys and apes. In L. Berkowitz (Ed.), *Recent advances in experimental social psychology.* New York: Academic Press, 1964, Pp. 277–305.

Mason, W. A. Social organization of the South American monkey, *Callicebus moloch:* A preliminary report. *Tulane Studies in Zoology,* 1966, **13,** 23–28.

Mason, W. A. Use of space by Callicebus groups. In P. C. Jay (Ed.), *Primates: Studies in adaptation and variability.* New York: Holt, 1968, Pp. 220–216.

Mason, W. A. Field and laboratory studies of social organization in *Saimiri* and *Callicebus.* In L. A. Rosenblum (Ed.), *Primate Behavior,* Vol. 2. New York: Academic Press, 1971, Pp. 107–137.

Mason, W. A., Davenport, R. K., & Menzel, E. W., Jr. Early experience and the social development of rhesus monkeys and chimpanzees. In G. Newton & S. Levine (Eds.), *Early experience and behavior.* Springfield, Illinois: Charles C Thomas, 1968. Pp. 1–41.

Mitchell, G. D. Attachment differences on male and female infant monkeys. *Child Development,* 1968, **39,** 611–620.

Mitchell, G. Abnormal behavior in primates. In L. A. Rosenblum (Ed.), *Primate behavior,* Vol. 1. New York: Academic Press, 1970, Pp. 195–249.

Morrison, J. A., & Menzel, E. W., Jr. Adaptation of a free-ranging rhesus monkey group to division and transplantation. *Wildlife Monographs,* November, 1972, No. 31.

Nadler, R. D., & Rosenblum, L. A. Factors influencing sexual behavior of male bonnet macaques *(Macaca radiata). Proceedings of the 3rd International Congress of Primatology, Zurich,* 1971, **3,** 100–107.

Nishida, T. A sociological study of solitary male monkeys. *Primates,* 1966, **7,** 141–204.

Reynolds, V., & Luscombe, B. Chimpanzee rank order and the function of displays. *Proceedings of the 2nd International Congress of Primatology, Atlanta, Ga.,* 1969, **1,** 81–86.

Ripley, S. Intertroop encounters among ceylon gray langurs *(Presbytis entellus).* In S. A. Altmann (Ed.), *Social communication among primates.* Chicago: University of Chicago Press, 1967, Pp. 237–253.

Rose, R. M., Gordon, T. P., & Bernstein, I. S. Plasma testosterone levels in the male rhesus: Influences of sexual and social stimuli. *Science,* 1972, **178,** 643–645.

Rose, R. M., Holaday, J. W., & Bernstein, I. S. Plasma testosterone, dominance rank and aggressive behaviour in male rhesus monkeys. *Nature (London),* 1971, **231,** 366–368.

Rosenblum, L. A., & Kaufman, I. C. Laboratory observations of early mother-infant relations in pigtail and bonnet macaques. In S. A. Altmann (Ed.), *Social communication among primates*. Chicago: University of Chicago Press, 1967. Pp. 33–41.

Rosenblum, L. A., Kaufman, I. C., & Stynes, A. J. Individual distance in two species of macaque. *Animal Behaviour*, 1964, **12**, 338–342.

Rowell, T. E., Forest living baboons in Uganda. *Journal of Zoology (London)*, 1966, **149**, 344–364. (a)

Rowell, T. E. Hierarchy in the organization of a captive baboon group. *Animal Behaviour*, 1966, **14**, 430–443. (b)

Rowell, T. E. Female reproductive cycles and the behavior of baboons and rhesus macaques. In S. A. Altmann (Ed.), *Social communication among primates*. Chicago: University of Chicago Press, 1967. Pp. 15–32. (a)

Rowell, T. E. A quantitative comparison of the behaviour of a wild and a caged baboon group. *Animal Behaviour*, 1967, **15**, 499–509. (b)

Rowell, T. E. The effect of temporary separation from their group on the mother infant relationship of baboons. *Folia Primatologica*, 1968, 9, 114–122.

Rowell, T. E. Organization of caged groups of Cercopithecus monkeys. *Animal Behaviour*, 1971, **19**, 625–645.

Rumbaugh, D. M. Learning skills of anthropoids. In L. A. Rosenblum (Ed.), *Primate behaviour*, Vol. 1. New York: Academic Press: 1970. Pp. 1–70.

Rumbaugh, D. M. Evidence of qualitative differences in learning processes among primates. *Journal of Comparative Physiological Psychology*, 1971, **76**, 250–255.

Sackett, G. P. Abnormal behavior in laboratory-reared rhesus monkeys. In M. W. Fox (Ed.), *Abnormal behavior in animals*. Philadelphia: W. B. Saunders, 1968, Pp. 293–331.

Sade, D. S. An algorithm for dominance relations among rhesus monkeys: Rules for adult females and sisters. Paper presented at the Annual Meeting of American Association of Physical Anthropologists, Mexico City, Mexico, 1969.

Sade, D. S. A longitudinal study of social behavior of rhesus monkeys. In R. Tuttle (Ed.), *The functional and evolutionary biology of primates*. Chicago: Aldine-Atherton, 1972, Pp. 378–398.

Sassenrath, E. N. Increased adrenal responsiveness related to social stress in rhesus monkeys. *Hormones & Behavior*, 1970, **1**, 283–298.

Schrier, A. M., Harlow, H. F., & Stollnitz, F. (Eds.) *Behavior of Nonhuman Primates*, Vol. 1. New York: Academic Press, 1965.

Singh, S. D. The effects of human environment on the social behavior of rhesus monkeys. *Primates*, 1966, **7**, 33–39.

Singh, S. D. Social interactions between the rural and urban monkeys, *Macaca mulatta*. *Primates*, 1968, **9**, 69–74.

Southwick, C. H. An experimental study of intragroup agonistic behavior in rhesus monkeys (*Macaca mulatta*). *Behaviour*, 1967, **28**, 182–209.

Southwick, C. H. Aggressive behaviour of rhesus monkeys in natural and captive groups. In S. Garattini & E. B. Sigg (Eds.), *Aggressive behaviour*. New York: Wiley, 1969, Pp. 32–43.

Struhsaker, T. T., & Gartlan, J. S. Observations on the behaviour and ecology of the Patas monkey (*Erythrocebus patas*) in the Waza Reserve, Cameroon. *Journal of Zoology (London)*, 1970, **161**, 49–63.

Sugiyama, Y. Behavioral development and social structure in two troops in hanuman langurs (*Presbytis entellus*). *Primates*, 1965, **6**, 213–247.

Sugiyama, Y. Social organization of hanuman langurs. In S. A. Altmann (Ed.), *Social communication among primates*. Chicago: University of Chicago Press, 1967, Pp. 221–236.

Thorington, R. W., Jr. Observations of squirrel monkeys in a Colombian forest. In L. A. Rosenblum & R. W. Cooper (Eds.), *The Squirrel Monkey*. New York: Academic Press, 1968, Pp. 69–85.

Tokuda, K., & Jensen, G. D. The leader's role in controlling aggressive behavior in a monkey group. *Primates*, 1968, 9, 319–222.

van Lawick-Goodall, J. *My friends the wild chimpanzees*. Washington, D.C.: National Geographic Society, 1967.

Vessy, S. H. Interactions between free-ranging groups of rhesus monkeys. *Folia Primatologica*, 1968, 8, 228–239.

Vessey, S. H. Free-ranging rhesus monkeys: Behavioural effects of removal, separation and reintroduction of group members. *Behaviour*, 1971, 40, 216–227.

Vogel, C. Behavioral differences of *Presbytis entellus* in two different habitats. *Proceedings of the 3rd International Congress of Primatology, Zurich*, 1971, 3, 41–47.

Warren, J. M. The comparative psychology of learning. *Annual Review of Psychology*, 1965, 16, 95–118. (a)

Warren, J. M. Primate learning in comparative perspective. In A. M. Schrier, H. F. Harlow, & F. Stollnitz (Eds.), *Behavior of nonhuman primates*, Vol. 1. New York: Academic Press, 1965. Pp. 249–281. (b)

Wilson, A. P., & Vessey, S. H. Behavior of free-ranging, castrated rhesus monkeys, *Folia Primatologica*, 1968, 9, 1–14.

Wolfheim, J. H., Jensen, G. D., & Bobbitt, R. A. Effects of group environment on the mother-infant relationship in pigtailed monkeys (*Macaca nemestrina*). *Primates*, 1970, 11, 119–124.

Yamada, M. A study of blood-relationship in the natural society of the Japanese macaque. *Primates*, 1963, 4, 43–65.

Yoshiba, K. Local and intertroop variability in ecology and social behavior of common Indian langurs. In P. C. Jay (Ed.), *Primates: Studies in adaptation and variability*. New York: Holt, 1968, Pp. 217–242.

Author Index

Numbers in *italics* refer to the pages on which the complete references are listed.

A

Abercrombie, M., 318, *326*
Ackerman, R. F., 232, *243*
Ackil, J. E., 175, *188*
Adám, G., 266, 269, *326*
Adelman, W. J., Jr., 291, *326*
Adolph, E. F., 297, 312, *326*
Adrian, E. D., 42, 45, 46, 236, *240*
Adrien, J., 322, *328*
Agayan, A. L., 132, 133, 135, *146*
Agranoff, B. W., 297, 319, *326*
Aidley, D. J., 291, *326*
Ajamone Marsan, C., 150, 151, *164*
Akert, K., 396, *405*
Albers, R. W., 297, 319, *326*
Albert, D. J., 269, *326*
Albiniak, B. A., 209, *216*
Aldrich-Blake, F. P. G., 427, *451*
Alexander, B. K., 448, *450*
Alexander, R. McN., 301, 316, *326*
Alland, A., Jr., 269, 274, *326*
Allison, T., 230, *240*, 397, *403*
Allman, J. M., 285, 288, *326, 336*, 396, 397, *403*
Altman, J., 269, 306, 323, *326*
Altmann, J., 446, 447, *450*
Altmann, S. A., 446, 447, *450*
Altner, H., 30, 35, *46*

Ambros, V. R., 135, *143*
Amsel, A., 180, *187*, 205, *211*
Anderson, S., 230, *240*
Andersen, H. T., 267, *326*
Anderson, O., 35, *48*
Andrew, R. J., 273, 301, *326*
Andy, O. J., 42, *51*, 230, 232, *242*
Anfinsen, C. B., 297, *326*
Angeletti, P., 306, *326*
Angevine, J. B., Jr., 157, *163*
Ansell, G. B., 269, *326*
Anthony, J., 22, *23*
Appel, S. H., 311, *341*
Ariëns-Kappers, C. U., 39, *46*, 115, 140, 141, 283, 285, 287, 289, *326*, 360, 362, 363, 365, 367, *378*, 399, *403*
Armstrong, J. A., 153, *163*
Armstrong, M. E., 322, *333*
Aronson, L. R., 128, *141*
Arumugasamy, N., 287, 320, *333*
Aschoff, J., 257, *326*
Ashton, E. H., 45, *49*
Asimov, I., 274, *327*
Atema, J., 34, *46*
Atkins, D. L., 229, *240*
Audubon, J. J., 38, *46*
Auerbach, A. A., 280, *327*
Ayala, F. J., 383, *390*
Ayers, H., 90, *104*

Subject Index